Aerospace Engineering: Design, Development and Applications

Aerospace Engineering: Design, Development and Applications

Edited by Stephen Baggins

CLANRYE
INTERNATIONAL
www.clanryeinternational.com

Clanrye International,
750 Third Avenue, 9th Floor,
New York, NY 10017, USA

ISBN: 978-1-63240-935-5

Cataloging-in-Publication Data

Aerospace engineering : design, development and applications / edited by Stephen Baggins.
    p. cm.
Includes bibliographical references and index.
ISBN 978-1-63240-935-5
1. Aerospace engineering. 2. Aeronautics. 3. Astronautics. 4. Engineering. I. Baggins, Stephen.
TL545 .A37 2020
629.1--dc23

For information on all Clanrye International publications
visit our website at www.clanryeinternational.com

CLANRYE
INTERNATIONAL

Contents

Preface

Over the recent decade, advancements and applications have progressed exponentially. This has led to the increased interest in this field and projects are being conducted to enhance knowledge. The main objective of this book is to present some of the critical challenges and provide insights into possible solutions. This book will answer the varied questions that arise in the field and also provide an increased scope for furthering studies.

Aerospace engineering is the discipline of engineering that deals with the development of aircraft and spacecraft. Aeronautical engineering and astronautical engineering are the two branches of aerospace engineering. Aeronautical engineering is concerned with the study, design and manufacturing of air flight capable machines, whereas astronautically engineering is linked to the field of rocket science. The interaction between technologies like aerodynamics, propulsion, avionics, materials science, structural analysis and manufacturing is characterised as aerospace engineering. It includes elements like radar cross section, flight test, aero acoustics, noise control, risk and reliability, solid mechanics etc. Theoretical physics is the basis of most of these elements that fall under aerospace engineering. This book unravels the recent studies in the field of aerospace engineering. Also included herein is a detailed explanation of the various concepts and applications of aerospace engineering. Through this book, we attempt to further enlighten the readers about the new concepts in this field.

I hope that this book, with its visionary approach, will be a valuable addition and will promote interest among readers. Each of the authors has provided their extraordinary competence in their specific fields by providing different perspectives as they come from diverse nations and regions. I thank them for their contributions.

Editor

Satellite Constellation Orbit Design Optimization with Combined Genetic Algorithm and Semianalytical Approach

Tania Savitri,[1] Youngjoo Kim,[1] Sujang Jo,[2] and Hyochoong Bang[1]

[1]Department of Aerospace Engineering, Korea Advanced Institute of Science and Technology (KAIST),
 Daejeon 305-701, Republic of Korea
[2]Korea Aerospace Research Institute (KARI), Daejeon 305-701, Republic of Korea

Correspondence should be addressed to Hyochoong Bang; hcbang@ascl.kaist.ac.kr

Academic Editor: Christian Circi

This paper focuses on maximizing the percent coverage and minimizing the revisit time for a small satellite constellation with limited coverage. A target area represented by a polygon defined by grid points is chosen instead of using a target point only. The constellation consists of nonsymmetric and circular Low Earth Orbit (LEO) satellites. A global optimization method, Genetic Algorithm (GA), is chosen due to its ability to locate a global optimum solution for nonlinear multiobjective problems. From six orbital elements, five elements (semimajor axis, inclination, argument of perigee, longitude of ascending node, and mean anomaly) are varied as optimization design variables. A multiobjective optimization study is conducted in this study with percent coverage and revisit time as the two main parameters to analyze the performance of the constellation. Some efforts are made to improve the objective function and to minimize the computational load. A semianalytical approach is implemented to speed up the guessing of initial orbital elements. To determine the best parametric operator combinations, the fitness value and the computational time from each study cases are compared.

1. Introduction

Compared to a single satellite, a constellation may provide better coverage and higher reliability under failure of some satellites, ensuring a higher rate of survival and mission success. A constellation can offer unique capabilities that are often difficult to achieve through different means, for instance, enhanced temporal coverage [1]. A satellite orbit constellation design usually relies on the Walker approach [2] or streets of coverage method [3, 4]. Besides, a ground track-based approach has been developed [5], and recently the sliding ground track concept applied to constellations composed of one or more orbital planes has been introduced [6].

Related to the coverage mission, the idea of satellite constellation designs for complex coverage was presented by Ulybyshev [7]. Indrikis and Cleave, in their work about SPACE EGGS or the Satellite Coverage Model for LEO (Low Earth Orbit) Constellations [8], also tried to access the effectiveness of global, regional, and area coverage for proliferated small satellite constellations in low altitude orbits, stressing the capability of conventional analytical techniques. The idea of staged deployments of satellite constellations in LEO was also proposed by including the uncertainty feature of the expected number of users and activity level [9]. In the process, a satellite's life cycle cost and its capacity is traded off to satisfy the minimum per-channel performance requirement. Another approach to optimize a satellite constellation design by using tiers of satellites with variations in the orbit's altitude and inclination parameters was proposed Razoumny et al. in [10]. This approach tries to reduce the redundancy in the Walker approach by dividing the regional coverage into several latitude regions that can be addressed by different pairs of satellites' altitude and inclination, allowing for a better revisit time or a reduced number of satellites.

Besides the analytic methods, the genetic algorithm (GA) is known for its robustness in obtaining a global optimum solution for nonlinear multivariable problems through its stochastic and heuristic search algorithms. In their thesis, Pegher and Parish [11] tried to compare the coverage optimization and the revisit time of sparse military satellite constellations using traditional approaches and GA. Another method of hybrid satellite constellation design called Genetic Satellite Constellation (GSC) was proposed in [12] by using single GA optimization. This approach demonstrates the performance of hybrid LEO/MEO, LEO/GEO, and MEO/GEO constellations. A study on Simulated Annealing and GA approaches to satellite constellation design for coverage of a limited latitude region was conducted by Crossley and Williams [13], where both methods outperformed the conventional Walker approach at low Earth central angles.

Multiobjective genetic algorithm (MOGA) is a variation of GA, which is known to be a robust technique to solve multiobjective optimization, resulting in a Pareto optimal set solution [14]. A class of fast and elite MOGA was introduced by Deb et al. in 2002 [15]. This new MOGA called NSGA-II stands for Nondominated Sorting GA, which uses a fast nondominated sorting and a crowding distance assignment in its algorithm. Several multiobjective optimization studies have been performed. For example, Ely used two-branch tournament GA for a constellation design with eccentric orbits, aiming to minimize the maximum constellation's orbit altitude and the total number of satellites [16]. Several works from Ferringer are also focused on multiobjective optimization of satellite constellation designs including the trade-off analysis [17–20]. Related to MOGA, Mason et al. introduced a type of MOGA approach, called the Modified Illinois Nondominated Sorting Genetic Algorithm (MINSGA), combined with Satellite Tool Kit (STK) software to produce several constellation designs that provides a continuous global coverage [21]. Another work was also done by Confessore et al. to optimize a satellite constellation's orbit design with GA aiming to minimize the number of satellites and maximize region coverage [22].

However, to the best of our knowledge, there is no specific unified design approach for a local continuous coverage or surveillance mission over a region. This paper addresses coverage and revisit time problems using a sparse satellite constellation design with limited coverage capability, as is often encountered in low-cost satellite missions. Some low-cost small satellites tend to use small cameras for image acquisition [23, 24]. Several recent developments made in this area are the BRITE mission, which is an explorer targeting bright stars, conducted by the Canadian Space Agency, partnering with the University of Vienna and Graz University of Technology from Austria and the Copernicus Astronomical Center from Poland [16], and the PRISM Earth-imaging validation mission developed by the University of Tokyo using a CMOS imager on a narrow-angle camera [25].

The contribution of this paper is twofold: the first is a method to address a complex coverage area by using sparse satellite constellation design with a limited coverage capability and the second is an efficient approach to lower the computational burden when performing the GA optimization. In this paper, a case study is conducted on a sparse satellite constellation using multiple circular LEO satellites equipped with imaging sensors. A GA-based optimization is chosen due to its ability to locate a global optimum solution for a wide class of nonlinear problems that may be applicable to the case of this study. Several constellations' figures of merits [26], such as the area percent coverage and revisit time, are selected to evaluate the constellation performance using a multiobjective optimization algorithm. As much as GA seems to be a compatible optimization method to solve such a problem, it is often avoided due to its high computational load [27]. To minimize the computational load, a semianalytical approach is also proposed, which brings about a significant savings in computational time. The proposed approach, a combination of GA and an efficient semianalytical approach, will be a viable option for optimum satellite constellation designs.

2. Problem Statement

2.1. Mathematical Formulation of a Target Area. The area of interest is defined by inputting latitude-longitude coordinates of the main cities located in the country of interest. From these points, a convex hull polygon is generated using a computational geometry method, as depicted in Figure 1. The convex hull of a set of points is the smallest convex set that contains the points [28]. The area of interest is divided into several effective grid points, which are then evaluated to determine the target access time.

The grid points are defined in geodetic latitude Φ_t and longitude λ_t, where $t = 1, \ldots, N$ and N is the total number of grid points. Furthermore, zero ellipsoidal height of the target points is assumed.

2.2. Orbit Design. Initial orbit design consists of six standard orbital elements: semimajor axis a, eccentricity e, inclination i, longitude of ascending node Ω, argument of perigee ω, and mean anomaly M.

The maximum Earth-centered half-angle θ and the maximum swath width of the satellite sensor occur when the satellite is at apogee $r_a = a(1 + e)$. As shown in Figure 2, a satellite footprint projection on the Earth's surface with elevation angle ε defines a coverage circle with a radius θ through the following equation [29]:

$$\theta_{\max} = -\alpha + \operatorname{asin}\left(\frac{a(1+e)}{R_e}\sin\alpha\right). \tag{1}$$

3. Semianalytical Initial Guess

Using a semianalytical initial guess, the time step for propagation could be adjusted in order to reduce the overall computational load. Comparison between this method and the numerical propagation method in terms of the area coverage difference is presented as well in this paper. This method is based upon the correlation between nonsingular orbital elements and the target Φ and Ψ. This approach is especially useful for observation over a repeating ground track orbit.

In comparison with numerical propagation, this approach allows for information about the satellite position at a specific time without having to propagate the satellite's

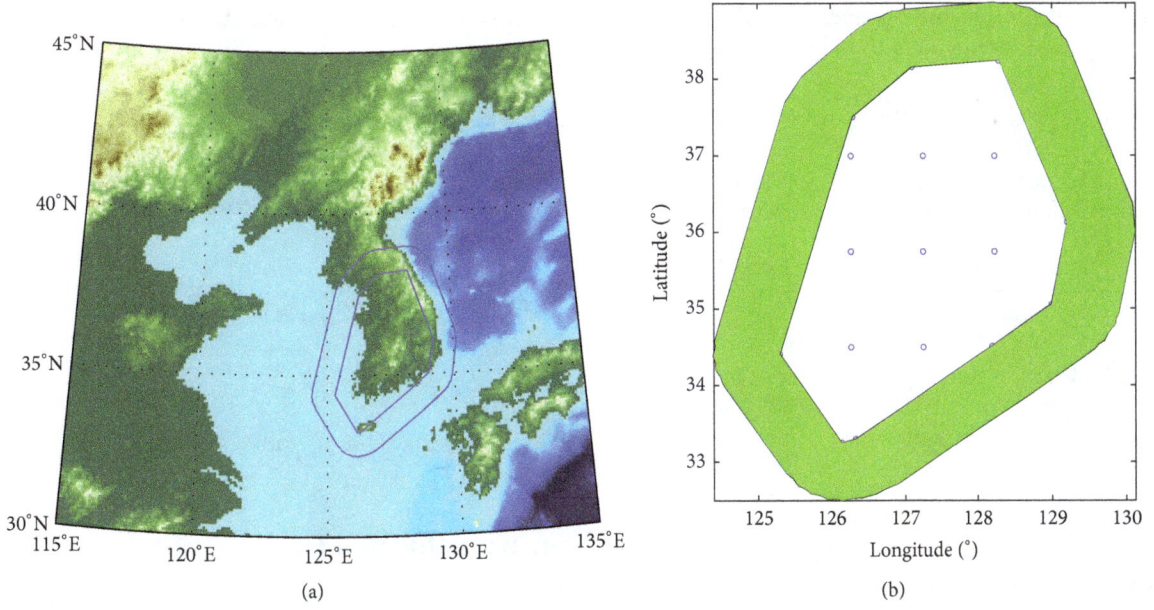

FIGURE 1: Example region (a) and the polygon made by convex hull mapping (b).

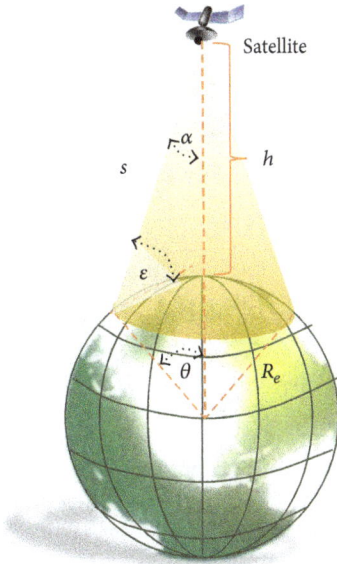

FIGURE 2: Satellite footprint projection.

orbital elements or the satellite's position and velocity vector over a time step. The accuracy of numerical propagation results depends on the size of the time step, which, on the other hand, conflicts with the size of available computing memory and computational load.

By the mean element theory [30], a satellite orbit is described using its mean orbital elements. The time rate of secular changes of mean Ω, ω, and M is given by [31]

$$\dot{\Omega} = -\frac{K}{a^{7/2}\left(1-e^2\right)^2}\cos i,$$

$$\dot{\omega} = \frac{K}{a^{7/2}\left(1-e^2\right)^2}\left(2-\frac{5}{2}\sin^2 i\right),$$

$$\dot{M} = n + \frac{K}{a^{7/2}\left(1-e^2\right)^{3/2}}\left(1-\frac{3}{2}\sin^2 i\right),$$

$$(2)$$

where $K = (3/2)J_2 R_e^2 \sqrt{\mu}$ and the mean motion n is given by

$$n = \sqrt{\frac{\mu}{a^3}}.$$

$$(3)$$

Drag perturbation is described using the following equations [32]:

$$\dot{a}_d = -\frac{\rho A C_d \sqrt{\mu a}}{m}\left(1-\left(\frac{\omega_e}{n}\right)\cos i\right)^2.$$

$$(4)$$

3.1. Latitude Access. The correlation between the satellite mean orbital elements and the access to a target region's latitude is defined in the following form [32]:

$$\phi - \frac{\pi}{2} + \cos^{-1}\sin i \cos\phi_s = 0,$$

$$(5)$$

$$\phi_s = \frac{3\pi}{2} + \omega_0 + \dot{\nu}\Delta t,$$

$$(6)$$

where ϕ_s represents the angular distance along the ground track measured from the ascending node at time = 0. The equation is only applied to an orbit with small eccentricity. On the assumption that the satellite in this study is equipped with a conical sensor, (6) becomes

$$\phi \mp \theta_{\max} - \frac{\pi}{2} + \cos^{-1}\left(\sin i \cos\phi_s\right) = 0.$$

$$(7)$$

For a noncircular orbit, the maximum swath width θ_{max} achieved on the perigee point is given by

$$\theta_{max} = -\alpha_{half} + \sin^{-1} \left(\frac{a(1+e)}{R_E} \sin \alpha_{half} \right). \tag{8}$$

It is recommended to use the time derivative of θ for a noncircular orbit with drag and J_2 perturbation included.

$$\dot{\theta} = \frac{(\sin \alpha_{half}/R_e)\left((r/a)\,\dot{a} + (r^2/p)\,e\sin(f)\,\dot{f}\right)}{\sqrt{1 - r\left(\sin \alpha_{half}/R_e\right)}}. \tag{9}$$

The latitude access time repeats for every period of mean argument of latitude or every $2\pi/(\dot{\omega} + \dot{M})$. $T_{lat\,i}$ is defined as a set of latitude access times such that

$$T_{lat\,i} = \left[p_{in\,i}, p_{out\,i} \right], \quad i = 1, \ldots, m. \tag{10}$$

3.2. Longitude Access. The correlation between satellite orbital elements and the area longitude is defined as

$$\Psi - (\Omega_0 + \omega_0 - \text{GMST}_0 \mp \theta_{max}) - \frac{(\dot{\Omega} + \dot{\omega} - \omega_E)\,l}{l} \tag{11}$$

$$= \text{atan2}\left(\cos i \sin l, \cos l\right),$$

where l is the mean argument of latitude and $\mp\theta_{max}$ represents the entrance and exit of the access. A similar approach using a straightforward expansion to solve this equation is explained in [29] with a difference in the usage of atan2 instead of the usual atan, since the mean argument of latitude lies between $-\pi$ and π. The analytical solution approach is described briefly by first defining several variables. Let A_1, A_2, and A_3 be defined as follows:

$$A_1 = \Psi - (\Omega_0 + \omega_0 - \text{GMST}_0 \mp \theta_{max}), \tag{12}$$

$$A_2 = \frac{(\dot{\Omega} + \dot{\omega} - \omega_E)}{l}, \tag{13}$$

$$A_3 = \cos i. \tag{14}$$

Substituting (13)–(15) into (12) results in the following equation:

$$A_1 - A_2 l - \text{atan2}\left(A_3 \sin l, \cos l\right) = 0. \tag{15}$$

Equation (16) can be solved by using a solver in MATLAB®. $\alpha = \text{atan2}(y, x)$ is a quadrant-sensitive inverse of $\tan \alpha = y/x$, which returns a value in the range $-\pi \leq \alpha \leq \pi$. Using Taylor series expansion, $\text{atan}(A_3 \sin l, \cos l)$ can be defined as

$$\text{atan}(x) = x - \frac{x^3}{3} + \frac{x^5}{5} - \frac{x^7}{7} + \cdots, \quad (-1 \leq x \leq 1), \tag{16}$$

$$x = A_3 \sin l, \cos l. \tag{17}$$

Let us put a value of l into the equivalent Kepler equation in terms of the mean element and derive Δt:

$$l = (\omega_0 + M_0) + (\dot{\omega} + \dot{M})\,\Delta t. \tag{18}$$

$T_{\text{lon}\,j}$ is defined as a set of latitude access time for which

$$T_{\text{lon}\,j} = \left[q_{in\,j}, q_{out\,j} \right], \quad j = 1, \ldots, n. \tag{19}$$

The time for intersection of a target point can be found from the time window intersection of each latitude and longitude access. The intersection time T_{access} consists of s access times, where

$$T_{access} = \bigcup_{i=1}^{m} \left[T_{lat\,i}, T_{lon\,i} \right] = \left[T_{in\,k}, T_{out\,k} \right], \tag{20}$$

$$k = 1, \ldots, s.$$

The flowcharts of the algorithm are depicted in Figures 3 and 4.

4. Multiobjective Optimization

4.1. Pareto Approach. Multiobjective optimizations are usually expressed by introducing different objectives in a cost function, for which each objective importance is represented by a weight w_i parameter for the i-th objective, as in [33].

$$J = \min_{x} \sum_{i=1}^{n} w_i f_i(x), \quad x \in C, \tag{21}$$

$$J = \min_{x} (1 - \alpha) f_1(x) + \alpha f_2(x), \quad x \in C, \tag{22}$$

where

$$\alpha = \frac{\sin \theta}{\cos \theta + \sin \theta}. \tag{23}$$

Using such a weighted sum approach, the objective functions should always be normalized or scaled so that their objective values are in similar magnitudes. However, according to [34], there is a certain drawback of minimizing the weighted sums of objectives for Pareto set generation in multicriteria optimization problems. The reason is that an evenly distributed set of weighting vectors cannot guarantee an evenly distributed representation of the Pareto optimal solutions.

4.2. Nondominated Sorting Genetic Algorithms Optimization (NSGA-II). As described in Figure 5, NSGA-II, which stands for Nondominated Sorting GA, uses a fast nondominated sorting and a crowding distance assignment [8]. GA optimization itself is a heuristic, stochastic global search based on the Darwinian concepts of evolution and natural selection and was first introduced by John Holland in mid-1960s [33]. An elitist GA favors individuals with a better fitness measure (rank). A controlled elitist GA, which is used in this algorithm, facilitates the optimization process to converge into a set of diverse solutions, even if the solution might consist of several lower performing individuals. NSGA-II is selected for this research because it has been shown to exhibit a better spread and convergence, relatively close to the true Pareto optimal front compared to PAES and SPEA [34].

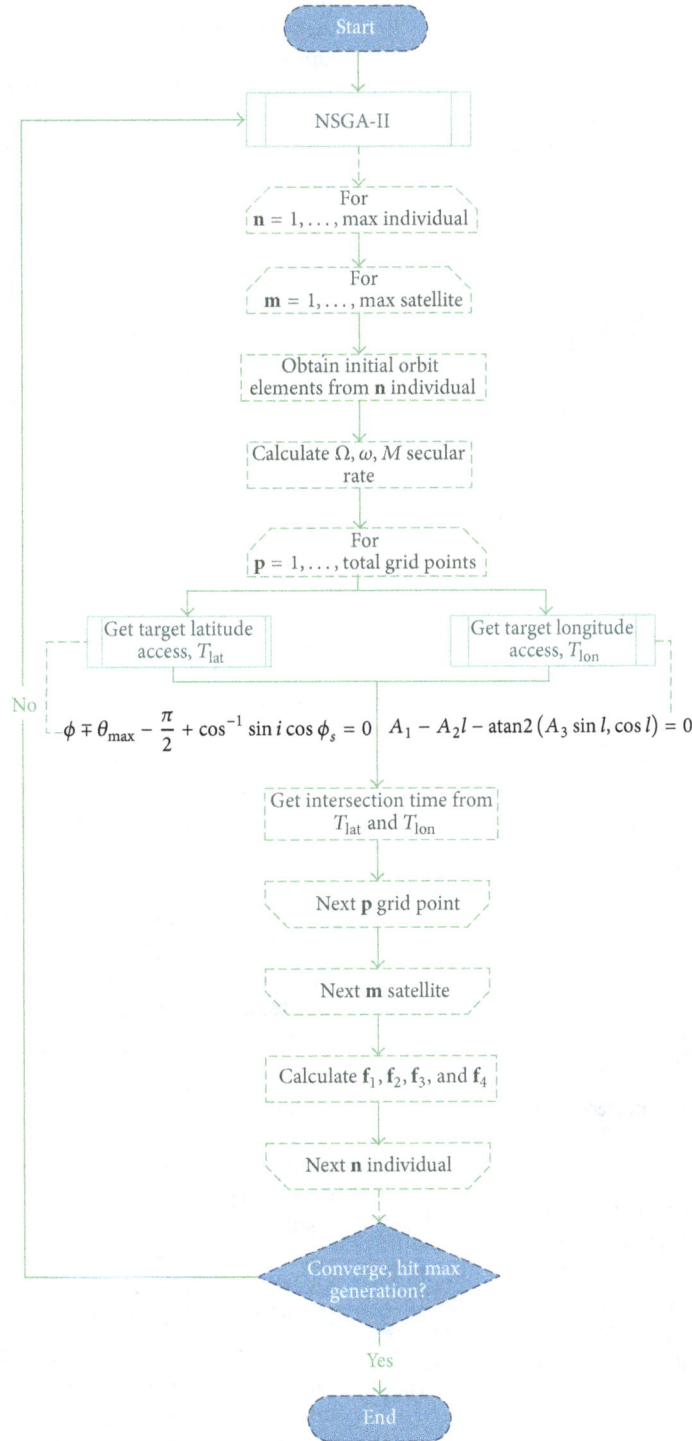

FIGURE 3: Latitude and longitude target access determination flowchart.

SPEA stands for Strength Pareto Evolutionary Algorithm, and PAES is Pareto Archived Evolution Strategy algorithm. PAES is a multiobjective optimizer approach, which uses an archive of previously found solutions in order to identify the dominance ranking of the current and candidate solution vectors. SPEA provides a form of elitism by using an archive of the nondominated set, which is maintained separately from the population of candidate solutions.

In the first generation, parent population P_t of I individuals creates an offspring population Q_t also in I individuals through the common tournament Selection, Crossover, and Mutation operators. Next, population R_t with a size of $2I$

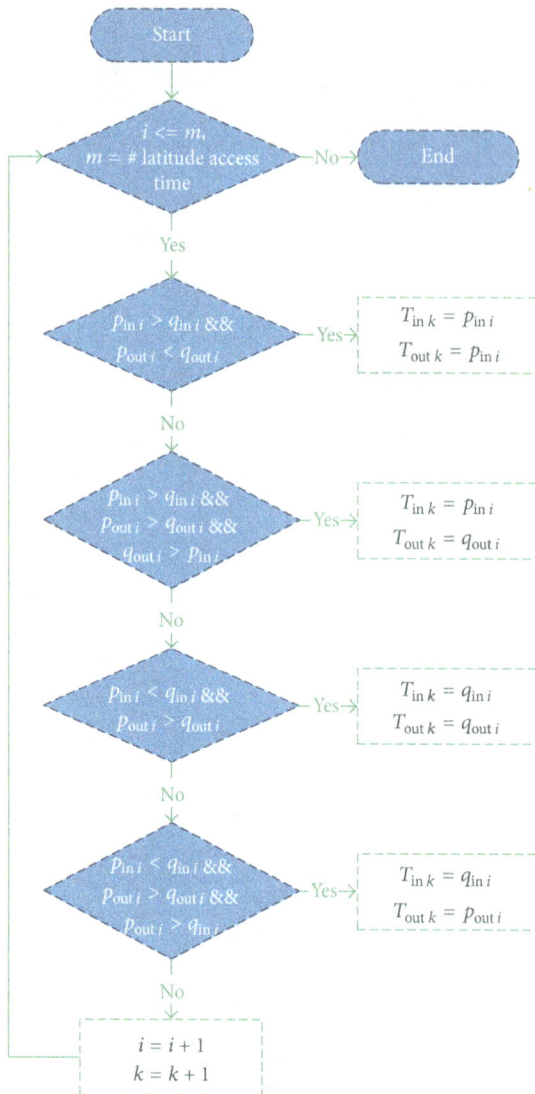

FIGURE 4: Target access time intersection flowchart.

individuals goes through a nondominated sorting. After the first generation, crowded tournament selection is used to select the top I individuals from R_t to form the next generation P_{t+1}. A solution that wins a tournament has either the highest rank or a better crowding distance parameter. The crowding distance parameter itself is defined as the largest cuboid surrounding a solution, for which no other solution is present [35].

Two parameters used in this algorithm to control the elitism are the Pareto fraction, which is the nondominated sorting using the Pareto set, and the distance function, which uses a crowding distance assignment. The Pareto fraction limits the number of individuals that can be put into the Pareto front (elite members). The distance function itself favors individuals who are located relatively far from the front since it can create a diverse, noncrowded set of solutions.

In this paper, the parametric operator is varied for the number of generations and population. The number of generations is usually set at the beginning of the optimization and

is usually used as a stopping criterion for the optimization. Another useful parameter is the stall generation, where the optimization stops if the best fitness function value does not stop after hitting several generations until the stall generation is met. Population, which has a constant value during the optimization, is the number of individual sets for every generation.

5. Simulations

A target region for the case South Korea, with its 37 cities' latitude and longitude positions, is used as an input for simulation. The area of interest itself is represented as a polygon of boundary points that includes all of the 37 cities. This area of the polygon is then divided into a number of grid points. N denotes the number of fast nondominated target points which is the number of grid points located inside the polygon, for which 10×10 grids dividing the example area result in a total of 50 grid points. The basic design criteria and constraints are as follows. A maximum satellite constellation of three satellites with a 10-degree half-angle of a conical sensor is used. This conical sensor is set as well so that only one satellite cannot cover the whole area of interest. Although from (1) it was mentioned that the maximum swath width occurs at the apogee, due to the circular orbit used in this simulation, this parameter is a matter of the satellite altitude only, assuming that the pointing error is neglected.

The constellation is placed into a circular orbit at an epoch date on August 1, 2020. Up to six hours of the satellite's lifetime is studied. In comparison with the simulation done in [29], the satellite's orbiting time period is much smaller; therefore, the argument of perigee and mean anomaly are included as optimization variables to compensate for the shorter observation time.

The 6-hour slot of the satellite lifetime was studied as an example of applying the proposed optimization algorithm to satellite constellation orbit design. The part of the satellite lifetime was selected to compare performance of the proposed semianalytical algorithm to that of typical genetic algorithms in a perspective of computational time. To get optimized results for whole time period of repeating ground track seemed unnecessary since the satellites in the constellation are on the same configuration every about 30 days and therefore much more time for simulation is needed.

Imaging sensors is assumed for this study. The illuminating conditions will change at every pass because the orbit is not Sun-synchronous. However, we believe it is feasible to use imaging sensors in such constellation as existing systems equipped with CMOS or thermal infrared camera.

The constraints for initial orbit elements used as optimization variables are defined in Table 1.

A comparison between the 4th order Runge-Kutta (4RK) numerical integration and the semianalytical approach is performed as well. The results match within an error of 0.15% to 4.33% over a 1.5- to approximately a 13-day orbit propagation time. In a single application, the semianalytical approach can save up to four times computational time compared to the numerical integration approach, using a

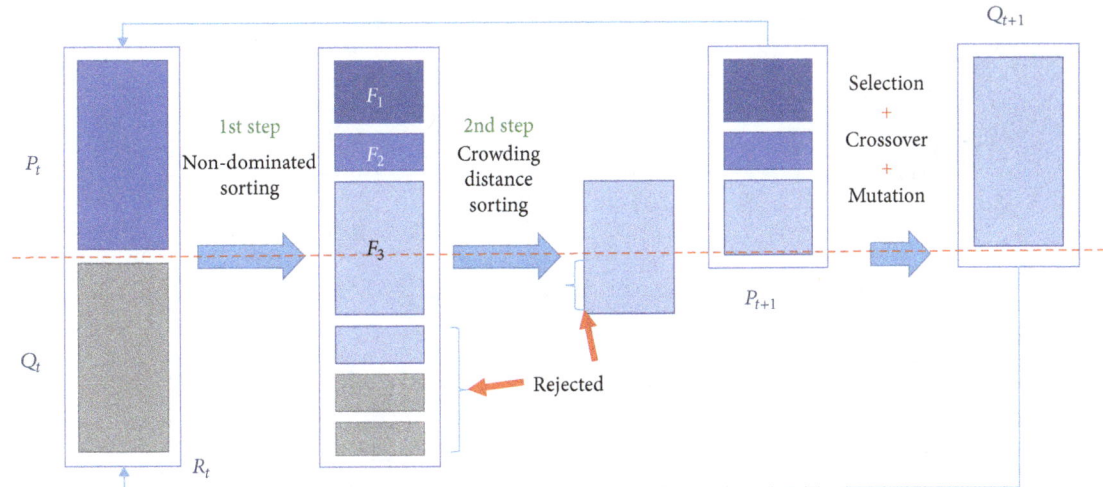

FIGURE 5: NSGA-II schema.

TABLE 1: Parameters variation for constellation visibility GA simulation.

Variation parameter	Value
Inclination	0–90 degrees
Semimajor axis	6538 to 6978 km, orbit height 160 to 600 km (LEO)
Longitude of ascending node	0–360 degrees
Argument of latitude	0–360 degrees

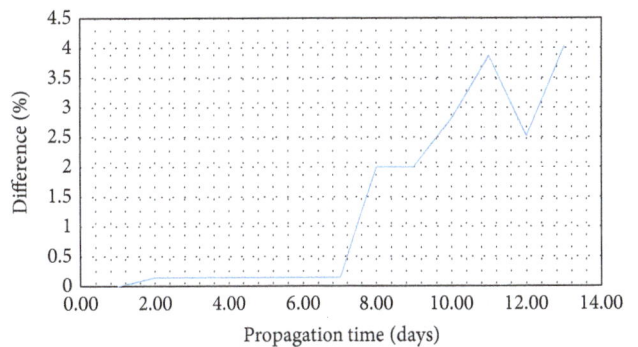

FIGURE 6: Coverage time comparison between 4th order Runge-Kutta propagation versus semianalytical approach.

TABLE 2: Drag parameters for a sample satellite (330 km LEO alt., $e = 0$, $i = 79°$).

Satellite parameter	Value
Mass (kg)	2.0
Area (m^2)	0.01
C_D	2.0

computer with an Intel ® Core™2 Duo CPU E7500 2.93 GHz processor and only 2.0 GB RAM. This test was performed using a random target site at 37° latitude and 14°E longitude. A satellite located in orbit with 0.005 eccentricity, $a = 6812.2$ km, $i = 37°$, and 30° Ω is propagated over five orbital periods. Figure 6 shows the error growth between the 4RK coverage results and the semianalytical approach, which can reach 4 minutes and 42 seconds during the longest 13-day propagation time. From the data, it is predicted that the difference will grow as the propagation time increases.

In overall application, the optimization's computational load can be reduced up to nine times using the semianalytical algorithm. If the satellite does not cover any target latitude, the algorithm automatically assigns a high value to the cost function and skips the access computation for longitude sections. Such bad performing individuals are automatically marked by the NSGA-II in the overall computation to exclude this individual in the next generation evaluation. This approach can speed up the entire computational time by skipping at least half of the calculation.

The fitness function of each individual is evaluated for every generation, and while the evaluation does not hit the stopping criteria, the selected parents continue to mate and reproduce new generations via Crossover and Mutation. The stopping criteria of the NSGA-II optimization algorithm used in this paper use MATLAB® default numbers, except for the TolFun or tolerance function being set at 10^{-3} (driven by the total number of grids and the definition of the fitness function formula). Optimum constellation design then covers as many grid points as possible and has a minimum average revisit time (ART), also with the lowest maximum revisit time.

Five out of six orbital elements, semimajor axis a, inclination i, longitude of ascending node Ω, argument of perigee ω, and mean anomaly M, are used as optimization design variables, while eccentricity e is kept as a constant. As a note, ω and M are included as design variables, despite the circular orbit assumption, to maximize the range of individual diversity. Using (4) and (5), the drag perturbation effects can be modeled and show increasing θ over propagation time because drag tends to lower the semimajor axis of a satellite. The data for drag is obtained from [35] in Table 2.

The effect of drag is presented in Figure 7.

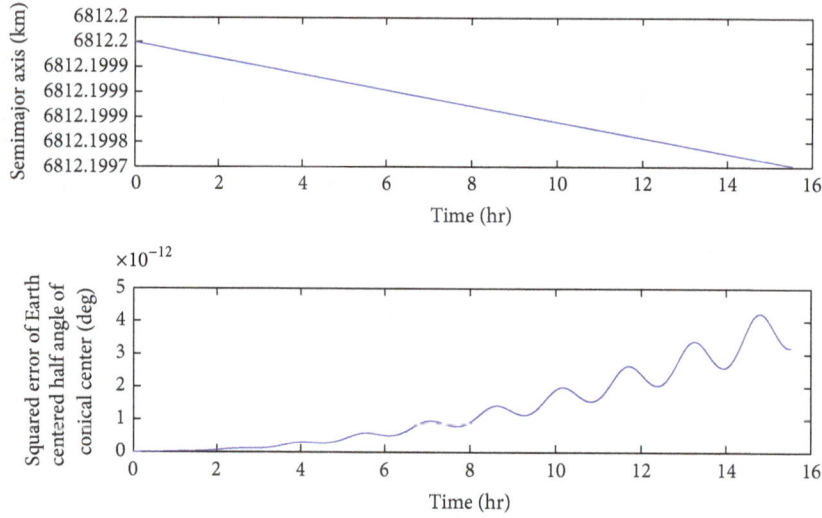

FIGURE 7: Effect of drag perturbation on the θ angle.

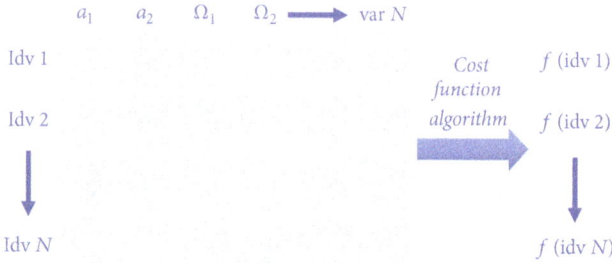

FIGURE 8: Schema for GA-based optimization algorithm.

There are four optimization objectives in this study, which is divided into two main case studies as follows: the coverage factor and the revisit time factor. The percent coverage for any point on the grid is described as the number of points covered, which is divided by the total number of points in the grid. The percent of coverage directly shows the effectiveness of satellite coverage. However, it cannot provide any information about the distribution of the gaps [32]. The coverage gap is defined here as the length of time where a point is not covered by any of the satellites in the constellation.

In the following, those two figures of merit of satellite constellations are represented in four objectives. Regarding the previous discussion about the convexity of the Pareto solution, the cost function in this study is represented in several objectives instead of using a weighted cost function. The area coverage is different from the time coverage since the time coverage is a summation of total satellite access time divided by the number of grids and the total simulation time. The following algorithm for GA-based optimization is depicted in Figure 8:

(1) Maximum area percent coverage

$$f_1 = 1 - \frac{\sum_{i=1}^{N} \text{coverage } idx}{N} + \text{reward}. \tag{24}$$

(2) Maximum average area time coverage

$$f_2 = 1 - \frac{\sum_{i=1}^{N} \text{coverage time}}{N \cdot \text{simulation period}}. \tag{25}$$

(3) Minimization of maximum coverage gap time

$$f_3 = \frac{\sum_{i=1}^{N} \text{max revisit}}{N \cdot 60}. \tag{26}$$

(4) Average coverage gap time

$$f_4 = \frac{\sum_{i=1}^{N} \text{revisit}}{N \cdot 60}. \tag{27}$$

The reward is defined as a negative value that is added to the cost function if 100% area coverage can be achieved by less than the maximum initial three satellites. Several case studies that were conducted are presented in Figure 9.

Using the semianalytical technique, the time of access accuracy is independent of the time step since it does not use numerical integration such as the Runge-Kutta orbit propagation. In this simulation, test cases are analyzed by varying two types of parameters: the optimization algorithm parameters (or the NSGA-II parameters in this case) and the constellation orbit design parameters. The NSGA-II parameters that can be varied include the population, generation, and the stall criteria. Other adjustments that could be made for Selection, Crossover, and Mutation parameters for the population mating and generation are kept at their default values this time since the effect of changing those parameters has been discussed previously [33, 36].

In Case 2, a different number of populations and generations are tried. By increasing the number of populations and generations, there is a greater probability to achieve a global optimum for any nonlinear optimization problem as the number of search points increases. The analogy for this can be seen in Figure 10, where the increment in the number

Case number	Optimized parameter					# of sat	NSGA-II parametric operators	
	a	i	Ω	ω	ν		Population	Generation
1						3	500	30
2						3	750	100
3						6	500	30
4						3	500	30
5						3	500	30
6						3	750	100
7		2 planes				3	500	30
8		Fuel penalty				3	500	30
9						3	500	100
10						3	750	30
11						3	500	30
12			2 planes			3	500	30

FIGURE 9: Parameters variations for percent coverage and ART by NSGA-II simulation.

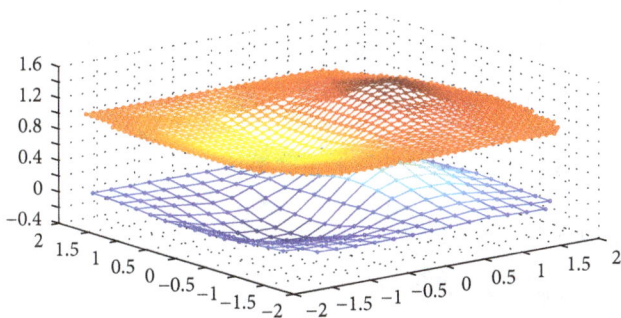

FIGURE 10: Mesh figure for nonlinear problem.

of populations looks like a mesh figure with an increasing number of mesh points. This results in a bigger search space throughout the overall optimization framework. In our case, it indicates that the Pareto front generated will have a better value, as can be seen in Case 2. However, Case 2 suffers from longest computational time due to the increased number of cost function counting.

Case 3 iterates the constellation orbit design parameters by increasing the number of satellites in the constellation from 3 to 6. The increment in the number of satellites can increase the area percent coverage and coverage time as well while also reducing both ART and maximum revisit time. However, the increment achieved is still lower than the results achieved in Case 2, even though the computational load is about two times lower. The role of the inclination angle in this satellite constellation design is to maximize the launch vehicle capability by choosing the inclination that is closest to the target latitude.

Cases 5, 6, and 7 all employ i and Ω only as optimized variables. In comparison with Case 5, the number of populations and generations is increased in Case 6, and only two orbital planes are used in Case 7. It can be seen from Figures 14–16 that the largest percent coverage from all four cases (a comparison with Case 1 included) is achieved by Case 6 on 88%

coverage. Both the lowest ART and maximum revisit time are achieved by Case 5 at 1.614 and 0.896 hours, respectively. Case 6 achieves the best overall performance in all four cases, although the CPU computational load significantly increases up to 4.5 times the average CPU computational load for all the three cases. To strengthen the results, Case 6 is compared with Case 2, both with 750 population and 100 generations. From these two cases, the results are almost identical in terms of the area percent coverage with only 2% difference. In terms of the time percent coverage, Case 6 exhibits slightly better coverage with 5 minutes difference, while in terms of ART, Case 2 achieves 16.8 minutes shorter time compared to Case 6. Figures 11–16 show the results for various cases.

Case 7 is conducted to seek an optimum design for a constellation of the same three satellites in two orbital planes in order to reduce the launch cost by using a combination of i, Ω, and ω. Seven variables are used in total for three satellites and two orbits, two slots for i, two for Ω, and three for ω variation. However, the separation between two satellites in the same plane is considered too small with only 13° difference in argument of latitude. Therefore, an additional case is attempted using five optimized variables in total, which are a, i, Ω, ω, and ν in Case 12. In Case 12, a constellation of three satellites distributed in two orbital planes is used. In comparison with Case 7, although the area coverage performance is similar, the time coverage achieved in Case 7 is about 15 minutes longer. Also, it can be seen that Case 7 has a greater number of solutions in the Pareto front with near optimum values. From the previous trends, regarding optimization results with respect to the number of optimized variables, one can see that the optimization using NSGA-II leads to a better set of solutions using a lower number of optimized variables (only i and Ω, compared to variations of a, i, Ω, ω, and ν) within a considerably lower number of populations and generations. Figures 17–19 show the results for Cases 7 and 12.

Cases 9 and 10 were conducted to see the results if the number of populations and generations is increased separately. In comparison with Cases 9 and 10, Case 2 yields

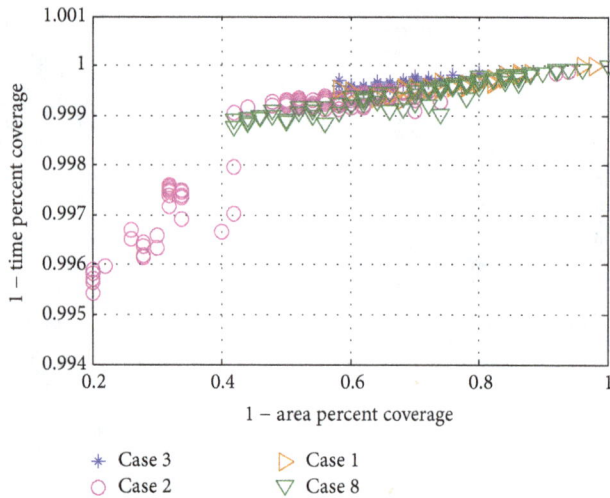

FIGURE 11: NSGA-II result for $f1$ and $f2$ objectives (Cases 1, 2, 3, and 8).

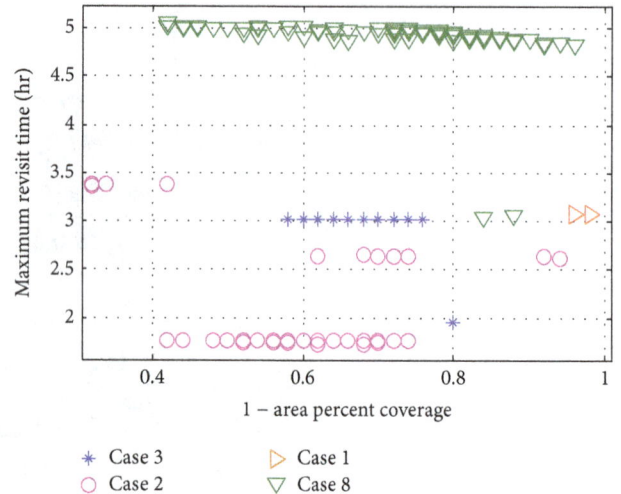

FIGURE 13: NSGA-II result for $f1$ and $f4$ objectives (Cases 1, 2, 3, and 8).

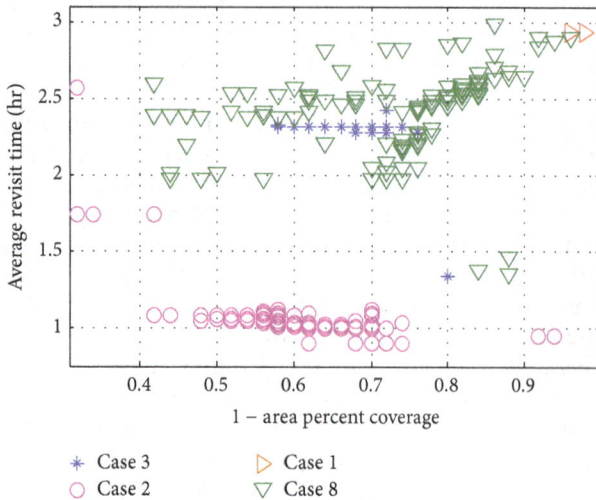

FIGURE 12: NSGA-II result for $f1$ and $f3$ objectives (Cases 1, 2, 3, and 8).

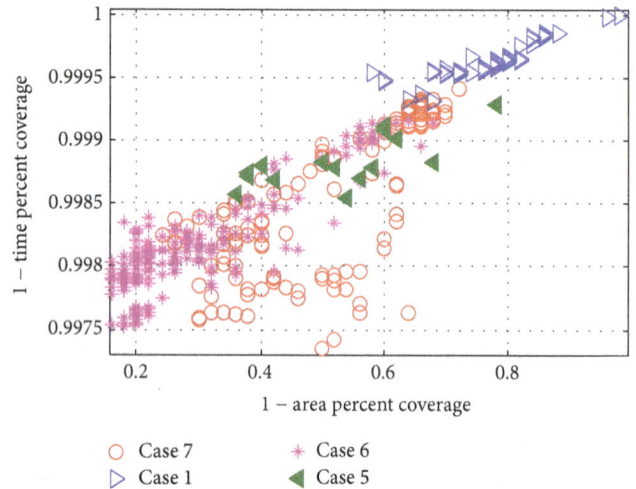

FIGURE 14: NSGA-II result for $f1$ and $f2$ objectives (Cases 1, 5, 6, and 7).

the best performance in all of the four optimized objectives, while the second-best performance is by Case 9. The results of the simulation are given in Table 3. From Table 4, similar or better performance with Case 9 is achieved by Cases 5, 7, and 12 with approximately 15% less computational load. By increasing the number of populations, the NSGA-II optimization approach allows a greater chance of finding the global optimum prior to converging at a cost of increased computational load. This argument explains why the results from Case 10 are similar to Case 9 despite only half of the function evaluations in Case 9. Figures 20–22 show the results for Cases 1, 2, 9, and 10.

Figures 23–25 present a comparison between Cases 1, 4, 5, and 11 results with a different number of orbital elements used for optimization variables. The comparison indicates that Case 5 yields the best results in terms of overall objectives (percent coverage and revisit time) followed by Case 11 with

the best area coverage at 56% and then Case 4 at 44%. However, in terms of both the maximum revisit time and ART, the second rank is held by Case 4, while Case 11 offers a longer maximum revisit time at 3.089 hours compared to 1.683 hours in Case 11. The least performing constellation is Case 1 with a variation of a, i, Ω, and ω as optimized variables, while the best performing constellation is Case 5 with only i and Ω as optimized variables. Figures 26–29 show satellites' full ground tracks for various conditions and Tables 5–8 summarize orbit elements for the solutions.

In comparison with the best area coverage result in Case 7, it can be seen that, by introducing ν as an optimized variable in Case 12, the satellites' ascending node in the same orbit plane (satellite 1 and 2) is separated up to 264°. Nonetheless, this separation does not create significant changes in terms of percent coverage nor the revisit time performance compared to the results of Case 7.

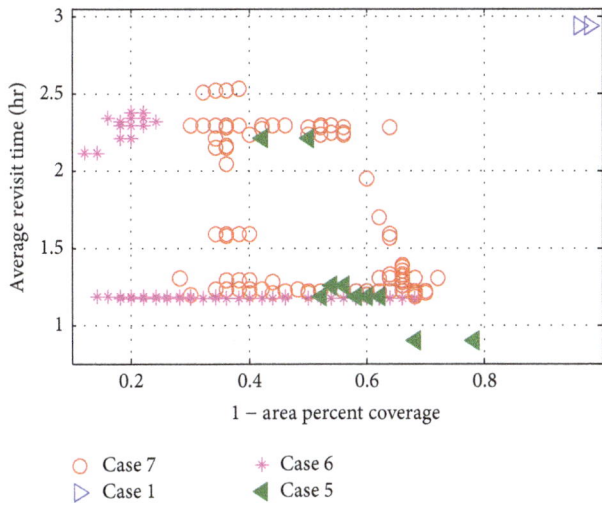

FIGURE 15: NSGA-II result for $f1$ and $f3$ objectives (Cases 1, 5, 6, and 7).

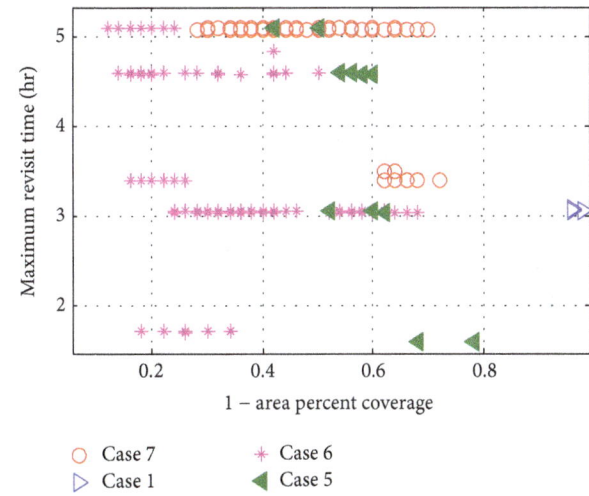

FIGURE 16: NSGA-II result for $f1$ and $f4$ objectives (Cases 1, 5, 6, and 7).

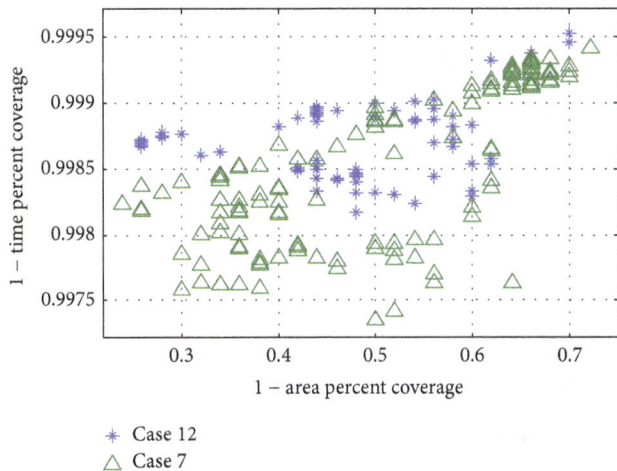

FIGURE 17: NSGA-II result for $f1$ and $f2$ objectives (Cases 7 and 12).

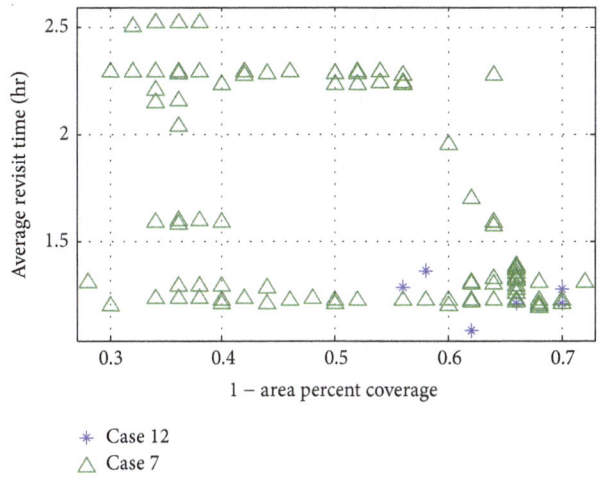

FIGURE 18: NSGA-II result for $f1$ and $f3$ objectives (Cases 7 and 12).

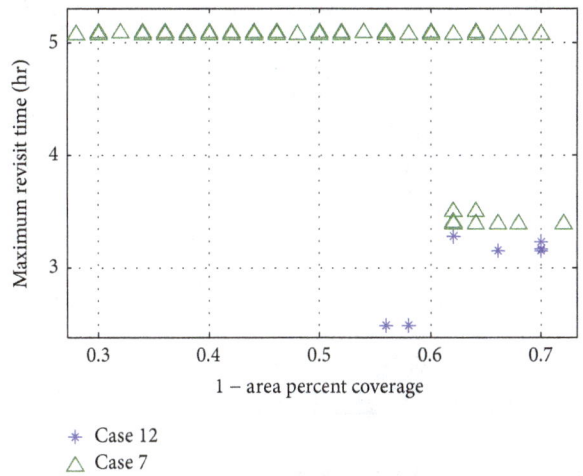

FIGURE 19: NSGA-II result for $f1$ and $f4$ objectives (Cases 7 and 12).

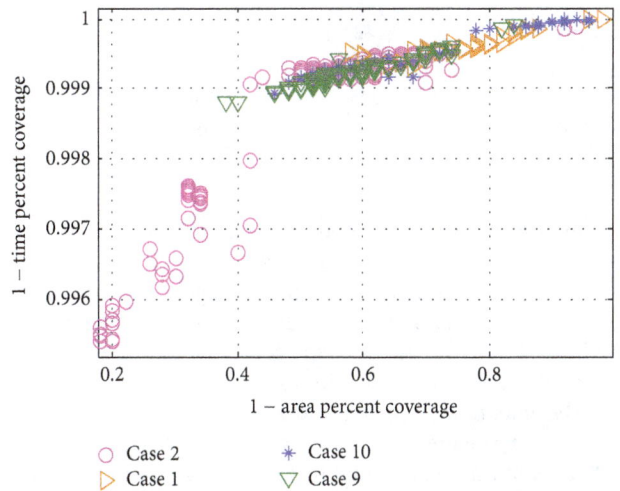

FIGURE 20: NSGA-II result for $f1$ and $f2$ objectives (Cases 1, 2, 9, and 10).

TABLE 3: Results for the NSGA-II simulation.

Case number	Σ points in Pareto front	Computation time (hr)
1	105	4.104
2	199	16.134
3	125	8.308
4	175	4.279
5	16	4.116
6	263	18.864
7	175	4.098
8	175	3.716
9	175	12.584
10	263	5.386
11	175	3.913
12	175	3.942

TABLE 4: Best result for all objective functions (f_1, f_2, f_3, f_4).

Case number	f_1 (%)	f_2 (min)	f_3 (hour)	f_4 (hour)
1	42	12.40	3.070	2.938
2	86	48.25	1.727	0.899
3	42	10.01	1.951	1.346
4	44	13.05	1.683	1.148
5	64	26.25	1.614	0.896
6	88	53.37	1.715	1.179
7	76	47.54	3.398	1.193
8	58	22.39	3.039	1.359
9	62	21.48	1.682	1.179
10	54	19.29	2.945	1.235
11	56	22.69	3.089	1.240
12	74	32.99	2.493	1.078

TABLE 5: Orbit elements for a solution in Case 7 (72% area coverage and 1 hour and 19 minutes of ART).

Orbit element	Sat #1	Sat #2	Sat #3
a (km)	6978	6978	6978
e	0	0	0
i (deg)	42	42	44
Ω (deg)	90	90	131
ω (deg)	324	326	284
ν (deg)	0	0	0

In summary, the overall objective function can be improved using several methods that are described in Table 9. This table was made by taking the average of performances from the cases that have either the same orbital elements as optimized variables or cases represented by the same number of populations and/or generations.

The maximum coverage time is around 53 minutes on average per satellite orbit, which was achieved in Case 6. Also, 88% maximum area coverage was achieved in Case 6. In terms of the revisit time, the minimum ART is 54 minutes in Case 5, while the lowest maximum gap time is around 1 hour and 37 minutes for Case 5.

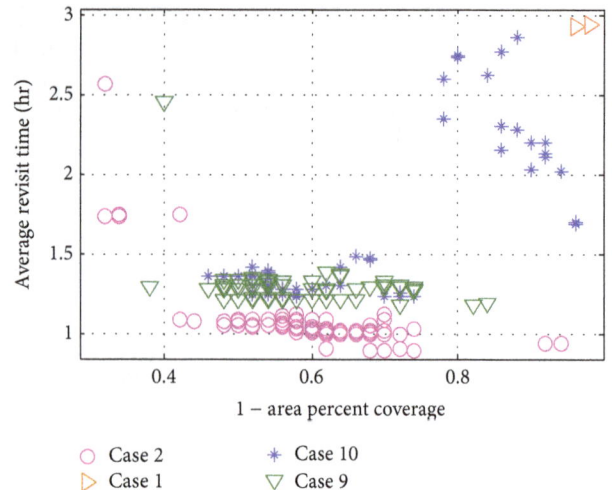

FIGURE 21: NSGA-II result for $f1$ and $f3$ objectives (Cases 1, 2, 9, and 10).

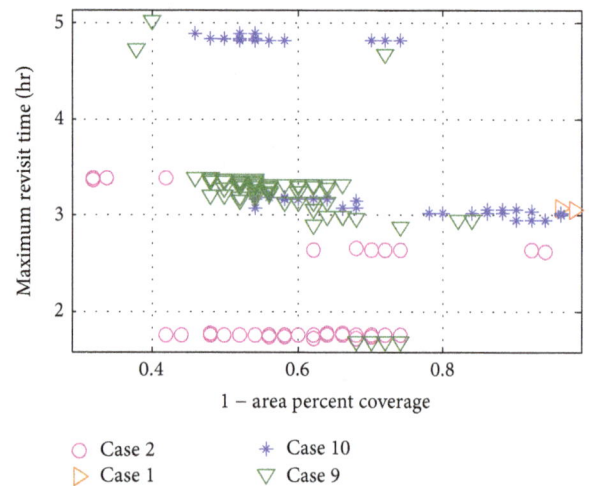

FIGURE 22: NSGA-II result for $f1$ and $f4$ objectives (Cases 1, 2, 9, and 10).

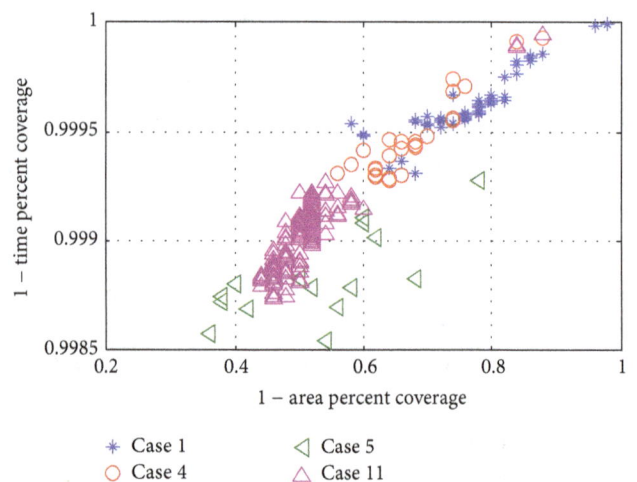

FIGURE 23: NSGA-II result for $f1$ and $f2$ objectives (Cases 1, 4, 5, and 11).

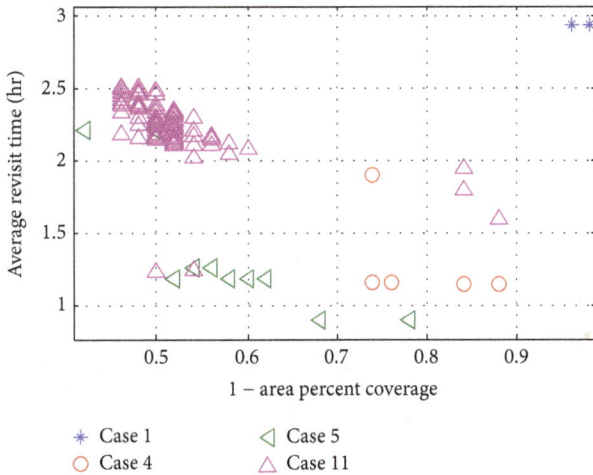

FIGURE 24: NSGA-II result for $f1$ and $f3$ objectives (Cases 1, 4, 5, and 11).

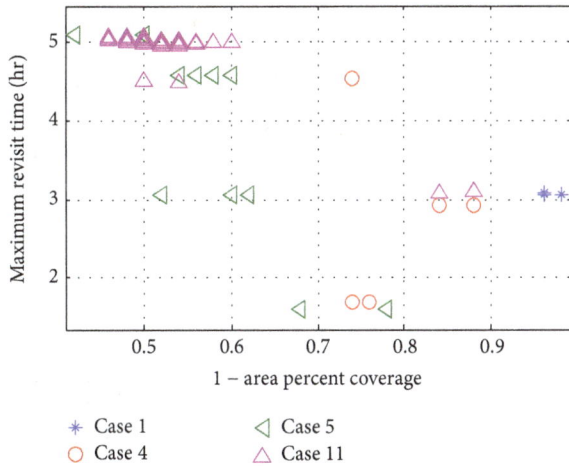

FIGURE 25: NSGA-II result for $f1$ and $f4$ objectives (Cases 1, 4, 5, and 11).

6. Conclusion and Future Work

By implementing a semianalytical approach, this study showed that computational load was reduced by over nine times within 0.5% error without having to numerically integrate satellite position or orbital elements over time to get the area coverage. Several figures of merits were chosen to analyze the performance of the optimized constellation orbit design, including maximum coverage, average coverage time, ART, and maximum revisit time over a target area. Circular LEO satellite constellation was used as a case study with a combination of a, i, Ω, and ω as optimization design variables. The satellites in this constellation design case were assumed to have a small conical sensor angle, which means that the satellites were subjected to a limited coverage. Therefore, the study cases were defined as well to achieve maximum coverage efficiency from the available number of satellites. Several cases were analyzed to see how

TABLE 6: Orbit elements for a solution in Case 7 with best maximum revisit time at 3 hours and 24 minutes and 32% area coverage.

Orbit element	Sat #1	Sat #2	Sat #3
a (km)	6978	6978	6978
e	0	0	0
i (deg)	42	42	53
Ω (deg)	154	154	136
ω (deg)	261	274	76
ν (deg)	0	0	0

TABLE 7: Orbit elements for a solution in Case 12 (best 74% area coverage and 9 hours of ART).

Orbit element	Sat #1	Sat #2	Sat #3
a (km)	6936	6936	6884
e	0	0	0
i (deg)	38	38	75
Ω (deg)	199	199	9
ω (deg)	251	159	158
ν (deg)	335	163	202

TABLE 8: Orbit elements for a solution in Case 12 with best maximum revisit time at 2 hours and 30 minutes and 44% area coverage.

Orbit element	Sat #1	Sat #2	Sat #3
a (km)	6938	6938	6871
e	0	0	0
i (deg)	39	39	78
Ω (deg)	198	198	118
ω (deg)	251	245	291
ν (deg)	333	342	23

the NSGA-II parameter or the orbital parameter changes might impact the optimizations results. To determine the best parametric operator combination, NSGA-II employed the Pareto concept to give a set of points that are able to satisfy all objectives.

From the conducted study cases, it was shown that results could be improved by increasing the number of populations and generations in NSGA-II or by increasing the number of satellites in the constellation. The latter option was considered inefficient since the results of those two cases were not much different, and the option of adding more satellite in the constellation may cost more. Also, by minimizing the number of optimization variables, better results were obtained since the complexity of the problem was decreased with the decreased number of variables. The percent of coverage was also improved by introducing a fuel penalty. Future work, including analysis of sensor pinpointing accuracy, may yield more meaningful results and make this problem closer to a real case.

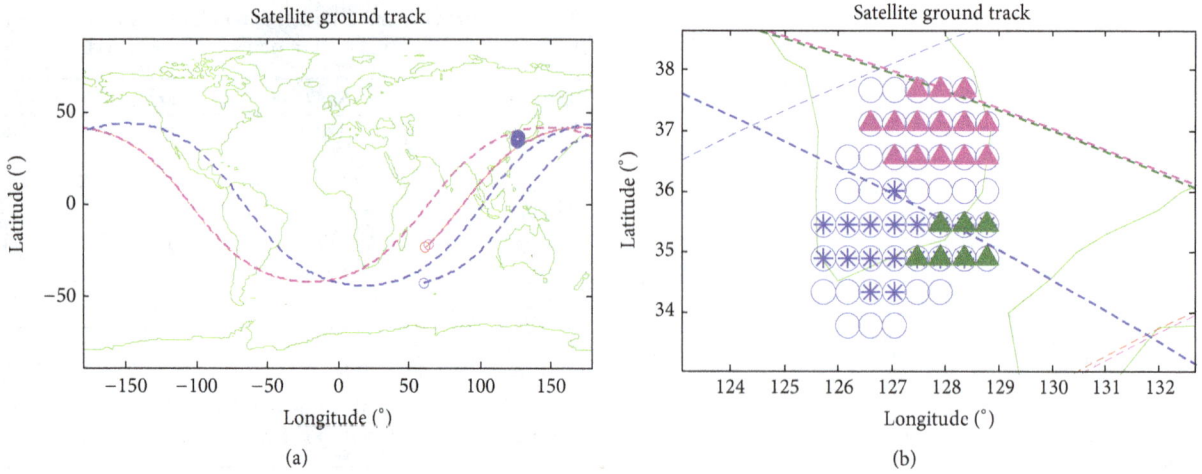

(a)   (b)

FIGURE 26: Satellites' full ground track (a) and close-up (b) for a solution in Case 7 with 72% area coverage and 1 hour and 19 minutes of ART. Red line: ground track of sat #1, pink line: ground track of sat #2, and blue line: ground track of sat #3. Green triangle: area covered by sat #1, pink triangle: area covered by sat #2, and blue star: area covered by sat #3.

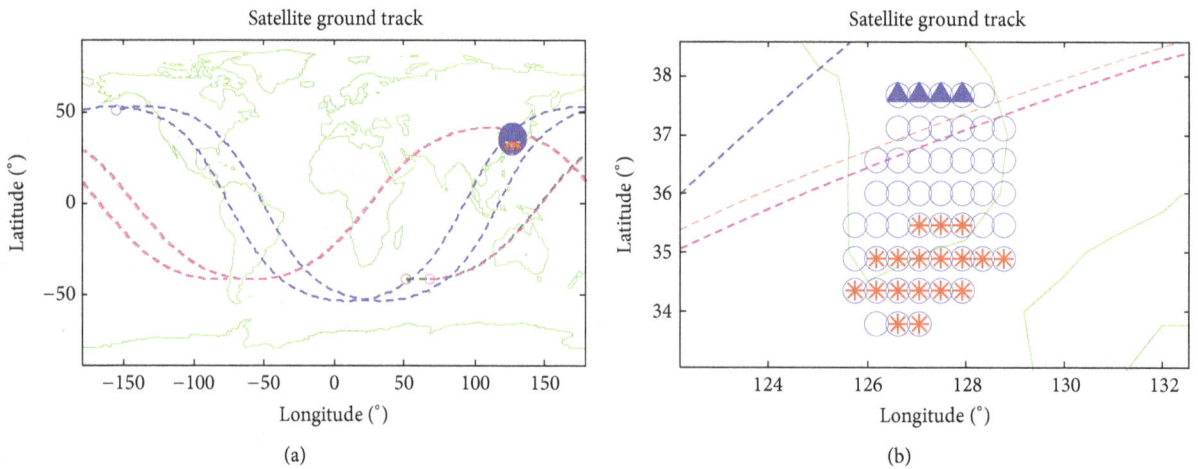

(a)   (b)

FIGURE 27: Satellites' full ground track (a) and close-up (b) for a solution in Case 7 with best maximum revisit time at 3 hours and 24 minutes and 32% area coverage. Red line: ground track of sat #1, pink line: ground track of sat #2, and blue line: ground track of sat #3. Red star: area covered by sat #1 and blue triangle: area covered by sat #3. In this case, area covered by satellites #1 and #2 is similar, so the paths are stacked on top of each other.

Nomenclature

A: Reference cross-sectional area of the satellite, m^2

a: Orbit semimajor axis, km

C_d: Drag coefficient of the satellite

e: Eccentricity

f_x: x-th cost function

h: Orbit altitude, km

h_{max}: Maximum number of satellites in the constellation

i: Orbit inclination, deg

J_2: Earth oblateness gravity harmonic coefficient, 1.0826×10^{-3}

l: Mean argument of latitude, deg

M: Mean anomaly, deg

m: Mass of the satellite, kg

N: Total number of target points

n: Orbit mean motion, deg/s

n_g: Number of grids

n_i: Number of individuals

naked(h^*): Number of uncovered grids by satellite denoted with h^* in percent ratio of all total available grids

R_e: Spherical radius of the Earth, 6378.1363 km

r_a: Satellite's apogee height, km

t: Time, s

α: Sensor half-angle of conical center, deg

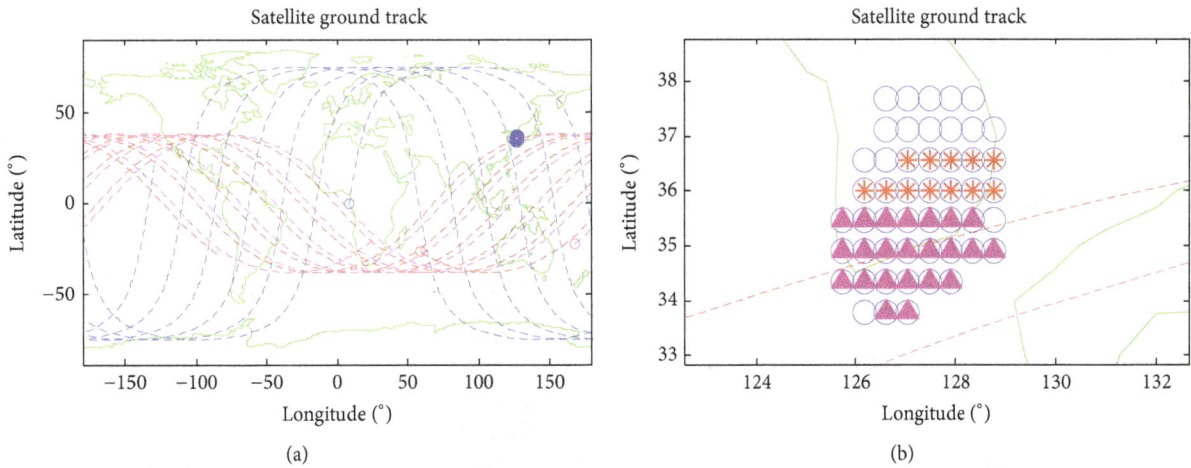

(a)   (b)

FIGURE 28: Satellites' full ground track (a) and close-up (b) for a solution in Case 12 (best 74% area coverage and 9 hours of ART). Red line: ground track of sat #1, pink line: ground track of sat #2, and blue line: ground track of sat #3. Pink triangle: area covered by sat #1 and red star: area covered by sat #2.

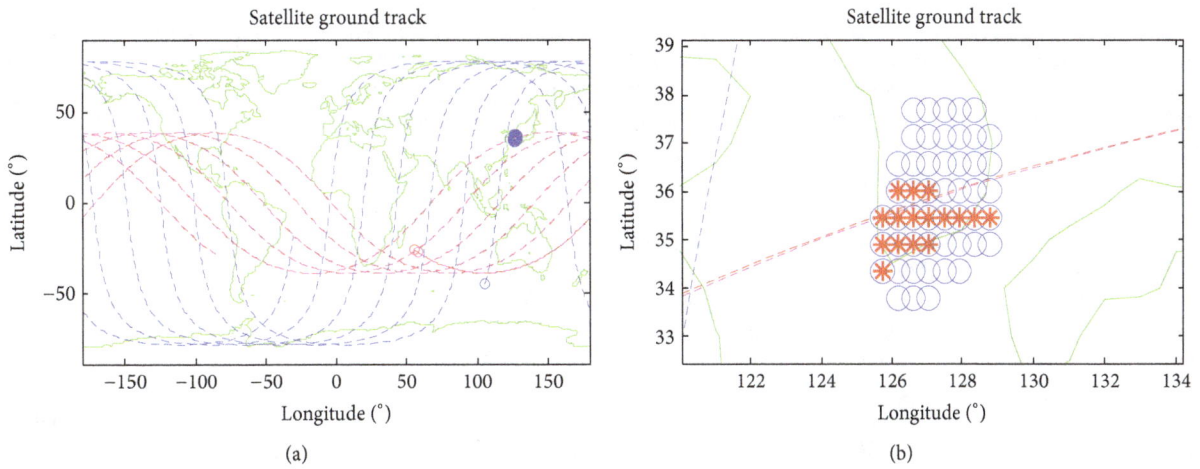

(a)   (b)

FIGURE 29: Satellites' full ground track (a) and close-up (b) for a solution in Case 12 with best maximum revisit time at 2 hours and 30 minutes and 44% area coverage. Purple line: ground track of sat #1 and blue line: ground track of sat #2. Red star: area covered by sat #1.

TABLE 9: Summary of parameter variation optimization results.

Variation	Percent coverage	Average revisit time	Computational load	Pareto front
Increasing number of populations	Better up to 29%	Better up to 54%	Worse up to 31%	Increasing number of solutions in Pareto front
Increasing number of generations	Better up to 48%	Better up to 60%	Worse up to 206%	N/A
Increasing number of populations & generations	Better up to 71%	Better up to 69%	Worse up to 293%	Increasing number of solutions in Pareto front
Using a variation of a, i, Ω, ω, and ν	Better up to 33%	Better up to 82%	N/A	N/A
Using a variation of a, i, and Ω	Better up to 5%	Better up to 61%	N/A	N/A
Using a variation of i and Ω	Better up to 29%	Better up to 70%	N/A	N/A
Fuel penalty used	Better up to 38%	Better up to 54%	N/A	N/A
Semisparse constellation (Cases 7 & 12)	Better up to 25%	Better up to 61%	N/A	N/A
Increasing number of satellites in constellation	Same value	Better up to 54%	Worse up to 102%	N/A

ε: Elevation angle, deg
λ_i: Target location longitude, deg, where $i = 1,\ldots,n_g$
μ: Gravitational parameter of the Earth, 398,600.4415 km^3/s^2
Ω: Longitude of ascending node, deg
ρ: Atmospheric density, kg/m^3
Φ_i: Target location latitude, deg, where $i = 1,\ldots,n_g$
θ: Earth-centered half-angle, deg
ω: Argument of perigee, deg
ω_E: The Earth's rotation rate, 7.2921×10^{-5} rad/s
v: True anomaly, deg.

Conflicts of Interest

The authors declare that there are no conflicts of interest.

References

[1] A. da Silva Curiel, A. Cawthorne, and M. Sweeting, "Progress in small satellite technology for Earth observation missions," in *Small Satellites for Earth Observation: Selected Proceedings of the 5th International Symposium of the International Academy of Astronautics, Berlin, April 4–8 2005*, p. 50, 2005.

[2] J. G. Walker, "Satellite constellations," *Journal of the British Interplanetary Society*, vol. 37, pp. 559–572, 1984.

[3] R. D. Luders, "Satellite networks for continuous zonal coverage," *ARS Journal*, vol. 31, no. 2, pp. 179–184, 1961.

[4] T. J. Lang and W. S. Adams, "A comparison of satellite constellations for continuous global coverage," in *Mission Design & Implementation of Satellite Constellations*, pp. 51–62, Springer, 1998.

[5] J. C. King, *Quantization and Symmetry in Periodic Coverage Patterns with Applications to Earth Observation*, 1975.

[6] C. Circi, E. Ortore, and F. Bunkheila, "Satellite constellations in sliding ground track orbits," *Aerospace Science and Technology*, vol. 39, pp. 395–402, 2014.

[7] Y. Ulybyshev, "Satellite constellation design for complex coverage," *Journal of Spacecraft and Rockets*, vol. 45, no. 4, pp. 843–849, 2008.

[8] J. Indrikis and R. Cleave, 'SPACE EGGS'-Satellite Coverage Model for Low Earth Orbit Constellations, 1991.

[9] O. L. De Weck, R. D. Neufville, and M. Chaize, "Staged deployment of communications satellite constellations in low earth orbit," *Journal of Aerospace Computing, Information and Communication*, vol. 1, pp. 119–136, 2004.

[10] Y. N. Razoumny, P. G. Kozlov, V. Y. Razoumny, and A. A. Moshnin, "On optimization of Earth coverage characteristics for compound satellite constellations based on orbits with synchronized nodal regression," in *Proceedings of the 2nd IAA Conference on Dynamics and Control of Space Systems*, Roma, Italy, March 2014.

[11] D. J. Pegher and J. A. Parish, *Optimizing Coverage and Revisit Time in Sparse Military Satellite Constellations: A Comparison of Traditional Approaches and Genetic Algorithms*, DTIC Document, 2004.

[12] M. Asvial, R. Tafazolli, and B. G. Evans, "Genetic hybrid satellite constellation design," *Parameters*, vol. 1, p. 100, 2003.

[13] W. A. Crossley and E. A. Williams, "Simulated annealing and genetic algorithm approaches for discontinuous coverage satellite constellation design," *Engineering Optimization*, vol. 32, no. 3, pp. 353–371, 2000.

[14] C. M. Fonseca and Fleming P. J., "Genetic algorithms for multiobjective optimization: formulation discussion and generalization," in *Proceedings of the 5th International Conference on Genetic Algorithms (ICGA '93)*, pp. 416–423, 1993.

[15] K. Deb, A. Pratap, S. Agarwal, and T. Meyarivan, "A fast and elitist multiobjective genetic algorithm: NSGA-II," *IEEE Transactions on Evolutionary Computation*, vol. 6, no. 2, pp. 182–197, 2002.

[16] T. Ely, "Satellite constellation design for zonal coverage using genetic algorithms," in *Proceedings of the 8th AAS/AIAA Space Flight Mechanics Meeting*, pp. 443–460.

[17] M. P. Ferringer and D. B. Spencer, "Satellite constellation design tradeoffs using multiple-objective evolutionary computation," *Journal of Spacecraft and Rockets*, vol. 43, no. 6, pp. 1404–1411, 2006.

[18] M. P. Ferringer, R. S. Clifton, and T. G. Thompson, "Efficient and accurate evolutionary multi-objective optimization paradigms for satellite constellation design," *Journal of Spacecraft and Rockets*, vol. 44, no. 3, pp. 682–691, 2007.

[19] M. P. Ferringer, R. S. Clifton, and T. G. Thompson, "Constellation design with parallel multi-objective evolutionary computation," in *Proceedings of AIAA/AAS Astrodynamics Specialist Conference and Exhibit*, Keystone, Colo, USA, 2006.

[20] M. P. Ferringer, D. B. Spencer, P. M. Reed, R. S. Clifton, and T. G. Thompson, "Pareto-hypervolumes for the reconfiguration of satellite constellations," in *Proceedings of the AIAA/AAS Astrodynamics Specialist Conference and Exhibit*, pp. 1–31, August 2008.

[21] W. J. Mason, V. Coverstone-Carroll, and J. Hartmann, *Optimal Earth Orbiting Satellite Constellations via A Pareto Genetic Algorithm*, University of Illinois at Urbana-Champaign, 2001.

[22] G. Confessore, M. Di Gennaro, and S. Ricciardelli, "A Genetic Algorithm to Design Satellite Constellations for Regional Coverage," *Operations Research Proceedings*, vol. 2001, pp. 35–41.

[23] N. Orr, J. Eyer, B. Larouche, and R. Zee, "Precision formation flight: the CanX-4 and CanX-5 dual nanosatellite mission," in *Proceedings of the 21st Annual AIAA/USU Conference on Small Satellites*, 2007.

[24] H. Ashida, K. Fujihashi, S. Inagawa et al., "Design of Tokyo Tech nano-satellite Cute-1.7+APD II and its operation," *Acta Astronautica*, vol. 66, no. 9-10, pp. 1412–1424, 2010.

[25] M. Komatsu and S. Nakasuka, "University of Tokyo Nano Satellite Project 'PRISM'," *Transactions of Space Technology Japan*, vol. 7, 2009.

[26] J. R. Wertz, *Mission Geometry; Orbit and Constellation Design and Management: Spacecraft Orbit and Attitude Systems*, vol. 13 of *Space Technology Library*, Microcosm, Boston, Mass, USA; Kluwer Academic Publishers, El Segundo, Calif, USA, 2001.

[27] E. S. H. Hou, N. Ansari, and H. Ren, "Genetic algorithm for multiprocessor scheduling," *IEEE Transactions on Parallel and Distributed Systems*, vol. 5, no. 2, pp. 113–120, 1994.

[28] D. Sunday, *The Convex Hull of a 2D Point Set or Polygon*, vol. 109, 2006.

[29] P. Sengupta, S. R. Vadali, and K. T. Alfriend, "Satellite orbit design and maintenance for terrestrial coverage," *Journal of Spacecraft and Rockets*, vol. 47, no. 1, pp. 177–187, 2010.

[30] Y. Kozai, "The motion of a close earth satellite," *The Astronomical Journal*, vol. 64, pp. 367–377, 1959.

[31] H. Schaub and J. L. Junkins, *Analytical Mechanics of Space Systems*, AIAA, 2003.

[32] J. R. Wertz, *Mission Geometry; Orbit and Constellation Design and Management*, Space Technology Library, 2001.

[33] H.-D. Kim, O.-C. Jung, and H. Bang, "A computational approach to reduce the revisit time using a genetic algorithm," in *Proceedings of the International Conference on Control, Automation and Systems (ICCAS '07)*, pp. 184–189, IEEE, Seoul, Republic of Korea, October 2007.

[34] I. Das and J. E. Dennis, "A closer look at drawbacks of minimizing weighted sums of objectives for Pareto set generation in multicriteria optimization problems," *Structural Optimization*, vol. 14, no. 1, pp. 63–69, 1997.

[35] D. Ilyas, *Orbital Propagation and Formation Flying of CubeSats within QB50 Constellation*, Science and Technology (SpaceMaster), 2011.

[36] T. Savitri, O. Jung, and H. Bang, "Optimization of grid points coverage by satellite constellation using genetic algorithm," in *Proceedings of the Asia-Pacific International Symposium*, Takamatsu, Japan, 2013.

2

Assessment of Helicopter Pilot-in-the-Loop Models

Massimo Gennaretti, Federico Porcacchia, Simone Migliore, and Jacopo Serafini

Department of Engineering, Roma Tre University, Rome, Italy

Correspondence should be addressed to Massimo Gennaretti; massimo.gennaretti@uniroma3.it

Academic Editor: Christopher J. Damaren

The aim of this paper is the evaluation of several pilot models found in the literature, suited for helicopter pilot-assisted and pilot-induced oscillations analyses. Three main topics are discussed: (i) sensitivity of rotorcraft-pilot couplings simulations on the application of the different pilot models available in the literature; (ii) effect of vehicle modeling on active pilot modeling; (iii) effects of interactions between active and passive pilot models. The focus is on hovering flight, where a specific adverse rotorcraft-pilot coupling phenomenon, the vertical bounce, may occur. Pilot models are coupled with a comprehensive aeroservoelastic model of a mid-weight helicopter. The numerical investigations are performed in frequency domain, in terms of eigenanalysis and frequency response analysis.

1. Introduction

Machine-pilot couplings are a wide class of phenomena potentially affecting every sort of vehicle, with consequences that may vary from discomfort to severe accidents. While being well documented in aeronautics (where they are usually called aircraft-pilot couplings, APC) since the beginning of the twentieth century [1], they were often ascribed to pilot faults, due to both lack of preparation or incorrect evaluation. Consequently, they received limited attention, mostly during training, when the pilot was instructed to perform particular actions, like to release controls, if an anomalous aircraft behavior was arising. On the contrary, APCs have been ignored during design process for decades. The result was a number of accidents (even recently, like that on the Japan Airlines flight JAL 706 [2, 3], in 1997) that has been reported as inability of the pilot to deal with unexpected aircraft oscillations (in that case arising when he took controls without disengaging the autopilot).

However, in the last decades, the scientific and technical communities have become more conscious about the importance of considering pilot and aircraft as a whole, in order to avoid such, potentially catastrophic, phenomena. In that sense, the role of the pilot in their insurgency has been partially reduced, while more attention has been devoted to the assessment of the conditions under which the aircraft-pilot coupling may yield undesired aircraft response. APC events have been defined alternatively as "involuntary trajectories and flight behaviors originating from an anomalous interaction between the pilot and the aircraft" or as "pilot-in-the-loop instabilities" [4], in both cases remarking the need of an active, though not necessarily voluntary, pilot participation. The attention devoted to APCs has allowed the identification of undiscovered potential instabilities in some cases related to dynamics traditionally considered stable. As an example, it has been observed that the short-period mode damping may consistently drop in presence of pilot in the loop [5].

Additional challenges in the analysis of human-vehicle interaction arise nowadays, mostly due to the increased presence of flight control systems in modern aircraft. Acting in parallel or in series with the pilot, they may lose effectiveness or become source of instabilities if their interaction with him is not carefully assessed during the design phase.

Although cases of rotorcraft-pilot couplings (RPCs) are reported since World War II [6], studies on RPC have lagged about thirty years with respect to fixed-wing counterpart, the pioneering works of Mayo [7] and Pharam [8, 9] being dated back to the last decade of the twentieth century. This fact, combined with the greater proneness of rotorcraft to instabilities, results in a wider area of relatively unexplored

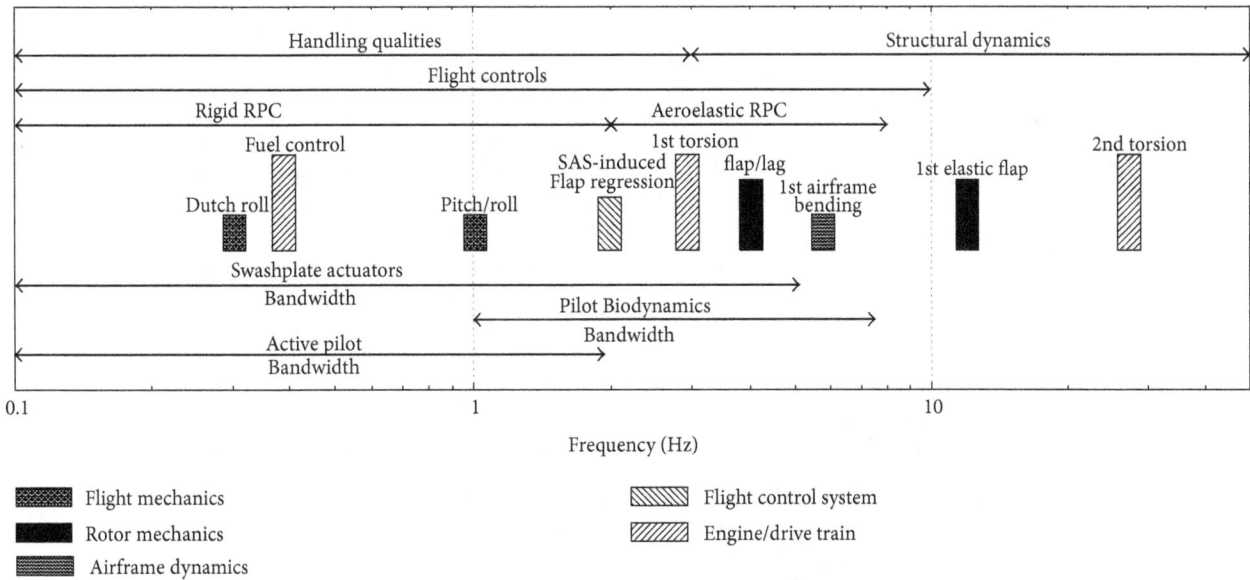

FIGURE 1: Typical rotorcraft frequencies (adapted from [23]).

research. In the last years, the EC-funded FP7 research project ARISTOTEL has been mainly focused on the analysis of RPCs insurgency and on means for their prevention. In that activity, the authors have been mostly involved in the assessment of aeroelastic triggers of RPCs, with particular emphasis on main rotor-airframe aeroelastic coupling [10–14].

Nevertheless, pilot modeling is still an open issue, usually addressed by separating active (voluntary, behavioral, and low-frequency) actions [15–17] from passive (involuntary, biodynamic, and mid-frequency) actions [7–9, 18–21], following the classification developed for fixed-wing aircraft. As a consequence, active pilot operations are more related to flight dynamics response phenomena, while passive pilot actions couple with aeroelastic responses.

The objective of the present work is the assessment of the following issues regarding pilot-in-the-loop rotorcraft response and stability analyses: (i) RPCs simulations sensitivity on the application of the different pilot models available in the literature; (ii) effect of vehicle modeling on active pilot modeling; (iii) effects of interactions between active-passive pilot models.

2. Rotorcraft-Pilot Coupling Classification

The RPC study requires competence in the modeling of several subsystems. Indeed, several different helicopter dynamics phenomena occur in the frequency range where RPCs events take place (as shown in Figure 1 for a medium-weight vehicle). These, combined with voluntary and involuntary pilot actions, leads to the definition of two main different kinds of RPCs: the so-called pilot-induced oscillations (or rigid-body RPCs), which are mostly related to active piloting and rigid-body dynamics, and the pilot-assisted oscillations (or aeroelastic RPCs) characterized by the involuntary feedback on controls due to cabin vibrations and mostly involving aeroelasticity and structural dynamics. The former involve

events in the frequency range from 0 to 2 Hz, while the latter occur in the frequency range from 2 to 8 Hz (above, both pilot biodynamics and controls actuator responses become negligible).

It is worth noting that this separation is quite arbitrary, since an overlap between active and passive pilot behavior exists. Moreover, unlike in the fixed-wing case, aeroelastic effects usually play a crucial role in rotorcraft flight dynamics, primarily due to the low-frequency main rotor deformation. However, the 2 Hz boundary is commonly accepted as valid from both the aeromechanics and the biodynamics point of view.

Another commonly used classification derived from fixed-wing field is that discerning the characteristic of the dynamics driving a RPC phenomenon and its trigger. Category I RPCs are caused by linear dynamics, while strong non-linearities (e.g., control saturations) determine categories II and III (in the latter, an external trigger like malfunctioning, gust, or pilot task is required). Finally, an informal category IV often overlapping with PAOs is associated with RPCs characterized by significant elastic deformations. While cat. I phenomena are usually investigated through eigenanalysis, cat. II and III require specific techniques (typically in the time domain) [4]. Cat. IV RPCs may be approached in different ways depending on the characteristics of the system: thus far, most of the analyses have been limited to linearizable systems, thus exploiting eigenanalysis (see, e.g., the investigation on vertical bouncing described in [11]).

Control designers have identified a number of retrofit solutions to alleviate RPCs, usually consisting of the introduction of small friction on controls. At the same time, the pilot community has identified piloting procedures to deal with specific RPCs. In the case of vertical bouncing, an instable loop involving collective control, main rotor coning, and vertical motion of pilot seat (see Figure 2), the procedure simply consists of loosening the grasp of the hand on the collective

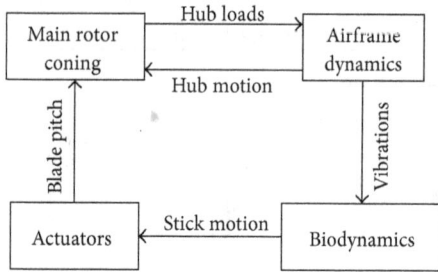

FIGURE 2: Vertical bouncing loop.

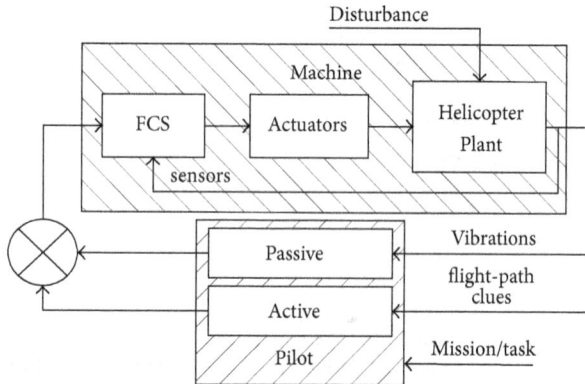

FIGURE 3: Pilot-in-the-loop representation.

lever, weakening the feedback. It is however evident that the capability to predict potential adverse couplings between pilot and aircraft since the design phase would represent a great advantage in the production of safer helicopters.

3. Pilot Modeling

A schematic representation of the pilot-in-the-loop system is shown in Figure 3. The pilot action is modeled as the combination of active and passive behavior, both given as feedback to the machine response. The commands exerted by the pilot (both voluntarily and involuntarily) are processed by FCSs (if present) and executed by actuators. These produce loads and consequently vibrations and rigid-body motion which, in turn, force pilot dynamics. Note that the inputs to the whole system are the specific mission/task to be performed and potential external disturbances.

In the following, a review of the pilot models used in this work is outlined.

3.1. Passive Pilot Modeling. Involuntary pilot actions on controls are determined by the complex dynamics of pilot body (including bones, muscles, and articulations) in response to the vibration transmitted from seat and controls. The most common approach to introduce passive pilot modeling in the loop is the use of transfer functions (usually of single-input/single-output type) between seat and control sticks motion. These are often determined with experimental campaigns, using a shaker to force the seat and measuring stick displacement [7, 8] for a number of pilots. While the fidelity

of these tests may be increased by positioning the pilot in a simulator with a realistic cockpit and visual feedback, it has been demonstrated that the response is heavily influenced by several factors. The most important of them are (i) pilot physical characteristics (weight, height, and tonus) [7, 19]; (ii) cockpit configuration; and (iii) piloting task. In order to deal with these issues, some authors proposed the direct biodynamic modeling through a multibody analysis [22], while others focused their work on the influence of workload on pilot response [20]. However, due to the variability of pilot skills and characteristics and of piloting tasks, a reliable and practical approach to RPCs should include statistical considerations to model those uncertainties.

Below, several passive pilot models proposed in literature are briefly outlined. They are all presented in terms of transfer functions between seat acceleration, a_s, and stick rotation, α, namely,

$$H(s) = \frac{\tilde{\alpha}}{\tilde{a}_s}. \tag{1}$$

Note that the number and position of poles and zeros of transfer functions change with the considered model.

3.1.1. Pilots Acting on Collective Control

Mayo's Transfer Function. In Mayo's experiment [7], the collective stick motion was recorded while the seat was perturbed using vertical, sinusoidal acceleration disturbances at frequencies ranging from 1 Hz to 5 Hz. The acceleration of pilot's wrist was recorded using three-axis accelerometers, which were mounted on the collective grip, in a classical configuration cockpit of helicopter. The experiment was performed in an open-loop way, in that the control input provided by the pilot did not affect the acceleration of the motion platform. In order to characterize the influence of the pilot body, mesomorphic (larger size) and ectomorphic (smaller size) test pilots were considered.

The following 4-poles and 2-zeros transfer functions were identified:

$$H_{\text{ecto}} = \frac{G}{r_{\text{st}}} \frac{1}{(s - p_1)(s - p_2)} \frac{-s^2 - 8.51s}{s^2 + 13.70s + 452.3},$$

$$H_{\text{meso}} = \frac{G}{r_{\text{st}}} \frac{1}{(s - p_1)(s - p_2)} \frac{-s^2 - 9.29s}{s^2 + 13.31s + 555.4}, \tag{2}$$

where r_{st} is the length of the collective lever and p_1 and p_2 are two pseudointegrators (usually set to 1 Hz in order to avoid nonphysical behavior at low frequency [11]), while G depends on the reference position of collective lever, as shown in Figure 4.

Practical Biodynamic Feedthrough Models. These transfer functions, proposed in [20], provide enhanced BDFT rotorcraft-pilot models of increased accuracy and complexity equivalent to Mayo's one. The main improvements concern the extended model frequency range and the inclusion of both the somatotype effect and the effect of task difficulty on neuromuscular admittance. Also the influence of the subject

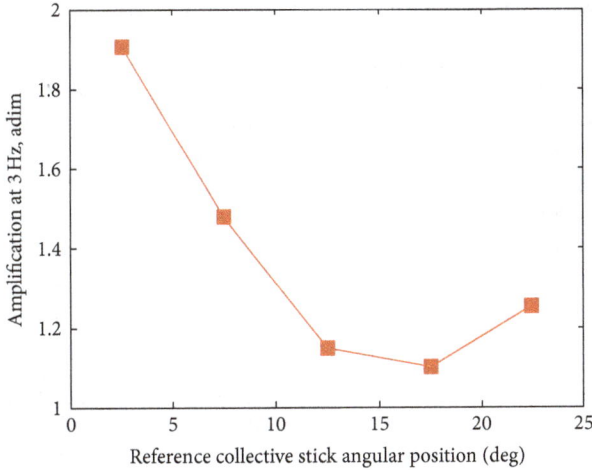

FIGURE 4: Influence of reference arm position on pilot transfer function.

TABLE 1: Poles, zeros, and gains of transfer functions identified at the Bibby flight simulation laboratory.

Pilot	Pole #1	Pole #2	Zero	Gain
#1, 10%	$9.8189 \pm 20.4374i$	$7.0661 \pm 31.2961i$	$2.6282 \pm 28.3482i$	-12.446
#1, 50%	$6.6574 \pm 19.3086i$	$4.9026 \pm 35.8785i$	$3.5630 \pm 27.6716i$	-6.86
#1, 90%	$4.6876 \pm 15.3775i$	$3.5824 \pm 36.1740i$	$7.3902 \pm 27.8659i$	-2.842
#2, 10%	$12.2048 \pm 19.8534i$	$5.0502 \pm 33.7910i$	$3.2423 \pm 30.9463i$	-12.348
#2, 50%	$5.9031 \pm 16.9689i$	$7.7169 \pm 38.3072i$	$5.7946 \pm 24.1660i$	-6.468
#2, 90%	$1.9331 \pm 12.6278i$	$6.1569 \pm 37.2060i$	$6.5938 \pm 18.3922i$	-3.332

variability is included. The data have been obtained by an experimental campaign exploiting the 6-DOF SIMONA Research Simulator at Delft University of Technology. In the tests, acceleration and force disturbances have been applied to the control devices. The subjects were instructed to perform three disturbance rejection tasks, namely, position task, relax task, and force task.

The first task consisted in resisting the force perturbations as much as possible, while maintaining the reference position of the collective lever; for the relax task it was required to relax the arms and passively undergo the perturbations, whereas in the last task the pilot was instructed to minimize the force applied to the collective lever, yielding to the perturbation force as much as possible [20]. In these experiments, the angle of deflection of collective lever and the applied force at the inceptor were measured. In order to characterize the bandwidth of the pilot, the 0.05–21.5 Hz frequency range was considered for vertical force and acceleration disturbances, actually extending that of Mayo's experiments (1–5 Hz).

The following three passive pilot transfer functions have been identified by the described BDFT modeling effort:

$$H_{\mathrm{BDFT}}^{\mathrm{FT}} = \frac{1}{0.7s\,(s + 3.26)} \left(\frac{554.00}{s^2 + 18.00s + 550.36} - 1 \right),$$

$$H_{\mathrm{BDFT}}^{\mathrm{RT}} = \frac{1}{0.7s\,(s + 5.06)} \left(\frac{597.00}{s^2 + 17.23s + 599.81} - 1 \right), \quad (3)$$

$$H_{\mathrm{BDFT}}^{\mathrm{PT}} = \frac{1}{0.7s\,(s + 5.57)} \left(\frac{447.56}{s^2 + 8.28s + 446.42} - 1 \right).$$

Bibby Transfer Functions. Further passive pilot models have been identified experimentally at the University of Liverpool's Bibby flight simulation laboratory, using a visual channel flight simulator mounted upon a six-axis motion [24]. The human subject was seated in the simulator cockpit with MTx motion sensors attached near to the wrist and the elbow. Because, ideally, most of the tests required the collective lever forces to be null, the collective lever friction was set to the

minimum available. A display was created with indication of the position of the control inceptors, in order to avoid drift phenomena due to the lack of feedback to the occupant. The controls (longitudinal and lateral cyclic stick, collective lever, and rudder pedals) were set to their nominal start positions and three reference positions of collective lever have been investigated (10%, 50%, and 90% of the full scale deflection). Axis excitation was a sinusoidal frequency sweep starting at 0.4 Hz that steps up in 0.2 Hz increments to the final value of 7 Hz. Also in this case a single-input/single-output (SISO) model has been identified, with the transfer function provided in the following zeros-poles form:

$$H_{\mathrm{Bibby}} = G \frac{(s - Z)\,(s - Z^*)}{(s - P_1)\,(s - P_1^*)\,(s - P_2)\,(s - P_2^*)}, \quad (4)$$

characterized by a fourth-order denominator and a second-order numerator (deriving from two pairs of complex conjugated poles and one pair of complex conjugated zeros). G is a gain dependent on the reference arm position. The identified values of G, P_1, P_2, and Z for two pilots for three reference positions of the collective lever are reported in Table 1.

Pilot Modeling for Aeroservoelastic RPC in ARISTOTEL. During ARISTOTEL European FP7 Research Project several BDFT tests have been conducted by the flight simulator at University of Liverpool [19], in order to identify pilots bio-dynamic response while being subject to vertical and lateral acceleration. The excitation consisted of colored noise signals, band-pass-filtered between 1 and 10 Hz, with zero mean value and 0.004 g rms.

The motion induced to the control inceptors by the oscillations imposed to the cockpit was measured, along with the motion induced to the limbs. No specific flight task was required, the occupant being requested to hold the controls without compensating the stick vibrations and to maintain the collective lever in the proximity of the reference position to avoid excessive drift. Also in this case the pilot has been considered as a SISO system (vertical acceleration as input and wrist acceleration as output).

TABLE 2: Coefficients of collective control transfer functions identified in ARISTOTEL.

	n_5	n_4	n_3	n_2	n_1	n_0
Berryman1	0	0	$-5.53 \cdot 10^3$	$3.34 \cdot 10^5$	$-1.03 \cdot 10^7$	$8.49 \cdot 10^8$
Berryman2	$-5.02 \cdot 10^1$	$2.40 \cdot 10^3$	$-1.55 \cdot 10^5$	$8.72 \cdot 10^6$	$3.02 \cdot 10^6$	$4.94 \cdot 10^9$
Berryman3	$-2.01 \cdot 10^1$	$3.94 \cdot 10^2$	$-2.12 \cdot 10^4$	$1.47 \cdot 10^6$	$3.53 \cdot 10^7$	$6.82 \cdot 10^8$
Berryman4	$-5.27 \cdot 10^1$	$2.29 \cdot 10^3$	$-1.18 \cdot 10^5$	$8.84 \cdot 10^6$	$3.32 \cdot 10^7$	$5.22 \cdot 10^9$
Berryman5	$-6.13 \cdot 10^2$	$3.03 \cdot 10^4$	$-1.82 \cdot 10^6$	$1.08 \cdot 10^8$	$2.80 \cdot 10^8$	$6.61 \cdot 10^{10}$
Berryman6	$-1.02 \cdot 10^2$	$3.95 \cdot 10^3$	$-1.05 \cdot 10^5$	$1.61 \cdot 10^7$	$1.07 \cdot 10^8$	$1.00 \cdot 10^{10}$
Mayer1		$2.11 \cdot 10^3$	$3.34 \cdot 10^4$	$1.28 \cdot 10^7$	$1.26 \cdot 10^8$	$1.25 \cdot 10^{10}$
Mayer2		$1.32 \cdot 10^2$	$3.57 \cdot 10^3$	$8.75 \cdot 10^5$	$1.43 \cdot 10^7$	$1.08 \cdot 10^9$
Cheyne		$1.36 \cdot 10^3$	$4.30 \cdot 10^4$	$6.04 \cdot 10^6$	$9.13 \cdot 10^7$	$2.73 \cdot 10^9$

	d_6	d_5	d_4	d_3	d_2	d_1	d_0
Berryman1		1.	$2.88 \cdot 10^2$	$1.93 \cdot 10^4$	$1.17 \cdot 10^6$	$2.82 \cdot 10^7$	$8.75 \cdot 10^8$
Berryman2	1.	$1.61 \cdot 10^2$	$1.08 \cdot 10^4$	$4.60 \cdot 10^5$	$1.61 \cdot 10^7$	$2.54 \cdot 10^8$	$4.94 \cdot 10^9$
Berryman3	1.	$6.42 \cdot 10^1$	$4.75 \cdot 10^3$	$1.61 \cdot 10^5$	$4.99 \cdot 10^6$	$7.66 \cdot 10^7$	$7.07 \cdot 10^8$
Berryman4	1.	$1.73 \cdot 10^2$	$1.11 \cdot 10^4$	$4.97 \cdot 10^5$	$1.70 \cdot 10^7$	$2.84 \cdot 10^8$	$5.3 \cdot 10^9$
Berryman5	1.	$1.59 \cdot 10^3$	$1.06 \cdot 10^5$	$5.36 \cdot 10^6$	$1.95 \cdot 10^8$	$3.38 \cdot 10^9$	$6.62 \cdot 10^{10}$
Berryman6	1.	$2.72 \cdot 10^2$	$1.74 \cdot 10^4$	$8.24 \cdot 10^5$	$2.9 \cdot 10^7$	$5.03 \cdot 10^8$	$1.00 \cdot 10^{10}$
Mayer1	1.	$7.4 \cdot 10^1$	$9.45 \cdot 10^3$	$3.59 \cdot 10^5$	$2.17 \cdot 10^7$	$3.5 \cdot 10^8$	$1.26 \cdot 10^{10}$
Mayer2		1.	$2.95 \cdot 10^2$	$1.21 \cdot 10^4$	$1.25 \cdot 10^6$	$2. \cdot 10^7$	$1.07 \cdot 10^9$
Cheyne	1.	$7.68 \cdot 10^1$	$7.1 \cdot 10^3$	$2.55 \cdot 10^5$	$9.84 \cdot 10^6$	$1.54 \cdot 10^8$	$2.73 \cdot 10^9$

Pilot transfer functions have been determined in a rational polynomial form of the type

$$H_A = \frac{1}{r_{st}(s - p_1)(s - p_2)} \left(\frac{n_5 s^5 + \cdots + n_0}{d_6 s^6 + \cdots + d_0} - 1 \right). \quad (5)$$

In Table 2 the transfer function coefficients identified for each pilot/test examined are presented (pseudointegrator poles, p_1 and p_2, are set equal to 1 Hz).

Analyzing these transfer functions, it is possible to infer that there are significantly damped biodynamic poles in the range between 3 and 10 Hz. Even for the same pilot, different tests are characterized by different responses. It is worth underlining that it seems impossible to clearly distinguish professional from nonexpert pilot behavior and that there is no clear dependence on biometric measures.

3.1.2. Pilots Acting on Cyclic Control

Parham Lateral and Longitudinal Transfer Functions. An aeroelastic analysis with longitudinal/lateral pilot in the loop has been performed by Parham [8], using data from the Osprey V-22 flight test program. The initial pilot model used in that analysis was a math model, relating the pilot lateral stick displacement expressed in inches, with lateral acceleration of pilot seat, measured in g. It has been refined and validated by examination of several sets of flight test data, as well as measured outputs from shake tests with pilot holding the controls. Several tests revealed a large variation of responses from different pilots and also a strongly nonlinear behavior. The following transfer functions, respectively, for lateral and longitudinal passive pilot models relating cyclic stick rotation to seat acceleration (expressed in m/s^2) have

been finally determined as an optimal fit of the large amount of responses examined:

$$H_{P_{lat}}(s) = \frac{0.0254}{9.81 r_{cyc}}$$
$$\cdot \frac{9.4487 \cdot 10^3 s - 2.8526 \cdot 10^5}{s^3 + 1.2641 \cdot 10^3 s^2 + 9.7102 \cdot 10^3 s + 3.8554 \cdot 10^5},$$
$$H_{P_{lon}}(s) = \frac{0.0254}{9.81 r_{cyc}} \quad (6)$$
$$\cdot \frac{-9.0227 \cdot 10^3 s^2 + 1.4602 \cdot 10^4 s + 5.7467 \cdot 10^7}{s^4 + 1308.5 s^3 + 75206 s^2 + 1.2590 \cdot 10^7 s + 3.0382 \cdot 10^7},$$

where r_{cyc} denotes the cyclic stick length.

Pilot Modeling for Aeroservoelastic RPC in ARISTOTEL. In the same experimental test campaign held at the University of Liverpool for the collective control passive pilot modeling (see Table 3), lateral control transfer functions have also been identified. Measured data revealed a good level of coherence between the lateral acceleration of the pilot seat and the lateral acceleration of the pilot wrist. Pilot's lateral responses have been fitted using a rational transfer function composed of a 6th-order polynomial denominator and 4th-order polynomial numerator, which reads

$$H_{A_{lat}} = \frac{1}{r_{cyc} s^2} \frac{n_4 s^4 + \cdots + n_0}{d_6 s^6 + \cdots + d_0}. \quad (7)$$

In Table 3 the transfer function coefficients identified for each pilot/test examined are presented.

These transfer functions significantly depend on stiffness, damping, and mass of the control inceptors. In every transfer

TABLE 3: Coefficients of lateral control transfer functions identified in ARISTOTEL.

	n_5	n_4	n_3	n_2	n_1	n_0
*Berryman*1	-2.73	$-1.55 \cdot 10^1$	$1.26 \cdot 10^4$	$-1.64 \cdot 10^5$	$4.4 \cdot 10^6$	$5.40 \cdot 10^7$
*Berryman*2	0	$-2.91 \cdot 10^2$	$1.53 \cdot 10^4$	$-2.48 \cdot 10^5$	$3.25 \cdot 10^6$	$4.33 \cdot 10^7$
*Mayer*1	0	0	-2.00	$7.41 \cdot 10^2$	$8.03 \cdot 10^3$	$4.29 \cdot 10^5$
*Mayer*2	0	0	$1.59 \cdot 10^1$	$3.02 \cdot 10^2$	$5.58 \cdot 10^3$	$2.86 \cdot 10^4$
*Cheyne*1	$1.78 \cdot 10^1$	$-1.12 \cdot 10^3$	$3.23 \cdot 10^4$	$-5.27 \cdot 10^5$	$4.69 \cdot 10^7$	$1.11 \cdot 10^9$
*Cheyne*2	$1.65 \cdot 10^1$	$-1.05 \cdot 10^3$	$2.75 \cdot 10^4$	$-6.19 \cdot 10^5$	$3.81 \cdot 10^7$	$9.23 \cdot 10^8$
	d_6	d_5	d_4	d_3	d_2	d_1	d_0
*Berryman*1	1	$4.74 \cdot 10^1$	$2.74 \cdot 10^3$	$5.67 \cdot 10^4$	$1.05 \cdot 10^6$	$9.51 \cdot 10^6$	$6 \cdot 10^7$
*Berryman*2	1	$4.2 \cdot 10^1$	$2.54 \cdot 10^3$	$4.89 \cdot 10^4$	$9.15 \cdot 10^5$	$7.86 \cdot 10^6$	$4.86 \cdot 10^7$
*Mayer*1	0	0	1	$4.27 \cdot 10^1$	$1.88 \cdot 10^3$	$2.56 \cdot 10^4$	$4.42 \cdot 10^5$
*Mayer*2	0	0	1	$2.66 \cdot 10^1$	$5.55 \cdot 10^2$	$6.04 \cdot 10^3$	$3.37 \cdot 10^4$
*Cheyne*1	1	$5.95 \cdot 10^1$	$5.82 \cdot 10^3$	$1.68 \cdot 10^5$	$7.14 \cdot 10^6$	$8.15 \cdot 10^7$	$1.34 \cdot 10^9$
*Cheyne*2	1	$5.66 \cdot 10^1$	$4.87 \cdot 10^3$	$1.46 \cdot 10^5$	$5.31 \cdot 10^6$	$6.69 \cdot 10^7$	$1.07 \cdot 10^9$

function there is one clear pair of dominant poles between 2 and 3 Hz. These are less damped than in case of collective control transfer functions.

3.2. Active Pilot Modeling. Voluntary pilot actions on controls have been extensively investigated since the half of XX century. The pioneer of the so-called control-theoretic pilot models (namely, the models considering the pilot as a feedback to vehicle dynamics) was Tustin, who suggested the first transfer function representing the human operator [25]. Thirty years later, McRuer put the basis of manual control theory [15], highlighting the correlation between pilot transfer function and controlled object dynamics (in the next paragraph, the McRuer crossover model is described with some details).

In the following years, the crossover model has inspired several quasi-linear models, where nonlinearities are introduced with a residual term (see, e.g., [15, 26–28]). Research work has been focused also on the effect of task difficulty on nonlinearities of pilot response, as well as on detailed modeling of pilot neuromuscular, proprioceptive, and visual sensory characteristics (see, among others, [29, 30]).

3.2.1. Crossover Model. The fundamentals of man-machine interaction theory were introduced by McRuer [15] during the 70s. His model starts from a simple consideration about human control from a skilled human on inanimate elements: when the stimulus forcing the system has random time-stationary characteristics, the human operator adapts his strategy control with actions having the same random and time-stationary properties. The simplest manual control, called single loop compensatory system with visual stimulus, is shown in Figure 5.

The forcing input $i(t)$ is a random function with stationary or quasi-stationary properties. The compensatory display shows to the human operator only the visual stimulus $e(t)$ (tracking error) defined as the difference between the system output $m(t)$ and $i(t)$. Usually, the pilot's task is to minimize $e(t)$, trying to keep the indicator superimposed to a stationary point or line. The error minimization is achieved by pilot control action, $c(t)$. This compensatory model is the paradigm for

TABLE 4: Dynamics of controlled elements.

Type	Equation of motion	Transfer function $Y_c(s)$
Proportional	$m(t) = K_c c(t)$	K_c
Rate of velocity	$\dot{m}(t) = K_c c(t)$	$\dfrac{K_c}{s}$
First-order lag	$\dot{m}(t) + am(t) = K_c c(t)$	$\dfrac{K_c}{s + a}$
Acceleration	$\ddot{m}(t) = K_c c(t)$	$\dfrac{K_c}{s^2}$

much more complicated configurations as multiloop system with visual stimuli and the pursuit mode in which the perception of system state is the input to the pilot model.

In order to obtain a simple description of the active pilot's behavior, McRuer asserted that the operator action strongly depends on the dynamics of controlled elements. The simplified model of manual control system relates the human operator to controlled element dynamics by a simple equation. In the tests, several dynamic models were controlled by a single highly trained operator. Table 4 shows the most relevant cases.

These dynamics were chosen in that they are very simple and have many applications in practical control systems. The human transfer function modeling the pilot behavior, $Y_p(s)$, is different for each controlled element described by the transfer function, $Y_c(s)$, but the loop transfer function, $Y_p(s) \cdot Y_c(s)$, is essentially the same. Moving from these considerations, the following crossover law, relating the operator and the controlled element, is introduced:

$$Y_p(s) Y_c(s) \doteq \frac{\omega_c e^{-s\tau_e}}{s}. \tag{8}$$

The above relation states that the pilot-vehicle system in open-loop behaves as an integrator around the crossover region. The model does not distinguish explicitly between controlled element and operator characteristics, although it is evident that the two parameters, ω_c and τ_e, depend on

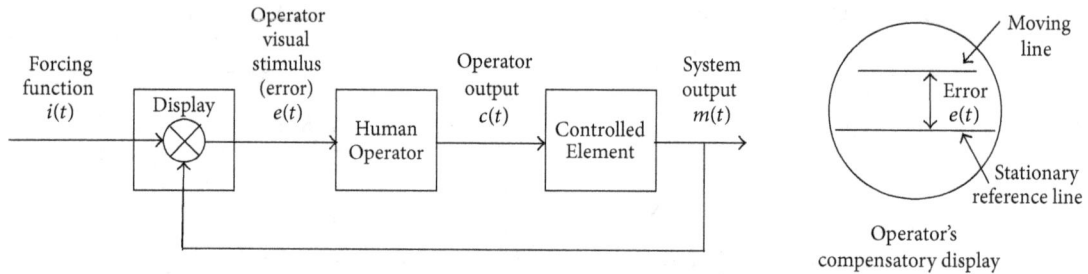

FIGURE 5: Single loop compensatory scheme. From [15].

controlled element dynamics, task variables, and other details which affect the operator.

A brief and intuitive interpretation of the crossover frequency can be made analyzing the closed-loop transfer function relating the input $\widetilde{i}(s)$ and the output $\widetilde{m}(s)$ that can be written as

$$\frac{\widetilde{m}(s)}{\widetilde{i}(s)} = \frac{G(s)}{1 + G(s)} = \frac{\omega_c e^{-s\tau}}{s + \omega_c e^{-s\tau}}, \tag{9}$$

where $G(s) = \omega_c e^{-s\tau}/s$. When $|G(s)| \gg 1$, then $\widetilde{m}(s)/\widetilde{i}(s) \to 1$, whereas when $|G(s)| \ll 1$, then $\widetilde{m}(s)/\widetilde{i}(s) \to G(s)$. At the same time, the transfer function relating the tracking error with the input signal, namely, $\widetilde{e}(s)/\widetilde{i}(s) = 1/(1 + G(s))$, tends to 0 when $|G(s)| \gg 1$, whereas it tends to 1 when $|G(s)| \ll 1$. The most significant property of feedback systems is that when the open-loop transfer function $|G(s)|$ is much greater than 1 (low frequency) the closed-loop response becomes similar to the input, $\widetilde{m}(s) \approx \widetilde{i}(s)$, whereas when $|G(s)|$ tends to 0 (high frequency), the closed-loop transfer function approaches the open-loop one. The crossover frequency, ω_c, defined as the intersection of the Bode diagram of $G(s)$ with the zero dB gain line divides the frequency range in two zones, defined as closed-loop behavior and no-feedback action. For stable applications the nondimensional parameter $\tau\omega_c$ ranges between 0 and $\pi/2$, whereas ω_c ranges from 3 to 6 rad/s, with a phase margin between 25° and 45°.

4. Results

The helicopter model considered in the present numerical investigation is representative of a lightweight helicopter with a hingeless rotor design, loosely inspired by the Bo-105. Airframe structural dynamics, fuselage and tail rotor aerodynamics, and main rotor aeroelastic responses (based on flap-lag-torsion blade dynamics) are modeled through a comprehensive aeromechanic formulation described in [11]. Suited transfer functions multiplied by the gearing ratio G describe the control system dynamics [11].

The airframe elastic deformation is taken into account by four normal vibration modes, reported in Table 5. The first mode involves significant bending of the airframe about the pitch axis, as well as out-of-phase relative vertical motion between the main rotor attachment and the cabin floor, thus possibly introducing nonnegligible interaction between the vertical oscillation of the rotorcraft and that of the pilot's seats.

TABLE 5: Airframe modes.

Mode	Frequency [Hz]
Tail boom vertical bending	6.
Tail boom lateral bending	8.
Fuselage vertical bending	11.5
Tail boom torsion	12.5

However, all mode shapes show some participation of hub and cabin floor displacements.

The model has been linearized about the trim state, obtaining a state space representation of the whole helicopter related to the entire set of aeroservoelastic states, and pilot control variables consisting of collective, longitudinal, and lateral cyclic pitches and tail rotor collective pitch ($\theta_0, \theta_s, \theta_c, \theta_p$, resp.). A multiblade variable transformation [31] has been used to get a linear time invariant approximation of the system. Finally, the fully coupled system sketched in Figure 3 is obtained including active and passive pilot models in parallel feedback to the helicopter dynamics (in this paper, no FCS has been considered).

4.1. Passive Pilot in the Loop

Collective Pilot Model. First, the analysis of the influence on hovering helicopter stability of in-the-loop application of the different passive collective control pilot models examined in Section 3.1 has been performed. The results of this investigation are shown in Figures 6 and 7 (which is an enlargement of the former, for the region containing the aeroelastic poles most relevant for RPC), in terms of root locus determined by increasing the gearing ratio, G, from 0 (out-of-the-loop pilot) to 3 (this variation is represented by markers of decreasing size). Complementary information may be obtained from Figure 8 that, representing the areas delimited by the aeroelastic poles for $G = 1, 2, 3$, gives an estimation of the scattering of helicopter poles with different pilots in the loop, for the same range of frequency of Figure 7.

It is worth noting that the different pilot models are not directly comparable, since some of them are obtained averaging on several subjects belonging to the same class (e.g., classes may be defined according to physical features, like Mayo's approach), while others are representative of the same subject in different operating conditions (for details, see Section 3.1).

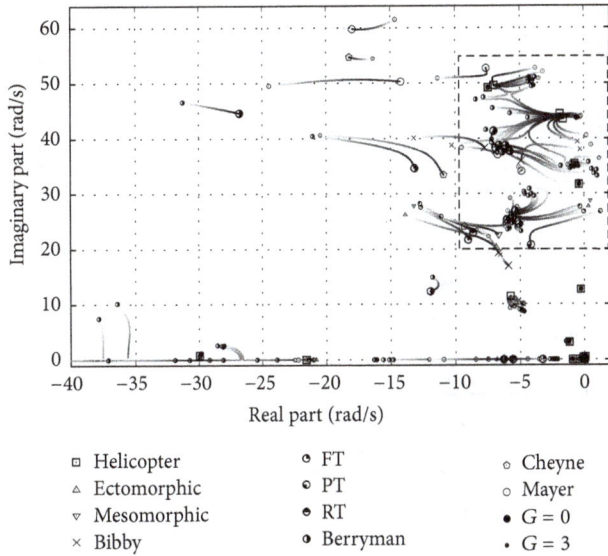

□ Helicopter	◉ FT	○ Cheyne
△ Ectomorphic	◉ PT	○ Mayer
▽ Mesomorphic	◉ RT	● G = 0
× Bibby	◉ Berryman	• G = 3

FIGURE 6: Collective control pilot-in-the-loop root loci, for gearing ratio increase. The area delimited by dashed rectangle is that represented in Figure 7.

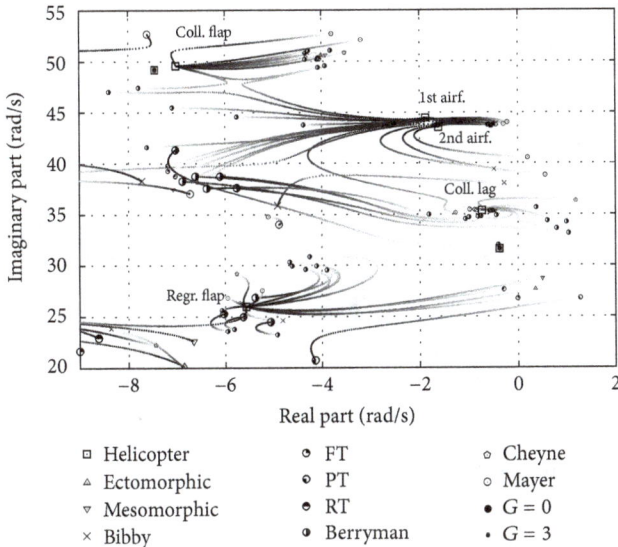

□ Helicopter	◉ FT	○ Cheyne
△ Ectomorphic	◉ PT	○ Mayer
▽ Mesomorphic	◉ RT	● G = 0
× Bibby	◉ Berryman	

FIGURE 8: Areas delimited by root loci, for $G = 1$ (violet), 2 (green), and 3 (red).

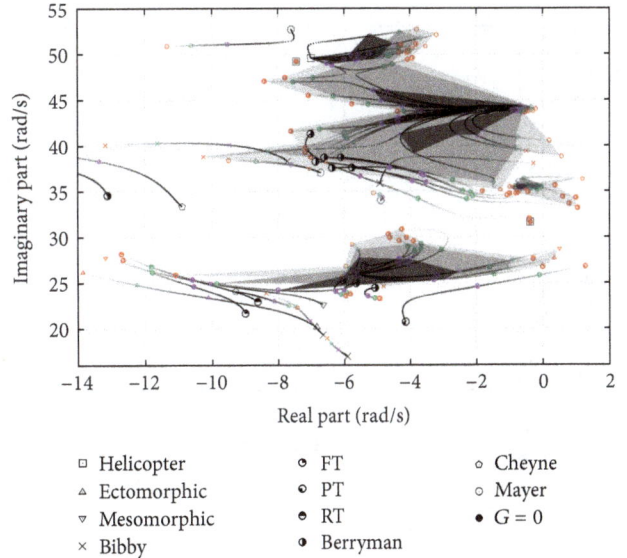

□ Helicopter	◉ FT	○ Cheyne
△ Ectomorphic	◉ PT	○ Mayer
▽ Mesomorphic	◉ RT	● G = 0
× Bibby	◉ Berryman	• G = 3

FIGURE 7: Collective control pilot-in-the-loop root loci, for gearing ratio increase, enlargement. The most relevant aeroelastic poles in RPC are indicated.

However, it is possible to draw some interesting considerations: In this scenario the task influence on pilot-helicopter closed-loop is present only in the BDFTs proposed in [20]; see Section 3.1. In that study, the authors concluded that the "position task" causes responses stronger than those from relax task (RT) and force task (FT), in the range of frequencies between 2 and 3 Hz. That behavior seems to be due to the stiffer response of the pilot when instructed to maintain the position of the control stick. Such effect affects the whole system, as shown in Figure 6 where the poles obtained using

PT-BDFT are more prone to became unstable with respect to those obtained by RT-BDFT or FT-BDFT. Similar deduction can be drawn considering the effect of Mayo's ectomorphic and mesomorphic pilot transfer function on the closed-loop stability. Both models give similar root loci, coherently with the fact that the tasks in the experiments were the same. It is worth reminding that Mayo's models take also into account the effect of arm position (through the coefficient G shown in Figure 4). However, that coefficient is determined once the linearization condition is set.

Other general considerations that may be drawn are as follows:

(i) Pilot in the loop significantly affects most of the aeroelastic poles of the helicopter in the 2–10 Hz frequency range, confirming that above this range no significant couplings arise (however, it is worth noting that also noncritical low-frequency poles are significantly affected).

(ii) While some aeroelastic poles show a well-defined behavior with the increase of gearing ratio, in other cases the behavior is more erratic: this suggests the need of a robust approach to PAO analysis.

(iii) The stability of some poles is strongly affected by the pilot model considered: for example, some poles become unstable only when a subset of pilots is considered.

(iv) The airframe elastic modes, the collective and regressive flap, and the collective lag blade modes are those most involved in collective bounce RPC. This is due both to their frequency (between 2 and 8 Hz, i.e., in the range of passive pilot actions) and to the fact that the corresponding modal shapes are strongly coupled with collective control and vertical motion.

(a) w versus θ_0

(b) p versus θ_c

(c) q versus θ_s

(d) r versus θ_p

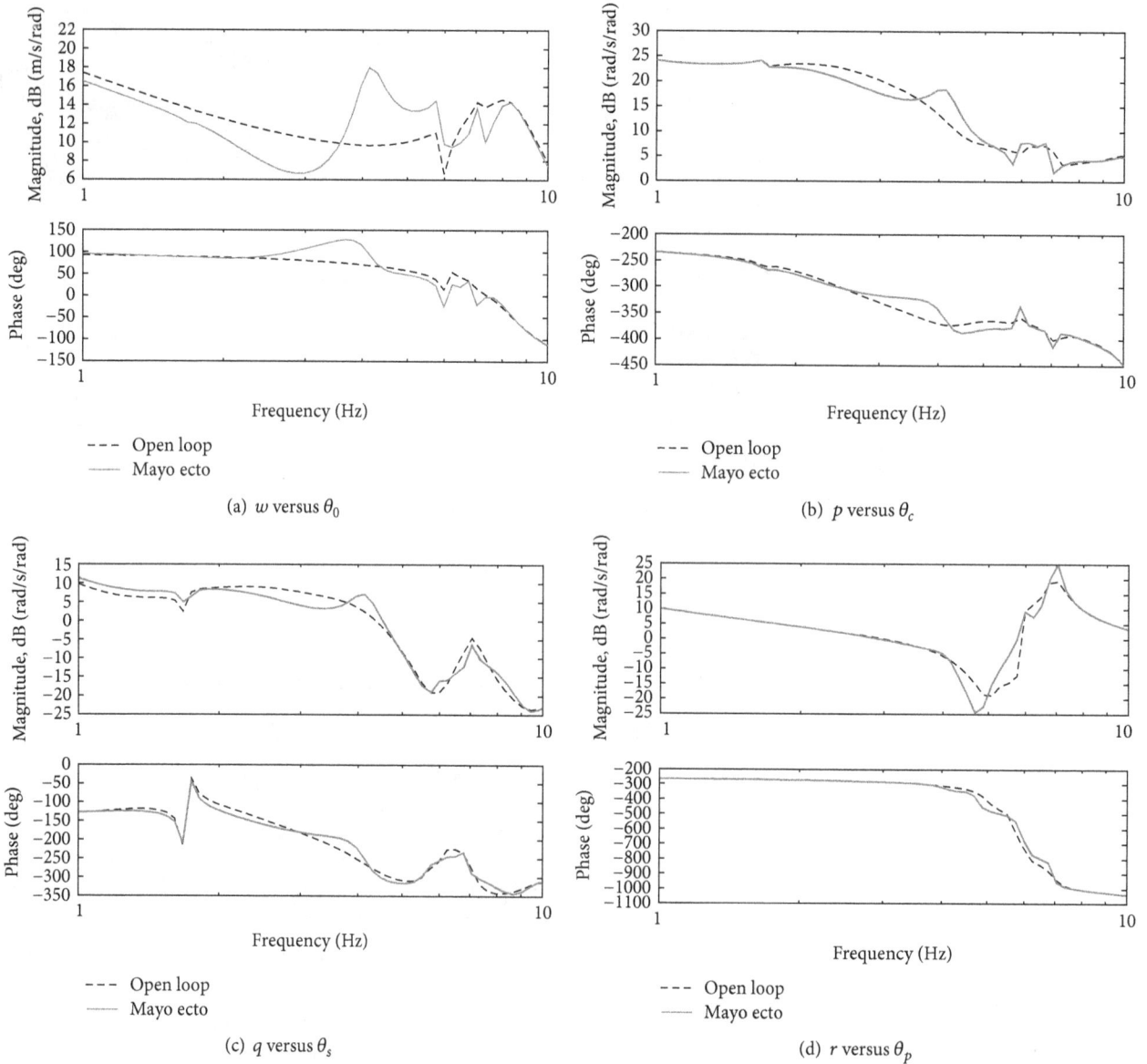

FIGURE 9: Helicopter transfer functions for in-the-loop and out-of-the loop collective control pilot.

Considering in detail, for instance, Mayo's ectomorphic pilot model in the loop, its effects on the most representative helicopter transfer functions are provided in Figures 9(a)–9(d). As expected, the influence is bigger on z-axis response (Figure 9(a)). Moreover, coherently with results given in [32–34], Mayo's pilot reduces the magnitude of w_{θ_0} in the frequency range below its characteristic pole, while in its proximity it causes increase of vehicle response.

Finally, Figure 10 shows the range of variability of the transfer function w versus θ_0 with respect to the applied collective control pilot model, for $G = 1.2$. As expected, the different number and positioning of collective pilot poles significantly modify the vertical response of the helicopter, in terms of both magnitude and phase. This is definitely true for model identified from different subjects responses and also for the models extracted from different tests performed by the same pilot (coherently with what has been obtained in terms of root loci).

Lateral Cyclic Pilot. Still considering a helicopter hovering condition, Figure 11 shows the root loci obtained changing the lateral control gearing ratio from 0 to 3, with the pilot models described in Section 3.1.2 included in the loop. Also in this case, some common trends are observed, but a number of poles have a clearly scattered behavior. In particular, only some of the pilot models make one of the aeroelastic system poles unstable for G approaching 2. Moreover, while collective pilots significantly affect only poles in the mid-frequency range, the influence of lateral pilot extends to flight dynamics (low) frequencies.

Figure 12 represents the transfer functions more affected by the presence of lateral pilots models in the loop. The effect

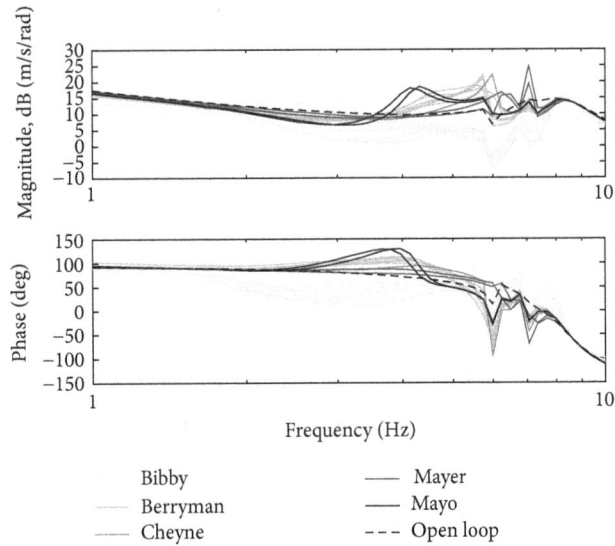

FIGURE 10: Influence on w versus θ_0 of different pilot models.

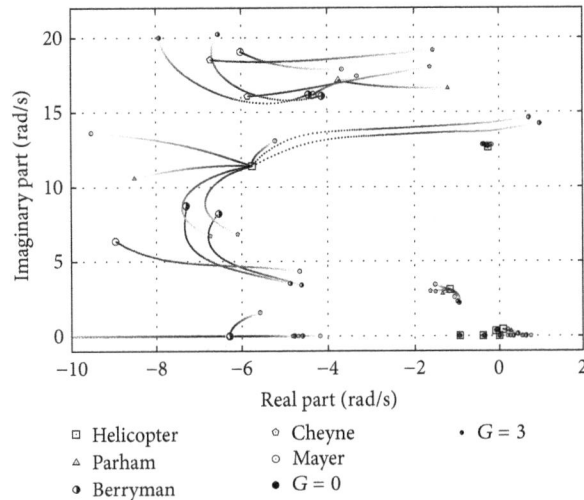

FIGURE 11: Lateral pilot-in-the-loop root loci, for gearing ratio increase.

of lateral pilots on helicopter dynamics is significantly smaller than that of vertical pilots but still appreciable, especially on lateral variables.

Finally, Figure 13 shows the influence of Parham's longitudinal pilot on lateral pilot-in-the-loop root loci. When lateral pilot is out of the loop, the poles of the fixed-control helicopter (thin squares) are significantly perturbed by the presence of longitudinal pilot (big squares, $G = 1$), only below 1 Hz. Comparing with Figures 11 and 13, it is possible to note that its influence is small also when lateral pilots enter the loop. In other words, couplings between the two pilot dynamics are negligible and longitudinal pilot is the least relevant for aeroelastic RPC analyses.

4.2. Active Pilot in the Loop. Concerning the examination of active pilot-in-the-loop effects, first the influence of different

helicopter dynamics models, Y_c, on the definition of the crossover active pilot transfer function is assessed. To this purpose, three rotorcraft models have been considered: a simple SISO model for the vertical velocity, w,

$$M\dot{w} + Z_w w = Z_{\theta_0}\theta_0 \qquad (10)$$

a 6-DOF rigid-body model, and a more complex model including blade and airframe elasticity effects.

The corresponding transfer functions relating vertical response, z, to collective pitch, θ_0, are presented in Figure 14, where it can be observed that the 1-DOF and 6-DOF models provide quite similar results, whereas the inclusion of the elastic-body effects particularly affects the mid/high frequency range (above 10 Hz). The phase plot presents relevant differences between results from rigid and aeroelastic models

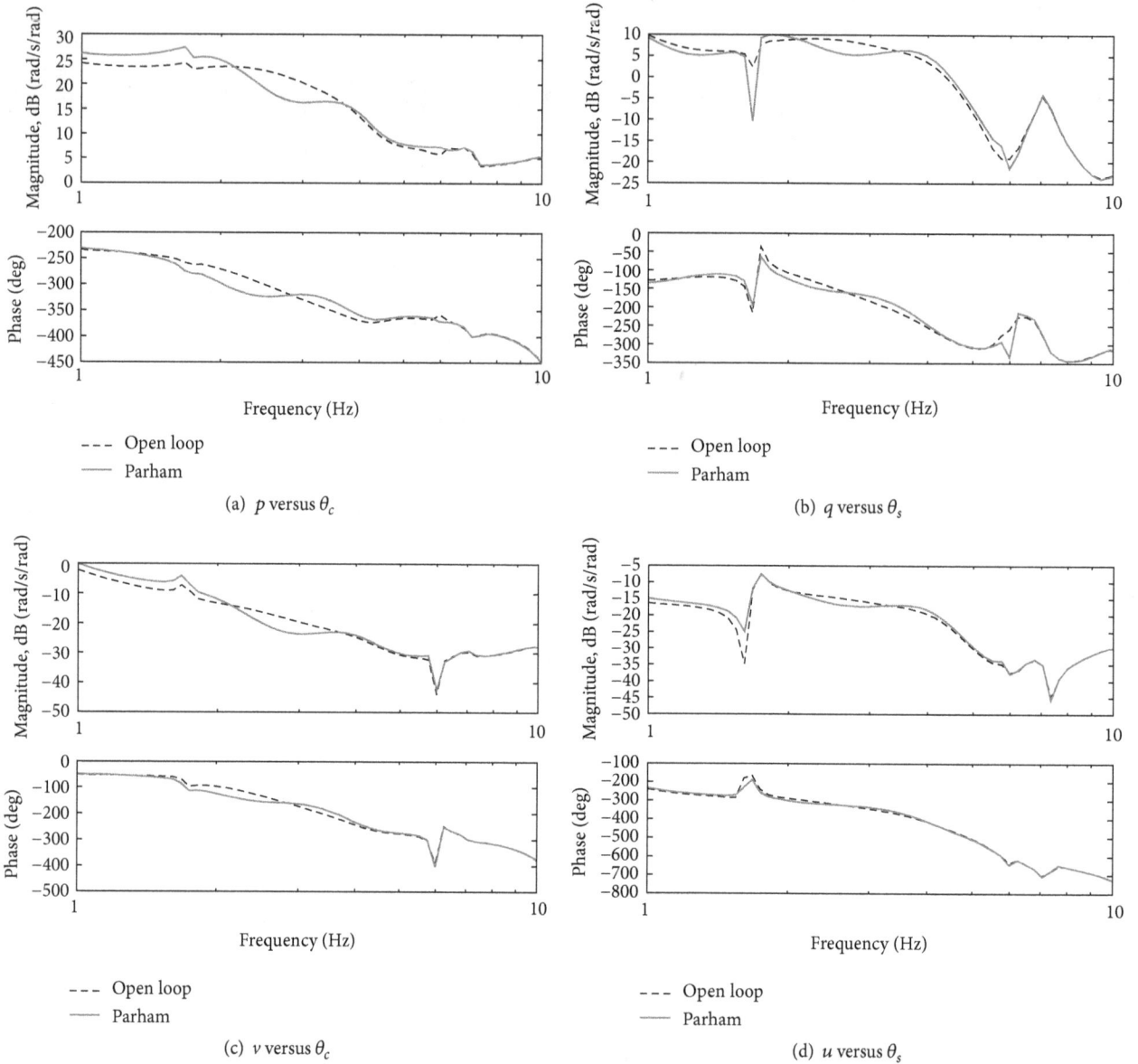

(a) p versus θ_c

(b) q versus θ_s

(c) v versus θ_c

(d) u versus θ_s

FIGURE 12: Helicopter transfer functions for in-the-loop and out-of-the loop lateral control pilot.

one also in proximity of 0.1 Hz (due to the low-frequency blade flapping dynamics effects). At high frequency, where the rigid response is dominated by inertial terms, the two rigid models become practically coincident.

The crossover pilot transfer functions derived from these helicopter models are depicted in Figure 15. Since active piloting actually affects low-frequency range dynamics (4π rad/s is a commonly accepted upper boundary), the crossover model has been modified by introducing two Butterworth filters in order to avoid nonphysical high frequency pilot response.

Figures 16 and 17 show the different helicopter spectral response resulting from the application of the three active pilot models depicted in Figure 15. Although the overall root loci are quite similar, significant discrepancies arise in terms of stability margin (low-frequency poles), with the active pilot

based on "elastic" rotorcraft model bringing to instability for gearing ratio 30% lower than that occurring for instability when Y_c consists of the simplest "1-DOF" rotorcraft model.

Note that, because of the inversion of the helicopter dynamics, $Y_c(s)$, required in (8) to determine the pilot transfer function, $Y_p(s)$, particular attention must be paid to the application of crossover pilot models based on vehicle dynamics characterized by zeros with positive real parts (nonminimum phase, stable systems), in that these would introduce unstable pilot poles.

5. Conclusions

Several pilot models have been examined for pilot-in-the-loop helicopter aeromechanics analysis. These have been

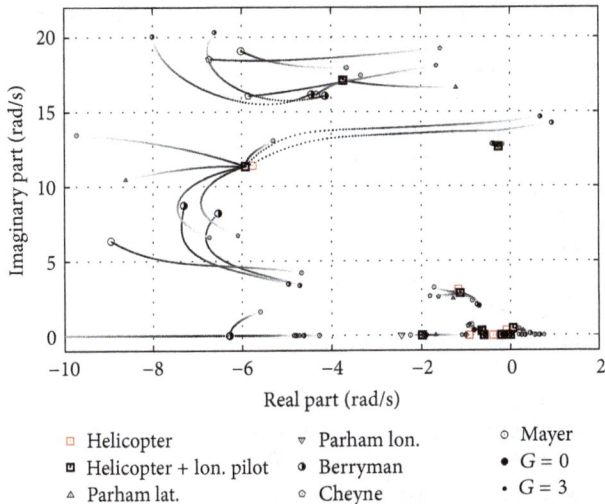

FIGURE 13: Effects of longitudinal pilot-in-the-loop on root loci, for gearing ratio increase.

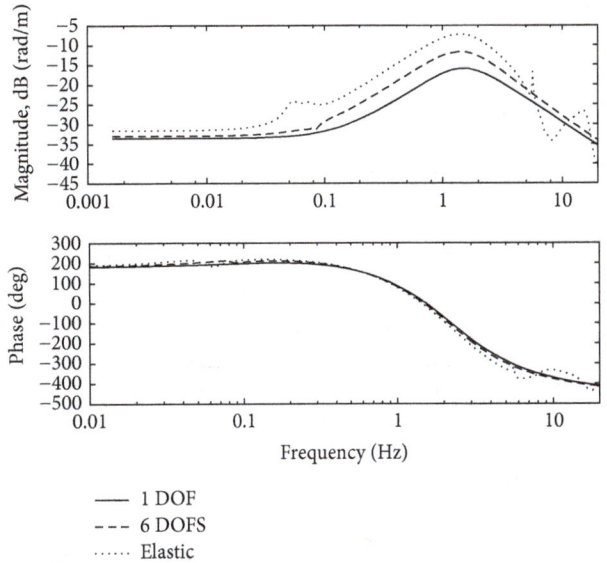

FIGURE 14: Transfer functions z versus θ_0 from different helicopter modeling.

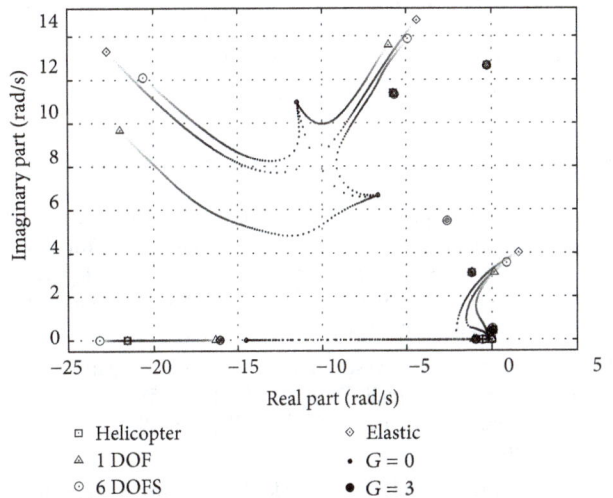

FIGURE 15: Active pilot models obtained using machine dynamics shown in Figure 14.

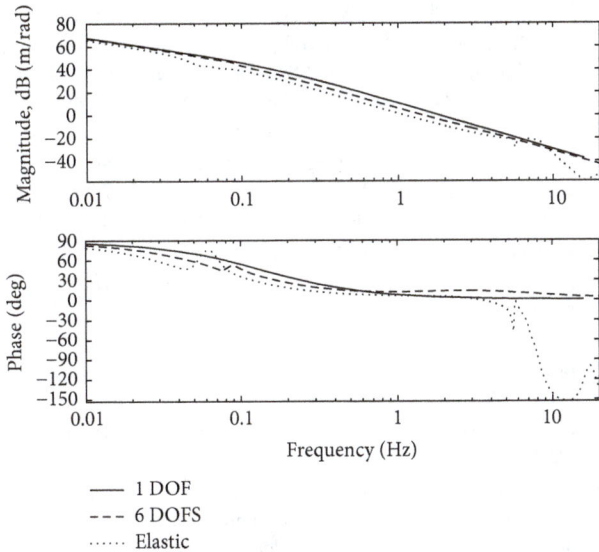

FIGURE 16: Influence of helicopter dynamic modeling on active pilot in the loop.

selected form the literature produced in the last 30 years and include collective, lateral, and longitudinal controls, as well as active and passive pilot behavior. The focus has been on the sensitivity of helicopter responses to pilot model variation (due to its intrinsic uncertainties), with particular attention on the aeromechanical stability of a lightweight helicopter. The investigation has highlighted several critical issues which may be summarized as follows:

(i) Passive pilot transfer functions on collective and lateral cyclic controls vary from pilot to pilot employed in the identification tests and depend on the specific pilot task considered. Literature provides several models which are often reported after averaging a number of tests. The observed considerable high

corresponding variance of the pilot-in-the-loop helicopter aeromechanic response suggests the search of a more robust modeling approach.

(ii) Helicopter dynamics are appreciably affected by lateral pilot responses and scarcely affected by longitudinal ones; couplings between longitudinal and lateral pilot responses are negligible.

(iii) The influence of the crossover active pilot model on helicopter dynamics is significantly affected by the rotorcraft model used in its definition; furthermore, critical responses are achieved when this model is represented by a transfer function with a real, positive zero.

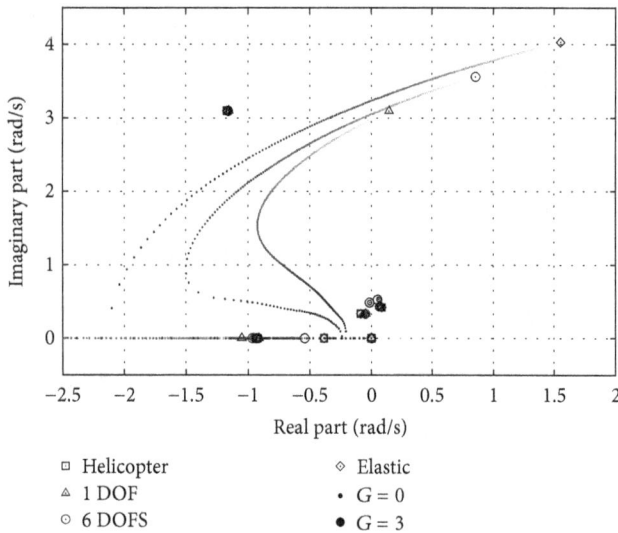

FIGURE 17: Influence of helicopter dynamic modeling on active pilot in the loop, enlargement.

Conflicts of Interest

The authors declare that there are no conflicts of interest regarding the publication of this paper.

References

[1] D. G. Mitchell and D. H. Klyde, "Recommended practices for exposing pilot-induced oscillations or tendencies in the development process," in *Proceedings of the USAF Developmental Test and Evaluation Summit*, AIAA 2004-6810, Woodland Hills, Calif, USA, November 2004.

[2] Aircraft Accident Investigation Commission, "Japan Airlines flight 706 McDonnell Douglas MD-11 JA8580 over Shima peninsula, Japan, 8 June 1997," Tech. Rep. TR-99-8, Ministry of Transport, 1999.

[3] D. G. Mitchell and D. H. Klyde, "Identifying a PIO signature—new techniques applied to an old problem," JAL 1500:11, 2006.

[4] D. T. McRuer, *Aviation Safety and Pilot Control: Understanding and Preventing Unfavorable Pilot-Vehicle Interactions*, National Research Council, National Academy Press, Washington, DC, USA, 1997.

[5] I. L. Ashkenas, H. R. Jex, and D. T. McRuer, "Pilot-induced oscillations: their cause and analysis," Tech. Rep., DTIC Document, 1964.

[6] R. D. Connor, "Wrecked rotors: understanding rotorcraft accidents, 1935–1945," in *Proceedings of the American Helicopter Society 66th Annual Forum*, Phoenix, Ariz, USA, May 2010.

[7] J. R. Mayo, "The involuntary partecipation of a human pilot in a helicopter collective control loop," in *Proceedings of the 15th European Rotorcraft Forum*, Association Industrie Aerospaziali, Amsterdam, The Netherlands, September 1989.

[8] T. Parham, "V-22 pilot-in-the-loop aeroelastic stability analysis," in *Proceedings of the 47th American Helicopter Society Annual Forum*, pp. 1307–1319, Phoenix, Ariz, USA, May 1991.

[9] T. Parham, "Aeroelastic and aeroscrvoelastic stability of the BA 609," in *Proceedings of the 25th European Rotorcraft Forum*, Rome, Italy, September 1999.

[10] M. D. Pavel, M. Jump, B. Dang-Vu et al., "Adverse rotorcraft pilot couplings—past, present and future challenges," *Progress in Aerospace Sciences*, vol. 62, pp. 1–51, 2013.

[11] M. Gennaretti, J. Serafini, P. Masarati, and G. Quaranta, "Effects of biodynamic feedthrough in rotorcraft/pilot coupling: collective bounce case," *Journal of Guidance, Control, and Dynamics*, vol. 36, no. 6, pp. 1709–1721, 2013.

[12] J. Serafini, M. M. Colella, and M. Gennaretti, "A finite-state aeroelastic model for rotorcraft-pilot coupling analysis," *CEAS Aeronautical Journal*, vol. 5, no. 1, pp. 1–11, 2014.

[13] G. Quaranta, A. Tamer, V. Muscarello et al., "Rotorcraft aeroelastic stability using robust analysis," *CEAS Aeronautical Journal*, vol. 5, no. 1, pp. 29–39, 2014.

[14] M. D. Pavel, P. Masarati, M. Gennaretti et al., "Practices to identify and preclude adverse Aircraft-and-Rotorcraft-Pilot Couplings—a design perspective," *Progress in Aerospace Sciences*, vol. 76, pp. 55–89, 2015.

[15] D. T. McRuer and E. S. Krendel, "Mathematical models of human pilot behavior," Tech. Rep., DTIC Document, 1974.

[16] M. M. Lone and A. K. Cook, "Review of pilot modelling techniques," in *Proceedings of the 48th AIAA Aerospace Sciences Meeting Including the New Horizons Forum and Aerospace Exposition*, Orlando, Fla, USA, January 2010.

[17] D. Yilmaz, M. Jump, L. Linghai, and M. Jones, "State-of-the-art pilot model for RPC prediction report," ARISTOTEL EU Project, ACPOGA-2010-266073, Technical Report D2.3, 2011.

[18] R. A. Hess, "Modeling biodynamic interference in helicopter piloting tasks," in *Proceedings of the AHS Aeromechanics Specialists Conference*, pp. 496–504, January 2010.

[19] P. Masarati, G. Quaranta, and M. Jump, "Experimental and numerical helicopter pilot characterization for aeroelastic rotorcraft-pilot coupling analysis," *Proceedings of the Institution of Mechanical Engineers Part G: Journal of Aerospace Engineering*, vol. 227, no. 1, pp. 125–141, 2013.

[20] J. Venrooij, M. D. Pavel, M. Mulder, F. C. T. van der Helm, and H. H. Buülthoff, "A practical biodynamic feedthrough model for helicopter," *CEAS Aeronautical Journal*, vol. 4, no. 4, pp. 421–432, 2014.

[21] L. Zaichik, P. Desyatnik, Y. Yashin et al., "Pilot modelling for aero-servo-elastic A/RPC," ARISTOTEL EU Project, ACPO-GA-2010-266073, Technical Report D3.6, 2013.

[22] P. Masarati and G. Quaranta, "Bioaeroservoelastic analysis of involuntary rotorcraft-pilot interaction," *Journal of Computational and Nonlinear Dynamics*, vol. 9, no. 3, Article ID 031009, 2014.

[23] W. A. Kuczynski, D. E. Copper, W. J. Twomey, and J. J. Howlett, "Influence of engine/fuel control design on helicopter dynamics and handling qualities," *Journal of the American Helicopter Society*, vol. 25, no. 2, pp. 26–34, 1980.

[24] M. Mattaboni, A. Fumagalli, M. Jump, P. Masarati, and G. Quaranta, "Biomechanical pilot properties identification by inverse kinematics/inverse dynamics multibody analysis," in *Proceedings of the 26th Congress of International Council of the Aeronautical Sciences (ICAS '08)*, pp. 3684–3698, September 2008.

[25] A. Tustin, "An investigation of the operators response in manual control of a power-driven gun," CS Memorandum 169, 1944.

[26] K. B. Zaychik, F. M. Cardullo, and G. George, "A conspectus on operator modeling: past, present and future," AIAA Paper 6625:2006, 2006.

[27] R. Heffley, "Use of a task-pilot-vehicle (TPV) model as a tool for flight simulator math model development," in *Proceedings of the AIAA Guidance, Navigation, and Control Conference*, Toronto, Canada, August 2010.

[28] M. M. Lone, N. Ruseno, and A. K. Cooke, "Towards understanding effects of non-linear flight control system elements on inexperienced pilots," *Aeronautical Journal*, vol. 116, no. 1185, pp. 1201–1206, 2012.

[29] R. A. Hess, "Unified theory for aircraft handling qualities and adverse aircraft-pilot coupling," *Journal of Guidance, Control, and Dynamics*, vol. 20, no. 6, pp. 1141–1148, 1997.

[30] R. Hosman and H. Stassen, "Pilot's perception in the control of aircraft motions," *Control Engineering Practice*, vol. 7, no. 11, pp. 1421–1428, 1999.

[31] W. Johnson, *Helicopter Theory*, Courier Corporation, 2012.

[32] J. Serafini, D. Muro, and M. Gennaretti, "Pilot-in-the-loop influence on controlled tiltrotor stability and gust response," in *Proceedings of the 27th International Congress of the Aeronautical Sciences*, Nice, France, 2010.

[33] D. Muro, M. M. Colella, J. Serafini, and M. Gennaretti, "An optimal control approach for alleviation of tiltrotor gust response," *Aeronautical Journal*, vol. 116, no. 1180, pp. 651–666, 2012.

[34] J. Serafini, L. Greco, and M. Gennaretti, "Rotorcraft-pilot coupling analysis through state-space aerodynamic modelling," *Aeronautical Journal*, vol. 119, no. 1219, pp. 1105–1122, 2015.

3

Wear Characteristics of the Material Specimen and Method of Predicting Wear in Floating Spline Couplings of Aero-Engine

Xiangzhen Xue,[1] Sanmin Wang,[1] Jie Yu,[2] and Liyun Qin[2]

[1]*School of Mechanical Engineering, Northwestern Polytechnical University, Xi'an 710072, China*
[2]*35th Department, Xi'an Space Engine Factory, Xi'an 710061, China*

Correspondence should be addressed to Xiangzhen Xue; a_zheny@163.com

Academic Editor: Nicolas Avdelidis

In order to reduce wear and design high-performance spline coupling, the friction coefficient, wear coefficient, and wear depth of 14 groups of material specimens were tested using multifunctional friction and wear tester. The effect of materials, loads, rotation speed, and surface treatment on friction coefficient, wear coefficient, and wear depth was investigated. A method using an Archard's equation based on the finite element method to calculate the wear depth of 14 groups of material specimens was proposed, and the results were consistent with the experimental results. Then, the wear of a floating involute spline coupling of aero-engine was predicted using this method. It can be concluded that carburizing and silvering can decrease the friction coefficient. The wear and wear coefficient decreased after carburizing. So, it is necessary to take 18CrNi4A with carburization and 32Cr3MoVA with nitridation as the material of the spline coupling in aero-engine to minimize wear. Furthermore, the method presented to predicate the wear of spline coupling in this work provided a good fundament for the fatigue prediction methodology of spline coupling.

1. Introduction

Two kinds of involute splines are widely used in aero-mechanical transmission systems including aero-reducer and aerospace engine turbine pump [1–4], and they are fixed spline and floating spline. For the first one, internal spline and external spline are relatively static. The failure of it was mainly caused by fretting wear due to the special load forms of variable torque, bending moment, and axial force during the taking-off, cruising, and landing processes of an aircraft [5–7]. So, these days, there have lots of relation studies on it. Madge et al. have studied the contact-evolution of spline couplings and analyzed the role of fretting wear on the fretting fatigue [8, 9]. Ding et al. proposed a finite element method to predict the fretting wear of spline couplings [10]. Houghton et al. presented an experimental method to simulate the multiaxial fretting conditions between spline teeth, basing on the concept of a simplified representative test [11]. Ratsimba et al. have investigated the friction and wear properties of spline couplings using a cylinder-on-flat fretting testing [12].

Moreover, McColl et al. also established a finite element model based on a modified Archard's equation for simulating both the wear and the evolution of fretting variables with number of wear cycles in a cylinder-on-flat fretting configuration [13]. Cuffaro et al. investigated the behavior of spline couplings in real working conditions and fretting wear phenomena particularly using a novel spline coupling test rig. Simultaneously, an estimation of the Ruiz parameter has been obtained [14, 15].

For the second one, mechanical sliding along axial direction is generated between the internal and external splines. Thus, the failure caused by mechanical (abrasive) wear is serious. But it can be gotten that there was almost no literature about the predicting wear of aero-engine floating spline coupling from the literature above.

According to the latest statistics, in the failure of the helicopter transmission system, the failure caused by these spline couplings accounts for about 10%~20%, and the failure caused by wear accounts for more than 90%. It affects the stability and safety of aero-engine transmission systems seriously. Therefore, it has a great demand of

TABLE 1: Chemical composition of alloy steels 18CrNi4A and 32Cr3MoVA.

Composition/wt%	C	Mn	Si	Ni	Cr	S	P	Al	V	Mo	Cu	Fe
18CrNi4A	0.17	0.47	0.15	3.93	0.90	0.005	0.009	0.055	—	—	—	94.311
32Cr3MoVA	0.34	0.60	0.38	0.06	3.07	0.002	—	—	0.26	0.99	0.07	94.228

TABLE 2: Mechanical properties of alloy steels 18CrNi4A and 32Cr3MoVA.

	18CrNi4A	32Cr3MoVA
E/GPa	205.1	211
G/GPa	79.9	82.4
$\sigma_{0.2}$/MPa	≥980	≥880
σ_b/MPa	1176–1274	1080–1280

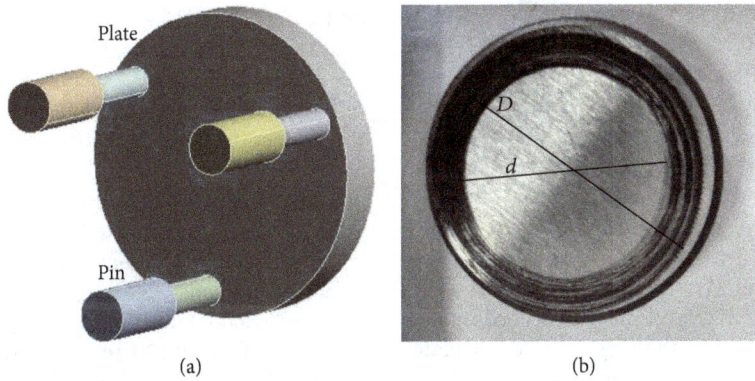

FIGURE 1: Experimental specimen configuration and worn trace: (a) scheme of pin-on-plate couple; (b) worn trace formed on plate specimen.

engineering background to investigate on the wear of floating spline coupling. Guo et al. investigated the gear-coupling contact and loads under considering the effects of the misalignment, torque, and friction [16]. Similarly, Leen et al. have studied the frictional contact in spline couplings [17, 18] and investigated the effect of axial profile modification on friction coefficient with considering the coupling which is under combined torque and axial loads [19]. Hu et al. analyzed the influences of the contact length, friction coefficient, spline wall thickness on the contact stress, and slip distribution of aviation involute spline couplings using a finite element method [20]. Though these studies provide some effective measures to decrease the wear damage by improving the loads and structure geometries, the failure of wear is mainly resulted from many factors: load condition, material pair, heat treatment, lubricating situation, geometric parameters, and surface temperature. So, a lot of efforts have been performed to decrease the wear damage of involute spline couplings by controlling other factors.

In order to improve the wear resistance and reliability of spline coupling, the influences of surface treatment, loads, and rotation on the friction and wear properties of the materials used in spline coupling were investigated experimentally in this work. A method was proposed to predict accurately the wear of floating spline coupling basing on an Archard's equation and finite element method.

2. Friction Experimental

2.1. Experimental Specimen. In the past, lots of carburizing steels were usually used to manufacture the aviation parts, such as 12CrNi3A, 14CrMnSiNi2MoA, 18Cr2Ni4WA, and 20CrNi3A. Later, more and more materials are introduced; for example, the carburizing steels are 16Cr3NiWMoVNbE, 16CrNi4MoA, and 18CrNi4A. The nitridation steels are 30Cr3MoA, 32Cr3MoVA, 38CrMoAlA, and so on. Alloy steels 18CrNi4A and 32Cr3MoVA were chosen as the investigated materials since they were widely used to manufacture spline coupling. The chemical compositions and mechanical properties of these two materials were listed in Tables 1 and 2, respectively. 18CrNi4A was one kind of nickel chromium alloy structural steel possessing good comprehensive performance. Alloy steel 32Cr3MoVA had high hardness, good comprehensive performance, and thermal stability.

The friction and wear properties of these materials were tested under a universal friction tester with a pin-on-flat configuration as shown in Figure 1. The material of pin specimen was alloy steel 18CrNi4A. The big end of pin specimen was clamped by the friction tester, whose diameter was 5 mm. Another end with the diameter of 3 mm was used to grind with the plate specimen. The length of these two parts was 10 mm. The dimension of plate specimen was Φ 30 mm × 5 mm, which was made of alloy steels 18CrNi4A or 32Cr3MoVA.

Furthermore, the pin and plate specimens were treated via carburization or nitridation processes for investigating the effect of surface treatment on the friction and wear properties. The thickness of carburization and nitridation layers was 0.5–0.7 mm and 0.4–0.6 mm, respectively. Moreover, some specimens were silvered on the carburization or nitridation layers.

2.2. Experimental Procedure.
Prior to the friction testing, the specimens were cleaned by an ultrasonic cleaning machine. The mass of the specimens was measured by a microbalance with the accuracy of 0.1 mg. During the testing, three same pins were arranged on one plate to form friction couples and then immersed into the lubricating oil to simulate the working environment of spline coupling. These friction couples were fixed on the friction and wear tester, the lever was levelled, and the load was applied. The testing time was set as 120 min, and the friction force was measured by a strain gauge. The room temperature and relative humidity were 20°C and 50 RH, respectively. Table 3 exhibits the experimental materials and testing conditions.

After every 20 minutes, the testing was interrupted to measure the mass loss with the microbalance of cleaned specimens and the length of D and d by a micrometer with the accuracy of 0.01 mm. Then, the friction coefficient of the materials was calculated by. The wear resistance was described by wear coefficient (k) presented as [7]. The wear depth (h) of plate specimen was calculated according to. Before and after the whole testing, the worn surface of plate specimen was observed by an optical microscope.

$$\mu = \frac{F_f}{F_N}, \tag{1}$$

$$k = \frac{m_w}{(\rho \cdot \delta \cdot F_N)}, \tag{2}$$

$$\delta = n_n \cdot t \cdot l_z, \tag{3}$$

$$l_z = \frac{\pi \cdot (D + d)}{2}, \tag{4}$$

$$h = \frac{m_w}{(\rho \cdot s_w)}, \tag{5}$$

$$s_w = \frac{\pi \cdot (D^2 - d^2)}{4}, \tag{6}$$

where F_f and F_N are the maximum friction force and applied normal load (N), m_w is the mass loss of specimen (g), ρ is the density of the materials (g/cm^3), δ is the sliding distance (mm), n_n is the rotation speed of pins (r/min), t is the testing time (120 min), l_z is the perimeter of worn trace formed on plate specimen (mm), D and d denote the outer and inner diameters of worn trace shown in Figure 1(b) (mm), s_w is the apparent area of worn trace (mm^2).

2.3. Results and Discussion

2.3.1. Friction Coefficient.
Tables 4, 5, and 6 show the results of the experiment, and Figure 2(a) shows the result of friction coefficients for different pairs of materials. As can be seen from Table 3, materials of the 2nd group and the 5th group are different, and materials of the 3rd group and the 6th group are different. Figure 2(a) shows that the plate made of 18CrNi4A (number 1 and 2 couples) has bigger fluctuating range than that of 32Cr3MoVA (number 3 and 4 couples). The friction coefficient of number 2 decreases with the increasing time, which has a higher average value of 0.149. Testing couple number 4 has low and steady friction coefficient, whose average friction coefficient is about 0.115 (Figure 2(a)). It always can be seen that the friction coefficient of couple numbers 3 and 4 increases with the increasing time, whereas the friction coefficient of couple numbers 1 and 2 decreases at last. It is attributed to the different hardness of 18CrNi4A and 32Cr3MoVA.

Figure 2(b) shows the result of friction coefficients for different surface treatment. As can be seen from Table 3, the surface treatments of the 5th group and the 7th group are different, and the surface treatments of the 6th group and the 8th group are different. From Figure 2(b), compared with the nontreated case (Figure 2(a)), the average friction coefficient is increased by the carburation or nitridation treatments of plate specimen. For example, the average friction coefficient is increased from 0.107 to 0.121 and from 0.120 to 0.127 for the plate made of carburized 18CrNi4A and nitrided 32Cr3MoVA, respectively. These results indicate that the friction coefficient of nontreated materials is larger than that of the carburized one. When the carburation or nitridation layers are silvered, the average friction coefficients present a decrease whereas the variations of friction coefficient with time are enhanced. Testing couple number 6 has a relatively low friction coefficient of 0.066. It means that silvering onto the carburization or nitridation layers can further decrease the friction coefficient.

Furthermore, the applied load and rotation speed have an important influence on the friction coefficient (Figures 2(c) and 2(d)). The friction coefficient decreases with the increasing load from 50 to 150 N. It is because that the friction properties of metal material are related to the deformation and interaction of friction surfaces. Under the lower normal load of 50 N, the interfacial shear stress on the friction surfaces is insufficient to deform the micro peaks and wear debris. The friction process is dominated by the interaction of uneven friction surfaces, leading to a higher friction coefficient. As the normal load increases, the deformation of micro peaks and compaction of wear debris are enhanced by the increased interfacial shear stress, which favors the formation of smooth surfaces thereby decreases the friction coefficient. It exhibits an increasing trend as the rotation speed increases from 150 to 260 r/min, whereas it decreases as the rotation speed increases to 360 r/min. It is because that there are boundary friction and fluid friction when the speed is slow. And as the speed increases, the dynamic pressure oil film is formed, and the friction coefficient decreases.

TABLE 3: Materials and testing conditions.

Testing couple	Pin (3)		Plate (1)		Load/N	Speed/r·min^{-1}
	Material	Treatment	Material	Treatment		
1	18CrNi4A	Carburization	18CrNi4A	Nontreatment	100	200
2	18CrNi4A	Nontreatment	18CrNi4A	Nontreatment	100	200
3	18CrNi4A	Carburization	32Cr3MoVA	Nontreatment	100	200
4	18CrNi4A	Nontreatment	32Cr3MoVA	Nontreatment	100	200
5	18CrNi4A	Carburization	18CrNi4A	Carburization	100	200
6	18CrNi4A	Carburization*	18CrNi4A	Carburization*	100	200
7	18CrNi4A	Carburization	32Cr3MoVA	Nitridation	100	200
8	18CrNi4A	Carburization*	32Cr3MoVA	Nitridation*	100	200
9	18CrNi4A	Carburization	18CrNi4A	Carburization	50	200
10	18CrNi4A	Carburization	18CrNi4A	Carburization	120	200
11	18CrNi4A	Carburization	18CrNi4A	Carburization	150	200
12	18CrNi4A	Carburization	18CrNi4A	Carburization	100	150
13	18CrNi4A	Carburization	18CrNi4A	Carburization	100	260
14	18CrNi4A	Carburization	18CrNi4A	Carburization	100	360

Note: the superscript "*" denotes that the material is silvered on the carburization or nitridation layers.

TABLE 4: Friction coefficient.

Groups	1	2	3	4	5	6	7
Friction coefficient	0.1208	0.1068	0.1494	0.1269	0.1196	0.1148	0.0661
Groups	8	9	10	11	12	13	14
Friction coefficient	0.0717	0.1307	0.1109	0.1059	0.1167	0.1340	0.1113

TABLE 5: Wear coefficient.

Groups	Wear weight (g)	Wear volume (cm^3)	Slide distance (mm)	Wear coefficient
1	0.0002	0.0256	1707657.6	0.0884×10^{-8}
2	0.0004	0.0513	1818436.8	1.6603×10^{-8}
3	0.0044	0.5641	1818436.8	18.2639×10^{-8}
4	0.0059	0.7564	1838784.0	24.2191×10^{-8}
5	0.0179	2.2948	1773974.4	76.1632×10^{-8}
6	0.0045	0.5769	1887768.0	17.9930×10^{-8}
7	0.0010	0.1282	1778496.0	4.2441×10^{-8}
8	0.0050	0.6410	1846320.0	20.4410×10^{-8}
9	0.0001	0.0128	1782264.0	0.4235×10^{-8}
10	0.0011	0.1410	1811654.4	4.5831×10^{-8}
11	0.0013	0.1667	1751366.4	5.6028×10^{-8}
12	0.0046	0.5897	1844812.8	18.8211×10^{-8}
13	0.0003	0.0385	1746091.2	1.2969×10^{-8}
14	0.0026	0.3333	1782264.0	11.0114×10^{-8}

2.3.2. Wear Coefficient and Depth. Figure 3 depicts the variations of wear coefficient with friction time under different conditions. From Figures 3(a)–3(d), it can be concluded that the wear coefficient increases with the increasing friction time. The material of pin specimens is carburized 18CrNi4A, and the plate made of nontreated 32Cr3MoVA (number 3

couple) possesses higher wear coefficient about 7.62×10^{-7} kPa^{-1} after 120 min testing. The plate made of nontreated 18CrNi4A (number 1 couple) has relatively low wear coefficient around 1.66×10^{-8} kPa^{-1} (Figure 3(a)). It relates to the hardness of the materials.

From Figure 3(b), the plate treated by carburization or nitridation has lower wear coefficient compared to the nontreated cases shown in Figure 3(a). After a silvering layer formed on the carburization or nitridation layers, the wear coefficient is increased (number 6 and 8 couples). The couple made of carburized 18CrNi4A (number 5 couple) possesses lower wear coefficient of 8.84×10^{-9} KPa^{-1} after 120 min testing. These suggest that the pins and plate are fabricated by the same material 18CrNi4A with carburization layer and without silvering.

Meanwhile, the wear coefficient increases with the increasing load, which exhibits an increasing tendency followed by a decrease as the rotation speed increases (Figures 3(c) and 3(d)).

In order to further understand the wear property, the depth of worn trace on the plate specimen was measured. Figure 4 shows the variations of wear depth. It can be seen that the wear depth increases with the increasing friction time. From Figure 4(a), when the pins are made of carburized 18CrNi4A, the wear depth of plate specimen made of nontreated 32Cr3MoVA is higher than that of nontreated 18CrNi4A. In the case of the nontreated pins, the wear depth of plate is opposite to the former. Among these couples, the

TABLE 6: Wear depth.

Groups	1	2	3	4	5	6	7
Wear area (mm^2)	273.6271	273.8906	134.1097	275.8176	224.7034	200.5754	200.0808
Wear depth (mm)	9.37×10^{-5}	18.72×10^{-5}	420.63×10^{-5}	274.24×10^{-5}	1021.28×10^{-5}	287.63×10^{-5}	64.08×10^{-5}
Groups	8	9	10	11	12	13	14
Wear area (mm^2)	261.5620	271.0527	223.4374	216.0019	221.3775	278.6471	256.2005
Wear depth (mm)	245.08×10^{-5}	4.73×10^{-5}	63.12×10^{-5}	77.16×10^{-5}	266.39×10^{-5}	13.80×10^{-5}	130.11×10^{-5}

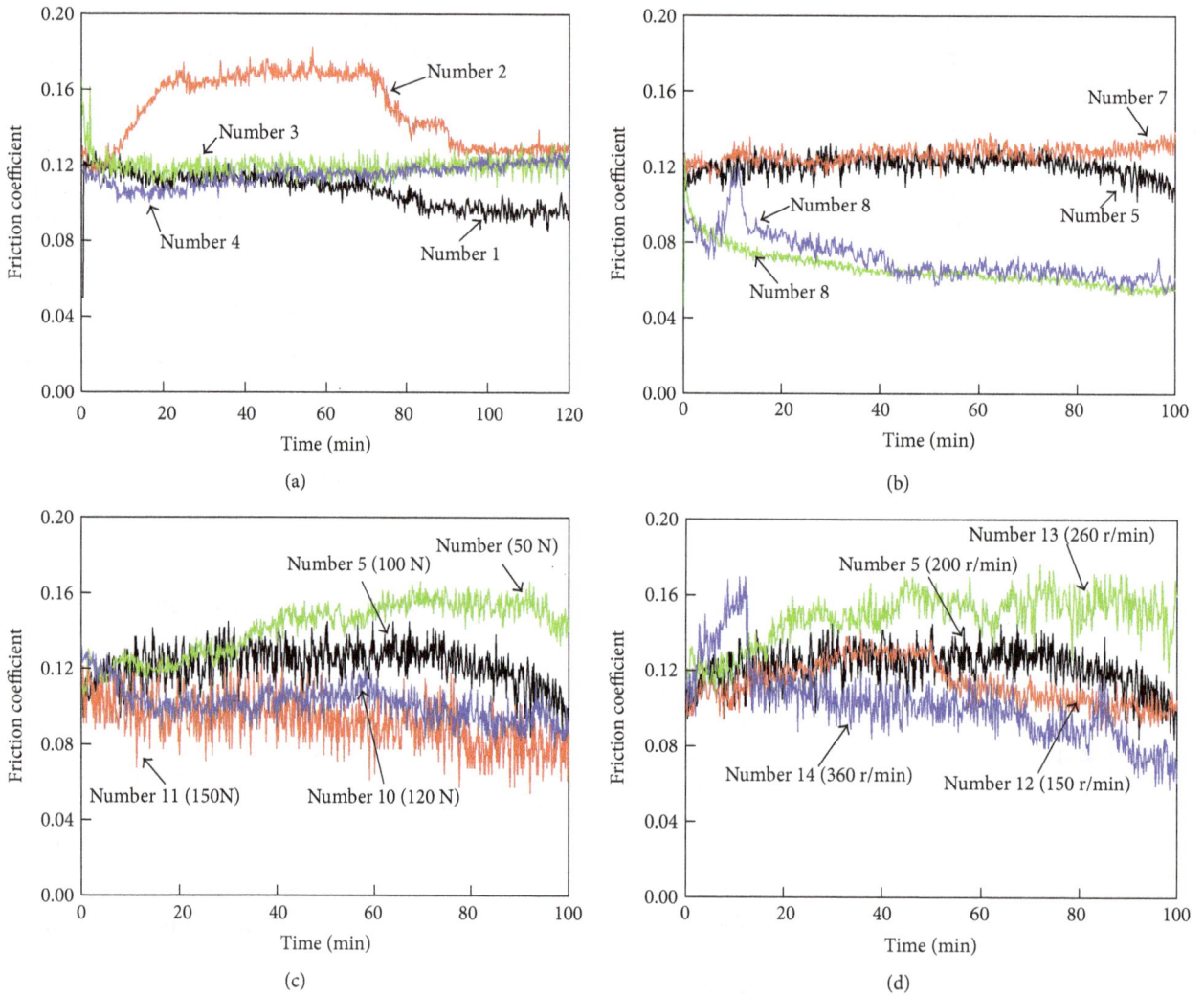

FIGURE 2: Variations of friction coefficient with time under different materials (a), surface treatments (b), loads (c), and rotation speeds (d).

plate of couple number 3 has higher wear depth, and its maximum value is 10.2 μm after 120 min testing. The reason for this phenomenon is that the hardness of the pins made of carburized 18CrNi4A is higher than the others. The plate of couple number 1 possesses relatively low wear depth. These suggest that alloy steel 18CrNi4A has better wear resistance than 32Cr3MoVA under the nontreatment state.

From Figure 4(b), the wear depth of plate is decreased via the carburization or nitridation treatments under the pins made of carburized 18CrNi4A. However, the wear depth increases after silvering layer formed on the carburization or nitridation layers. The treated plate has higher wear resistance than the nontreated one. Among these friction couples, couple number 5 made of carburized 18CrNi4A has relatively low wear depth of about 0.09 μm after 120 min testing. Also, Figure 4(c) shows that the wear depth increases with the increasing load. As the rotation speed increases, the wear depth exhibits a decreasing trend followed by an increase as

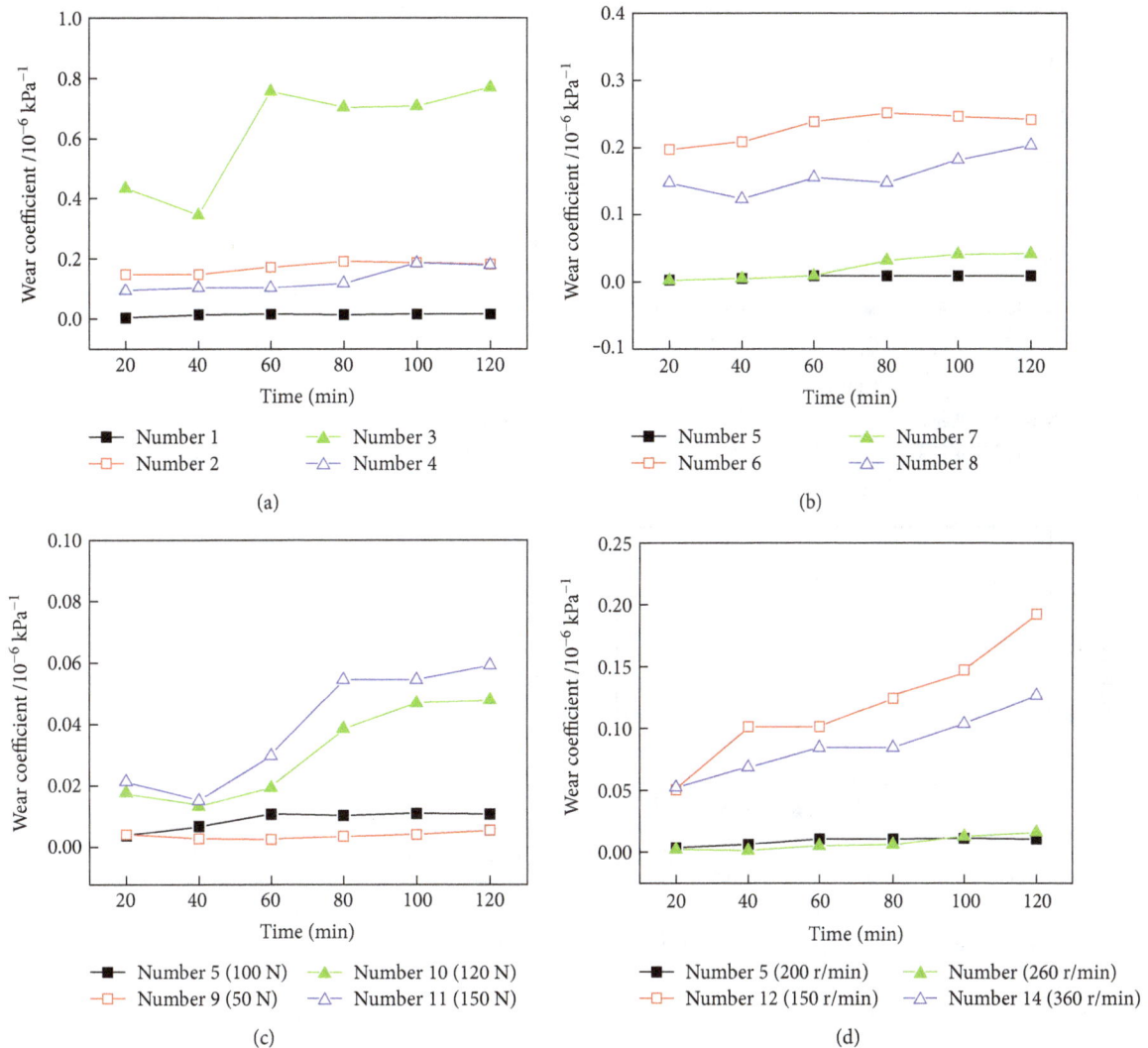

FIGURE 3: Variations of wear coefficient with time under different materials (a), surface treatments (b), loads (c), and rotation speeds (d).

shown in Figure 4(d). After 120 min testing, the wear depth reaches the minimum of about 0.09 μm at 200 r/min and the maximum of about 2.66 μm at 150 r/min.

In general, from Figures 2–4, it can be gotten that there are some inconsistencies between the results of friction coefficient and the results of wear depth. It is caused by the different mechanical properties and hardness of pins and plates. Meanwhile, the pins have higher wear resistance than the plates because of the differences in the geometry. Thus, the contact stress and slide distance can be calculated from above parameters.

3. Prediction of Wear of Specimen

Prediction of the wear is very significant for designing aero-engine spline couplings with high performance. In order to predict the wear in spline couplings accurately, a good method is necessary. Here, a method was proposed to predict the wear of the material specimens with an Archard's

equation as well as finite element method. Based on it, the wear of the floating spline couplings was predicted.

After the solid model established using PRO/E was imported into ANSYS, the global coordinate system should be converted into column coordinate system. The physical and mechanical properties used for building the finite element models were shown in Table 7, and then the finite element models of the pin-on-plate friction couple were established as shown in Figure 5. And, at this time, the nodal coordinate systems of each node should also be converted into column coordinate system. The target and contact surfaces correspond to the internal and external splines. The selected contact stiffness factor FKN is 1.0 [21], the contact pairs were established. According to the moving rules of pin-on-plate friction couple in the test, the plate surface was constrained completely, and the freedom of three pins was constrained completely too except for the freedom degrees of circumferential rotation degrees. Next, the loads were applied according to the experiment conditions listed in Table 3.

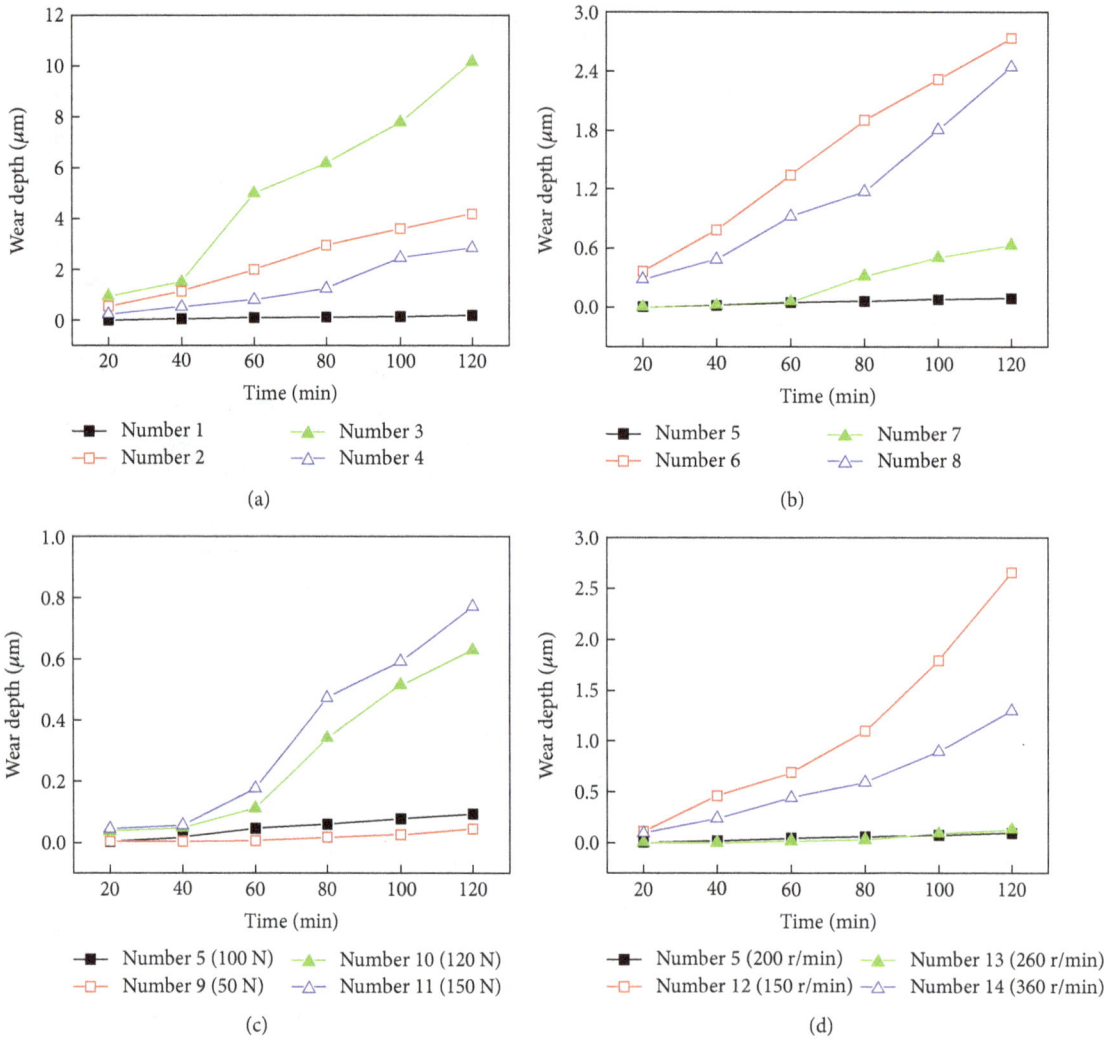

FIGURE 4: Variations of wear depth with time under different materials (a), surface treatments (b), loads (c), and rotation speeds (d).

TABLE 7: Physical and mechanical properties for building the finite element models.

Unit type	Modulus of elasticity	Density	Friction coefficient	Poisson ratio
Solid185	210 GPa	7.800 g/cm^3	0.12	0.28

The wear depth of the plate specimen is predicted using Archard's equation (7) [22]. Basing on this equation, the wear depth of one minor cycle can be expressed by (8).

$$h(x) = 2k \cdot s(x) \cdot p(x), \qquad (7)$$

$$\Delta h_j(x) = 2k \cdot s(x) \cdot p(x), \qquad (8)$$

where $h(x)$ is the wear depth in one cycle of x; k is the wear coefficient obtained from friction testing; $s(x)$ is the slide distance of x; $p(x)$ is the contact stress of x; j ($j = 1, 2, 3, \ldots, t$) is the number of load cycle; $s(x) = l \times z$.

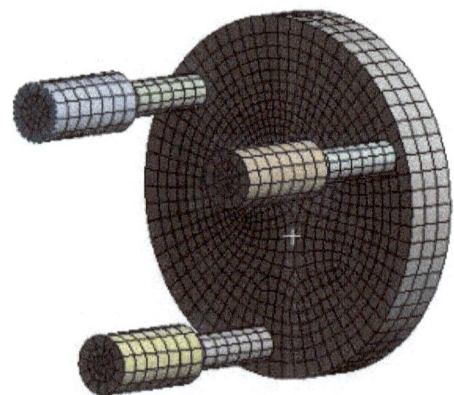

FIGURE 5: Finite element model of pin-on-plate friction couple.

The model of pin-on-plate friction couple is modified according to $\Delta h_j(x)$ in PRO/E. Therefore, the finite element model and contact couple can be renewed. The contact stress and sliding distance are calculated once again as

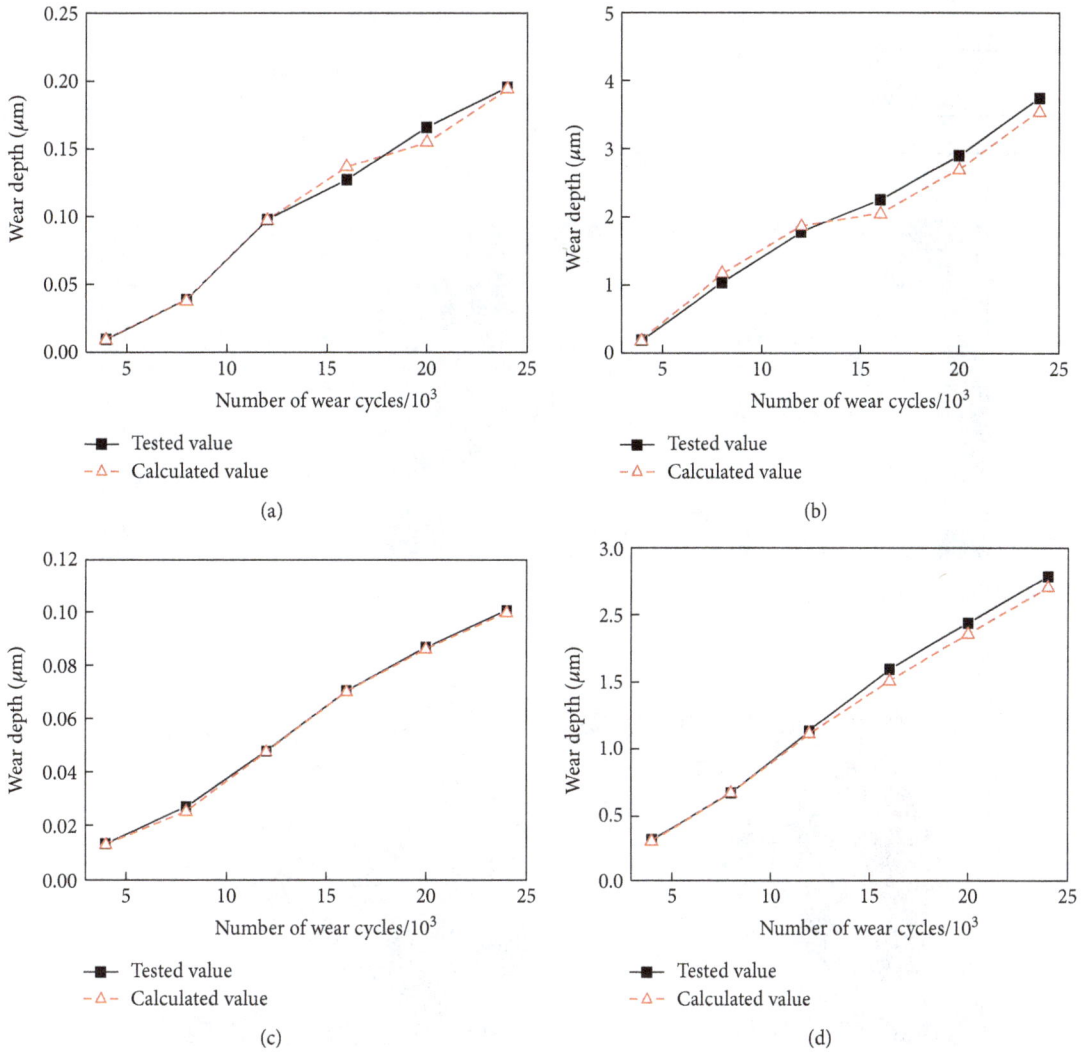

FIGURE 6: Wear depth of the friction couple number 1 (a), number 2 (b), number 5 (c), and number 7 (d).

well as the wear depth of one minor cycle in the modified model. After the cycles of N_t-1 iteration, the wear depth can be expressed by

$$h = \sum_{j=1}^{N_t} \Delta h_j. \qquad (9)$$

Actually, the wear depth of one minor cycle can be neglected since the value is too small to modify the model. It is necessary to select ΔN as the increment of wear cycle [18]. Therefore, the wear depth of one wear cycle is as follows:

$$\Delta h_{\Delta N} = 2k \cdot \Delta N \cdot s(x) \cdot p(x). \qquad (10)$$

From (10), the total wear depth of friction couple can be described using

$$h = \sum_{i=1}^{N_t/\Delta N} \Delta h_i, \qquad (11)$$

where i is the number of wear cycle increment, $i = 1, 2, 3, \ldots, N_t/\Delta N$, ΔN is 4000 under the $N_t = 24000$, and the wear coefficient k is the average value obtained in Figure 2.

Figure 6 depicts the calculated and tested wear depths. It can be seen that the wear depth calculated via finite element models is nearly the same to the tested value. From Figure 6, the wear depth of the friction of couple numbers 1 and 5 is less than that of couple number 2 and couple number 7. The wear depth of the former two couples is approximate to zero when the number of wear cycle is less than 6000. As the number of wear cycle increases, the wear depth presents an increasing trend. When the number of wear cycle is increased to 24000, the wear depth of the former two couples (numbers 1 and 5) is less than 0.19 μm. Especially for couple number 5, its wear depth reaches the minimum of about 0.09 μm. It also can be obtained from Figure 6 that the results of calculated value are consistent with the results of tested value. This indicates that the method of calculating wear proposed here is feasible and can provide an accurate basis for the predicting wear of floating spline coupling.

TABLE 8: Geometric parameters of floating involute spline.

Number of teeth (z)	Press anger (a)	Contact length (L)	Thickness of tooth top (B)	Pitch diameter (D)
22	30°	27.5 (mm)	2.4 (mm)	55 (mm)

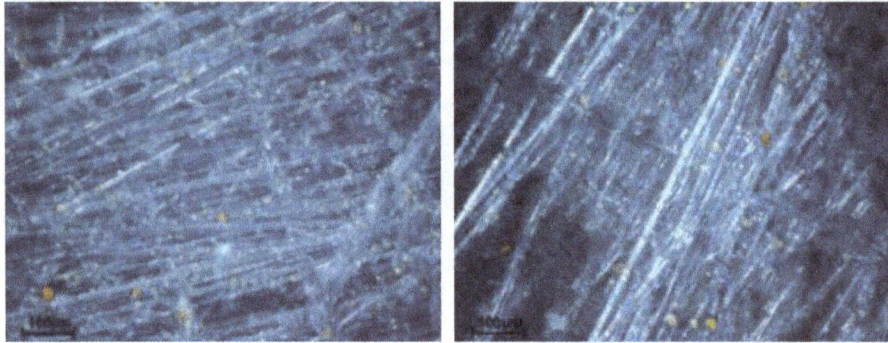

FIGURE 7: Surface topography by industrial microscopes of worn material specimen.

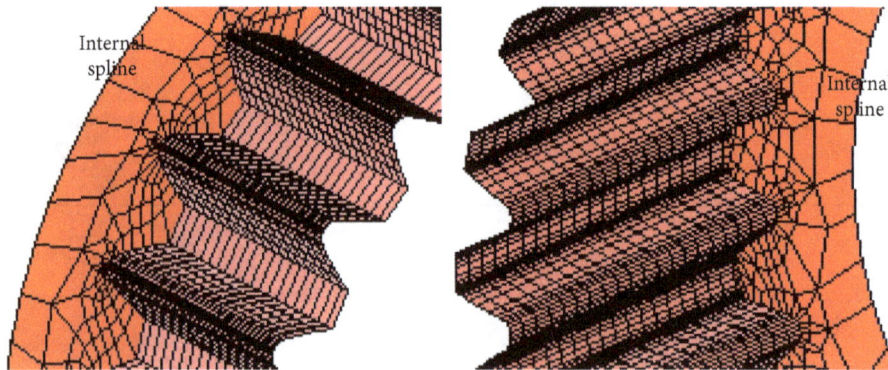

FIGURE 8: Finite element model of spline coupling.

4. Prediction of Wear of Floating Involute Spline

4.1. Prediction of Contact Stress. Table 8 shows the geometric parameters of the wear of a floating involute spline. The axial sliding distance of this floating spline coupling is 2 mm. The surface morphology of worn spline teeth was observed by an industrial microscope to analyze the wear form. From the morphology of worn spline teeth measured by industrial microscopes (Figure 7), many furrows and wear debris can be found on the worn surface because of the long range reciprocating sliding ($>100 \, \mu$m). It is a representative characteristic of abrasive wear (mechanical wear), and it implies that the damage in floating spline coupling is caused mainly by abrasive wear. In Section 2, the specimens were doing the mechanical sliding, and its mechanism is mechanical wear and similar to that of the floating spline coupling because they all suffer the long range reciprocating sliding ($>100 \, \mu$m). Moreover, the method proposed in Section 3 for calculating the wear of specimens is consistent

with testing results of Section 2. Therefore, this method can be used to predict the wear depth of floating involute spline.

The process of establishing the finite element model (Figure 8) and contact pair of floating involute spline coupling is similar as pin-on-plate friction couple. The differences are the ways to constrain and load.

According to the moving mode of spline coupling, the external cylindrical surface of internal spline was constrained completely. The degree of freedom of external spline coupling was constrained except for the Y direction. Meanwhile, the torque is expressed as follows:

$$T(t) = T_m(1 + \varepsilon_T \cdot \cos(\omega_T \cdot t)), \tag{12}$$

where T_m is the average torque; ε_T is the fluctuation coefficient of torque, $\varepsilon_T = 0.1$; ω_T is the angular velocity of system; n is the rotation speed of the system. The torque of one minor cycle (0–2π) is separated into 5 steps T_i ($i = 1, 2, 3, 4,$ and 5). According to the equation $\omega_T = 2n\pi/60$, the one minor cycle is $\tau = 60/n$. The load applied to the model is calculated using (7).

$$F_i = \frac{T_i}{(r \cdot N)}, \tag{13}$$

where D is the pitch circle diameter of external spline, r is the hole radius of the external spline ($r = 0.8D/2$), and N is the node number in the hole of the external spline.

The force calculated using (7) is applied on the internal cylindrical surface of the external spline coupling along the y direction, and displace of s (given by designer) along the x direction is applied too. Because the distance s caused by axial movement of floating spline coupling is much bigger than the distance caused by fretting movement, the distance caused by fretting movement is neglected here. And then, the model is solved, and the contact stress is achieved.

4.2. The Wear of Floating Spline Coupling. For simplifying the wear of spline coupling, the floating distance is set as 2 mm. The elastic distortion and damage caused by fretting wear are all neglected. The wear of floating spline coupling in one minor cycle is defined as follows [10]:

$$\Delta h_j(x) = 2k \sum_{i=1}^{5} s \cdot p(x)_i. \tag{14}$$

Here, $s = 2$ mm. When the number of load cycle is N_t, the wear of the floating spline coupling is defined as follows:

$$h = \sum_{j=1}^{N_t} \Delta h_j. \tag{15}$$

Then, in one wear cycle incremental (ΔN) [17], the wear of floating spline coupling is as follows [10]:

$$\Delta h_{\Delta N_j}(x) = 2k\Delta N \sum_{i=1}^{5} s \cdot p(x)_i. \tag{16}$$

The total wear depth of floating spline coupling is as follows:

$$h = \sum_{l=1}^{N_t/\Delta N} \Delta h_l. \tag{17}$$

In the model, N_t and ΔN are 120000 and 60000, respectively. The wear coefficient k is the average value of each group tested in Section 2.2. The calculation of the wear of the spline coupling geometric parameters is shown in Table 3, and the materials and heat treatment are listed in Section 2.1. When the input power P is 1015 kW, the rotation speed n is 5915 rpm. When the axial distances of floating are 2.0 mm, 2.5 mm, and 3.0 mm, the calculated results of wear are shown in Figure 9.

5. Conclusions

This work investigates the effect of materials, loads, rotation speed, and surface treatment on friction coefficient, wear coefficient, and wear depth, and the wear prediction for

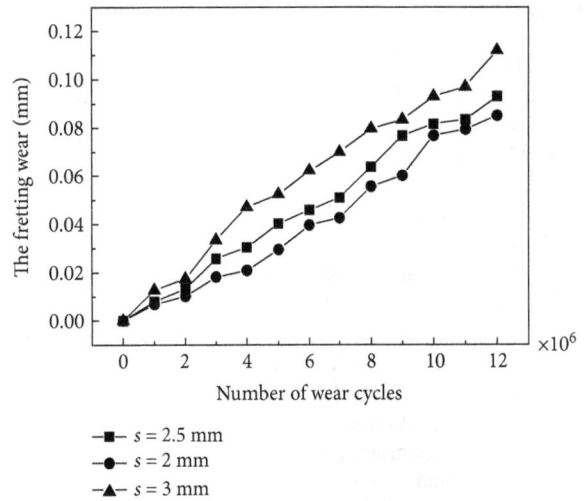

FIGURE 9: Wear depth results of the spline coupling.

floating spline coupling is proposed. From the work, it can be gotten that the friction coefficient of the specimen can be decreased by carburizing and silvering. The material couple has an impact on the friction process. The friction coefficient decreases when the load is increasing, whereas it increases when the rotation is increasing. Carburization or nitridation can decrease wear coefficient, whereas silvering on the carburization or nitridation layers can increase wear coefficient. And, carburization or nitridation treatments can decrease wear depth, whereas silvering on the carburization or nitridation layers can increase wear depth. It also can be gotten that there are some inconsistencies between the results of friction coefficient and results of wear depth. It is mainly caused by the different mechanical properties and hardness of pins and plates. Meanwhile, the pins have higher wear resistance than the plates because of the differences in the geometry. In other words, 18CrNi4A with carburization and 32Cr3MoVA with nitridation should be the material of the spline coupling in aero-engine.

At the same time, the method proposed here to calculate the wear of floating involute spline was feasible. From the result, it can be concluded that the wear depth increases with the axial floating distance increasing. So, the distance of axial floating must be designed as small as possible. It provides a basis for predicting the fatigue life of the spline coupling. It also gives an accurate value based on the design of spline coupling.

Conflicts of Interest

The authors declare that there is no conflict of interests regarding the publication of this paper.

Acknowledgments

The authors gratefully acknowledge the financial support of the National Natural Science Foundation of China (Grant no. 51175422).

References

[1] S. Medina and A. V. Olver, "An analysis of misaligned spline coupling," *Journal of Engineering Tribology*, vol. 216, pp. 269–279, 2002.

[2] C. H. Wink and M. Nakandakar, "Influence of gear loads on spline coupling," *Power Transmission Engineering*, vol. 3, pp. 41–49, 2014.

[3] G. N. D. S. Sudhakar and A. S. Sekhar, "Coupling misalignment in rotating machines: modelling, effects and monitoring," *Noise & Vibration worldwide*, vol. 8, pp. 17–39, 2009.

[4] D. C. H. Yang and S. H. Tong, "On the profile design of transmission splines and keys," *Mechanism and Machine Theory*, vol. 42, pp. 82–87, 2007.

[5] W. S. Sum, S. B. Leen, E. J. Williams, R. Sabesan, and I. R. McColl, "Efficient finite element modelling for complex shaft couplings under non-symmetric loading," *The Journal of Strain Analysis for Engineering Design*, vol. 40, pp. 655–673, 2005.

[6] S. Fouvry, "Shakedown analysis and wear response under gross slip condition," *Wear*, vol. 251, pp. 1320–1331, 2001.

[7] T. R. HYDE, S. B. LEEN, and I. R. MCCOLL, "A simplified fretting test methodology for complex shaft couplings," *Fatigue & Fracture of Engineering Materials & Structures*, vol. 28, pp. 1047–1067, 2005.

[8] J. J. Madge, S. B. Leen, I. R. McColl, and P. H. Shipway, "Contact-evolution based prediction of fretting fatigue life: effect of slip amplitude," *Wear*, vol. 262, pp. 1159–1170, 2007.

[9] J. J. Madge, S. B. Leen, and P. H. Shipway, "The critical role of wear in the analysis of fretting fatigue," *Wear*, vol. 263, pp. 542–551, 2007.

[10] J. Ding, I. R. McColl, and S. B. Len, "The application of fretting wear modelling to a spline coupling," *Wear*, vol. 262, pp. 1205–1216, 2007.

[11] D. Houghton, P. M. Wavish, E. J. Williams, and S. B. Leen, "Multiaxial fretting fatigue testing and prediction for splined couplings," *International Journal of Fatigue*, vol. 31, pp. 1805–1815, 2009.

[12] C. H. H. Ratsimba, I. R. McColl, E. J. Williams, S. B. Leen, and H. P. Soh, "Measurement, analysis and prediction of wear damage in a representative aeroengine spline coupling," *Wear*, vol. 257, pp. 1193–1206, 2004.

[13] I. R. McColl, J. Ding, and S. B. Leen, "Finite element simulation and experimental validation of wear," *Wear*, vol. 256, pp. 1114–1127, 2004.

[14] V. Cuffaro, F. Cura, and A. Mura, "Test rig for spline coupling working in misaligned conditions," *Journal of Tribology*, vol. 136, pp. 1–7, 2014.

[15] V. Cuffaro, F. Cura, and A. Mura, "Damage identification on spline coupling teeth by means of roughness parameters," *Theoretical and Applied Fracture Mechanics*, vol. 82, pp. 9–16, 2016.

[16] Y. Guo, S. Lambert, R. Wallen, R. Errichello, and J. Keller, "Theoretical and experimental study on gear-coupling contact and loads considering misalignment, torque, and friction influences," *Mechanism and Machine Theory*, vol. 98, pp. 242–262, 2016.

[17] S. B. Leen, T. R. Hyde, E. J. Williams et al., "Development of a representative test specimen for frictional contact in spline joint couplings," *The Journal of Strain Analysis for Engineering Design*, vol. 35, pp. 521–544, 2000.

[18] S. B. Leen, I. J. Richardson, I. R. McColl, E. J. Williams, and T. R. Hyde, "Macroscopic fretting variables in a splined coupling under combined torque and axial load," *The Journal of Strain Analysis for Engineering Design*, vol. 36, pp. 481–497, 2001.

[19] S. B. Leen, T. H. Hyde, C. H. H. Ratsimba, E. J. Williams, and I. R. McColl, "An investigation of the fatigue and fretting," *The Journal of Strain Analysis for Engineering Design*, vol. 37, pp. 565–583, 2003.

[20] Z. G. Hu, R. P. Zhu, G. H. Jin, and D. Ni, "Analysis of fretting frictional contact parameters of aviation involute spline coupling," *Journal of Central South University (Science and Technology)*, vol. 44, pp. 1822–1828, 2013.

[21] J. Hong, D. Talbot, and A. Kahraman, "Load distribution analysis of clearance-fit spline joints using finite elements," *Mechanism and Machine Theory*, vol. 74, pp. 42–57, 2014.

[22] A. Barrot, M. Paredes, and M. Sartor, "Extended equations of load distribution in the axial direction in a spline coupling," *Engineering Failure Analysis*, vol. 16, pp. 200–211, 2009.

Adaptive Reduced Dimension Fuzzy Decoupling Control Method with its Application to a Deployable Antenna Panel

Zhi-yong Liu (ID),[1,2] **Jing-li Du** (ID),[1,3] **Hong Bao** (ID),[1,3] **and Qian Xu**[4]

[1]*The Key Laboratory of Electronic Equipment Structure Design of Ministry of Education, Xidian University, Xian 710071, China*
[2]*Xian Yang Vocational & Technical College, Xianyang 712000, China*
[3]*Collaborative Innovation Center of Information Sensing and Understanding, Xidian University, Xian 710071, China*
[4]*Xinjiang Astronomical Observatories, CAS, Urumqi 830011, China*

Correspondence should be addressed to Hong Bao; hbao@xidian.edu.cn

Academic Editor: Paul Williams

This paper addresses the multiple-input multiple-output (MIMO) control problem with an active deformation adjustment mechanism on a 5-metre deployable antenna panel. An adaptive reduced-dimension fuzzy decoupling (ARDFD) control strategy based on fusion functions is proposed to eliminate the effects uncertainties due to coupled subsystems by designing and incorporating a fuzzy decoupling disturbance observer. Moreover, the MIMO system can be decomposed into a number of single-input single-output (SISO) systems using the unit diagonal matrix decoupling theory. Based on Lyapunov stability theory, it was shown that the error of fuzzy interference observer and the error between the fuzzy rule post parameter matrix and optimal post parameter matrix converge to a small region. Results of the simulation also show that the ARDFD control strategy can significantly eliminate adverse effects of the coupled subsystems.

1. Introduction

To achieve long-term continuous meteorological observations, 5-meter deployable antennas can be arranged in geosynchronous orbit [1, 2]. The reflector can significantly affect meteorological observation performance due to deformation of the antenna caused by external factors. To offset the impact of the environmental load on the antenna panel deformation, an active control strategy must be adopted, which can control the surface deformation down to microscale precision.

In this paper, a high precision active control method is studied to overcome the problem of antenna surface deformation on space reflecting antennas. An active adjustment system for a 5-meter deployable antenna is proposed, as shown in Figure 1. Because the reflecting panel is supported by the more active actuators, the supporting actuators interact with one another during the adjustment process. Therefore, it is necessary to develop decoupling control methods for complex multivariable systems, in particular, to realize

high precision active adjustments of the reflecting panel of 5-meter deployable antennas.

In real-world multiple-input multiple-output (MIMO) systems, a variety of factors may cause many parameters to change at runtime. Traditional fuzzy theory can be directly applied to MIMO systems; however, due the large number of relationships, the "rule explosion" problem occurs. That is, the number of fuzzy rules increases exponentially with the number of fuzzy input variables [3]. In the last decade, a great deal of progress has been made in relation to the control of MIMO systems, and numerous control strategies have been introduced [4–7]. Decoupling controllers are based on nonlinear dynamic inverse and decentralized methods, but cannot guarantee robustness when certain system characteristics are unknown [8]. The automatic adjustment of a proportional–integral–derivative (PID) decoupling controller using a neural network was previously proposed; however, its precision for MIMO systems with strong coupling cannot be guaranteed [9]. Intelligent control methods for MIMO systems have been put forward combining fuzzy control theory

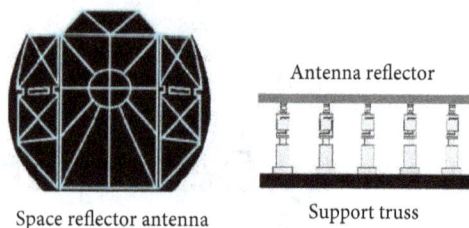

FIGURE 1: The active adjustment system of a 5-meter deployable antenna.

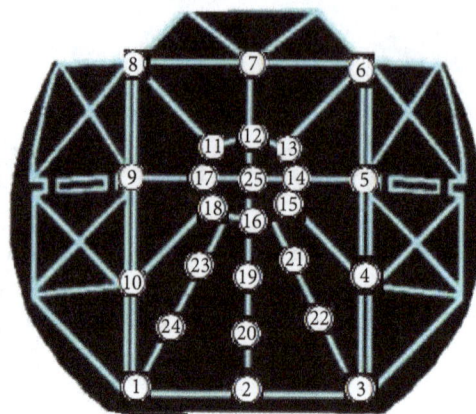

FIGURE 2: Distribution of piezoelectric ceramic actuators on the middle reflector of the 5 m deployable antenna.

with artificial intelligence and adaptive control [10–12]. However, such methods are only applicable to cases wherein the output is a state quantity. An adaptive fuzzy controller was previously applied to a multivariable system with uncertainties and demonstrated some practical value [13]. Interferences were approximated using a specifically designed disturbance observer, and a single precise controller was successfully implemented on a system of multiple variables. Efficient observers for the decentralized control of uncertain MIMO systems have also been studied [14, 15]. Tian et al. proposed a novel inverted fuzzy decoupling scheme [16]; moreover, the Smith predictor (SP) based on a fuzzy decoupling scheme was proposed for a MIMO chemical process with multiple time delays [17]. The latter two schemes were achieved using fuzzy logic, thereby avoiding the reliance on model-based analyses. Further to this, Hamdy et al. proposed another fuzzy decoupling-based PI controller [18]. Fuzzy decoupling schemes are simple to design and easy to implement and demonstrate good decoupling capabilities.

For the active adjustment system of the 5-meter deployable antenna, the relationship between the input and output is unclear and previously adopted control strategies cannot effectively deal with the coupling of variables and influence of uncertainty factors.

This paper presents an adaptive reduced-dimension fuzzy decoupling (ARDFD) control strategy based on a number of fuzzy observers. The efficiency of the method was demonstrated via simulations. Moreover, compared to the aforementioned controllers designed for MIMO systems, the ARDFD controller has the following advantages:

(i) MIMO systems can be decomposed into a number of single-input single-output (SISO) systems using the unit diagonal matrix decoupling method.

(ii) The dimension of multidimensional systems can be reduced using fusion function theory. Furthermore, the controller can be constructed by introducing a fuzzy decoupling disturbance observer.

(iii) The value of disturbances can be approximately obtained based on the Lyapunov stability and using the fuzzy observer. Further, disturbances are considered by the addition of compensation signals to the ARDFD controller, which eliminates the influence of uncertainties due to coupled subsystems.

2. Problem Statement

Twenty-five actuators are arranged on the 5-meter space deployable antenna reflector. The distribution of the actuators is shown in Figure 2, and each actuator is labelled with a number between 1–25.

For active adjustment system of the 5-meter deployable antenna, the positions of the zero-displacement datum of the 25 actuators were chosen to directly represent the system state.

According to [16], the mathematical model of the active adjustment mechanism can be described as follows:

$$Z(k + 1) = Z(k) + B * U(k),$$
$$Y(k) = Z(k), \tag{1}$$

where $Z(k)$ denotes the displacement before the kth adjustment, $U(k)$ is the external force of the kth adjustment, and B represents the transfer matrix of the external force and actual displacement.

Using existing control strategies to adjust the active adjustment system of the 5-meter deployable antenna, the following problems are encountered:

(i) Mapping relation is unclear between the input and output of each subsystem; therefore, strong coupling effects are caused.

(ii) The "rule explosion" problem occurs if traditional fuzzy methods are used for the active adjustment system of the 5-meter deployable antenna.

(iii) With existing control methods, it is difficult to determine the values of the coupling term and unknown disturbances.

To solve the above problems, an ARDFD control strategy based on the fusion function is proposed. An important aspect of the ARDFD control strategy is that the strongly coupled MIMO system can be deconstructed into 25 independent subsystems using the unit diagonal matrix method, thereby reducing the dimensions of the MIMO system and solving the "rule explosion" problem. Moreover, by designing the fuzzy decoupling observer, estimated values of the disturbance terms can be obtained based on the input and output information. Then, estimated values can be incorporated into the decoupling controller by adding a compensation signal to eliminate the uncertainties due to the coupled subsystems.

3. Adaptive Reduced-Dimension Fuzzy Decoupling Controller Design

3.1. Adjustment System Decoupling. Based on (1), the discrete multivariable control model with uncertainties can be described as

$$Z(k+1) = Z(k) + (B + \Delta B) * U(k),$$
$$Y(k) = Z(k),$$
(2)

where ΔB represents the uncertainty of the system. In the active adjustment system, any changes in inputs will affect the outputs, because there is a strong coupling between the input and output. To optimize the control, (1) can be decoupled using the unit diagonal matrix decoupling method.

For any MIMO system with a full rank transfer matrix, a compensator can be designed to decouple it [17].

Assuming $W(s)$ is the transfer function matrix of (1), then $W(s) = C(SI - A)^{-1}B$, where C and A are unit diagonal matrices and $W(s)$ is the full rank matrix. Thus, a compensator can be designed to decouple the MIMO system.

Based on the decoupling principle for a unit diagonal matrix, (3) can be written as

$$W_p(s)W(s) = I,$$
(3)

where $W_p(s)$ is the transfer function matrix of the feedforward compensator.

According to (3), we can derive

$$W_p(s) = W^{-1}(s) = B^{-1}(SI - A)C^{-1}.$$
(4)

From (4), the output matrix of the feed-forward compensator is B^{-1}; then, the output equation of the feed-forward compensator can be written as

$$X(k) = B^{-1}Z(k).$$
(5)

Rearranging (5), we obtain

$$Z(k) = BX(k).$$
(6)

Substituting (6) into (2), we obtain

$$BX(k+1) = BX(k) + (B + \Delta B) * U(k).$$
(7)

Then, multiplying both sides of (7) by B^{-1}, the equation can be expressed as

$$X(k+1) = X(k) + U(k) + B^{-1}\Delta B * U(k),$$
(8)

where $X(k)$ is the state vector of the decoupled system, $X(k) = [x_1(k), x_2(k), \ldots, x_{25}(k)]$, and $B^{-1}\Delta B$ is an unknown vector representing the unmodeled disturbance matrix of the MIMO system.

Here, assuming $G = B^{-1}\Delta B$, (8) can be rewritten as

$$X(k+1) = X(k) + U(k) + GU(k),$$
(9)

where $G = [G_1^T, G_2^T, \ldots, G_{25}^T]^T$.

From (9), it can be seen that the discrete multivariable decoupled system can be decomposed into 25 independent single variable state equations. The coupling effects caused by other subsystems and unknown terms are regarded as unknown quantities G. Thus, the active adjustment coupling system of the 5-meter deployable antenna is decoupled.

Since the unknown disturbance G cannot be obtained, the disturbance observer can be constructed to estimate the unknown disturbance using the fuzzy strategy. Furthermore, since the active adjustment system of the 5-meter deployable antenna is a MIMO system in this case, directly adopting the traditional fuzzy control method leads to the "rule explosion" problem. That is, the fuzzy rules grow exponentially with the number of fuzzy inputs, which is unfavorable for controlling the active adjustments of a coupled system. However, using the fusion function method, the input state variables can be merged to reduce the dimensions of the system.

3.2. Fusion Function Design. Controlling this type of MIMO system using multidimensional fuzzy control strategies is difficult; therefore, a multilevel control method is adopted. Fusion function theory can be used to convert the complex fuzzy control strategy into a more simplified nested control strategy [18, 19], as follows:

$$X_i(k) \xrightarrow{f_1()} \tilde{X}_i(k) \xrightarrow{f_2()} Y.$$
(10)

Equation (10) shows that the multivariable system is preliminary processed by $f_1()$. Then, based on the output, $f_1()$ and $f_2()$ control the active adjustment of the coupled system. If the dimension of the output vector of $f_1()$ is less than the dimension of $X_i(k)$, the problem has been simplified.

In this paper, the reduction in dimensions can be achieved by transforming the 25 state variables $x_i(k)$ into 5 synthetic errors E_i, as follows:

$$E_1 \triangleq [k_1 \, k_2 \, k_3 \, k_4 \, k_5][x_1(k) \, x_2(k) \, x_3(k) \, x_4(k) \, x_5(k)]^T,$$
$$E_2 \triangleq [k_6 \, k_7 \, k_8 \, k_9 \, k_{10}][x_6(k) \, x_7(k) \, x_8(k) \, x_9(k) \, x_{10}(k)]^T,$$
$$E_3 \triangleq [k_{11} \, k_{12} \, k_{13} \, k_{14} \, k_{15}][x_{11}(k) \, x_{12}(k) \, x_{13}(k) \, x_{14}(k) \, x_{15}(k)]^T,$$
$$E_4 \triangleq [k_{16} \, k_{17} \, k_{18} \, k_{19} \, k_{20}][x_{16}(k) \, x_{17}(k) \, x_{18}(k) \, x_{19}(k) \, x_{20}(k)]^T,$$
$$E_5 \triangleq [k_{21} \, k_{22} \, k_{23} \, k_{24} \, k_{25}][x_{21}(k) \, x_{22}(k) \, x_{23}(k) \, x_{24}(k) \, x_{25}(k)]^T,$$
(11)

where $k_1, k_2, \ldots, k_{25}$ represent the comprehensive coefficients.

A number of assumptions can be made:

$$
\begin{aligned}
K_1 &= [k_1 k_2 k_3 k_4 k_5], \\
K_2 &= [k_6 k_7 k_8 k_9 k_{10}], \\
K_3 &= [k_{11} k_{12} k_{13} k_{14} k_{15}], \\
K_4 &= [k_{16} k_{17} k_{18} k_{19} k_{20}], \\
K_5 &= [k_{21} k_{22} k_{23} k_{24} k_{25}], \\
Y_1 &= [x_1(k)\ x_2(k)\ x_3(k)\ x_4(k)\ x_5(k)]^T, \\
Y_2 &= [x_6(k)\ x_7(k)\ x_8(k)\ x_9(k)\ x_{10}(k)]^T, \\
Y_3 &= [x_{11}(k)\ x_{12}(k)\ x_{13}(k)\ x_{14}(k)\ x_{15}(k)]^T, \\
Y_4 &= [x_{16}(k)\ x_{17}(k)\ x_{18}(k)\ x_{19}(k)\ x_{20}(k)]^T, \\
Y_5 &= [x_{21}(k)\ x_{22}(k)\ x_{23}(k)\ x_{24}(k)\ x_{25}(k)]^T.
\end{aligned}
\tag{12}
$$

The output of the fusion function can then be expressed as

$$
f_1(x) =
\begin{bmatrix} E_1 \\ E_2 \\ E_3 \\ E_4 \\ E_5 \end{bmatrix}
=
\begin{bmatrix}
K_1 & & & & \\
& K_2 & & & \\
& & K_3 & & \\
& & & K_4 & \\
& & & & K_5
\end{bmatrix}
\cdot
\begin{bmatrix}
Y_1 & & & & \\
& Y_2 & & & \\
& & Y_3 & & \\
& & & Y_4 & \\
& & & & Y_5
\end{bmatrix}.
\tag{13}
$$

By designing the feedback gain matrix using the linear quadratic regulator (LQR) method, the optimization index can be defined as

$$
J = \frac{1}{2} \int_0^\infty \left(x^T(k) Q x(k) + u^T(k) R u(k) \right) dk,
\tag{14}
$$

where Q is a positive semidefinite matrix and R represents a symmetrical and positive definite matrix.

The Riccati equation can be solved as follows:

$$
-PA - AP + PBR^T B^T P - Q = 0,
\tag{15}
$$

where P is a positive definite matrix of constants.

Then, the feedback matrix can be obtained as

$$
K^T = R^{-1} B^T P [k_1\ k_2\ k_3\ k_4\ k_5 \cdots k_{25}] = \frac{K^T}{\|K\|}.
\tag{16}
$$

It can be seen that the linear combination reduces the dimensions of the MIMO system, which greatly reduces the number of fuzzy rules.

3.3. ARDFD Controller Design. In this section, the 1st decoupled subsystem is taken as an example to introduce the ARDFD controller. According to (9), the state equation of the 1st decoupled subsystem can be written as

$$
x_1(k+1) = x_1(k) + u_1(k) + G_1 U(k).
\tag{17}
$$

Here, we define the following set:

$$
G_1 U(k) = d_1(k).
\tag{18}
$$

Substituting (18) into (17), the decoupled subsystem, which takes into account disturbances caused by other subsystems, can be further expressed as

$$
x_1(k+1) = x_1(k) + u_1(k) + d_1(k), \quad d_1(k) = \sum_{j=1}^{25} G_{1,j} u_j(k),
\tag{19}
$$

where $d_1(k)$ are the disturbances due to the coupled subsystems and contains the uncertainty terms. In this paper, the $d_1(k)$ values of the 5-meter deployable antenna are unknown. Equation (19) represents an SISO system with disturbances and a one-to-one relationship between the input and output of the 1st decoupled subsystem exists.

Based on fuzzy theory, the fuzzy decoupling disturbance observer can be constructed to obtain the approximate disturbance values of the 1st decoupled subsystem. Moreover, an adaptive law to adjust the consequent parameter based on fuzzy rules can be simultaneously obtained.

Since $d_1(k)$ is a nonlinear function of arbitrary accuracy, the real matrix $\theta_1^*(k)$ exists [20, 21] and can be defined as

$$
d_1(k) = \theta_1^{*T}(k)\xi(k) + \varepsilon,
\tag{20}
$$

where $\theta_1^*(k)$ is the consequent parameter matrix of the fuzzy rules, ε represents a nonnegative constant, $1 > \varepsilon > 0$, and $\xi(k)$ is the basis function of the fuzzy rules.

Subsequently, the fuzzy disturbance observer can be constructed, and the estimated values of the disturbances can be obtained.

Suppose the 1st fuzzy rule is given as follows:

if $x_1(k)$ is A_1^i and $x_2(k)$ is A_2^i and $\cdots$ and $x_5(k)$ is A_{25}^i

then $\hat{d}_1(k) = a_1^i(k)$.

$$
\tag{21}
$$

Then, the output of the fuzzy system can be expressed as

$$
\hat{d}_1(k) = \frac{\sum_{j=1}^{r} a_1^i(k) \prod_{i=1}^{5} \mu_{A_j^i}(x_j(k))}{\sum_{j=1}^{r} \prod_{i=1}^{5} \mu_{A_j^i}(x_j(k))},
\tag{22}
$$

where r is the number of fuzzy rules and $\mu_{A_j^i}$ represents the membership functions of the input variables.

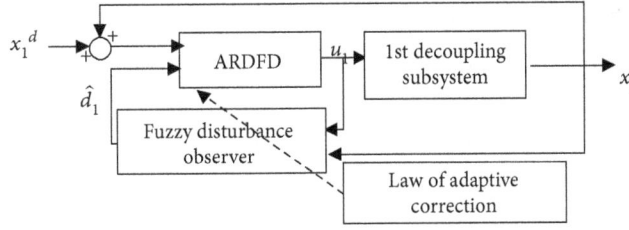

FIGURE 3: The closed-loop model of the 1st decoupled subsystem.

Now, suppose the basic function $\xi_i(k)$ of the ith fuzzy rule is expressed as

$$\xi_i(k) = \frac{\prod_{i=1}^{5} \mu_{A_i^j}(x_i(k))}{\sum_{j=1}^{r} \prod_{i=1}^{5} \mu_{A_i^j}(x_i(k))}. \tag{23}$$

Then, the basic function vector of the fuzzy rules can be defined as

$$\xi(k) = [\xi_1(k), \xi_2(k), \ldots, \xi_r(k)]^T. \tag{24}$$

Suppose the fuzzy posterior parameter matrix is

$$\theta_i(k) = [a_i^1(k), a_i^2(k), \ldots, a_i^r(k)]^T. \tag{25}$$

Based on (24) and (25), the output of the fuzzy system can be defined as

$$\widehat{d}_i(k) = \theta^T_i(k)\xi(k). \tag{26}$$

Then, the fuzzy disturbance observer for the 1st decoupled subsystem can be defined as

$$\sigma_1(k+1) = \sigma_1(k) + u_1(k) - \gamma_1\sigma_1(k) + \gamma_1 x_1(k) + \widehat{d}_1(k),$$
$$\widehat{d}_1(k) = \theta^T_1(k)\xi(k), \tag{27}$$

where $d_1(k)$ is the estimated value of the disturbance, γ_1 represents a parameter of the fuzzy interference term, and $\sigma_1(k)$ is the state of the fuzzy disturbance observer.

According to $\sigma_1(k)$ and $x_1(k)$, the observation error can be defined as

$$e_1(k) = x_1(k) - \sigma_1(k). \tag{28}$$

The error between the fuzzy rule posterior parameter matrix and the optimal fuzzy posterior parameter matrix can be expressed as

$$\tilde{\theta}_1(k) = \theta_1^*(k) - \theta_1(k). \tag{29}$$

Furthermore, the control law based on the fuzzy disturbance observer can be designed as

$$u_1(k) = -hx_1(k) - \widehat{d}_1(k), \tag{30}$$

where h is the feedback control law parameter.

Substituting (30) into (19), the 1st decoupled subsystem can be rewritten as

$$x_1(k+1) = x_1(k) - hx_1(k) - \widehat{d}_1(k) + d_1(k)$$
$$= (1-h)x_1(k) - \widehat{d}_1(k) + d_1(k). \tag{31}$$

Then, substituting (20) and (26) into (30), the equation can be rewritten as

$$x_1(k+1) = x_1(k) - hx_1(k) - \widehat{d}_1(k) + d_1(k)$$
$$= (1-h)x_1(k) - \theta_1^T(k)\xi(k) + \theta_1^{*T}(k)\xi(k) \tag{32}$$
$$= (1-h)x_1(k) + \tilde{\theta}_1^T(k)\xi(k).$$

According to (17), (18), and (27), the observation error equation of an unknown disturbance can be represented as

$$e_1(k+1) = e_1(k) - \gamma_1 e_1(k) + \tilde{\theta}_1^T(k)\xi(k). \tag{33}$$

Then, based on (19), (27), and (30) for the 1st decoupled subsystem, the adaptive law of the fuzzy post parameter matrix can be written as

$$\theta_1(k+1) - \theta_1(k) = \rho_1 x_1(k) p_1 \xi(k) + \rho_1(k) e_1(k)\xi(k), \tag{34}$$

where p_1 is a positive constant, ρ_1 is the adaptive parameter, and $\rho_1 > 0$.

Now, the closed-loop model of the 1st decoupled subsystem can be constructed, as shown in Figure 3.

Therefore, the fuzzy system can be used to obtain approximate values of the disturbances based on the input and output information, which can be added to the decoupling controller as a compensation signal to eliminate the uncertainties due to the coupled subsystems.

Similarly, control laws for the other 24 decoupled subsystems can be defined and the fuzzy disturbance estimator can easily be obtained for each.

4. Stability Analysis

For the mathematical model expressed in (1), the fuzzy disturbance observer of (27) and the control law based on the fuzzy disturbance observer presented as (29), when $k = 1, 2, \ldots, n$, the error of the fuzzy disturbance observer and the error between the fuzzy rule posterior parameter matrix and optimal fuzzy posterior parameter matrix converge uniformly with discrete time k.

The Lyapunov candidate function is defined as

$$y(k) = \frac{1}{2}p_1[x_1(k)]^2 + \frac{1}{2}[e_1(k)]^2 + \frac{1}{2\rho_1}\tilde{\theta}_1^T(k)\tilde{\theta}_1(k), \qquad (35)$$

where $p_1 > 0$, $\rho_1 > 0$, $y(k) > 0$, and $y(k)$ is a positive definite matrix.

According to (35),

$$\Delta y(k) = y(k+1) - y(k). \qquad (36)$$

Substituting (35) into (36), we obtain

$$\Delta y(k) = y(k+1) - y(k) = \frac{1}{2}p_1[x_1(k+1)]^2$$
$$+ \frac{1}{2}[e_1(k+1)]^2 + \frac{1}{2\rho_1}\tilde{\theta}_1^T(k+1)\tilde{\theta}_1(k+1)$$
$$- \frac{1}{2}p_1[x_1(k)]^2 - \frac{1}{2}[e_1(k)]^2 - \frac{1}{2\rho_1}\tilde{\theta}_1^T(k)\tilde{\theta}_1(k). \qquad (37)$$

Based on (32), the following can be derived:

$$\frac{1}{2}p_1[x_1(k+1)]^2 - \frac{1}{2}p_1[x_1(k)]^2$$
$$= p_1 x_1(k)\tilde{\theta}_1^T(k)\tilde{\theta}_1(k)$$
$$\quad + \frac{1}{2}p_1\left\{ (h^2 - 2h)[x_1(k)]^2 - 2hx_1(k)\tilde{\theta}_1^T(k)\xi(k) \right.$$
$$\quad \left. + \left[\tilde{\theta}_1^T(k)\xi(k)\right]^2 \right\}$$
$$= p_1 x_1(k)\tilde{\theta}_1^T(k)\tilde{\theta}_1(k) + \frac{1}{2}p_1 \begin{bmatrix} x_1(k) \\ \tilde{\theta}_1^T(k)\tilde{\theta}_1(k) \end{bmatrix}$$
$$\cdot \begin{bmatrix} h^2 - 2h & -h \\ -h & 1 \end{bmatrix} \begin{bmatrix} x_1(k) \\ \tilde{\theta}_1^T(k)\tilde{\theta}_1(k) \end{bmatrix}. \qquad (38)$$

According to (33), the following can be derived:

$$\frac{1}{2}[e_i(k+1)]^2 - \frac{1}{2}[e_i(k)]^2$$
$$= e_i(k)\tilde{\theta}_i^T(k)\tilde{\theta}_i(k)$$
$$\quad + \frac{1}{2}\left\{ (\gamma^2 - 2\gamma)[x_i(k)]^2 - 2\gamma x_i(k)\tilde{\theta}_i^T(k)\xi(k) + \left[\tilde{\theta}_i^T(k)\xi(k)\right]^2 \right\}$$
$$= e_i(k)\tilde{\theta}_i^T(k)\tilde{\theta}_i(k) + \frac{1}{2}\begin{bmatrix} e_i(k) \\ \tilde{\theta}_i^T(k)\tilde{\theta}_i(k) \end{bmatrix}\begin{bmatrix} \gamma^2 - \gamma & -\gamma \\ -\gamma & 1 \end{bmatrix}$$
$$\cdot \begin{bmatrix} e_i(k) \\ \tilde{\theta}_i^T(k)\tilde{\theta}_i(k) \end{bmatrix}^2. \qquad (39)$$

Furthermore, based on (33), we can derive the following:

$$\frac{1}{2\rho_i}\tilde{\theta}_i^T(k+1)\tilde{\theta}_i(k+1) - \frac{1}{2\rho_i}\tilde{\theta}_i^T(k)\tilde{\theta}_i(k)$$
$$= \frac{1}{2\rho_i}\left[2\tilde{\theta}_i(k) - \rho_i\Delta\right][-\rho_i\Delta] \qquad (40)$$
$$= -\tilde{\theta}_i^T(k) + \frac{\rho_i}{2}\Delta^T\Delta,$$

where

$$\Delta = x_i(k)p_i\xi(k) + e_i(k)\xi(k). \qquad (41)$$

Substituting (38), (39), and (40) into (37), the equation can be rewritten as

$$\Delta y(k) = \frac{1}{2}p_i \begin{bmatrix} x_i(k) \\ \tilde{\theta}_i^T(k)\tilde{\theta}_i(k) \end{bmatrix} \begin{bmatrix} h^2 - 2h & -h \\ -h & 1 \end{bmatrix} \begin{bmatrix} x_i(k) \\ \tilde{\theta}_i^T(k)\tilde{\theta}_i(k) \end{bmatrix}$$
$$+ \frac{1}{2}\begin{bmatrix} e_i(k) \\ \tilde{\theta}_i^T(k)\tilde{\theta}_i(k) \end{bmatrix} \begin{bmatrix} \gamma^2 - \gamma & -\gamma \\ -\gamma & 1 \end{bmatrix} \begin{bmatrix} e_i(k) \\ \tilde{\theta}_i^T(k)\tilde{\theta}_i(k) \end{bmatrix}$$
$$+ \frac{\rho_i}{2}[x_i(k)p_i\xi(k) + e_i(k)\xi(k)]^T[x_i(k)p_i\xi(k) + e_i(k)\xi(k)]. \qquad (42)$$

In the case of $p_i > 0$, $\rho_i > 0$, $\gamma_i > 0$, we have

$$\frac{1}{2}p_i\begin{bmatrix} h^2 - 2h & -h \\ -h & 1 \end{bmatrix} = -hp_i < 0,$$
$$\frac{1}{2}p_i\begin{bmatrix} \gamma^2 - \gamma & -\gamma \\ -\gamma & 1 \end{bmatrix} = -\gamma_i < 0. \qquad (43)$$

Then, $\Delta y(k) < 0$.

According to the Lyapunov stability theorem of discrete systems, $x_i(k)$, $e_i(k)$, and $\theta_i(k)$ uniformly converge with discrete time k. In a similar way, the stability analysis of the other 24 decoupled subsystems can be performed.

5. Simulation Verification

To test the validity of the ARDFD controller, a simulation analysis was performed. During the simulation, control matrix B was obtained using finite element software ANSYS (version 14.5).

$$B = \begin{bmatrix} B_1 & & & & \\ & B_2 & & & \\ & & B_3 & & \\ & & & B_4 & \\ & & & & B_5 \end{bmatrix}, \qquad (44)$$

where

$$B_1 = \begin{bmatrix} 90.775 & 0.907 & 4.138 & -1.646 & 0.616 \\ 0.907 & 90.740 & 0.613 & 0.029 & 4.123 \\ 4.138 & 0.613 & 6.313 & 0.005 & 0.280 \\ -1.643 & 0.032 & 0.008 & 3.647 & -0.002 \\ 7.791 & 11.482 & 7.636 & 7.330 & 13.690 \end{bmatrix},$$

$$B_2 = \begin{bmatrix} -3.762 & -2.267 & -7.351 & -9.914 & -4.350 \\ 5.088 & 174.390 & 10.081 & 9.752 & 96.525 \\ 4.138 & 0.613 & 6.313 & 0.005 & 0.280 \\ -1.643 & 0.032 & 0.008 & 3.647 & -0.002 \\ 7.791 & 11.482 & 7.636 & 7.330 & 13.690 \end{bmatrix},$$

$$B_3 = \begin{bmatrix} 0.773 & 0.184 & 0.003 & -0.002 & -0.031 \\ 0.187 & 115.940 & 0.002 & -1.924 & 0.047 \\ 0.003 & -0.015 & 0.772 & 0.185 & 0.030 \\ 0.002 & -1.923 & 0.189 & 115.930 & 0.049 \\ 0.032 & 0.046 & 0.032 & 0.048 & 29.912 \end{bmatrix},$$

$$B_4 = \begin{bmatrix} 25.795 & -0.004 & 0.059 & 0.017 & 0.174 \\ -0.009 & 25.683 & -0.012 & 0.058 & 0.175 \\ 0.054 & 0.013 & 21.254 & -0.389 & -0.022 \\ 0.017 & 0.064 & -0.384 & 21.260 & 0.028 \\ 0.169 & 0.176 & 0.023 & 0.024 & 1.114 \end{bmatrix},$$

$$B_5 = \begin{bmatrix} 11.516 & 17.269 & -0.115 & -0.985 & 0.341 \\ 17.268 & 54.711 & -0.518 & -2.334 & 0.867 \\ -0.112 & -0.514 & 76.493 & -0.500 & 0.160 \\ -0.980 & -2.310 & -0.500 & 440.780 & -8.235 \\ 0.340 & 0.870 & 0.160 & -8.239 & 286.420 \end{bmatrix}.$$

$$(45)$$

The initial value $Z(k)$ of the discrete state vector of the 5-meter deployable antenna adjustment system is

$$Z(k) = [-1.141 \quad 12.668 \quad 12.656 \quad -3.987 \quad 4.032 \quad 8.982 \quad 0.676$$
$$-3.183 \quad -12.143 \quad -11.927 \quad 0.297 \quad 0.045 \quad 0.465$$
$$-9.303 \quad -14.889 \quad 2.119 \quad 2.003 \quad 0.194 \quad 11.079$$
$$0.631 \quad 1.194 \quad 1.094 \quad -9.187 \quad 3.839 \quad 2.257]^{\mathrm{T}}.$$

$$(46)$$

Units are in μm.

First, the decomposed state vector $X(k)$ was given based on (5) and (6).

Then, the semipositive definite matrix was designed according to the actual requirements as follows:

$$Q = \mathrm{diag} [80 \quad 20 \quad 100 \quad 80 \quad 150 \quad 50 \quad 200 \quad 100 \quad 300$$
$$200 \quad 200 \quad 150 \quad 500 \quad 200 \quad 200 \quad 150 \quad 100$$
$$500 \quad 200 \quad 300 \quad 150 \quad 100 \quad 500 \quad 200 \quad 300].$$

$$(47)$$

The symmetric matrix is $R = 1$. Further, the state feedback matrix K was obtained as

$$K = [10 \quad 5.58 \quad 11.05 \quad 10 \quad 13.29 \quad 8.14 \quad 15.18$$
$$11.05 \quad 3018.35 \quad 15.18 \quad 15.18 \quad 13.29 \quad 23.38$$
$$15.18 \quad 15.18 \quad 13.29 \quad 11.05 \quad 23.38 \quad 15.18$$
$$18.34 \quad 13.28 \quad 11.04 \quad 23.38 \quad 15.18 \quad 18.35].$$

$$(48)$$

The comprehensive coefficient was then obtained based on (16) as follows:

$$K_1 = \begin{bmatrix} k_1 & k_2 & k_3 & k_4 & k_5 \end{bmatrix}$$
$$= [0.132 \quad 0.074 \quad 0.146 \quad 0.132 \quad 0.175],$$
$$K_2 = \begin{bmatrix} k_6 & k_7 & k_8 & k_9 & k_{10} \end{bmatrix}$$
$$= [0.107 \quad 0.2001 \quad 0.146 \quad 0.242 \quad 0.2001],$$
$$K_3 = \begin{bmatrix} k_{11} & k_{12} & k_{13} & k_{14} & k_{15} \end{bmatrix}$$
$$= [0.2001 \quad 0.175 \quad 0.308 \quad 0.2001 \quad 0.2001],$$
$$K_4 = \begin{bmatrix} k_{16} & k_{17} & k_{18} & k_{19} & k_{20} \end{bmatrix}$$
$$= [0.175 \quad 0.146 \quad 0.308 \quad 0.2001 \quad 0.242],$$
$$K_5 = \begin{bmatrix} k_{21} & k_{22} & k_{23} & k_{24} & k_{25} \end{bmatrix}$$
$$= [0.175 \quad 0.146 \quad 0.308 \quad 0.2001 \quad 0.242].$$

$$(49)$$

Based on fusion theory, the dimensions of $X(k)$ were reduced to

$$X(k) = \begin{bmatrix} Y_1 & & & & \\ & Y_2 & & & \\ & & Y_3 & & \\ & & & Y_4 & \\ & & & & Y_5 \end{bmatrix}, \quad (50)$$

where

$$Y_1 = [-0.1911 \quad 0.1311 \quad 4.0682 \quad -2.4169 \quad 0.2832]^{\mathrm{T}},$$
$$Y_2 = [0.0399 \quad 0.0366 \quad -0.0467 \quad -0.0584 \quad 0.0189]^{\mathrm{T}},$$
$$Y_3 = [0.1054 \quad 0.1334 \quad 0.7029 \quad -0.044 \quad -0.5334]^{\mathrm{T}},$$
$$Y_4 = [-0.0338 \quad 0.162 \quad -0.3691 \quad 0.4334 \quad 0.9486]^{\mathrm{T}},$$
$$Y_5 = [0.0604 \quad -0.0098 \quad 0.0272 \quad -0.0805 \quad -0.0541]^{\mathrm{T}}.$$

$$(51)$$

The uncertainty matrix ΔB was randomly generated in the range of $\pm 5\%$ based on the 2-norm of control matrix B, defined as

$$\Delta B = \begin{bmatrix} \Delta B_1 & & & & \\ & \Delta B_2 & & & \\ & & \Delta B_3 & & \\ & & & \Delta B_4 & \\ & & & & \Delta B_5 \end{bmatrix}, \qquad (52)$$

where

$$\Delta B_1 = \begin{bmatrix} 0.057 & 0.046 & -0.040 & -0.027 & -0.061 \\ 0.073 & 0.044 & 0.032 & -0.055 & 0.053 \\ -0.067 & -0.019 & 0.028 & -0.045 & -0.034 \\ 0.074 & 0.028 & -0.061 & 0.021 & 0.005 \\ 0.024 & -0.059 & -0.069 & -0.005 & -0.060 \end{bmatrix},$$

$$\Delta B_2 = \begin{bmatrix} 0.074 & -0.029 & -0.037 & 0.068 & -0.043 \\ -0.057 & 0.072 & 0.044 & 0.009 & 0.019 \\ -0.043 & -0.024 & -0.056 & 0.022 & 0.038 \\ 0.074 & 0.028 & -0.061 & 0.021 & 0.005 \\ -0.064 & -0.070 & 0.034 & 0.016 & -0.050 \end{bmatrix},$$

$$\Delta B_3 = \begin{bmatrix} -0.083 & 0.020 & 0.029 & -0.071 & -0.084 \\ 0.069 & 0.021 & 0.003 & -0.023 & 0.011 \\ 0.074 & 0.065 & 0.085 & -0.054 & 0.069 \\ 0.053 & 0.055 & 0.027 & -0.002 & 0.030 \\ -0.072 & 0.014 & 0.054 & -0.029 & -0.056 \end{bmatrix},$$

$$\Delta B_4 = \begin{bmatrix} 0.073 & 0.084 & -0.044 & 0.019 & -0.007 \\ 0.068 & 0.006 & -0.050 & -0.055 & 0.029 \\ 0.057 & -0.031 & 0.030 & 0.043 & 0.049 \\ -0.043 & -0.071 & 0.062 & -0.046 & -0.027 \\ 0.017 & 0.020 & -0.028 & 0.075 & 0.029 \end{bmatrix},$$

$$\Delta B_5 = \begin{bmatrix} -0.017 & 0.046 & 0.052 & -0.001 & 0.043 \\ -0.009 & -0.046 & 0.081 & -0.056 & 0.015 \\ -0.067 & -0.019 & 0.028 & -0.045 & -0.034 \\ 0.074 & 0.034 & 0.031 & -0.063 & 0.030 \\ 0.023 & -0.025 & -0.011 & -0.080 & -0.075 \end{bmatrix}.$$

$$(53)$$

The deformation was adjusted by the LQR controller, and the adjustment curve is presented in Figure 4.

After adding the uncertainty matrix to the actuator, the displacement control was adjusted by LQR controller

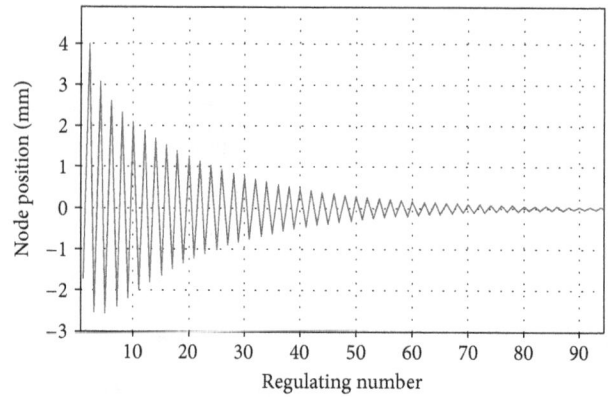

FIGURE 4: LQR adjustment curve of Node 1.

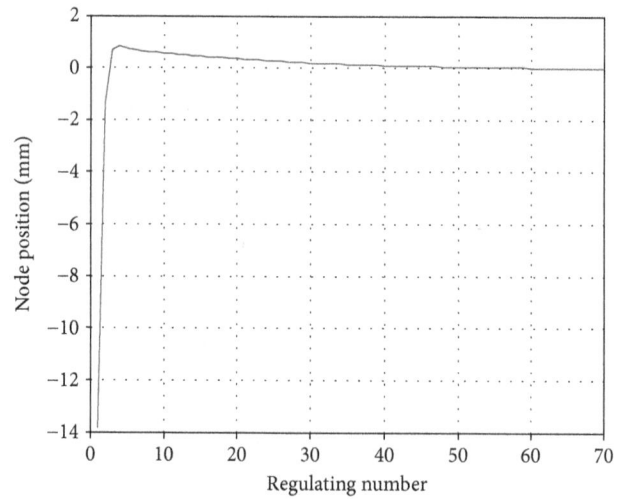

FIGURE 5: ARDFD adjustment curve for Node 1.

according to the curve shown in Figure 4. Uncertainty led to excessive overshooting of the system, which caused the curve converge.

Based on the same amount of interference, the 1st actuator was adjusted by the ARDFD controller, as shown in Figure 5. To assess the ability of the ARDFD controller to deal with the uncertainty of an observer, we have the following constraints on the parameters. The parameters of the fuzzy disturbances $\gamma_i \in [0.3, 1.2]$, $(i = 1, 2, 3, \ldots, 25)$, the adaptive algorithm parameters of the fuzzy observers $\rho_i = 0.01$, $p_i = 1$, $(i = 1, 2, 3, \ldots, 25)$, the parameters of the feedback control laws $h_i \in [0.6, 1.0]$, $(i = 1, 2, 3, \ldots, 25)$. Specific parameters can be adjusted according to actual simulation. The parameters of the fuzzy disturbances for the 1st to 25th decoupled subsystems were analyzed.

$$\gamma = [\gamma_1\, \gamma_2\, \gamma_3 \cdots \gamma_{25}]$$
$$= [0.4 \quad 0.6 \quad 0.3 \quad 0.6 \quad 1.2 \quad 0.4 \quad 0.6 \quad 0.3 \quad 0.6 \quad 1.2$$
$$0.4 \quad 0.6 \quad 0.3 \quad 0.6 \quad 1.2 \quad 0.4 \quad 0.6 \quad 0.3 \quad 0.6 \quad 1.2$$
$$0.4 \quad 0.6 \quad 0.3 \quad 0.6 \quad 1.2]^T.$$

$$(54)$$

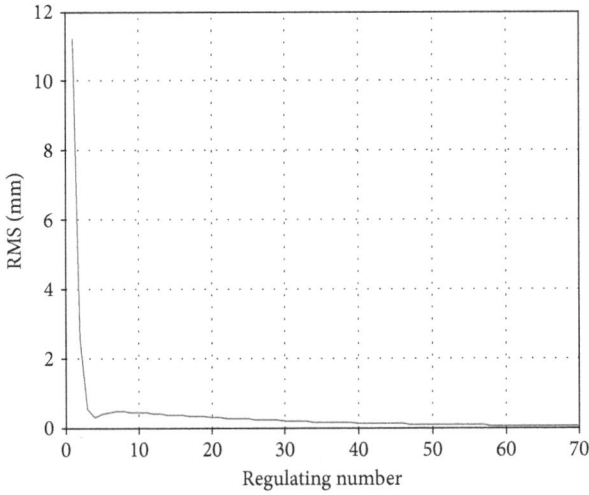

FIGURE 6: Root mean square (RMS) based on 25 points.

Adaptive algorithm parameters of the fuzzy observers were derived:

$$\rho = [\rho_1\, \rho_2\, \rho_3\, \cdots\, \rho_{25}]$$

$$= [0.01\quad 0.01\quad 0.01\quad 0.01\quad 0.01\quad 0.01\quad 0.01\quad 0.01\quad 0.01$$

$$0.01\quad 0.01\quad 0.01\quad 0.01\quad 0.01\quad 0.01\quad 0.01\quad 0.01\quad 0.01$$

$$0.01\quad 0.01\quad 0.01\quad 0.01\quad 0.01\quad 0.01\quad 0.01],$$

$$p = [p_1\, p_2\, p_3\, \cdots\, p_{25}]$$

$$= [1\quad 1\quad 1\quad 1\quad 1\quad 1\quad 1\quad 1\quad 1\quad 1\quad 1\quad 1\quad 1\quad 1\quad 1$$

$$1\quad 1\quad 1\quad 1\quad 1\quad 1\quad 1\quad 1\quad 1].$$

$$(55)$$

Finally, parameters of the feedback control laws were obtained:

$$h = [h_1\, h_2\, h_3\, \cdots\, h_{25}]$$

$$= [0.8\quad 0.7\quad 0.7\quad 1.0\quad 0.6\quad 0.8\quad 0.7\quad 0.7\quad 1.0\quad 0.6$$

$$0.8\quad 0.7\quad 0.7\quad 1.0\quad 0.6\quad 0.8\quad 0.7\quad 0.7\quad 1.0\quad 0.6$$

$$0.8\quad 0.7\quad 0.7\quad 1.0\quad 0.6\quad 0.8\quad 0.7\quad 0.7\quad 1.0\quad 0.6]^T.$$

$$(56)$$

As seen in Figure 5, when the uncertainty matrix ΔB is added to the 1st actuator, the curves do not show large fluctuations with the ARDFD controller and the system adjustments are very small. At the 10th step, the system reaches stability. Because the ARDFD controller uses a fuzzy system to realize the disturbance observer and estimate the uncertainty values of each coupled subsystem, the system can be controlled with high precision.

The root mean square (RMS) error curve for the 25 actuators is presented in Figure 6. The RMS value is 0.124 mm before the adjustment. After adjustment using the ARDFD controller, the displacements of all adjustment actuators are below 10^{-3} mm after the 45th adjustment. Accuracy is shown to be in the micron scale and demonstrates a tendency to narrow.

6. Conclusions

In this paper, the ARDFD controller for the active deformation adjustment mechanism of a 5-meter deployable antenna panel was studied. First, based on the unit diagonal matrix decoupling method, the MIMO system is decomposed into a number of SISO systems. Then, the ARDFD controller is constructed by introducing the fuzzy decoupling disturbance observer and using fusion function theory. Finally, the results of the simulation demonstrate that the proposed method for active deformation adjustment of a 5-meter deployable antenna panel is effective.

Furthermore, once the rule structure of the fuzzy system is defined during the design stage, it cannot be modified. Future research will consider a self-organized fuzzy system to improve the approximation capabilities of fuzzy systems.

Conflicts of Interest

The authors declare that there is no conflict of interest regarding the publication of this paper.

Acknowledgments

This work was supported in part by the National Natural Science Foundation of China through Grants 51675398 and 51775401, the National Key Basic Research Program of China through Grant 2015CB857100, and the CAS "Light of West China" Program (no. 2016-QNXZ-A-7).

References

[1] G. Hong and W. J. Zhang, "Development and prospects of Chinese meteorological satellite and application," *Meteorological Monthly*, vol. 34, no. 9, pp. 3–9, 2008.

[2] J. Yang, "Development and applications of China's Fengyun (FY) meteorological satellite," *Spacecraft Engineering*, vol. 17, no. 3, pp. 23–28, 2008.

[3] H. G. Zhang, "Application of fuzzy adaptive control theory," Bei Hang University Press, 2003.

[4] G. V. S. Raja, J. Zhou, and R. A. Kiser, "Fuzzy logic control for steam generator feedwater control," in *Proceedings of American Control Conference*, pp. 1491–1493, San Diego, CA, USA, 1990.

[5] G. R. Liu, "Multivariable systems fuzzy/neural network adaptive control," Science Press, 2012.

[6] H. A. Yousef and M. Hamdy, "Observer-based adaptive fuzzy control for a class of nonlinear time-delay systems," *International Journal of Automation and Computing*, vol. 10, no. 4, pp. 275–280, 2013.

[7] W. Siwakosit, S. A. Snell, and R. A. Hess, "Robust flight control design with handling qualities constraints using scheduled linear dynamic inversion and loop-shaping," *IEEE Transactions on Control Systems Technology*, vol. 8, no. 3, pp. 483–494, 2000.

[8] X. F. Zing, L. Zhang, and G. Z. Sheen, "Design of control systems for missiles based on dynamic inversion and decentra-

lized control," *Journal of Beijing University of Aeronautics and Astronautics*, vol. 33, no. 11, pp. 1303–1307, 2007.

[9] J. Zhang, J. G. Zou, and W. X. Li, "PID neural network for decoupling control of multi-input and multi-output system," *Journal of Harbin Engineering University*, vol. 21, no. 5, p. 60, 2000.

[10] Y. Li, C. Ren, and S. Tong, "Adaptive fuzzy backstepping output feedback control for a class of MIMO time-delay nonlinear systems based on high-gain observer," *Nonlinear Dynamics*, vol. 67, no. 2, pp. 1175–1191, 2012.

[11] W. J. Hao, G. L. Liu, S. Y. Wang, and W.-Y. Qiang, "Observer-based fuzzy adaptive control for a class of MIMO nonlinear systems," in *The Proceedings of the Multiconference on "Computational Engineering in Systems Applications"*, pp. 239–244, Beijing, China, 2006, IEEE Press.

[12] Y. Li, S. Tong, and Y. Li, "Observer-based adaptive fuzzy backstepping control of MIMO stochastic nonlinear strict-feedback systems," *Nonlinear Dynamics*, vol. 67, no. 2, pp. 1579–1593, 2012.

[13] C. Bait, F. Yao, Z. Rent, and P. Yang, "Adaptive decoupling control of a MIMO system based on fuzzy neural networks," *BEI Hang University Journal*, vol. 41, no. 11, pp. 2131–2136, 2015.

[14] A. Tornambe and P. Valigi, "A decentralized controller for the robust stabilization of a class of MIMO linear systems," *Systems and Control Letters*, vol. 18, no. 5, pp. 383–390, 1992.

[15] A. Tornambè and P. Valigi, "A decentralized controller for the robust stabilization of a class of MIMO dynamical systems," *Journal of Dynamic Systems, Measurement, and Control*, vol. 116, no. 2, pp. 293–304, 1994.

[16] L. Tian, H. Bao, M. Wang, and X. Duan, "Modeling and control of the redundant parallel adjustment mechanism on a deployable antenna panel," *Sensors*, vol. 16, no. 10, p. 1632, 2016.

[17] M. Hamdy, A. Ramadan, and B. Abozalam, "Comparative study of different decoupling schemes for TITO binary distillation column via PI controller," *IEEE/CAA Journal of Automatica Sinica*, vol. 5, no. 4, pp. 869–877, 2018.

[18] M. Hamdy, A. Ramadan, and B. Abozalam, "A novel inverted fuzzy decoupling scheme for MIMO systems with disturbance: a case study of binary distillation column," *Journal of Intelligent Manufacturing*, pp. 1–13, 2016.

[19] M. Hamdy and A. Ramadan, "Design of smith predictor and fuzzy decoupling for MIMO chemical processes with time delays," *Asian Journal of Control*, vol. 19, no. 1, pp. 57–66, 2017.

[20] B. Liu and W. S. Tang, *Modern Control Theory*, China Machine Press, 2006.

[21] S. Hue, *Automatic Control Principle*, Beijing Science Press, 2001.

5

An Improved Fuzzy Neural Network Compound Control Scheme for Inertially Stabilized Platform for Aerial Remote Sensing Applications

Xiangyang Zhou [1,2] **Yating Li,**[1] **Yuan Jia,**[3] **and Libo Zhao** [2]

[1]*School of Instrumentation Science & Opto-Electronics Engineering, Beihang University, Beijing 100191, China*
[2]*State Key Laboratory for Manufacturing Systems Engineering, Xi'an Jiaotong University, Xi'an 710049, China*
[3]*China Aerospace Academy of Electronic Technology Beijing Institute of Aerospace Micro-Electromechanical Technology, Beijing 100094, China*

Correspondence should be addressed to Xiangyang Zhou; xyzhou@buaa.edu.cn and Libo Zhao; libozhao@mail.xjtu.edu.cn

Academic Editor: Kenneth M. Sobel

An improved fuzzy neural network (FNN)/proportion integration differentiation (PID) compound control scheme based on variable universe and back-propagation (BP) algorithms is proposed to improve the ability of disturbance rejection of a three-axis inertially stabilized platform (ISP) for aerial remote sensing applications. In the design of improved FNN/PID compound controller, the variable universe method is firstly used for the design of the fuzzy/PID compound controller; then, the BP algorithm is utilized to finely tune the controller parameters online. In this way, the desired performances with good ability of disturbance rejection and high stabilization accuracy are obtained for the aerial ISP. The simulations and experiments are, respectively, carried out to validate the improved FNN/PID compound control method. The results show that the improved FNN/PID compound control scheme has the excellent capability in disturbance rejection, by which the ISP's stabilization accuracy under dynamic disturbance is improved significantly.

1. Introduction

For a high-resolution aerial remote sensing system, it needs the inertially stabilized platform (ISP) to isolate the attitude changes of an aircraft in the directions of three axes and to reject the multisource disturbances in real time whether they are inside or outside of the aircraft body; therefore, the ISP is a key component for an aerial remote sensing system, which is mainly used to hold and control the line of sight (LOS) of the imaging sensors keeping steady in the inertial space [1–5]. The first fundamental objective of an ISP is to help imaging sensors to obtain high-resolution images of the target. Therefore, the most critical performance metric for an ISP is the disturbance rejection.

Disturbances that affect the pointing vector arise from platform angular motion or maneuvers and external loads such as wind and airstream inducing torque. Disturbances arise from diverse sources; for example, the linear motion and vibration of the aircraft platform generate disturbance torques due to mass imbalance and gimbal geometry. Among these disturbances, friction, imbalance, vehicle motion kinematic coupling, and sensor noise are predominant [6, 7]. It is a principal issue for the control system of the ISP to minimize the effects of disturbances on the ISP. In [5], a compound scheme on parameter identification and adaptive compensation of nonlinear friction disturbance is proposed to improve the stabilization accuracy of the ISP. In [6], a dual-rate-loop control method based on a disturbance observer (DOB) of angular acceleration is proposed to improve the control accuracy and stabilization of the ISP. In [7], the common disturbances in the ISP are summarized systematically. In [8], a composite control method based on the adaptive radial basis function neural network (NN) feedback control and the extended state observer is applied to a two-axis ISP system. In [9], an adaptive decoupling control for three-axis gyro stabilized platform based on the NN is proposed. In [10], a three-closed-loop compound controller for a two-axis ISP with multisensors is proposed and validated by experiments. The best-known application of ISPs is stabilization and control of payloads such as electro-optical sensors and laser beams [11]. In [12], to realize a balance between high performance and complexity, a dual-stage inertial stabilization system based on frequency-domain analysis is established. In [13, 14], a high-precision control scheme based on active disturbance rejection control (ADRC) and a model reference adaptive control (MRAC)/ PID command scheme are, respectively, proposed for the disturbance rejection of the ISP.

Compared with other methods, the fuzzy control is an adaptive approach of intelligent control, which can deal with the nonlinear, complex, and sometimes mathematically intangible dynamic systems. It has been applied for a wide range [15–22]. It provides a convenient method for constructing nonlinear controllers via the use of heuristic information [23, 24]. The fuzzy control can ensure that the system maintains a small overshoot with fast response, which has a strong adapted ability to the change of control parameters. For the nonlinearity or complexity of the control object, it has the advantages of good robustness and strong anti-interference ability. Since it is not necessary to establish the mathematical model of the controlled object, therefore, it is very suitable for improving the immunity and stability of the ISP. However, there is an inevitable steady-state error when the fuzzy control is used in a control process. So, the fuzzy control and other control methods need to be used in combination so as to achieve a high stabilization accuracy control. Among these methods, the PID controller takes the error as an input, which can satisfy the different requirements of the PID parameter self-tuning [25]. Therefore, if the fuzzy and the PID control methods were combined to establish a fuzzy/PID compound controller, the PID parameters can be adjusted in real time on the basis of the fuzzy controller. Thus, the dynamic performance and disturbance rejection ability of the ISP control system can be improved.

In general, the fuzzy/PID compound controller is experimentally designed based on the fixed universes. In order to improve the control accuracy, the number of fuzzy partitions should be increased. But the increase in the number of fuzzy subsets will lead to the so-called rule explosion problem which makes the controller design difficult. The variable universe can make use of the contraction-expansion factors to achieve the equivalent results resembling the increase in the number of fuzzy subsets [26]. Since there is no typical generalization capability, the design of the variable universe fuzzy/PID compound controller is relying on the experts' knowledge and experience so that more parameters need to be adjusted further [27]. With the universal approximation ability, the neural network (NN) has become a powerful approximation tool for system control, which generates their own rules by learning [28]. The stabilization of a gimbal platform for optical sensor acquisitions in topographic applications using mobile vehicles is investigated, and an

NN-based approach was developed to stabilize the gimbal platform [29]. Since the NN is able to deliver highly desired learning faculties while the fuzzy/PID compound controller just lacks this capability, it could provide a highly desired functional skeleton for the fuzzy/PID compound controller [30]. In addition, due to the self-learning ability, the back-propagation (BP) algorithm can fine-tune the weight coefficient value of the system, as well as reducing the dependence of the controller performance on the initial value of the parameters and the tracking error under the interference. In [31], an improved BP neural network algorithm to detect anomaly network traffic with adjusted correlation rules is developed.

In this paper, an improved FNN/PID compound control scheme is proposed to improve the ability of disturbance rejection and the stabilization accuracy of an aerial ISP. In the scheme, the variable universe method is used for the design of the fuzzy/PID compound controller to improve the convergence at an early period, and the BP algorithm is used to fine-tune the controller parameters online. To verify the scheme, the simulations and experiments are carried out, respectively. Compared to our previous publications [13, 14], the novel and significant contribution of this paper is to propose an improved FNN/PID compound control scheme, in which the idea of variable universe and the BP algorithm are combined to improve the ability of disturbance rejection.

2. Background

2.1. Aerial Remote Sensing System. Figure 1 shows the schematic diagram of an aerial remote sensing system. Generally, an aerial remote sensing system consists of four main components, a three-axis ISP, a remote sensing sensor, a position and orientation system (POS), and an aviation platform. When the aviation platform rotates or jitters, the control system of the ISP gets the high-accuracy attitude reference information measured by the POS and then routinely controls the LOS of the imaging sensor to achieve accurate pointing relative to ground level and flight track [13, 14].

The function of the ISP is to act as an intelligent and physical interface between the remote sensing sensor and the aircraft. With the help of ISP, the influences of various disturbances either inside or outside the aircraft on the remote sensing sensor are initiatively isolated, hence leading to high-resolution images. The POS, which is mainly composed of three main components, inertial measurement unit (IMU), GPS receiving antenna, and data processing system [32], is used to provide an accurate reference of position and attitude in the inertial space for the control system of the ISP and imaging sensor through measuring the angular movement of the imaging sensor. The IMU is mounted on the top of the imaging sensor phase center [5].

2.2. Operating Principle of the Three-Axis ISP System. Figure 2 shows the schematic diagram of the three-axis ISP. We can see that the ISP consists of three gimbals, which are azimuth gimbal, pitch gimbal, and roll gimbal, respectively. G_x, G_y, and G_z, respectively, stand for the rate gyro that measures the inertial angular rate of the three gimbals.

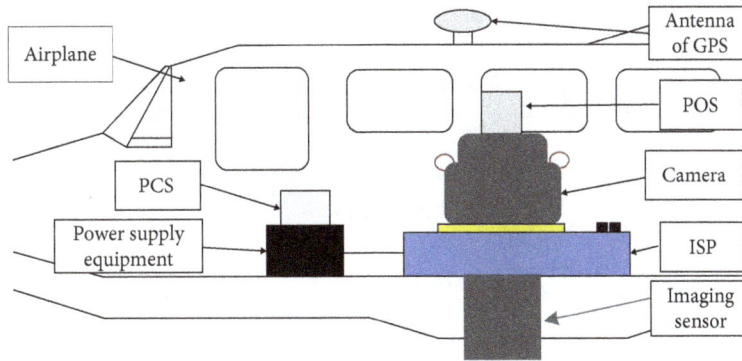

FIGURE 1: Schematic diagram of an aerial remote sensing system.

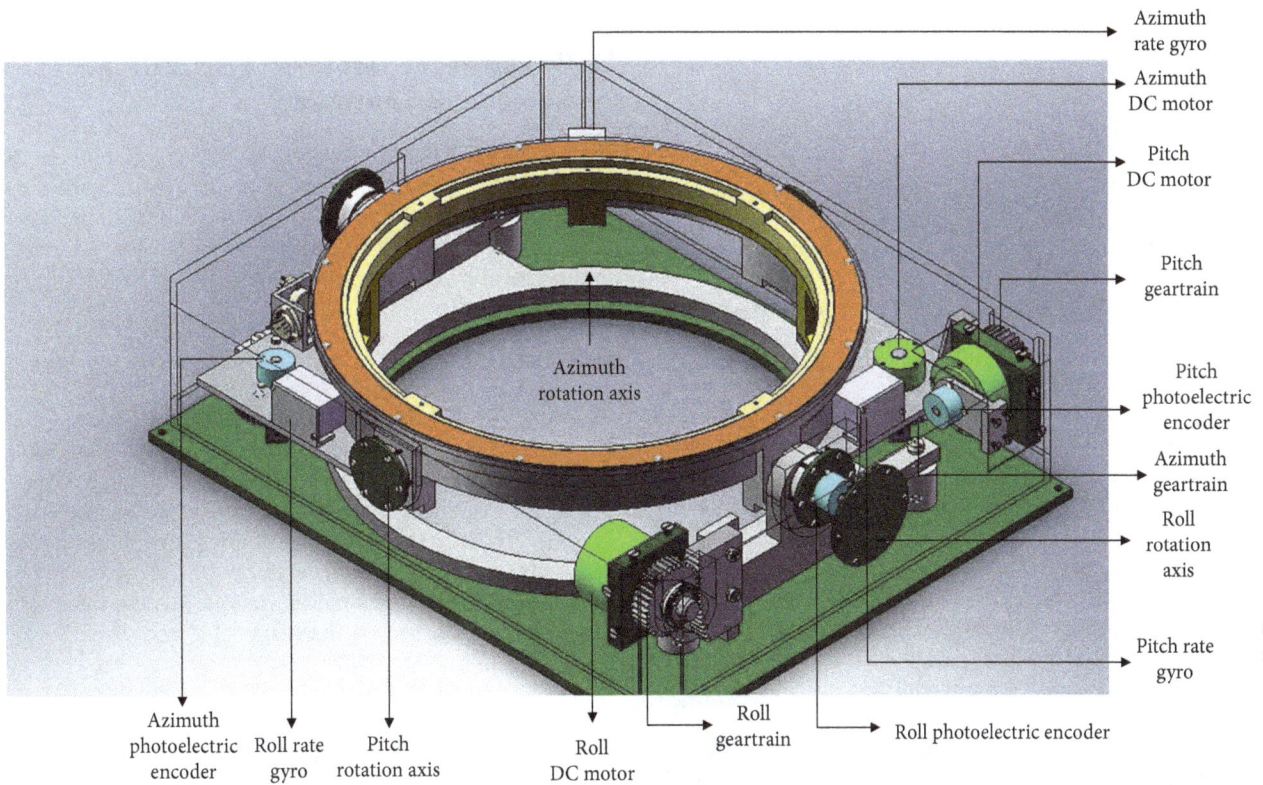

FIGURE 2: Schematic diagram of the three-axis ISP.

E_x, E_y, and E_z, respectively, stand for the photoelectric encoder which measures the relative angular rate between gimbals. M_r, M_p, and M_a, respectively, stand for the gimbal servo motor of the three gimbals [13, 14].

2.3. Three-Closed-Loop Compound Controller.

Figure 3 shows the block diagram of a traditional three-loop control system for the ISP [13, 14]. G-pos, G-spe, and G-cur separately represent the controllers in the tracking loop, stabilization loop, and current loop. PWM represents the power amplification used to amplify the current to drive the torque motor. L represents the inductance of the torque motor, and R represents the resistance. k_T represents the torque coefficient of

the motor, and N is the transition ratio from the torque motor to the gimbals. J_m represents the moment of inertia of the motor, and J_l represents the moment of inertia of the gimbals along the rotation axis.

For all of the three gimbals, the control systems are the same three-loop compound control structure, as shown in Figure 3. The specific functions of the three loops can be summarized as follows: the inner current loop is used to reduce the influence of voltage fluctuation from power supply or motor back electromotive force; the middle stabilization loop uses a rate gyro to measure the angular rate of each gimbal in the inertial space, which is used to compensate the difference between the rate command input and the angular rate of the gimbal and then improve the steady-

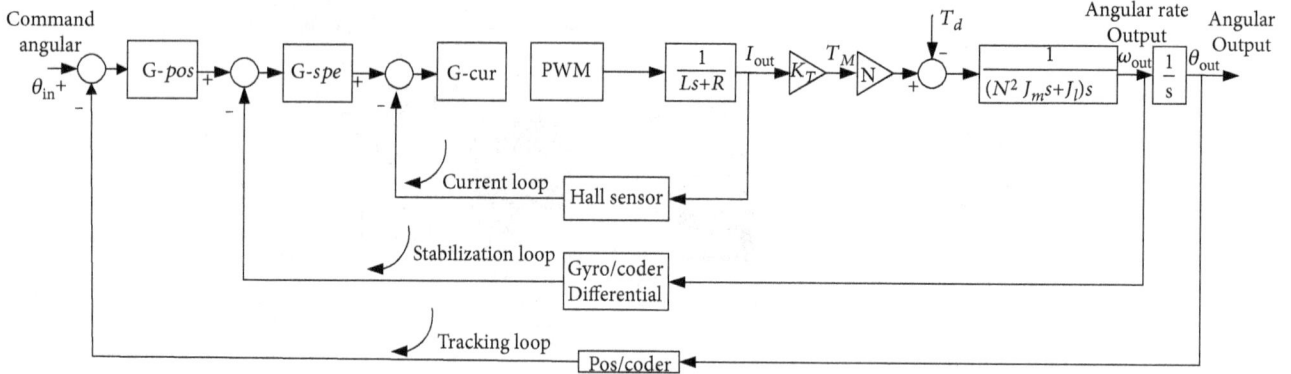

FIGURE 3: A block diagram of a traditional three-loop control system for the ISP.

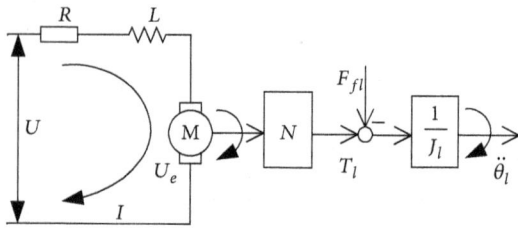

FIGURE 4: Block diagram for a simplified gear drive system with fixed transmission ratio.

state precision; and as to the main feedback path, the outer tracking loop takes the attitude angle measured by POS as accurate references to ensure the accurate pointing of the LOS [5].

To realize the high-drive torque of the load under the limitation of small size, the motor and gear train with high transmission ratio are usually used in the design of ISPs. If the influence of the backlash is ignored, the system model can be simplified with fixed transmission ratio, as shown in Figure 4 [5].

According to the current balance equation and the torque balance equation, at the side of the motor output axis, the following equations are satisfied:

$$U = IR + L\frac{dI}{dt} + k_b\omega_m,$$
$$J_m\ddot{\theta}_m + b_m\omega_m = T_m - \tau_l - F_{fm}, \tag{1}$$
$$T_m = k_T I,$$

where U, I, R, and L are the armature voltage, current, resistance, and inductance of the torque motor, respectively; k_b is the back EMF constant of the motor; ω_m is the angular velocity of the motor; J_m is the rotational inertia of the motor; θ_m is the angular displacement of the motor; b_m is the equivalent damping of the motor; T_m is the electromagnetic torque of the motor; τ_l is the load torque at the side of the motor output axis; F_{fm} is the inner friction torque of the motor; and k_T is the torque coefficient of the motor.

3. Design of the Improved FNN/PID Compound Controller

In the design of the improved FNN/PID compound controller, firstly, the variable universe method is used for the design of the fuzzy/PID compound controller, and then, the BP algorithm is used to fine tune the controller parameters online to improve the performance of the FNN/PID compound controller.

3.1. Fuzzy/PID Compound Controller Based on Variable Universe. Figure 5 shows the structure of the fuzzy/PID compound controller. In this work, the fuzzy/PID compound controller makes use of the nonlinear mapping of applied fuzzy logic to establish the angle position error parameters e and ec, as well as three PID parameters. On the basis of the input change, it adjusts the control gain in real time so as to obtain the fast-response ability to large deviation and the fine adjustment to small deviation. The combination of the parameters based on the fuzzy logic and the PID initial value is as follows:

$$K_p = K_{p0} + \text{Fuzzy}\left(\Delta K_p\right),$$
$$K_i = K_{i0} + \text{Fuzzy}(\Delta K_i), \tag{2}$$
$$K_d = K_{d0} + \text{Fuzzy}(\Delta K_d),$$

where K_{p0}, K_{i0}, and K_{d0} represent the initial parameters of the fuzzy/PID compound controller.

Fuzzification is an accurate numerical classification, which sets a fuzzy set corresponding to each file. The subsets of fuzzy languages are selected as negative big, negative middle, negative small, zero, positive small, positive middle, and positive big and also expressed as {NB, NM, NS, ZE, PS, PM, PB}, respectively [33]. The input universes are taken as {−6, −5, −4, −3, −2, −1, 0, 1, 2, 3, 4, 5, 6}. Based on simulation analysis, the membership function of nonlinear distribution is finally used [34].

Under the normal circumstances, a combination of the fuzzy control and PID control in the ISP relies on increasing the fuzzy rules to obtain higher accuracy, but it easily leads to the regular explosion. Therefore, the variable universe method is introduced in the controller design, which can

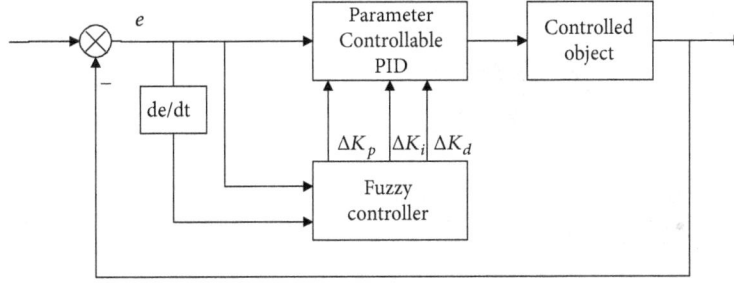

FIGURE 5: Fuzzy/PID compound controller structure.

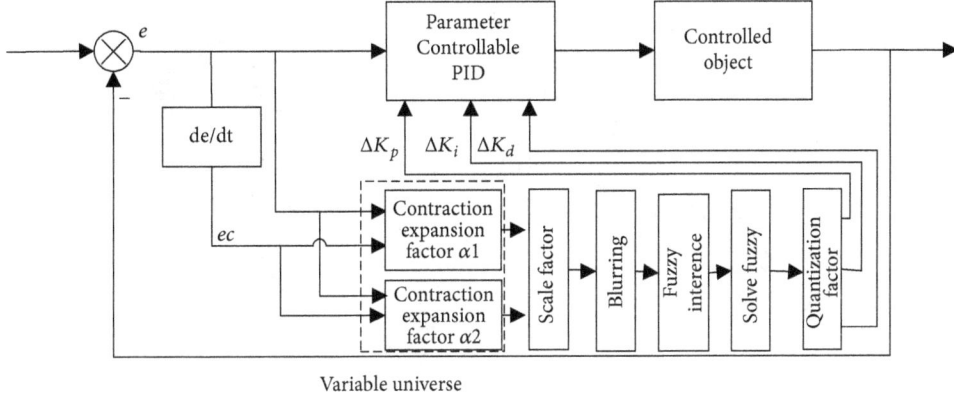

FIGURE 6: Variable universe fuzzy/PID compound controller structure.

improve the control effect under the same number of rules. In particular, the method is to introduce the contraction expansion factors, α_1 and α_2, on the basis of the fuzzy/PID compound controller, so that the universe of e and ec is $\{-U_1{}^*\alpha_1, U_1{}^*\alpha_1\}$ and $\{-U_2{}^*\alpha_2, U_2{}^*\alpha_2\}$, respectively. Thus, the corresponding output amounts can change according to different contraction expansion factors. In the experiments, the contraction expansion factors, α_1 and α_2, are chosen optimally as 0.25 and 0.5, respectively, so that the controller can converge quickly and have small stabilization error. Figure 6 shows the diagram of the variable universe fuzzy/PID compound controller.

3.2. FNN/PID Compound Control Scheme. The fuzzy control and the NN are frequently combined in a compound scheme called the FNN [35, 36]. The general FNN-combined PID controller is composed of two parts: the PID/NN and the FNN, which are the front part and the latter part of the controller, respectively. Through the learning offline and online, the real-time adaptive capability of the NN is trained so that the part of the fuzzy logic can be adjusted.

In the FNN using the second-order method, the fuzzy layers follow the two input nodes of the first layer. The two input nodes of the first layer are named as the deviation and the deviation rate, that is, e and ec, respectively. Since the number of nodes in the fuzzy layer depends on the number of fuzzy subsets divided by e and ec, the membership function, which is the activation

function in the network, can be written by using the Gaussian functions:

$$\mu_{Aij} = \exp\left[\frac{-(X_i - c)^2}{\sigma^2}\right], \tag{3}$$

where X_i, c, and σ represent the input variable, the activation function center, and the activation function width, respectively. The subscript A represents the fuzzy subset of the fuzzy variable. The number of nodes in the fuzzy rule layer is the number of fuzzy rules, which is shown as follows:

$$O_k = \mu_{A1k}(x_1) * \mu_{A2k}(x_2), \tag{4}$$

where O_k represents the degree of activation about the rule k. The weighted average method is usually used in the defuzzy layer. The front structure of the FNN is shown in Figure 7.

The input and output relationship of each node is shown in the following formula. The first-layer input (input layer) is

$$I_i^{(1)} = X_i, \quad i = 1, 2. \tag{5}$$

The first-layer output is

$$O_i^{(1)} = I_i^{(1)}, \quad i = 1, 2. \tag{6}$$

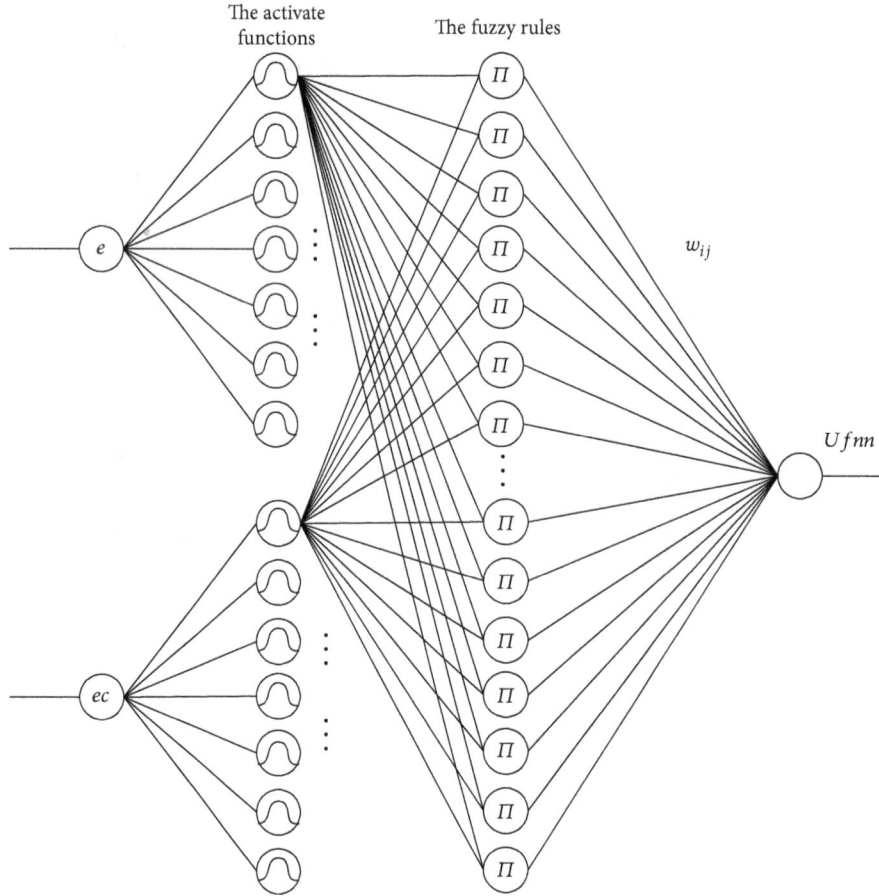

FIGURE 7: The front topological structure of the FNN.

The second-layer input (fuzzy layer) is

$$I_{ij}^{(2)} = \frac{-\left(O_{ij}^{(1)} - c\right)^2}{\sigma^2}, \quad i = 1, 2, j = 1, 2, \ldots, n. \tag{7}$$

The second-layer output is

$$O_{ij}^{(2)} = \exp\left(I_{ij}^{(2)}\right), \quad i = 1, 2, j = 1, 2, \ldots, n. \tag{8}$$

The third-layer input (fuzzy rule layer) is

$$I_{(j-1)n+k}^{(3)} = O_{1j}^{(2)} O_{2k}^{(2)}, \quad j = 1, 2, \ldots, n, k = 1, 2, \ldots, n. \tag{9}$$

The third-layer output is

$$O_i^{(3)} = I_i^{(3)}, \quad i = 1, 2, \ldots, n^2. \tag{10}$$

The fourth-layer input (defuzzy layer) is

$$I^{(4)} = \sum_{i=1}^{n^2} O_i^{(3)} wf_i, \tag{11}$$

where wf_i is the connected weight of the output layer.

The output of the fourth layer is

$$O_i^{(4)} = \frac{I^{(4)}}{\sum_{i=1}^{n^2} O_i^{(3)}}, \tag{12}$$

where n is the number of fuzzy subsets about the deviation and the deviation change rate.

The deviation e and deviation change rate ec are taken as the input of the fuzzy controller, adjusting the PID parameters K_p, K_i, and K_d, according to fuzzy control rules, to reduce the impact of control model time variation [37]. On the basis of the FNN, the introduction of neurons of P, I, and D makes the network become a dynamic network, which can enhance the information processing ability of the network and can reflect the dynamic behaviour of the system better. The PID topological structure of the FNN is shown in Figure 8.

The NN is introduced into the fuzzy control to achieve the adaptive adjustment of fuzzy rules, by which the NN structure is used to complete the fuzzy system function, including the extraction and the adjustment of the rule. Since the output of the front part network is the same as the input of the back part, in which the indirect control parameters (the nonlinear mapping of error and error change) are

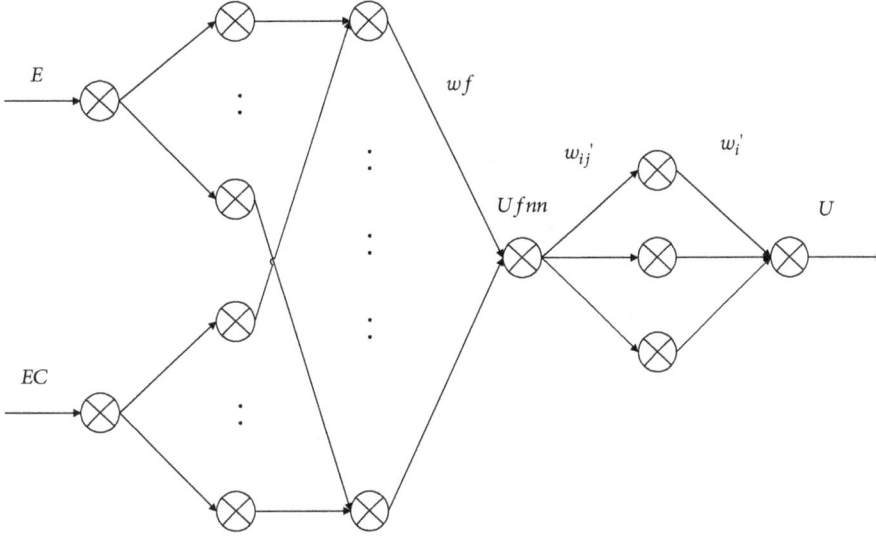

FIGURE 8: The PID topological structure of the FNN.

weighted and then input to the hidden layer of the back part network, the neurons of the hidden layer produce the PID control effect on the system. In this way, the control parameters are improved in real time, which integrate the advantages of the fuzzy, the NN, and the PID.

By using the incremental PID control algorithm, the output formula of the controller is derived as follows:

$$U(k) = \mathbf{w}_1' \mathbf{w}_{11}' U_{\text{fnn}}(k) + \mathbf{w}_2' \mathbf{w}_{21}' [U_{\text{fnn}}(k) + U_{\text{fnn}}(k-1)] \\ + \mathbf{w}_3' \mathbf{w}_{31}' [U_{\text{fnn}}(k) - U_{\text{fnn}}(k-1)], \tag{13}$$

where U is the final output of the network, which is the system controlled variable; $\mathbf{w}_i'$ is the weight coefficient matrix about the PID layer and output layer; $\mathbf{w}_{ij}'$ is the weight coefficient matrix about the PID layer and input layer; and U_{fnn} is the output of the FNN, which is actually a nonlinear mapping of error.

The main control principle of the FNN/PID compound control scheme is that on the basis of the error and error changes of the input, the controller learning algorithm can adjust and update their weight constantly, which output a more suitable control amount for the current state of the system so that the output of the system is closer to the desired input [38].

3.3. BP Algorithm Fine-Tuning Online.

After obtaining the offline suboptimal controller parameters, the BP algorithm is used to adjust the online simulations. The BP algorithm is adjusted on the basis of the initial parameters; therefore, the result of the offline acquisition is more close to the optimal value; it can more avoid the divergence or convergence to the suboptimal solution, but not the convergence to the optimal solution. Hence, only by offline, get better initial suboptimal parameters in order to better real-time online

adjustment. The online BP algorithm adjustment is adjusted according to the following formula:

$$c(n+1) = \frac{(c(n) + \text{xite}^* \partial E)}{(\partial c + \eta^* \Delta c(n))},$$

$$\sigma(n+1) = \frac{(\sigma(n) + \text{xite}^* \partial E)}{(\partial \sigma + \eta^* \Delta \sigma(n))}, \tag{14}$$

$$w_{\text{all}}(n+1) = \frac{(w_{\text{all}}(n) + \text{xite}^* \partial E)}{(\partial w_{\text{all}} + \eta^* \Delta w_{\text{all}}(n))},$$

where w_{all} contains all the weights that need to be adjusted in the controller, xite represents the learning factor, and η is the momentum factor. E is the mean square error (MSE), which is expressed as $E = (1/2)(\text{rin}(k)\text{-yout}(k))^2$.

3.4. Improved FNN/PID Compound Control Scheme.

For the existing FNN/PID compound control scheme, the initial parameter value has a great effect on the controller convergence. Since there are so many parameters, it is hard to find good initial values to realize the good control effects.

Combined with the fuzzy/PID, the FNN/PID has the advantages of small initial value dependency and easy convergence; besides, it also inherits the adaptive characteristics of the NN by the FNN/PID; the good control parameters can be obtained, such as high accuracy, small overshoot, and fast response speed.

In the improved FNN/PID controller, the input layer is composed of two nodes, e and ec, respectively. And the role of the fuzzy layer is designed to deal with the higher input and then start reasonable fuzzy segmentation; the number of nodes is equal to the number of subsets of variables.

$$f_1(i) = X = [x_1, \ldots, x_n], \quad n = 2. \tag{15}$$

When the improved FNN/PID compound control scheme is applied to the ISP, the accuracy should be taken into account; the fuzzy languages of each input variable are

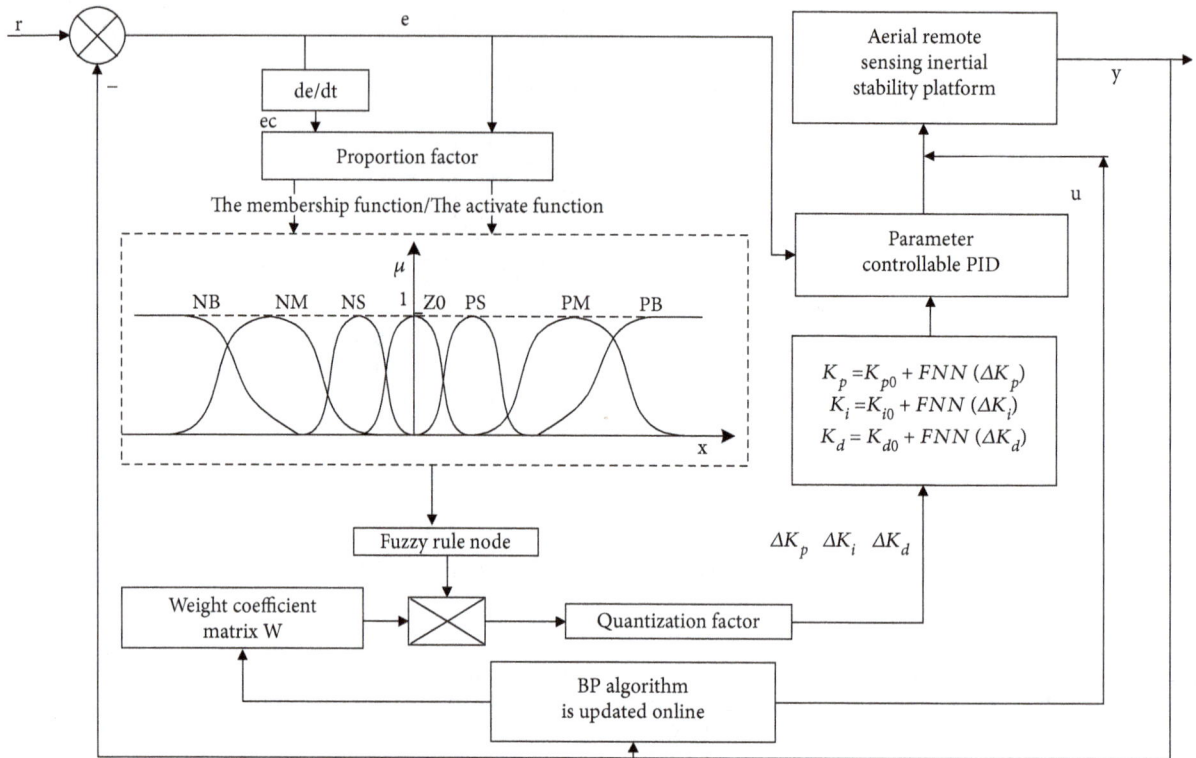

FIGURE 9: Schematic diagram of an improved FNN/PID compound control structure.

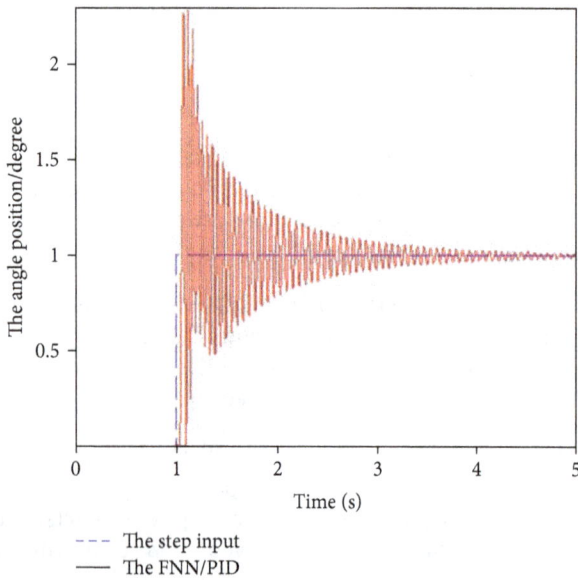

--- The step input
—— The FNN/PID

FIGURE 10: The response diagram of the FNN/PID controller under the step input.

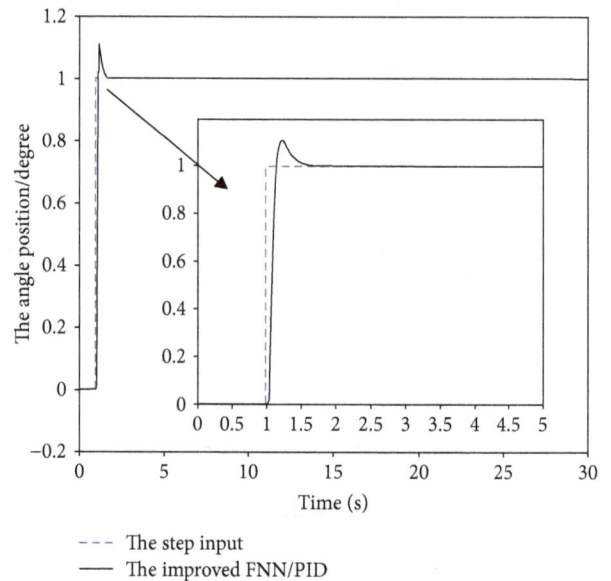

--- The step input
—— The improved FNN/PID

FIGURE 11: System response curve between the improved FNN/PID compound control scheme and step input.

TABLE 1: The accuracy of the FNN/PID compound control scheme in different time periods.

Time range (s)	0–30	5–10	3–30
RMS error (°)	0.0857	0.0015	0.0066

TABLE 2: Comparison of the accuracy of each method step response in different time periods.

Method	0–30 s/RMS error (°)
PID	0.0931
Fuzzy/PID	0.0770
Improved FNN/PID	0.0432

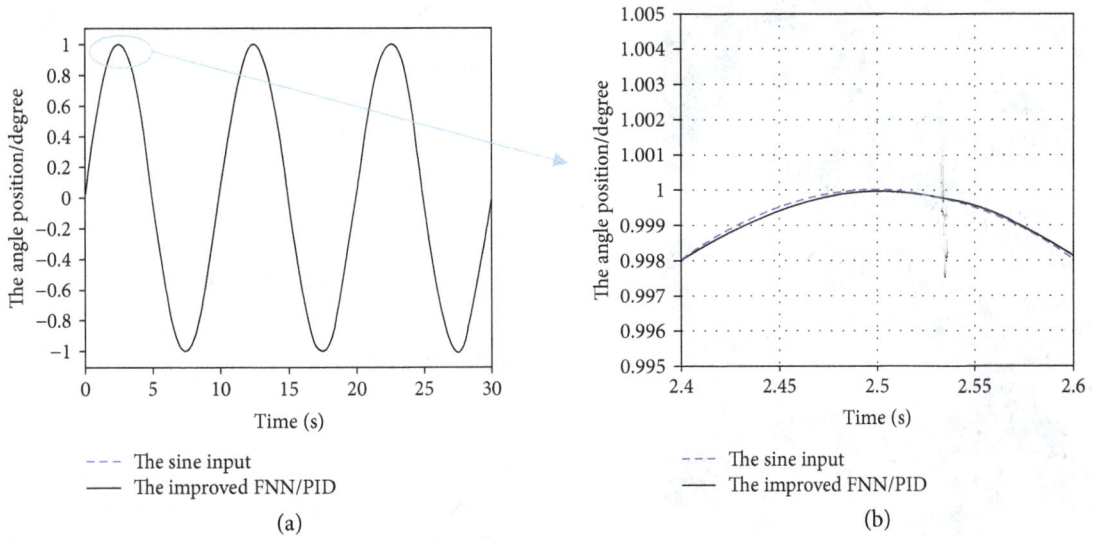

(a)

(b)

FIGURE 12: The sine response curve under the friction disturbance: (a) improved FNN/PID compound control schemes and (b) the partial enlargement.

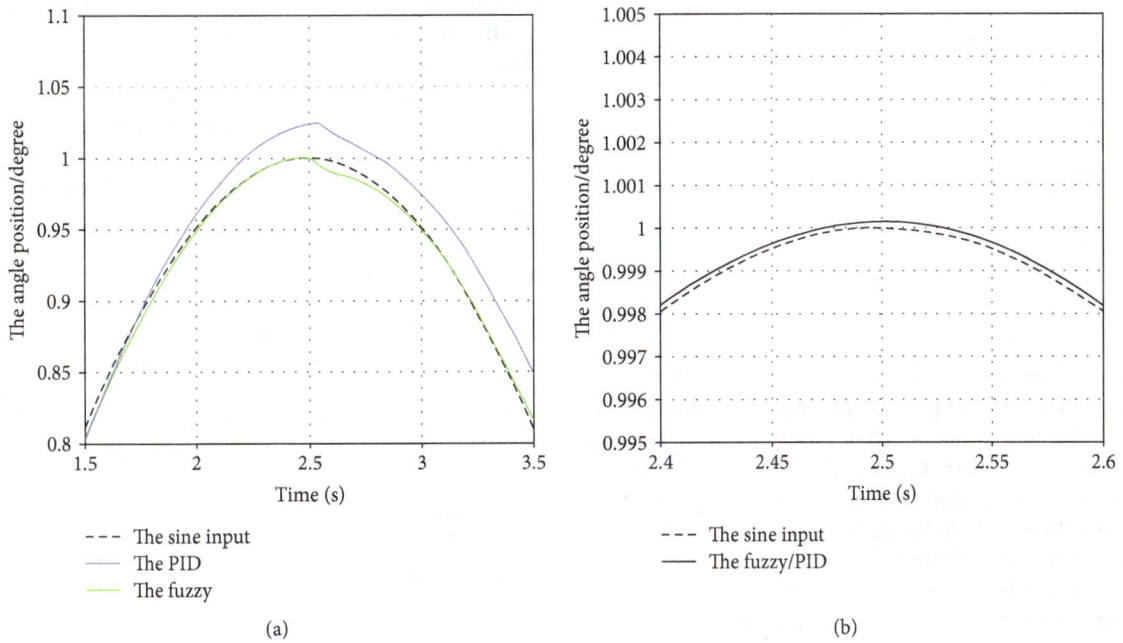

(a)

(b)

FIGURE 13: The comparison of system responses to sine input among different methods under the friction disturbance: (a) PID and fuzzy controller and (b) fuzzy/PID compound controller.

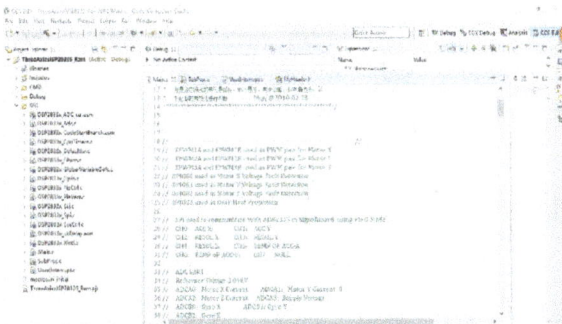

FIGURE 14: The control software.

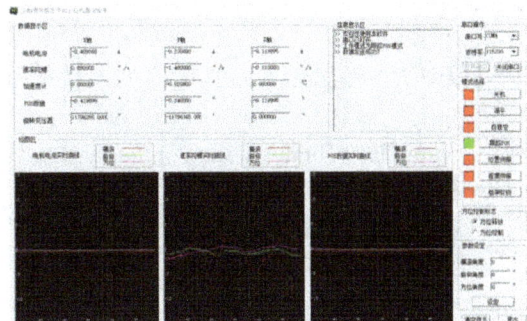

FIGURE 15: The upper computer software.

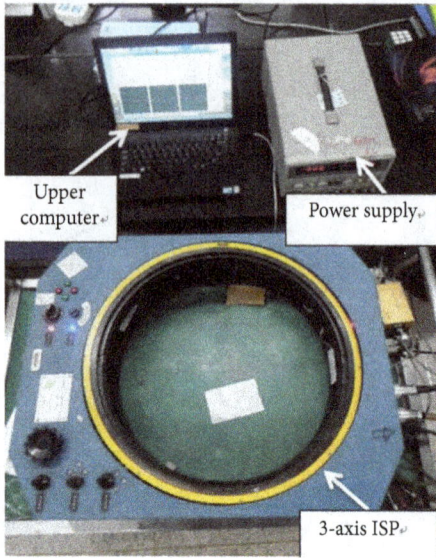

FIGURE 16: The picture of the three-axis ISP system in levelling static experiments.

divided into seven segments. According to the working principle of the ISP, in different fuzzy subsets, use different Gaussian functions and bell functions to represent.

$$f_2(i,j) = \mu_{Aij} = \exp\left[\frac{-(X_i - c_{ij})^2}{\sigma_{ij}^2}\right], \quad i = 1, 2, j = 2, 3, 4, 5, 6,$$

$$f_2(i,j) = \mu_{Aij} = \frac{1}{\left(1 + |X_i - c_{ij}|/a_{ij}\right)^{2\sigma_{ij}^2}}, \quad i = 1, 2, j = 1, 7.$$

(16)

Different from the simple activation function that is set by the general FNN/PID control algorithm, the parameters of the membership function about the improved method are different, which gain the experience about the fuzzy/PID compound controller from the membership function design. Therefore, it is more suitable for the system characteristics of the ISP. Since its parameters are no longer updated in the BP algorithm in real time, the blindness during the parameter updating is avoided and the calculation resources are saved, which is helpful to improve the control effect.

Correspondingly, the fuzzy rule layer has 49 nodes, and the fuzzy reasoning is calculated as

$$f_3(j) = \prod_{j=1}^{N} f_2(i,j) = \mu_{A1k}(x_1) * \mu_{A2k}(x_2), \quad N = \prod_{i=1}^{n} N_i. \quad (17)$$

For the output layer, omit the steps of the weighted average method and multiply the output of the upper layer directly by the connection weight matrix; the formula is shown as

$$f_4(j) = w \cdot f_3 = \sum_{j=1}^{N} w(i,j) \cdot f_3(j). \quad (18)$$

The output of the improved FNN/PID compound control scheme is taken as the compensation of the constant PID parameters:

$$\begin{aligned} K_p &= K_{p0} + \text{FNN}(\Delta K_p), \\ K_i &= K_{i0} + \text{FNN}(\Delta K_i), \\ K_d &= K_{d0} + \text{FNN}(\Delta K_d). \end{aligned} \quad (19)$$

So, the output formula of the controller is

$$u(k) = k_p[e(k) - e(k-1)] + k_i \sum_{i=1}^{k} e(i) + k_d[e(k) - e(k-1)]. \quad (20)$$

In the improved FNN/PID compound control scheme, the FNN only computes a small amount of change and reduces the dependency on the initial value, so the adjustment time becomes shorter. The control of the entire system has good stability as it is not entirely dependent on the output of the adaptive adjustment part. The improved FNN/PID compound control scheme structure is shown in Figure 9.

In the improved FNN/PID compound control scheme, since the number of fuzzy subsets, the membership function, the quantization factor, and constant PID parameters are used, therefore, it needs only to determine the weight of the controller. Since the output of the method is small, the effect of the initial value of the weight coefficient on its output is limited.

4. Simulations Analysis

To analyse the performance of the proposed methods, the simulations are conducted under the MATLAB/Simulink software environment. It should be noted that in both simulations and experiments of this work, since the influences of the interactions between gimbals are very small compared to other disturbances [39], they are ignored when simulations and experiments are conducted.

The LuGre friction model described by (21) is added into the PID model to exhibit the effects of nonlinear friction disturbances on the performance of the control system. Also, to show the effects of the proposed FNN/PID compound controller on nonlinear disturbance rejection, the LuGre friction model is also added into the FNN/PID controller to exhibit its disturbance rejection ability.

The LuGre friction model of the ISP can be established as follows [5]:

$$\begin{aligned} \frac{dz}{dt} &= \omega_f - \frac{|\omega_f|}{g(\omega_f)} z, \\ \sigma_0 g(\omega_f) &= T_c + (T_s - T_c) e^{-(\omega_f/\omega_s)^2}, \\ F_{fl} &= \sigma_0 z + \sigma_1 \frac{dz}{dt} + \sigma_2 \omega_f, \end{aligned} \quad (21)$$

where z is the average deflection of the bristles, ω_f is the velocity between the two surfaces in contact, $g(\omega_f)$ is the

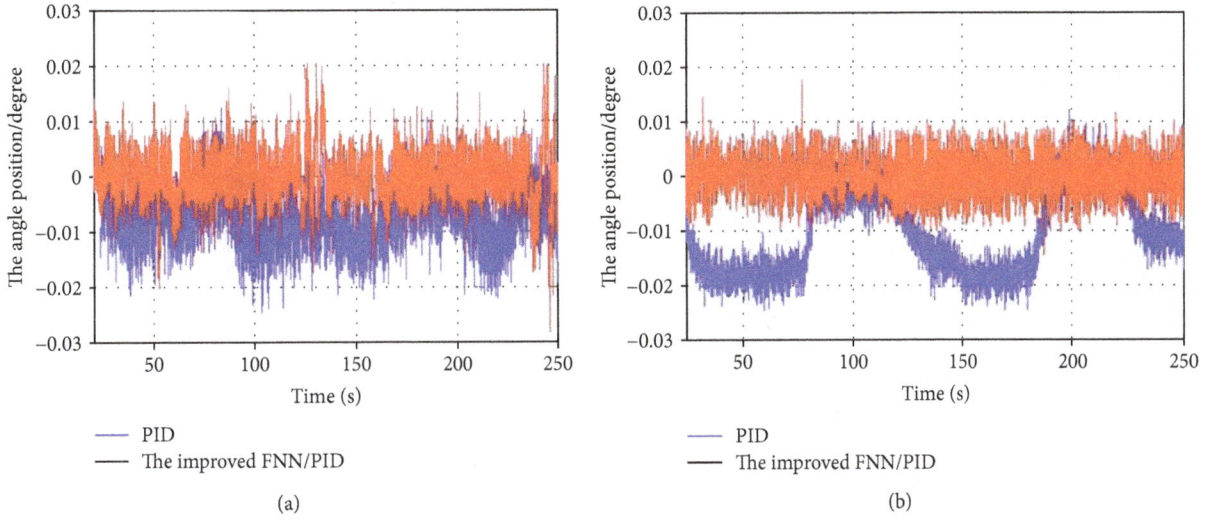

FIGURE 17: Static levelling experiment results: (a) rolling and (b) pitching.

TABLE 3: Error comparison of static levelling results.

| Frame | RMS errors (°) | | Improved extent than PID (%) | RMS errors (°) Improved FNN/PID | Improved extent than PID (%) |
	PID	Fuzzy/PID			
Rolling	0.0132	0.0065	50.758	0.0055	58.333
Pitching	0.0129	0.0035	72.868	0.0030	76.744

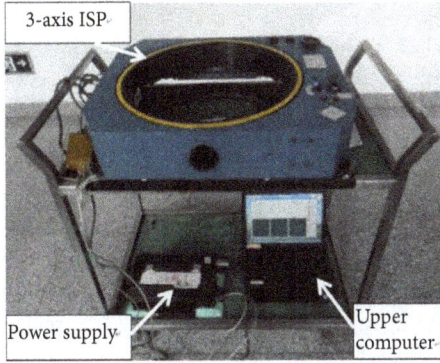

FIGURE 18: The picture of the three-axis ISP in moving base experiments.

velocity-dependent function, T_c is the coulomb friction torque, T_s is the maximum static friction torque, ω_s is the critical Stribeck speed, σ_0 is the bristle stiffness coefficient of the contact surface, σ_1 is the damping coefficient of the bristles, and σ_2 is the viscous friction coefficient.

4.1. FNN/PID Compound Control Scheme. Figure 10 shows the response curve of the FNN/PID compound control scheme under the step input. As can be seen from this figure, although the angle position errors of the controller determined by the trial and error method can be converged, and the response at the beginning stage is not stable, which has big overshoot errors. However, because of the self-adaptability itself, the controller can achieve the high convergence accuracy eventually.

Table 1 shows the accuracy for the FNN/PID compound control scheme in different periods (root mean square (RMS)). It can be seen that by using the trial and error method, the initial value that makes the system not diverged make the system keep high stabilization accuracy. However, the convergence speed is low and the overshoot is large; meanwhile, since there is an oscillation state in the output process at the early adjustment stage, it is very unfavourable for the work of the motor. In the actual application, it may cause the system to crash in some serious situations. Therefore, it is necessary to optimize the initial values of the parameters in the actual application.

4.2. Improved FNN/PID Compound Control Scheme. As shown in Figure 11, with the step input, the system response curve of the improved FNN/PID compound control scheme has the fast response speed, high stability, and low overshoot. Moreover, the method has a low degree of dependence on the initial value of the weight coefficient.

The RMS errors of the PID, the fuzzy/PID, and the improved FNN/PID compound control scheme are shown in Table 2. The improved FNN/PID compound control scheme has higher accuracy than the previous methods described above. In the table, the RMS errors of the three different methods to the step input are compared at different time stages. We can see that in the whole time stages of 0–30 s, the RMS error of the fuzzy/PID is 0.0770°, which is decreased up to 17.29% than that of the PID, and the

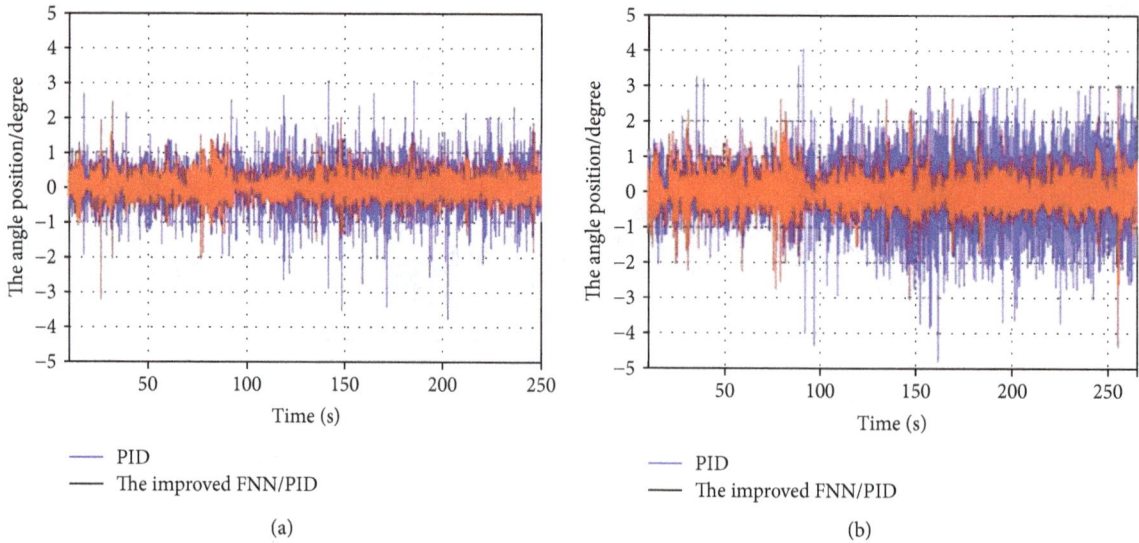

FIGURE 19: Moving base experiment results: (a) rolling and (b) pitching.

TABLE 4: Result comparison among three methods under dynamic levelling experiments.

| Frame | RMS errors (°) | | Improved extent than PID (%) | RMS errors (°) Improved FNN/PID | Improved extent than PID (%) |
	PID	Fuzzy/PID			
Rolling	0.5942	0.4808	19.084	0.4098	31.033
Pitching	0.9491	0.6590	30.566	0.5493	42.124

RMS error of the improved FNN/PID is 0.0432°, which is decreased up to 53.6% than that of the PID.

Compared with the traditional FNN/PID method, the improved FNN/PID compound control scheme has little dependence on the initial value of the weight coefficient, which has high stability. For the improved FNN/PID compound control scheme with higher stability, the sine signal with an amplitude of 1 and a period of 10 s is input into the system. The sine response curve under the friction disturbance of the system is shown in Figure 12. As can be seen in the figure, compared with the PID (0.0307° RMS error), the RMS error of the improved FNN/PID is decreased up to 99.71%, illustrating its excellent tracking accuracy.

Figure 13 shows the comparison of system responses to sine input among different methods under the friction disturbance. As seen in the figure, three methods are compared with the improved FNN/PID of Figure 12(b), which are the PID, the fuzzy control, and the fuzzy/PID, respectively. Compared with other methods, the improved FNN/ PID compound control scheme has better response speed and stabilization accuracy and has a greater ability for disturbance suppression compensation.

5. Experimental Validation

The improved FNN/PID compound control scheme is applied to the actual equipment of the ISP to conduct the static levelling experiment. The hardware is mainly composed of circuit boards, sensors, and actuators. The circuit

boards include the main control board, high-precision analog signal acquisition board, motor power driver board, power board, accelerometer signal processing board, and gyro signal processing board, respectively. Among them, the control board with DSP is the core of the entire circuit system, which can realize digital signal processing, high-speed ADC analog-digital conversion, and communication and output motor control PWM signal to the motor power drive unit. The main sensors and actuators include 3 Hall current sensors, 3 rate gyros, 3 photoelectrical encoders, 2 accelerometers, and 3 torqued motors, respectively. The software includes the control software and upper computer software, which are all developed by VC++6.0, as shown in Figures 14 and 15, respectively. The control software is used as the control program of the ISP. The upper computer software is used for monitoring the signal outputs of POS, gyro, and current.

5.1. Static Base Experiments. Figure 16 shows the picture of the three-axis ISP system in levelling static experiments. The weight of the ISP is 40 kg and the weight of the simulated load is 20 kg, and the largest weight that the ISP can carry is 80 kg. The maximum levelling rotation angle range is ±5°; the maximum heading rotation angle range is ±25°.

Figure 17 shows the static levelling experiment results. The stabilization errors (RMS) of the roll and pitch control system are 0.0055° and 0.0030°, which are, respectively, decreased up to 58.333% and 76.744% compared with 0.0132° and 0.0129° of the PID. The comparison results of static levelling are listed in Table 3. We see that the

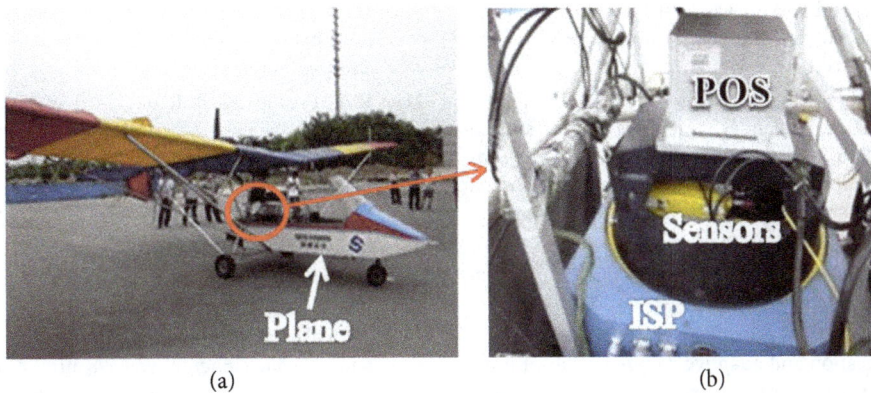

FIGURE 20: Aerial remote sensing system: (a) aerial remote sensing system and (b) integrated remote sensing system in a plane.

FIGURE 21: High-resolution surveying and mapping maps with a 1 : 500 scale.

stabilization accuracy has been significantly improved, which is in good accordance with the simulation results, illustrating the effectiveness of the improved FNN/PID compound control scheme.

5.2. Moving Base Experiments. Figure 18 shows the picture of the three-axis ISP in moving base experiments. In order to further validate the performance of the proposed method under the movable base, the dynamic experiments are conducted on a movable cart under uneven ground while changing the moving direction. The experimental results are shown in Figure 19. The comparison results among three methods under dynamic levelling experiments are listed in Table 4. It can be known that the stabilization errors (RMS) of the roll and pitch control systems are 0.4098° and 0.5493°, which are, respectively, decreased up to 31.033% and 42.124% than those of the PID, meaning that the disturbance rejection ability of the improved FNN/PID compound control scheme is significant. Figures 20 and 21 show the applications of the ISP in a real aerial remote sensing system, by which the high-resolution surveying and mapping maps with a scale of 1 : 500 were successfully obtained.

6. Conclusions

In this paper, to improve the ability of disturbance rejection and the stabilization accuracy of an aerial ISP, an improved FNN/PID compound control scheme based on variable universe and BP algorithms is proposed. In the compound controller design, the idea of variable universe and the BP algorithm are, respectively, used to improve the capability of the FNN/PID compound control scheme in disturbance rejection and the stabilization accuracy of an aerial ISP. To verify the improved FNN/PID compound control method, the simulations and experiments are carried out, respectively. The results show that the improved FNN/PID compound control scheme has excellent capability in disturbance rejection, by which the ISP stabilization accuracy under dynamic disturbance is improved significantly. Compared to the PID method, the RMS errors of the roll and pitch gimbal systems under the dynamic base obtained by the FNN/PID compound control scheme are decreased up to 31% and 42%, respectively.

Conflicts of Interest

The authors declare no conflict of interest.

Acknowledgments

This project is supported in part by the National Natural Science Foundation of China (Grant nos. 51775017 and 51375036), by the Beijing Natural Science Foundation (Grant no. 3182021), by the Open Research Fund of the State Key Laboratory for Manufacturing Systems Engineering (sklms2018005). The aerial remote sensing system shown in Figures 20 and 21 is supported by the National High-Tech Research and Development Program of China (Grant no. 2008AA121302) and the National Basic Research Program of China (Grant no. 2009CB724001).

References

[1] J. M. Hilkert, "Inertially stabilized platform technology concepts and principles," *IEEE Control Systems*, vol. 28, no. 1, pp. 26–46, 2008.

[2] X. Zhou, H. Zhang, and R. Yu, "Decoupling control for two-axis inertially stabilized platform based on an inverse system and internal model control," *Mechatronics*, vol. 24, no. 8, pp. 1203–1213, 2014.

[3] J. M. Hilkert and D. A. Hullender, "Adaptive control system techniques applied to inertial stabilization systems," in *Proceedings of SPIE, Acquisition, Tracking, and Pointing IV*, vol. 1304, pp. 190–206, Orlando, FL, USA, September 1990.

[4] P. J. Kennedy and R. L. Kennedy, "Direct versus indirect line of sight (LOS) stabilization," *IEEE Transactions on Control Systems Technology*, vol. 11, no. 1, pp. 3–15, 2003.

[5] X. Zhou, B. Zhao, and W. Liu, "A compound scheme on parameters identification and adaptive compensation of nonlinear friction disturbance for the aerial inertially stabilized platform," *ISA Transactions*, vol. 67, no. 1, pp. 293–305, 2017.

[6] X. Zhou, Y. Jia, and Q. Zhao, "Dual-rate-loop control based on disturbance observer of angular acceleration for a three-axis aerial inertially stabilized platform," *ISA Transactions*, vol. 63, no. 7, pp. 288–298, 2016.

[7] M. K. Masten, "Inertially stabilized platforms for optical imaging systems," *IEEE Control Systems*, vol. 28, no. 1, pp. 47–64, 2008.

[8] X. Lei, Y. Zou, and F. Dong, "A composite control method based on the adaptive RBFNN feedback control and the ESO for two-axis inertially stabilized platforms," *ISA Transactions*, vol. 59, no. 4, pp. 424–433, 2015.

[9] J. Fang, R. Yin, and X. Lei, "An adaptive decoupling control for three-axis gyro stabilized platform based on neural networks," *Mechatronics*, vol. 27, no. 4, pp. 38–46, 2015.

[10] X. Zhou, Y. Jia, Q. Zhao, and R. Yu, "Experimental validation of a compound control scheme for a two-axis inertially stabilized platform with multi-sensors in an unmanned helicopter-based airborne power line inspection system," *Sensors*, vol. 16, no. 3, p. 366, 2016.

[11] H. G. Wang and T. C. Williams, "Strategic inertial navigation systems - high-accuracy inertially stabilized platforms for hostile environments," *IEEE Control Systems*, vol. 28, no. 1, pp. 65–85, 2008.

[12] Z. Lin, K. Liu, and L. Zhang, "Coupling effect and control strategies of the maglev dual-stage inertially stabilization system based on frequency-domain analysis," *ISA Transactions*, vol. 64, no. 4, pp. 98–112, 2016.

[13] X. Zhou, C. Yang, B. Zhao, L. Zhao, and Z. Zhu, "A high-precision control scheme based on active disturbance rejection control for a three-axis inertially stabilized platform for aerial remote sensing applications," *Journal of Sensors*, vol. 2018, Article ID 7295852, 9 pages, 2018.

[14] X. Zhou, C. Yang, and T. Cai, "A model reference adaptive control/PID command scheme on disturbance rejection for an aerial inertially stabilized platform," *Journal of Sensors*, vol. 2016, Article ID 7964727, 11 pages, 2016.

[15] H. Nasser, E. H. Kiefer-Kamal, and H. Hu, "Active vibration damping of composite structures using a nonlinear fuzzy controller," *Composite Structures*, vol. 94, no. 4, pp. 1385–1390, 2012.

[16] P. Li and F. Jin, "Adaptive fuzzy control for unknown nonlinear systems with perturbed dead-zone inputs," *Acta Automatica Sinica*, vol. 36, no. 4, pp. 573–579, 2009.

[17] M. Marinaki, Y. Marinakis, and G. E. Stavroulakis, "Fuzzy control optimized by PSO for vibration suppression of beams," *Control Engineering Practice*, vol. 18, no. 6, pp. 618–629, 2010.

[18] D. Wang, X. Lin, and Y. Zhang, "Fuzzy logic control for a parallel hybrid hydraulic excavator using genetic algorithm," *Automation in Construction*, vol. 20, no. 5, pp. 581–587, 2011.

[19] R.-E. Precup, M.-B. Rădac, M. L. Tomescu, E. M. Petriu, and S. Preitl, "Stable and convergent iterative feedback tuning of fuzzy controllers for discrete-time SISO systems," *Expert Systems with Applications*, vol. 40, no. 1, pp. 188–199, 2013.

[20] S. Tong and G. Liu, "Real-time simplified variable domain fuzzy control of PEM fuel cell flow systems," *European Journal of Control*, vol. 14, no. 3, pp. 223–233, 2008.

[21] J. N. Lygouras, P. N. Botsaris, and J. Vourvoulakis, "Fuzzy logic controller implementation for a solar air-conditioning system," *Applied Energy*, vol. 84, no. 12, pp. 1305–1318, 2007.

[22] U. Zuperl, F. Cus, and M. Milfelner, "Fuzzy control strategy for an adaptive force control in end-milling," *Journal of Materials Processing Technology*, vol. 164-165, no. 2, pp. 1472–1478, 2005.

[23] L. A. Zadeh, "Fuzzy sets," *Information and Control*, vol. 8, no. 3, pp. 338–353, 1965.

[24] C. Radu and R. Wilkerson, "Using fuzzy set theory," *IEEE Potentials*, vol. 14, no. 5, pp. 33–35, 1995.

[25] Y. M. Li, Y. B. Wang, and Q. F. Yan, "The technology study of fuzzy control system for asynchronous motor," in *2013 3rd International Conference on Consumer Electronics, Communications and Networks*, pp. 710–713, Xianning, China, November 2013.

[26] J. Zhang, S. Zhang, and Z. Dan, "Variable universe fuzzy PID control for multi-level gas tank pressure," in *2014 IEEE 17th International Conference on Computational Science and Engineering*, pp. 1900–1904, Chengdu, China, December 2014.

[27] S. Xu, Y. Jing, and X. Chen, "Design of fuzzy-CMAC neural-network complex controller for gas mixing process," in *2011 Second International Conference on Mechanic Automation and Control Engineering*, pp. 1191–1194, Hohhot, China, July 2011.

[28] V. N. A. L. Dasilva and R. S. Zubulum, "An integration of neural networks and fuzzy logic for power systems diagnosis," in *Proceedings of International Conference on Intelligent System Application to Power Systems*, pp. 237–241, Orlando, FL, USA, February 1996.

[29] M. Michele and M. Giuseppe, "A gimbal platform stabilization for topographic applications," in *ICNAAM-2014: International Conference on Numerical Analysis and Applied Mathematics 2014*, vol. 1648, pp. 780011-1–780011-4, Rhodes, Greece, September 2014.

[30] W. Pedrycz and M. H. Smith, "Fuzzy inference networks: an introduction," in *Proceedings of International Conference on Neural Networks (ICNN'97)*, pp. 2342–2346, Houston, TX, USA, June 1997.

[31] Y. Cui, X. Li, and Z. Liu, "Application of improved BP neural network with correlation rules in network intrusion detection," *International Journal of Security and its Applications*, vol. 10, no. 4, pp. 423–430, 2016.

[32] T. Kang, J. Fang, and W. Wang, "In-flight calibration approach based on quaternion optimization for POS used in airborne remote sensing," *IEEE Transactions on Instrumentation and Measurement*, vol. 62, no. 11, pp. 2882–2889, 2013.

[33] H. Yang, D. Dong, and Y. Ren, "Study on fuzzy PID control in double closed-loop DC speed regulation system," in *2011 Third International Conference on Measuring Technology and*

Mechatronics Automation, pp. 465–469, Shangshai, China, January 2011.

[34] X. Zhou, L. Li, and Y. Jia, "Adaptive fuzzy/proportion integration differentiation (PID) compound control for unbalance torque disturbance rejection of aerial inertially stabilized platform," *International Journal of Advanced Robotic Systems*, vol. 13, no. 5, 2016.

[35] J. Cervantes, W. Yu, and S. Salazar, "Takagi–Sugeno dynamic neuro-fuzzy controller of uncertain nonlinear systems," *IEEE Transactions on Fuzzy Systems*, vol. 25, no. 6, pp. 1601–1615, 2016.

[36] S. Kara, S. Dasb, and P. K. Ghoshb, "Applications of neuro fuzzy systems: a brief review and future outline," *Applied Soft Computing*, vol. 15, no. 1, pp. 243–259, 2014.

[37] W. Wang, X. Huang, and W. Huang, "Design and simulation of heat substation controller based on neural network-fuzzy PID control," in *2011 Third International Conference on Intelligent Human-Machine Systems and Cybernetics*, pp. 3–6, Zhejiang, China, August 2011.

[38] J. Li and J. Yu, "An intelligent control system based on recurrent neural fuzzy network and its application to CSTR," *Journal of Systems Science and Complexity*, vol. 18, no. 1, pp. 43–54, 2005.

[39] X. Zhou, G. Gong, J. Li, H. Zhang, and R. Yu, "Decoupling control for a three-axis inertially stabilized platform used for aerial remote sensing," *Transactions of the Institute of Measurement and Control*, vol. 37, no. 9, pp. 1135–1145, 2015.

6

Adaptive Neural Control of Hypersonic Vehicles with Actuator Constraints

Changxin Luo [ID], Humin Lei, Dongyang Zhang, and Xiaojun Zou

Air and Missile Defence College, Air Force Engineering University, Xi'an 710051, China

Correspondence should be addressed to Changxin Luo; 1710794652@qq.com

Academic Editor: Yue Wang

An adaptive neural control method is proposed in this paper for the flexible air-breathing hypersonic vehicle (AHV) with constraints on actuators. This scheme firstly converts the original control problem with input constraints into a new control problem without input constraints based on the control input saturation function. Secondly, on the basis of the implicit function theorem, the radial basis function neural network (RBFNN) is introduced to approximate the uncertain items of the model. And the minimal-learning-parameter (MLP) technique is adopted to design the adaptive law for the norm of network weight vector, which significantly reduces calculations. Meanwhile, the finite-time convergence differentiator (FD) is introduced, through which the model state variables and their derivatives are accurately estimated to ensure the control effect. Finally, it is theoretically proved that the closed-loop control system is stable. And the effectiveness of the designed controller is verified by simulation.

1. Introduction

Air-breathing hypersonic vehicle (AHV) is a new type of aircraft flying in the near space at a speed of more than Mach 5. It has outstanding advantages in terms of high speed, high maneuverability, and large flight envelope that traditional aircrafts do not have, which possesses potential applications in both civilian and military fields [1–3]. But it also has more complex coupling, stronger nonlinearity, more severe elastic vibration, and tighter control input constraints than ordinary aircraft due to the flat and slender body design of the AHV and the integration of scramjet engine, which make the flight control of AHV a frontier issue in today's control field [4, 5].

In the design process of the control law for AHV, by introducing a neural network to approximate the model's unknown dynamics or control laws that are difficult to be directly implemented, the nonlinear, strong coupling, and uncertainties of the model can be handled well to ensure the robust performance of the control law [6, 7]. For the rigid body model of AHV, the continuous and discrete adaptive neural controllers are designed, respectively, in [7, 8] by expressing original model as a strict feedback form and introducing RBFNN to approximate the model's unknown function, which guarantee that the closed-loop signals are uniformly ultimately bounded but fail to consider the influence of flexible states. Different from references [7, 8], a nonsingular direct neural control is proposed in [9] based on backstepping method by firstly designing the ideal backstepping control law for each subsystem of AHV and then using the RBFNN to approximate the designed ideal control law online instead of the unknown function of the model. However, the design process and form of the control law in [9] are cumbersome and complex, which restrict its applications in engineering. Two novel neural backstepping control strategies are proposed in [10, 11], which are designed with improved backstepping methods, respectively. Simulations show that the proposed methods in [10, 11] have strong robustness and superior control effects, but both of them assume that the model is affine for control input.

On the other hand, considering that AHV controls the height and attitude of longitudinal motion with elevators, as the flying height increases, the efficiency of the elevator will decrease significantly [12]. In addition, the AHV can be affected by unknown airflows such as gust and turbulence during the flight process. Therefore, it is easy to encounter the phenomenon of elevator saturation when flying at high altitude. Once the actuators reach saturation, it may cause failure of the control system [13]. Thence, it requires urgent development of antisaturation control studies for AHV. However, the related research in this respect is still relatively little. Only considering the throttle setting constraint of the engine, a neural control method based on time-scale decomposition is designed in [14]. In [13], the neural network is used to approximate the saturation characteristics of the control law, which effectively solves the problem of control input constraint. However, the neural network weight parameters strongly depend on the model and are difficult to select. The tracking error is corrected through the designed auxiliary error compensation system in [15]. Although the simulation result shows that it has certain feasibility in dealing with the problem of actuator constraints, it cannot theoretically guarantee that the tracking error is bounded. The method of [15] is further extended to AHV flight control where both control input and flight attitudes are constrained. However, in the actual project, AHV has no corresponding actuators to directly limit its flight attitude.

There are two shortcomings in the above studies. First, forcing AHV's nonaffine motion model into an affine model will inevitably result in the loss of certain key dynamic characteristics. The control law designed based on the simplified affine model will have the risk of partial or complete failure. The second is that the difficulty of the controller design is increased to some extent by introducing the auxiliary system or using the neural network to approach the actuator saturation characteristic when dealing with the actuator constraints of the AHV.

Based on the above shortcomings, it is imperative to design a simple and effective nonaffine control law to solve the control problem of AHV when the actuators are constrained. This paper studies the control problem of AHV with actuator constraints and proposes an adaptive neural control method. By designing the hyperbolic tangent input saturation function, the original control problem with input constraints is transformed into a control problem without input constraints. Different from the above references, the AHV model is regarded as a pure feedback system with nonaffine control input that is closer to the actual situation of AHV. Based on the implicit function theorem, the RBFNN is introduced to accurately approximate the unknown function of the model. The MLP technique is adopted to adaptively adjust the norm of the weight vector, which greatly reduces the amount of adaptive calculations. And through employing FD to achieve effective estimation of the system state variables, the control accuracy is ensured. Simulation examples verify the effectiveness and superiority of the design method.

2. AHV Model and Preliminaries

2.1. Model Description. Parker, a scholar at the US Air Force Research Laboratory, based on the study of AHV models by Bolender and Doman [16] and by neglecting some slow dynamics and weak coupling of AHV, establishes the following AHV longitudinal motion model [17]:

$$\dot{V} = \frac{T \cos{(\theta - \gamma)} - D}{m} - g \sin \gamma, \tag{1}$$

$$\dot{h} = V \sin \gamma, \tag{2}$$

$$\dot{\gamma} = \frac{L + T \sin{(\theta - \gamma)}}{mV} - \frac{g}{V} \cos \gamma, \tag{3}$$

$$\dot{\theta} = Q, \tag{4}$$

$$\dot{Q} = \frac{M + \tilde{\psi}_1 \ddot{\eta}_1 + \tilde{\psi}_2 \ddot{\eta}_2}{I_{yy}}, \tag{5}$$

$$k_1 \ddot{\eta}_1 = -2\zeta_1 \omega_1 \dot{\eta}_1 - \omega_1^2 \eta_1 + N_1 - \tilde{\psi}_1 \frac{M}{I_{yy}} - \frac{\tilde{\psi}_1 \tilde{\psi}_2 \ddot{\eta}_2}{I_{yy}}, \tag{6}$$

$$k_2 \ddot{\eta}_2 = -2\zeta_2 \omega_2 \dot{\eta}_2 - \omega_2^2 \eta_2 + N_2 - \tilde{\psi}_2 \frac{M}{I_{yy}} - \frac{\tilde{\psi}_2 \tilde{\psi}_1 \ddot{\eta}_1}{I_{yy}}, \tag{7}$$

where

$$
\begin{aligned}
k_1 &= 1 + \frac{\tilde{\psi}_1}{I_{yy}}, \\
k_2 &= 1 + \frac{\tilde{\psi}_2}{I_{yy}}, \\
\tilde{\psi}_1 &= \int_{-L_f}^{0} \hat{m}_f \xi \phi_f(\xi) \mathrm{d}\xi, \\
\tilde{\psi}_2 &= \int_{0}^{L_a} \hat{m}_a \xi \phi_a(\xi) \mathrm{d}\xi,
\end{aligned}
\tag{8}
$$

where velocity V, altitude h, flight-path γ, pitch angle θ, and pitch rate Q are the five rigid body states; m and I_{yy} represent vehicle mass and moment of inertia, respectively; the four flexible modes η_1, $\dot{\eta}_1$, η_2, and $\dot{\eta}_2$ denote the first two bending modes of the fuselage; ζ_i and $\omega_i (i = 1, 2)$ represent the damping ratio and natural frequency for flexible modes, respectively; L_f and L_a represent the length of forward beam and aft beam; $\hat{m}_f$ and $\hat{m}_a$ are the mass distribution of forward beam and aft beam, respectively; and $\phi_f(\cdot)$ and $\phi_a(\cdot)$ are structural mode shapes [16]. The force map of an AHV model is shown in Figure 1. The approximations of thrust T, drag D, lift L, pitching moment M, and the generalized forces $N_i (i = 1, 2)$ are expressed as [17]

$$T \approx C_T^{\alpha^3} \alpha^3 + C_T^{\alpha^2} \alpha^2 + C_T^{\alpha} \alpha + C_T^0,$$

$$D \approx \bar{q} S \left(C_D^{\alpha^2} \alpha^2 + C_D^{\alpha} \alpha + C_D^{\delta_e^2} \delta_e^2 + C_D^{\delta_e} \delta_e + C_D^0 \right),$$

$$L \approx \bar{q} S \left(C_L^{\alpha} \alpha + C_L^{\delta_e} \delta_e + C_L^0 \right),$$

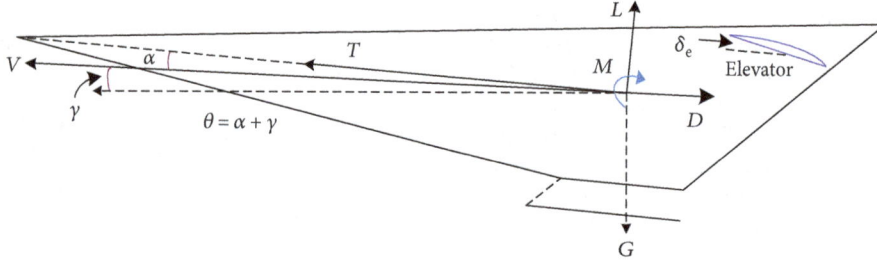

FIGURE 1: Force map of an AHV model.

$$M \approx z_T T + \bar{q} S \bar{c} \left[C_{M,\alpha}^{\alpha^2} \alpha^2 + C_{M,\alpha}^{\alpha} \alpha + C_{M,\alpha}^0 + c_e \delta_e \right],$$

$$N_1 \approx N_1^{\alpha^2} \alpha^2 + N_1^{\alpha} \alpha + N_1^0,$$

$$N_2 \approx N_2^{\alpha^2} \alpha^2 + N_2^{\alpha} \alpha + N_2^{\delta_e} \delta_e + N_2^0,$$

$$C_T^{\alpha^3} = \beta_1(h, \bar{q}) \Phi + \beta_2(h, \bar{q}),$$

$$C_T^{\alpha^2} = \beta_3(h, \bar{q}) \Phi + \beta_4(h, \bar{q}),$$

$$C_T^{\alpha} = \beta_5(h, \bar{q}) \Phi + \beta_6(h, \bar{q}),$$

$$C_T^0 = \beta_7(h, \bar{q}) \Phi + \beta_8(h, \bar{q}),$$

$$\bar{q} = \frac{1}{2} \bar{\rho} V^2,$$

$$\bar{\rho} = \bar{\rho}_0 \exp\left(\frac{h_0 - h}{h_s}\right), \tag{9}$$

where $\alpha = \theta - \gamma$ is the angle of attack; fuel equivalence ratio Φ and elevator angular deflection δ_e are control inputs; $\bar{c}$ and S are the aerodynamic chord and reference area, respectively; $\bar{q}$ represents dynamic pressure; $\bar{\rho}$ is the air density at height h; z_T means thrust moment arm; c_e is the elevator coefficient; h_0 and $\bar{\rho}_0$ represent nominal altitude and air density at the altitude h_0, respectively; $1/h_s$ is the air density decay rate; $C_T^{\alpha^i}(i = 1, 2, 3)$ is the ith order coefficient of α in T; $C_D^{*^i}(i = 1, 2; * = \alpha, \delta_e)$ represents the ith order coefficient of $*$ in D; $C_L^*(* = \alpha, \delta_e)$ means coefficient of $*$ in L; $C_*^0(* = T, D, L)$ is the constant coefficient in $*$; $C_{M,\alpha}^{\alpha^i}(i = 0, 1, 2)$ represents the ith order coefficient of α in M; $N_j^{\alpha^i}(i = 0, 1, 2; j = 1, 2)$ is the ith order contribution of α to N_j; and $N_2^{\delta_e}$ is the contribution of δ_e to N_2; $\beta_i(h, \bar{q})(i = 1, 2, \dots, 8)$ is the ith trust fit parameter. For more detailed definitions of the model geometric parameters and coefficients, the reader could refer to [17].

2.2. Input Constraint Problem and Model Conversion. According to [18] and combining (1), (2), (3), (4), (5), and (9), we know that velocity V is mainly controlled by equivalence ratio Φ since the thrust T is directly affected by Φ. On the other hand, altitude h is mainly controlled by elevator

angular deflection δ_e since δ_e directly affects the pitch rate Q and then affects the pitch angle θ and flight-path γ, ultimately controlling the change of h. Therefore, we can firstly decompose the AHV model into a velocity subsystem (1) and altitude subsystem ((2), (3), (4), and (5)) and then design the control law separately [19].

Taking the actual situation into account, the executable ranges of Φ and δ_e have certain limits, which are described as follows:

$$\Phi \in [\Phi_{min}, \Phi_{max}],$$
$$\delta_e \in [\delta_{e\,min}, \delta_{e\,max}], \tag{10}$$

where Φ_{min} and $\Phi_{max} \geq 0$ are the respective upper and lower bound of Φ; $\delta_{e\,min}$ and $\delta_{e\,max}(\delta_{emin} = -\delta_{e\,max})$ stand for the upper and lower bound of δ_e, respectively.

In the above situation of actuator constraints, for velocity subsystem and altitude subsystem, the control objective is to design the respective control law Φ and δ_e under the constraints of (10) such that V and h track their reference commands V_{ref} and h_{ref}.

Obviously, Φ and δ_e have saturation characteristics under the constraints of (10). In order to convert the original control problem with input constraints into a new control problem without input constraints, here we use hyperbolic tangent function to approximate Φ and δ_e

$$\Phi(\Phi^*) = \frac{\Phi_{max} - \Phi_{min}}{2} \tanh\left(\frac{2\Phi^*}{\Phi_{max} - \Phi_{min}} - 1\right)$$
$$+ \frac{\Phi_{max} + \Phi_{min}}{2}, \tag{11}$$

$$\delta_e(\delta_e^*) = \delta_{e\,max} \tanh\left(\frac{\delta_e^*}{\delta_{e\,max}}\right),$$

where Φ^* and $\delta_e^* \in \mathbf{R}$ are the actual input of actuator saturation loop. By (11) and the character of hyperbolic tangent function, we can know that no matter what values Φ^* and δ_e^* take, both Φ and δ_e satisfy the constraint of (10).

At this point, the original control objective can be converted to design unconstrained Φ^* and δ_e^* such that V and h track their reference commands V_{ref} and h_{ref}.

Remark 1. In order to maintain the normal operating mode of scramjet engine, a restriction is imposed on the value range of Φ, which is generally taken as $\Phi \in [0.05, 1.5]$; taking the physical limit of the deflection angle of elevator into account, usually $\delta_e \in [-20°, 20°]$ [20].

Remark 2. Although the problem of actuator saturation is often considered when planning the trajectory, AHV can be affected by unknown airflows such as gusts and turbulences as well as sudden actuator failures during the flight process, which will cause the actuator to be saturated instantly such that the actual executable range of it will be even smaller than its theoretical value [13].

3. Adaptive Neural Controller Design

3.1. RBFNN Approximation. RBFNN has the characteristics of simple structure, strong learning, and fault tolerance and has the ability of globally approximating nonlinear continuous functions [21]. It can be expressed as a mapping from input to output.

$$y = \mathbf{W}^T \mathbf{h}(\mathbf{X}), \tag{12}$$

where $\mathbf{X} = [X_1, X_2, \ldots, X_n]^T \in \mathbf{R}^n$ is the input vector (n represents the dimension of the input vector), $\mathbf{W} = [w_1, w_2, \ldots, w_N]^T \in \mathbf{R}^N$ stands for the weight vector (N is the number of hidden layer nodes), $\mathbf{h}(\mathbf{X}) = [h_1(\mathbf{X}), h_2(\mathbf{X}), \ldots, h_N(\mathbf{X})]^T \in \mathbf{R}^N$, $h_i(\mathbf{X})$, denotes the activation function. Here, $h_i(\mathbf{X})$ is chosen as the following Gaussian function:

$$h_i(\mathbf{X}) = \exp\left(-\frac{\|\mathbf{X} - \mathbf{c}_i\|^2}{2b_i^2}\right), \quad i = 1, 2, \ldots, N, \tag{13}$$

where $\mathbf{b} = [b_1, b_2, \ldots, b_N]^T \in \mathbf{R}^N$, b_i, is the width of the ith Gaussian function; $\mathbf{c} = [\mathbf{c}_1, \mathbf{c}_2, \ldots, \mathbf{c}_N] \in \mathbf{R}^{n \times N}$ ($\mathbf{c}_i$ represents the center of the ith Gaussian function), which can be expressed as follows:

$$\mathbf{c} = \begin{bmatrix} c_{11} & \cdots & c_{1N} \\ \vdots & \ddots & \vdots \\ c_{n1} & \cdots & c_{nN} \end{bmatrix}. \tag{14}$$

For any unknown nonlinear continuous function $F(\mathbf{X})$, using the RBFNN and by selecting enough nodes (selecting a sufficiently large N), there must be an ideal weight vector $\mathbf{W}^* = [w_1^*, w_2^*, \ldots, w_N^*]^T \in \mathbf{R}^N$ that satisfies [21]

$$F(\mathbf{X}) = \mathbf{W}^{*T} \mathbf{h}(\mathbf{X}) + \mu, \quad |\mu| \le \mu_M, \tag{15}$$

where $\mu \in \mathbf{R}$ is the approximation error and $\mu_M \in \mathbf{R}^+$ stands for the upper bound of approximation error. When taking N large enough, μ_M can be arbitrarily small.

Remark 3. Since $\exp(\cdot)$ is a strictly monotonically increasing and positive function and $-\|\mathbf{X} - \mathbf{c}_i\|^2/(2b_i^2) \le 0$, there is

$0 < h_i(\mathbf{X}) \le h_i(0) = 1$. Therefore, there must be a bounded constant $\bar{h} \in \mathbf{R}^+$ such that $\|\mathbf{h}(\mathbf{X})\| \le \bar{h}$.

3.2. Controller Design for Velocity Subsystem. Based on the research conclusion of [7], here velocity subsystem is further expressed as a more general nonaffine form of control input

$$\dot{V} = F_V(V, \Phi). \tag{16}$$

Combined with (11), the above formula can be rewritten as

$$\dot{V} = F_V(V, \Phi^*), \tag{17}$$

where $F_V(V, \Phi^*)$ is a completely unknown continuously differentiable function.

In order to design the control law Φ^*, the following theorem is given firstly.

Theorem 1. *For any* $(V, \Phi^*) \in \Omega_V \times \mathbf{R}$, *the following inequality is established:*

$$\frac{\partial F_V(V, \Phi^*)}{\partial \Phi^*} > 0, \tag{18}$$

where Ω_V *is a controllable area.*

Proof 1. From (11), (16), and (17), we can know

$$\begin{aligned} \frac{\partial F_V(V, \Phi^*)}{\partial \Phi^*} &= \frac{\partial F_V(V, \Phi^*)}{\partial \Phi} \cdot \frac{\partial \Phi}{\partial \Phi^*} \\ &= \frac{((\partial F_V(V, \Phi))/\partial \Phi)}{\cosh^2((2\Phi^*/(\Phi_{\max} - \Phi_{\min})) - 1)}. \end{aligned} \tag{19}$$

According to [17], there is

$$\frac{\partial F_V(V, \Phi)}{\partial \Phi} > 0. \tag{20}$$

Therefore,

$$\frac{\partial F_V(V, \Phi^*)}{\partial \Phi^*} > 0. \tag{21}$$

Define velocity tracking error

$$\tilde{V} = V - V_{\text{ref}}. \tag{22}$$

Taking time derivative along (22) and using (17) yield

$$\dot{\tilde{V}} = \dot{V} - \dot{V}_{\text{ref}} = F_V(V, \Phi^*) - \dot{V}_{\text{ref}}. \tag{23}$$

Let

$$F_V^*(V, \Phi^*) = F_V(V, \Phi^*) - k_V \Phi^*. \tag{24}$$

where $k_V \in \mathbf{R}^+$ is the design parameter.

Combine (23) and (24)

$$\dot{\tilde{V}} = \dot{V} - \dot{V}_{\text{ref}} = F_V^*(V, \Phi^*) + k_V \Phi^* - \dot{V}_{\text{ref}}. \quad (25)$$

Design the control law Φ^* as

$$\Phi^* = k_V^{-1}(\Phi_0^* - \Phi_1^*), \quad (26)$$

where

$$\Phi_0^* = -k_{V1}\tilde{V} - k_{V2}\int_0^t \tilde{V}d\tau + \dot{V}_{\text{ref}}. \quad (27)$$

k_{V1} and $k_{V2} \in \mathbf{R}^+$ stand for design parameters and Φ_1^* is the neural control law to be designed to counteract the influence of uncertain term $F_V^*(V, \Phi^*)$.

In order to facilitate the subsequent design process, the implicit function theorem is introduced here [22].

Theorem 2. *Assume that the implicit function* $\Psi : \mathbf{R}^l \times \mathbf{R}^r \to \mathbf{R}^l$ *is continuously differentiable at each point* (ς, σ) *on the open set* $Y \subset \mathbf{R}^l \times \mathbf{R}^r$. (ς_0, σ_0) *is the point in* Y; $\Psi(\varsigma_0, \sigma_0) = 0$ *and the Jacobian matrix* $(\partial\Psi/\partial\varsigma)(\varsigma_0, \sigma_0)$ *is nonsingular. Then, for any* $\sigma \in G$, *a neighborhood* $U \subset \mathbf{R}^l$ *of* ς_0 *and a neighborhood* $G \subset \mathbf{R}^r$ *of* σ_0 *can make the equation* $\Psi(\varsigma, \sigma) = 0$ *which has a unique solution* $\varsigma \in U$, *and the solution can be expressed as* $\varsigma = g_0(\sigma)$, *where* $g_0(\cdot)$ *is a continuously differentiable function on* $\sigma = \sigma_0$.

Remark 4. Theorem 2 shows that once the implicit function $\Psi(\varsigma, \sigma)$ satisfies all conditions in the theorem, ς can be expressed as a continuously differentiable function of σ, that is, $\varsigma = g_0(\sigma)$. At this time, using RBFNN to approximate $\Psi(\varsigma, \sigma)$, only σ, instead of ς, is used as the input signal of the neural network to obtain a satisfactory approximation effect. This is where the special meaning of the implicit function theorem lies.

Let

$$\Gamma(V, \Phi_0^*, \Phi_1^*) \triangleq F_V^*(V, \Phi^*) - \Phi_1^* \\ = F_V^*(V, k_V^{-1}(\Phi_0^* - \Phi_1^*)) - \Phi_1^*. \quad (28)$$

To illustrate that $\Gamma(V, \Phi_0^*, \Phi_1^*)$ satisfies the implicit function theorem, the following theorem is given.

Theorem 3. *Define*

$$k_V > \frac{1}{2}\frac{\partial F_V(V, \Phi^*)}{\partial \Phi^*}. \quad (29)$$

Then there are a controllable area $\Omega_V \subset \mathbf{R}$ *and a unique* Φ_1^* *for any* $(V, \Phi_1^*) \in \Omega_V \times \mathbf{R}$; Φ_1^* *satisfies*

$$\Gamma(V, \Phi_0^*, \Phi_1^*) = 0. \quad (30)$$

Proof 2. According to [23], a sufficient condition for the existence of Φ_1^* is that the following inequality holds:

$$\left|\frac{\partial F_V^*(V, \Phi^*)}{\partial \Phi_1^*}\right| < 1. \quad (31)$$

Consider (18), (24), (26), and (29), there are

$$\begin{aligned}
\left|\frac{\partial F_V^*(V, \Phi^*)}{\partial \Phi_1^*}\right| &= \left|\frac{\partial[F_V(V, \Phi^*) - k_V\Phi^*]}{\partial \Phi_1^*}\right| \\
&= \left|\frac{\partial[F_V(V, \Phi^*) - k_V\Phi^*]}{\partial \Phi^*}\frac{\partial \Phi^*}{\partial \Phi_1^*}\right| \\
&= \left|\left[\frac{\partial F_V(V, \Phi^*)}{\partial \Phi^*} - k_V\right]\frac{1}{k_V}\right| \\
&= \left|\frac{1}{k_V}\frac{\partial F_V(V, \Phi^*)}{\partial \Phi^*} - 1\right| < 1.
\end{aligned} \quad (32)$$

Therefore, Φ_1^* exists.

Further,

$$\begin{aligned}
\frac{\partial\Gamma(V, \Phi_0^*, \Phi_1^*)}{\partial \Phi_1^*} &= \frac{\partial[F_V^*(V, \Phi^*) - \Phi_1^*]}{\partial \Phi_1^*} \\
&= \frac{\partial[F_V(V, \Phi^*) - k_V\Phi^*]}{\partial \Phi_1^*} - 1 \\
&= \frac{\partial[F_V(V, \Phi^*) - k_V\Phi^*]}{\partial \Phi^*}\frac{\partial \Phi^*}{\partial \Phi_1^*} - 1 \\
&= \left[\frac{\partial F_V(V, \Phi^*)}{\partial \Phi^*} - k_V\right]\left(-\frac{1}{k_V}\right) - 1 \\
&= -\frac{1}{k_V}\frac{\partial F_V(V, \Phi^*)}{\partial \Phi^*},
\end{aligned} \quad (33)$$

where $\Phi^* = k_V^{-1}(\Phi_0^* - \Phi_1^*)$.

Combine (18) and (33)

$$\frac{\partial\Gamma(V, \Phi_0^*, \Phi_1^*)}{\partial \Phi_1^*} < 0. \quad (34)$$

According to Theorem 3 and (34), $\Gamma(V, \Phi_0^*, \Phi_1^*)$ satisfies the implicit function theorem. Therefore, Φ_1^* can be regarded as a function of Φ_0^* and V, and further $F_V^*(V, \Phi^*)$ can be regarded as a function of Φ_0^* and V.

Define $\mathbf{X}_1 = [V, \Phi_0^*]^T \in \mathbf{R}^2$ as the input vector of RBFNN and introduce RBFNN to approximate $F_V^*(V, \Phi^*)$

$$F_V^*(V, \Phi^*) = \mathbf{W}_V^{*T}\mathbf{h}(\mathbf{X}_1) + \varepsilon, \quad |\varepsilon| \leq \varepsilon_M, \quad (35)$$

where $\mathbf{W}_V^{*T} = [w_{V1}^*, w_{V2}^*, \ldots, w_{VN_1}^*]^T \in \mathbf{R}^{N_1}$ is the weight vector, N_1 is the number of nodes, and ε and ε_M are approximation errors and its upper bounds, respectively. And $\mathbf{h}(\mathbf{X}_1) = [h_1(\mathbf{X}_1), h_2(\mathbf{X}_1), \ldots, h_{N_1}(\mathbf{X}_1)]^T \in \mathbf{R}^{N_1}$, where $h_i(\mathbf{X}_1)$ is the activation function; here, $h_i(\mathbf{X}_1)$ is chosen as the Gaussian function.

The following is based on the MLP technique to adaptively adjust the norm of the RBFNN weight vector.

Define $\omega_V = \|\mathbf{W}_V^*\|^2$ and design Φ_1^* as

$$\Phi_1^* = \frac{1}{2}\tilde{V}\hat{\omega}_V\mathbf{h}^{\mathrm{T}}(\mathbf{X}_1)\mathbf{h}(\mathbf{X}_1), \qquad (36)$$

where $\hat{\omega}_V$ is the estimation of ω_V, and its adaptive law is designed as

$$\dot{\hat{\omega}}_V = \frac{\mu_V}{2}\tilde{V}^2\mathbf{h}^{\mathrm{T}}(\mathbf{X}_1)\mathbf{h}(\mathbf{X}_1) - 2\hat{\omega}_V, \qquad (37)$$

where $\mu_V \in \mathbf{R}^+$ is the design parameter.

Substitute (27) and (36) into (26) and finally get the control law Φ^*

$$\Phi^* = k_V^{-1}\left[-k_{V1}\tilde{V} - k_{V2}\int_0^t\tilde{V}d\tau + \dot{V}_{\mathrm{ref}} - \frac{1}{2}\tilde{V}\hat{\omega}_V\mathbf{h}^{\mathrm{T}}(\mathbf{X}_1)\mathbf{h}(\mathbf{X}_1)\right]. \tag{38}$$

Remark 5. Different from [24] in which the weight vector of the neural network is directly adjusted online, this paper regards $\mathbf{W}_V^*$ as a whole based on the MLP technique. Adaptive adjustment of ω_V requires only one online learning parameter $\hat{\omega}_V$, and the computational complexity of the approximation algorithm is significantly reduced.

3.3. Controller Design for Altitude Subsystem. Define altitude tracking error

$$\tilde{h} = h - h_{\mathrm{ref}}. \qquad (39)$$

Taking time derivative along (39) and using (2) yield

$$\dot{\tilde{h}} = V\sin\gamma - \dot{h}_{\mathrm{ref}}. \qquad (40)$$

The reference trajectory of γ is chosen as

$$\gamma_d = \arcsin\left(\frac{-k_{\gamma 1}\tilde{h} - k_{\gamma 2}\int_0^t\tilde{h}d\tau + \dot{h}_{\mathrm{ref}}}{V}\right), \qquad (41)$$

where $k_{\gamma 1}$, and $k_{\gamma 2} \in \mathbf{R}^+$ are design parameters. If $\gamma \to \gamma_d$, known by (40) and (41), the dynamic response of $\tilde{h}$ is

$$\ddot{\tilde{h}} + k_{\gamma 1}\dot{\tilde{h}} + k_{\gamma 2}\tilde{h} = 0. \qquad (42)$$

Performing Laplace transformation on both sides of (42) can get its characteristic equation

$$s^2 + k_{\gamma 1}s + k_{\gamma 2} = 0. \qquad (43)$$

The two characteristic roots $(-k_{\gamma 1} - \sqrt{k_{\gamma 1}^2 - 4k_{\gamma 2}})/2$ and $(-k_{\gamma 1} + \sqrt{k_{\gamma 1}^2 - 4k_{\gamma 2}})/2$ of (43) are negative real numbers, so the system (42) is stable and $\tilde{h}$ is exponentially convergent.

Therefore, as long as $\gamma \to \gamma_d$, h can track h_{ref} stably. In this way, the control objective of the altitude subsystem turns to ensure that γ tracks γ_d.

Define $x_1 = \gamma$, $x_2 = \theta$, and $x_3 = Q$. Based on the research conclusion of [7], the remaining part of the altitude subsystem ((3), (4), and (5)) is further expressed as the following more general nonaffine form:

$$\begin{aligned}\dot{x}_1 &= f_1(x_1, x_2),\\\dot{x}_2 &= x_3,\\\dot{x}_3 &= f_3(\mathbf{x}, \delta_e).\end{aligned} \qquad (44)$$

Considering (11), the above formula can be rewritten as

$$\begin{aligned}\dot{x}_1 &= f_1(x_1, x_2),\\\dot{x}_2 &= x_3,\\\dot{x}_3 &= f_3(\mathbf{x}, \delta_e^*),\end{aligned} \qquad (45)$$

where $\mathbf{x} = [x_1, x_2, x_3]^{\mathrm{T}}$. $f_1(x_1, x_2)$ and $f_3(\mathbf{x}, \delta_e^*)$ are completely unknown continuously differentiable functions.

Remark 6. Due to the strong coupling between the rigid body states and the flexible modes in the AHV model ((1), (2), (3), (4), (5), (6), and (7)), here the dynamic characteristics of strong nonlinearity and strong coupling in the original model are regarded as completely unknown continuous differentiable functions ((16) and (44)) by referring to the method of [25]. The proposed method in this paper will use RBFNN to accurately approximate these continuous differentiable functions and then complete the design of the control law and ensure the stability of the closed-loop control system.

In order to design the control law δ_e^*, the following theorem is given firstly.

Theorem 4. For any $(\mathbf{x}, \delta_e^*) \in \Omega_x \times \mathbf{R}$, the following inequalities are established:

$$\begin{aligned}\frac{\partial f_1(x_1, x_2)}{\partial x_2} &> 0,\\[2mm]\frac{\partial f_3(\mathbf{x}, \delta_e^*)}{\partial\delta_e^*} &> 0,\end{aligned} \qquad (46)$$

where Ω_x is a controllable area.

Proof 3. According to the results in [17], we can know

$$\begin{aligned}\frac{\partial f_1(x_1, x_2)}{\partial x_2} &> 0,\\[2mm]\frac{\partial f_3(\mathbf{x}, \delta_e)}{\partial\delta_e} &> 0.\end{aligned} \qquad (47)$$

Consider (11), (44) and (45), there is

$$\frac{\partial f_3(\mathbf{x}, \delta_e^*)}{\partial \delta_e^*} = \frac{\partial f_3(\mathbf{x}, \delta_e^*)}{\partial \delta_e} \cdot \frac{\partial \delta_e}{\partial \delta_e^*} = \frac{(\partial f_3(\mathbf{x}, \delta_e))/\partial \delta_e}{\cosh^2(\delta_e^*/\delta_{emax})}. \quad (48)$$

Combining (47) and (48), we can see that (46) is true.

In order to avoid the cumbersome and complicated design process of the traditional backstepping method in designing the control law, the following equivalent transformation is performed on the system (45).

Step 1. Let $z_1 = x_1 = \gamma$ and $z_2 = \dot{z}_1 = f_1(x_1, x_2)$. From (45), the time derivative of z_2 is derived as

$$\begin{aligned}
\dot{z}_2 &= \frac{\partial f_1(x_1, x_2)}{\partial x_1} \dot{x}_1 + \frac{\partial f_1(x_1, x_2)}{\partial x_2} \dot{x}_2 \\
&= \frac{\partial f_1(x_1, x_2)}{\partial x_1} f_1(x_1, x_2) + \frac{\partial f_1(x_1, x_2)}{\partial x_2} x_3 \\
&\triangleq f_{h1}(\mathbf{x}).
\end{aligned} \quad (49)$$

Step 2. Let $z_3 = \dot{z}_2 = f_{h1}(\mathbf{x})$. From (45), the time derivative of z_3 is derived as

$$\begin{aligned}
\dot{z}_3 &= \frac{\partial f_{h1}(\mathbf{x})}{\partial x_1} \dot{x}_1 + \frac{\partial f_{h1}(\mathbf{x})}{\partial x_2} \dot{x}_2 + \frac{\partial f_{h1}(\mathbf{x})}{\partial x_3} \dot{x}_3 \\
&= \frac{\partial f_{h1}(\mathbf{x})}{\partial x_1} f_1(x_1, x_2) + \frac{\partial f_{h1}(\mathbf{x})}{\partial x_2} x_3 + \frac{\partial f_{h1}(\mathbf{x})}{\partial x_3} f_3(\mathbf{x}, \delta_e^*) \\
&\triangleq f_{h2}(\mathbf{x}, \delta_e^*).
\end{aligned} \quad (50)$$

After the above transformation, system (45) is as follows:

$$\begin{aligned}
\dot{z}_1 &= z_2, \\
\dot{z}_2 &= z_3, \\
\dot{z}_3 &= f_{h2}(\mathbf{x}, \delta_e^*),
\end{aligned} \quad (51)$$

where $f_{h2}(\mathbf{x}, \delta_e^*)$ is a completely unknown continuously differentiable function.

Remark 7. From (46), (49) and (50), there is

$$\begin{aligned}
\frac{\partial f_{h2}(\mathbf{x}, \delta_e^*)}{\partial \delta_e^*} &= \frac{\partial f_{h1}(\mathbf{x})}{\partial x_3} \frac{\partial f_3(\mathbf{x}, \delta_e^*)}{\partial \delta_e^*} \\
&= \frac{\partial f_1(x_1, x_2)}{\partial x_2} \frac{\partial f_3(\mathbf{x}, \delta_e^*)}{\partial \delta_e^*} > 0.
\end{aligned} \quad (52)$$

Define flight-path tracking error e and error function E

$$e = \gamma - \gamma_d = z_1 - \gamma_d,$$

$$E = \left(\frac{d}{dt} + \lambda\right)^3 \int_0^t e d\tau = \ddot{e} + 3\lambda \dot{e} + 3\lambda^2 e + \lambda^3 \int_0^t e d\tau, \quad (53)$$

where $\lambda \in \mathbf{R}^+$ stands for the design parameter. Since $(s + \lambda)^3$ is a Hurwitz polynomial, when E is bounded, e must be bounded.

According to (51) and (53), we obtain

$$\begin{aligned}
\dot{e} &= \dot{z}_1 - \dot{\gamma}_d = z_2 - \dot{\gamma}_d, \\
\ddot{e} &= \dot{z}_2 - \ddot{\gamma}_d = z_3 - \ddot{\gamma}_d, \\
\dddot{e} &= \dot{z}_3 - \dddot{\gamma}_d = f_{h2}(\mathbf{x}, \delta_e^*) - \dddot{\gamma}_d.
\end{aligned} \quad (54)$$

Considering that z_2 and z_3 are unknown, we can know that $z_2 = \dot{\gamma}$ and $z_3 = \ddot{\gamma}$ from the previous model transformation process. A new finite-time convergence differentiator (FD) will be used below to accurately estimate differential signal. By taking γ as the input signal of FD (take $n = 4$), we can get the estimated values of z_2 and z_3, which are expressed as $\hat{z}_2$ and $\hat{z}_3$, respectively. Similarly, the estimations of $\dot{\gamma}_d$, $\ddot{\gamma}_d$, and $\dddot{\gamma}_d$ can be obtained via taking γ_d as the input signal of FD (take $n = 4$), which are expressed as $\hat{\dot{\gamma}}_d$, $\hat{\ddot{\gamma}}_d$, and $\hat{\dddot{\gamma}}_d$.

Therefore, the estimations of the first three derivatives of e can be expressed as

$$\begin{aligned}
\hat{\dot{e}} &= \hat{z}_2 - \hat{\dot{\gamma}}_d, \\
\hat{\ddot{e}} &= \hat{z}_3 - \hat{\ddot{\gamma}}_d, \\
\hat{\dddot{e}} &= f_{h2}(\mathbf{x}, \delta_e^*) - \hat{\dddot{\gamma}}_d.
\end{aligned} \quad (55)$$

From (53) and (55), we obtain the estimation of E

$$\hat{E} = \hat{\ddot{e}} + 3\lambda \hat{\dot{e}} + 3\lambda^2 e + \lambda^3 \int_0^t e d\tau. \quad (56)$$

Let

$$F_h(\mathbf{x}, \delta_e^*) = f_{h2}(\mathbf{x}, \delta_e^*) - k_h \delta_e^*, \quad (57)$$

where $k_h \in \mathbf{R}^+$ is the design parameter.
Combine (55), (56), and (57)

$$\dot{\hat{E}} = k_h \delta_e^* + F_h(\mathbf{x}, \delta_e^*) - \hat{\dddot{\gamma}}_d + 3\lambda \hat{\ddot{e}} + 3\lambda^2 \hat{\dot{e}} + \lambda^3 e. \quad (58)$$

Design the control law δ_e^* as

$$\delta_e^* = k_h^{-1}(\delta_{e0}^* - \delta_{e1}^*), \quad (59)$$

where $\delta_{e0}^* = -k_{h1}\hat{E} + \hat{\dddot{\gamma}}_d - 3\lambda \hat{\ddot{e}} - 3\lambda^2 \hat{\dot{e}} - \lambda^3 e$, $k_{h1} \in \mathbf{R}^+$ represents the design parameter, and δ_{e1}^* is the neural control law to be designed to counteract the influence of the uncertain term $F_h(\mathbf{x}, \delta_e^*)$.

Let

$$\begin{aligned}
H(\mathbf{x}, \delta_{e0}^*, \delta_{e1}^*) &\triangleq F_h(\mathbf{x}, \delta_e^*) - \delta_{e1}^* \\
&= F_h(\mathbf{x}, k_h^{-1}(\delta_{e0}^* - \delta_{e1}^*)) - \delta_{e1}^*.
\end{aligned} \quad (60)$$

Similar to velocity subsystem, to illustrate that $H(\mathbf{x}, \delta_{e0}^*, \delta_{e1}^*)$ satisfies the implicit function theorem, the following theorem is given.

Theorem 5. *If*

$$k_h > \frac{1}{2} \frac{\partial f_{h2}(\mathbf{x}, \delta_e^*)}{\partial \delta_e^*}, \qquad (61)$$

then there are a controllable area $\Omega_h \subset \mathbf{R}^3$ and a unique δ_{e1}^* for any $(\mathbf{x}, \delta_{e1}^*) \in \Omega_h \times \mathbf{R}$; δ_{e1}^* satisfies

$$H(\mathbf{x}, \delta_{e0}^*, \delta_{e1}^*) = 0. \qquad (62)$$

Proof 4. From [23], a sufficient condition for the existence of δ_{e1}^* is that the following inequality holds:

$$\left| \frac{\partial F_h(\mathbf{x}, \delta_e^*)}{\partial \delta_{e1}^*} \right| < 1. \qquad (63)$$

Consider (52), (59) and (61), there are

$$
\begin{aligned}
\left| \frac{\partial F_h(\mathbf{x}, \delta_e^*)}{\partial \delta_{e1}^*} \right| &= \left| \frac{\partial [f_{h2}(\mathbf{x}, \delta_e^*) - k_h \delta_e^*]}{\partial \delta_{e1}^*} \right| \\
&= \left| \frac{\partial [f_{h2}(\mathbf{x}, \delta_e^*) - k_h \delta_e^*]}{\partial \delta_e^*} \frac{\partial \delta_e^*}{\partial \delta_{e1}^*} \right| \\
&= \left| \left[\frac{\partial f_{h2}(\mathbf{x}, \delta_e^*)}{\partial \delta_e^*} - k_h \right] \frac{1}{k_h} \right| \\
&= \left| \frac{1}{k_h} \frac{\partial f_{h2}(\mathbf{x}, \delta_e^*)}{\partial \delta_e^*} - 1 \right| < 1.
\end{aligned} \qquad (64)
$$

Therefore, δ_{e1}^* exists.

Further,

$$
\begin{aligned}
\frac{\partial}{\partial \delta_{e1}^*} H(\mathbf{x}, \delta_{e0}^*, \delta_{e1}^*) &= \frac{\partial}{\partial \delta_{e1}^*} [F_h(\mathbf{x}, \delta_e^*) - \delta_{e1}^*] \\
&= \frac{\partial}{\partial \delta_{e1}^*} [f_{h2}(\mathbf{x}, \delta_e^*) - k_h \delta_e^*] - 1 \\
&= \frac{\partial}{\partial \delta_e^*} [f_{h2}(\mathbf{x}, \delta_e^*) - k_h \delta_e^*] \frac{\partial \delta_e^*}{\partial \delta_{e1}^*} - 1 \\
&= \left[\frac{\partial f_{h2}(\mathbf{x}, \delta_e^*)}{\partial \delta_e^*} - k_h \right] \left(-\frac{1}{k_h} \right) - 1 \\
&= -\frac{1}{k_h} \frac{\partial f_{h2}(\mathbf{x}, \delta_e^*)}{\partial \delta_e^*}.
\end{aligned}
$$

$$(65)$$

Combine (52) and (65)

$$\frac{\partial}{\partial \delta_{e1}^*} H(\mathbf{x}, \delta_{e0}^*, \delta_{e1}^*) < 0 \qquad (66)$$

According to Theorem 5 and (66), $H(\mathbf{x}, \delta_{e0}^*, \delta_{e1}^*)$ satisfies the implicit function theorem. Therefore, δ_{e1}^* can be regarded as a function of δ_{e0}^* and $\mathbf{x}$, and further $F_h(\mathbf{x}, \delta_e^*)$ can be regarded as a function of δ_{e0}^* and $\mathbf{x}$.

Define $\mathbf{X}_2 = [\mathbf{x}^T, \delta_{e0}^*]^T \in \mathbf{R}^4$ as the input vector of RBFNN and introduce RBFNN to approximate $F_h(\mathbf{x}, \delta_e^*)$

$$F_h(\mathbf{x}, \delta_e^*) = \mathbf{W}_h^{*T} \mathbf{h}(\mathbf{X}_2) + \iota, \quad |\iota| \leq \iota_M, \qquad (67)$$

where $\mathbf{W}_h^{*T} = [w_{h1}^*, w_{h2}^*, \dots, w_{hN_2}^*]^T \in \mathbf{R}^{N_2}$ is the weight vector, N_2 is the number of nodes, and ι and ι_M stand for approximation errors and its upper bounds, respectively. $\mathbf{h}(\mathbf{X}_2) = [h_1(\mathbf{X}_2), h_2(\mathbf{X}_2), \dots, h_{N_1}(\mathbf{X}_2)]^T \in \mathbf{R}^{N_2}$, $h_i(\mathbf{X}_2)$, is the activation function which is chosen as the Gaussian function here.

Define $\omega_h = \|\mathbf{W}_h^*\|^2$ and design δ_{e1}^* as

$$\delta_{e1}^* = \frac{1}{2} \hat{E} \hat{\omega}_h \mathbf{h}^T(\mathbf{X}_2) \mathbf{h}(\mathbf{X}_2), \qquad (68)$$

where $\hat{\omega}_h$ is the estimation of ω_h, and its adaptive law is designed as

$$\dot{\hat{\omega}}_h = \frac{\mu_h}{2} \hat{E}^2 \mathbf{h}^T(\mathbf{X}_2) \mathbf{h}(\mathbf{X}_2) - 2 \hat{\omega}_h, \qquad (69)$$

where $\mu_h \in \mathbf{R}^+$ is the design parameter.

3.4. Finite-Time Convergence Differentiator (FD)

Theorem 6. *Consider the following FD [26]:*

$$
\begin{aligned}
\dot{\xi}_1 &= \xi_2, \\
\dot{\xi}_2 &= \xi_3, \\
&\vdots \\
\dot{\xi}_{n-1} &= \xi_n, \\
\dot{\xi}_n &= R^n \left[-a_1 \arctan(\xi_1 - v(t)) - a_2 \arctan\left(\frac{\xi_2}{R}\right) \right. \\
&\qquad \left. - \cdots - a_n \arctan\left(\frac{\xi_n}{R^{n-1}}\right) \right],
\end{aligned} \qquad (70)
$$

where R, a_i $(i = 1, 2, \dots, n) \in \mathbf{R}^+$ stand for the design parameters. There are $\phi > 0$ and $\iota\phi > n$ so that

$$\xi_i - v^{(i-1)}(t) = O\left(\left(\frac{1}{R}\right)^{\iota\phi-i+1} \right), \quad i = 1, 2, \dots, n, \qquad (71)$$

where $O((1/R)^{\iota\phi-i+1})$ denotes the approximation order between ξ_i and $v^{(i-1)}(t)$, $\phi = (1-\vartheta)/\vartheta$, $\vartheta \in (0, \min\{\iota/(\iota + n), 1/2\})$, $n \geq 2$.

Proof 5. The detailed proof process of Theorem 6 can be found in [26].

Remark 8. In (70), ξ_i $(i = 1, 2, \ldots, n)$ is the state variable of the system, ξ_1 is the estimation of $v(t)$, and ξ_i $(i = 2, 3, \ldots, n)$ is the estimation of the $i - 1$th derivative of $v(t)$. At the same time, (71) shows that the estimation error is a high-order infinitesimal of $(1/R)^{\tau\phi - i + 1}$.

Use the above FD to estimate z_2, z_3, $\dot{\gamma}_d$, $\ddot{\gamma}_d$, and $\dddot{\gamma}_d$ of (54)

$$\dot{\hat{z}}_1 = \hat{z}_2,$$

$$\dot{\hat{z}}_2 = \hat{z}_3,$$

$$\dot{\hat{z}}_3 = \hat{z}_4,$$

$$\dot{\hat{z}}_4 = R_1^4 \left[-a_{11} \arctan\left(\hat{z}_1 - \gamma\right) - a_{12} \arctan\left(\frac{\hat{z}_2}{R_1}\right) \right. \\ \left. - a_{13} \arctan\left(\frac{\hat{z}_3}{R_1^2}\right) - a_{14} \arctan\left(\frac{\hat{z}_4}{R_1^3}\right) \right],$$

$$\left(\hat{\gamma}_d\right)' = \dot{\hat{\gamma}}_d,$$

$$\left(\dot{\hat{\gamma}}_d\right)' = \ddot{\hat{\gamma}}_d,$$

$$\left(\ddot{\hat{\gamma}}_d\right)' = \dddot{\hat{\gamma}}_d,$$

$$\left(\dddot{\hat{\gamma}}_d\right)' = R_2^4 \left[-a_{21} \arctan\left(\hat{\gamma}_d - \gamma_d\right) - a_{22} \arctan\left(\frac{\dot{\hat{\gamma}}_d}{R_2}\right) \right. \\ \left. - a_{23} \arctan\left(\frac{\ddot{\hat{\gamma}}_d}{R_1^2}\right) - a_{24} \arctan\left(\frac{\dddot{\hat{\gamma}}_d}{R_2^3}\right) \right],$$

(72)

where R_i, a_{ij} $(i = 1, 2; j = 1, 2, 3, 4) \in \mathbf{R}^+$ are design parameters; $\hat{z}_1$, $\hat{z}_2$, $\hat{z}_3$, and $\hat{z}_4$ represent the estimations of $z_1(\gamma)$, $z_2(\dot{\gamma})$, $z_3(\ddot{\gamma})$, and $z_4(\dddot{\gamma})$, respectively; and $\hat{\gamma}_d$, $\dot{\hat{\gamma}}_d$, $\ddot{\hat{\gamma}}_d$, and $\dddot{\hat{\gamma}}_d$ stand for the estimations of γ_d, $\dot{\gamma}_d$, $\ddot{\gamma}_d$, and $\dddot{\gamma}_d$.

4. Stability Analysis

Theorem 7. *For the velocity subsystem of AHV (17), considering the saturation characteristic of Φ (11), and adopting the control law (26) and the adaptive law (37) under the premise of Theorem 1, the closed-loop control system is semiglobally uniformly ultimately bounded.*

Proof 6. Define the estimation error of ω_V

$$\tilde{\omega}_V = \hat{\omega}_V - \omega_V.$$

(73)

Substitute (26), (27), (35), and (36) into (25)

$$\dot{\tilde{V}} = \Phi_0^* - \Phi_1^* + F_V^*(V, \Phi^*) - \dot{V}_{\text{ref}} \\ = -k_{V1}\tilde{V} - k_{V2}\int_0^t \tilde{V}d\tau - \frac{1}{2}\tilde{V}\hat{\omega}_V \mathbf{h}^{\text{T}}(\mathbf{X}_1)\mathbf{h}(\mathbf{X}_1) \\ + \mathbf{W}_V^{*\text{T}}\mathbf{h}(\mathbf{X}_1) + \varepsilon.$$

(74)

Choose the following Lyapunov function candidate:

$$L_V = \frac{\tilde{V}^2}{2} + \frac{k_{V2}}{2}\left(\int_0^t \tilde{V}d\tau\right)^2 + \frac{\tilde{\omega}_V^2}{2\mu_V}.$$

(75)

Taking time derivative of (75) and invoking (37), (73) and (74) yield

$$\dot{L}_V = \tilde{V}\dot{\tilde{V}} + k_{V2}\tilde{V}\int_0^t \tilde{V}d\tau + \frac{\tilde{\omega}_V\dot{\hat{\omega}}_V}{\mu_V} \\ = \tilde{V}\left[-k_{V1}\tilde{V} - k_{V2}\int_0^t \tilde{V}d\tau - \frac{1}{2}\tilde{V}\hat{\omega}_V \mathbf{h}^{\text{T}}(\mathbf{X}_1)\mathbf{h}(\mathbf{X}_1) + \mathbf{W}_V^{*\text{T}}\mathbf{h}(\mathbf{X}_1) + \varepsilon\right] \\ + k_{V2}\tilde{V}\int_0^t \tilde{V}d\tau + \frac{\tilde{\omega}_V}{\mu_V}\left(\frac{\mu_V}{2}\tilde{V}^2\mathbf{h}^{\text{T}}(\mathbf{X}_1)\mathbf{h}(\mathbf{X}_1) - 2\hat{\omega}_V\right) \\ = -k_{V1}\tilde{V}^2 - \frac{1}{2}\tilde{V}^2\omega_V\mathbf{h}^{\text{T}}(\mathbf{X}_1)\mathbf{h}(\mathbf{X}_1) + \tilde{V}\mathbf{W}_V^{*\text{T}}\mathbf{h}(\mathbf{X}_1) + \tilde{V}\varepsilon - \frac{2\tilde{\omega}_V\hat{\omega}_V}{\mu_V}.$$

(76)

Notice that

$$\tilde{V}\mathbf{W}_V^{*\text{T}}\mathbf{h}(\mathbf{X}_1) \le \frac{\tilde{V}^2}{2}\left\|\mathbf{W}_V^{*\text{T}}\mathbf{h}(\mathbf{X}_1)\right\|^2 + \frac{1}{2},$$

$$\tilde{V}\varepsilon \le \frac{\tilde{V}^2}{4} + \varepsilon_M^2, \quad 2\tilde{\omega}_V\hat{\omega}_V \ge \tilde{\omega}_V^2 - \omega_V^2.$$

(77)

Also from Cauchy-Schwarz inequality

$$\left\|\mathbf{W}_V^{*\text{T}}\mathbf{h}(\mathbf{X}_1)\right\| \le \left\|\mathbf{W}_V^*\right\|\left\|\mathbf{h}(\mathbf{X}_1)\right\|.$$

(78)

Further,

$$\tilde{V}\mathbf{W}_V^{*\text{T}}\mathbf{h}(\mathbf{X}_1) \le \frac{\tilde{V}^2}{2}\left\|\mathbf{W}_V^*\right\|^2\left\|\mathbf{h}(\mathbf{X}_1)\right\|^2 + \frac{1}{2} \\ = \frac{\tilde{V}^2}{2}\omega_V\mathbf{h}^{\text{T}}(\mathbf{X}_1)\mathbf{h}(\mathbf{X}_1) + \frac{1}{2}.$$

(79)

Then (76) becomes

$$\dot{L}_V \le -\left(k_{V1} - \frac{1}{4}\right)\tilde{V}^2 - \frac{\tilde{\omega}_V^2}{\mu_V} + \frac{1}{2} + \varepsilon_M^2 + \frac{\omega_V^2}{\mu_V}.$$

(80)

Let $k_{V1} > 1/4$, and define the following compact sets:

$$\Omega_{\tilde{V}} = \left\{ \tilde{V} \mid |\tilde{V}| \leq \sqrt{\frac{(1/2) + \varepsilon_M^2 + (\omega_V^2/\mu_V)}{(k_{V1} - (1/4))}} \right\},$$

$$\Omega_{\tilde{\omega}_V} = \left\{ \tilde{\omega}_V \mid |\tilde{\omega}_V| \leq \sqrt{\frac{(1/2) + \varepsilon_M^2 + (\omega_V^2/\mu_V)}{(1/\mu_V)}} \right\}. \tag{81}$$

If $\tilde{V} \notin \Omega_{\tilde{V}}$ or $\tilde{\omega}_V \notin \Omega_{\tilde{\omega}_V}$, then $\dot{L}_V < 0$. Therefore, the closed-loop control system is semiglobally uniformly ultimately bounded. Further, these error signals $\tilde{V}$ and $\tilde{\omega}_V$ are semiglobally uniformly ultimately bounded and can be invariant to the following sets $\Omega_{\tilde{V}}$ and $\Omega_{\tilde{\omega}_V}$. The radiuses of $\Omega_{\tilde{V}}$ and $\Omega_{\tilde{\omega}_V}$ can be made arbitrarily small by choosing k_{V1} that is big enough and μ_V that is small enough, and the tracking errors $\tilde{V}$ and $\tilde{\omega}_V$ can also be arbitrarily small.

Theorem 8. *For the altitude subsystem of AHV (51), considering the saturation characteristic of δ_e (11) and adopting the control law (59), the adaptive law (69) and FD (72) under the premise of Theorem 4, the closed-loop control system is semiglobally uniformly ultimately bounded.*

Proof 7. Define the estimation error of ω_h

$$\tilde{\omega}_h = \widehat{\omega}_h - \omega_h. \tag{82}$$

Then, define the estimation error of FD

$$\begin{aligned}
\chi_1 &= \widehat{z}_2 - z_2, \\
\chi_2 &= \widehat{z}_3 - z_3, \\
\chi_3 &= \dot{\widehat{\gamma}}_d - \dot{\gamma}_d, \\
\chi_4 &= \ddot{\widehat{\gamma}}_d - \ddot{\gamma}_d, \\
\chi_5 &= \dddot{\widehat{\gamma}}_d - \dddot{\gamma}_d.
\end{aligned} \tag{83}$$

According to Theorem 6, there is $\chi_{iM} \in \mathbf{R}^+$ $(i = 1, 2, \dots, 5)$

$$\chi_i \leq \chi_{iM}. \tag{84}$$

Substitute (59), (67), and (68) into (58)

$$\dot{\widehat{E}} = -k_{h1}\widehat{E} - \frac{1}{2}\widehat{E}\widehat{\omega}_h \mathbf{h}^T(\mathbf{X}_2)\mathbf{h}(\mathbf{X}_2) + \mathbf{W}_h^{*T}\mathbf{h}(\mathbf{X}_2) + \iota. \tag{85}$$

Choose the following Lyapunov function candidate:

$$L_h = \frac{1}{2}\widehat{E}^2 + \frac{\tilde{\omega}_h^2}{2\mu_h}. \tag{86}$$

Taking time derivative of (86) and invoke (69), (82), and (85) yield

$$\begin{aligned}
\dot{L}_h &= \widehat{E}\dot{\widehat{E}} + \frac{\tilde{\omega}_h \dot{\widehat{\omega}}_h}{\mu_h} \\
&= \widehat{E}\left[-k_{h1}\widehat{E} - \frac{1}{2}\widehat{E}\widehat{\omega}_h \mathbf{h}^T(\mathbf{X}_2)\mathbf{h}(\mathbf{X}_2) + \mathbf{W}_h^{*T}\mathbf{h}(\mathbf{X}_2) + \iota \right] \\
&\quad + \frac{\tilde{\omega}_h}{\mu_h}\left[\frac{\mu_h}{2}\widehat{E}^2 \mathbf{h}^T(\mathbf{X}_2)\mathbf{h}(\mathbf{X}_2) - 2\widehat{\omega}_h \right] \\
&= -k_{h1}\widehat{E}^2 - \frac{1}{2}\widehat{E}^2 \omega_h \mathbf{h}^T(\mathbf{X}_2)\mathbf{h}(\mathbf{X}_2) + \widehat{E}\mathbf{W}_h^{*T}\mathbf{h}(\mathbf{X}_2) \\
&\quad + \widehat{E}\iota - \frac{2\tilde{\omega}_h\widehat{\omega}_h}{\mu_h}.
\end{aligned} \tag{87}$$

Since

$$\begin{aligned}
\widehat{E}\mathbf{W}_h^{*T}\mathbf{h}(\mathbf{X}_2) &\leq \frac{\widehat{E}^2}{2}\left\|\mathbf{W}_h^{*T}\mathbf{h}(\mathbf{X}_2)\right\|^2 + \frac{1}{2}, \\
\frac{2\tilde{\omega}_h\widehat{\omega}_h}{\mu_h} &\geq \frac{\tilde{\omega}_h^2}{\mu_h} - \frac{\omega_h^2}{\mu_h}, \\
\widehat{E}\iota &\leq |\widehat{E}\iota| \leq \frac{\widehat{E}^2}{4} + \iota_M^2.
\end{aligned} \tag{88}$$

According to Cauchy-Schwarz inequality,

$$\left\|\mathbf{W}_h^{*T}\mathbf{h}(\mathbf{X}_2)\right\| \leq \|\mathbf{W}_h^*\|\|\mathbf{h}(\mathbf{X}_2)\|. \tag{89}$$

Further,

$$\begin{aligned}
\widehat{E}\mathbf{W}_h^{*T}\mathbf{h}(\mathbf{X}_2) &\leq \frac{\widehat{E}^2}{2}\|\mathbf{W}_h^*\|^2\|\mathbf{h}(\mathbf{X}_2)\|^2 + \frac{1}{2} \\
&= \frac{\widehat{E}^2}{2}\omega_h \mathbf{h}^T(\mathbf{X}_2)\mathbf{h}(\mathbf{X}_2) + \frac{1}{2}.
\end{aligned} \tag{90}$$

Then (87) becomes

$$\dot{L}_h \leq -\left(k_{h1} - \frac{1}{4}\right)\widehat{E}^2 - \frac{\tilde{\omega}_h^2}{\mu_h} + \frac{1}{2} + \iota_M^2 + \frac{\omega_h^2}{\mu_h}. \tag{91}$$

Let $k_{h1} > 1/4$, and define the following compact sets:

$$\Omega_{\widehat{E}} = \left\{ \widehat{E} \mid |\widehat{E}| \leq \sqrt{\frac{(1/2) + \iota_M^2 + (\omega_h^2/\mu_h)}{k_{h1} - (1/4)}} \right\},$$

$$\Omega_{\tilde{\omega}_h} = \left\{ \tilde{\omega}_h \mid |\tilde{\omega}_h| \leq \sqrt{\frac{(1/2) + \iota_M^2 + (\omega_h^2/\mu_h)}{1/\mu_h}} \right\}. \tag{92}$$

If $\widehat{E} \notin \Omega_{\widehat{E}}$ or $\tilde{\omega}_h \notin \Omega_{\tilde{\omega}_h}$, then $\dot{L}_h < 0$. Thus, these error signals $\widehat{E}$ and $\tilde{\omega}_h$ are semiglobally uniformly ultimately bounded and can be invariant to the following sets $\Omega_{\widehat{E}}$ and $\Omega_{\tilde{\omega}_h}$. The radiuses of $\Omega_{\widehat{E}}$ and $\Omega_{\tilde{\omega}_h}$ can be made arbitrarily

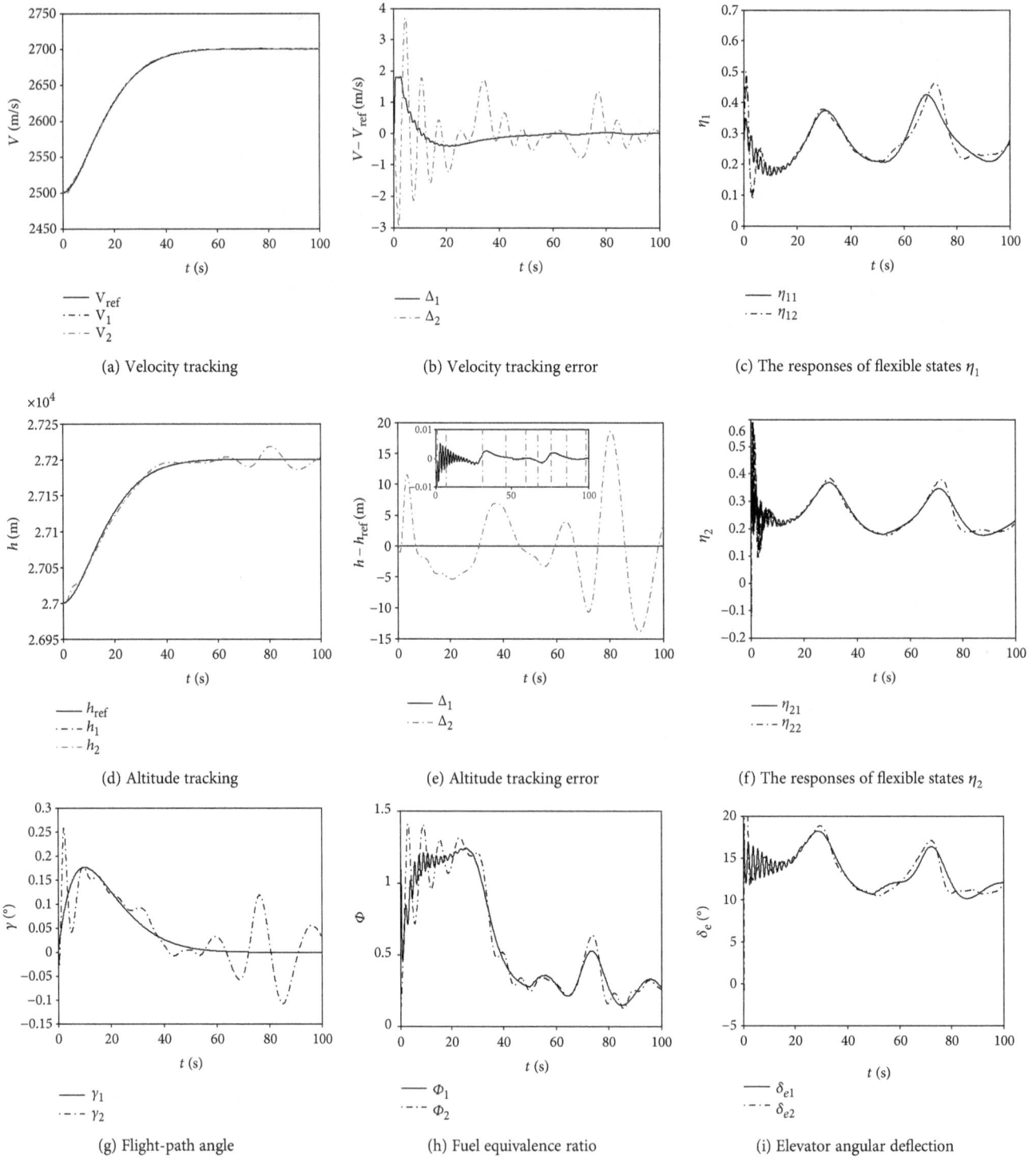

FIGURE 2: Simulation result of Case 1.

small by choosing big enough k_{h1} and small enough μ_h, and the tracking errors $\widehat{E}$ and $\tilde{\omega}_h$ can also be arbitrarily small.

Combine (53), (54), (55) and (56) and (83)

$$E = \widehat{E} + \left(z_3 - \widehat{z}_3 + \ddot{\widehat{\gamma}}_d - \ddot{\gamma}_d\right) + 3\lambda\left(z_2 - \widehat{z}_2 + \dot{\widehat{\gamma}}_d - \dot{\gamma}_d\right)$$
$$= \widehat{E} + \chi_4 - \chi_2 + 3\lambda(\chi_3 - \chi_1). \tag{93}$$

Consider (84), then (93) can become

$$E \leq \widehat{E} + \chi_{4M} + \chi_{2M} + 3\lambda(\chi_{3M} + \chi_{1M}). \tag{94}$$

Therefore, E and e are also bounded, then the closed-loop control system is semiglobally uniformly ultimately bounded.

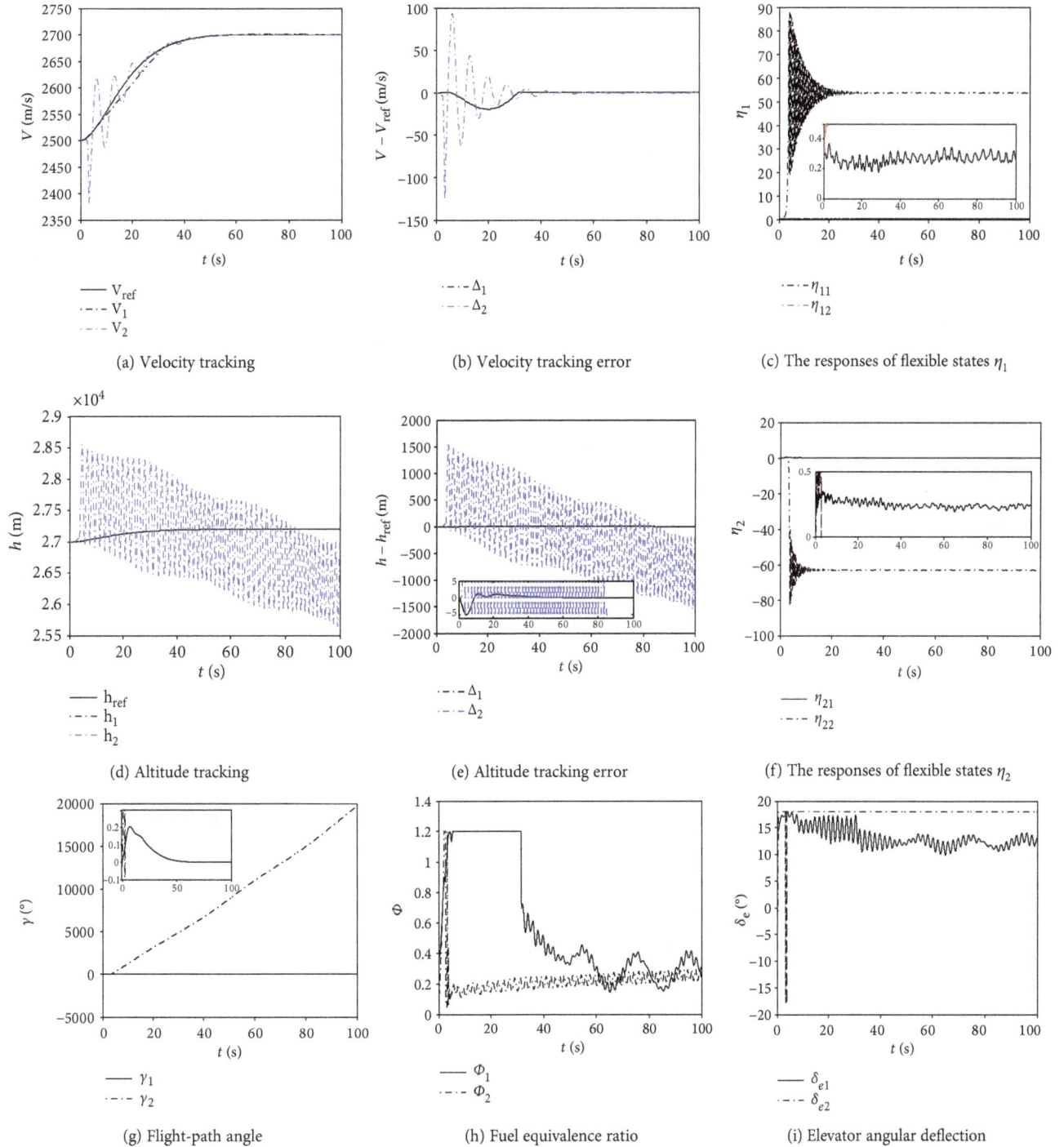

(a) Velocity tracking

(b) Velocity tracking error

(c) The responses of flexible states η_1

(d) Altitude tracking

(e) Altitude tracking error

(f) The responses of flexible states η_2

(g) Flight-path angle

(h) Fuel equivalence ratio

(i) Elevator angular deflection

FIGURE 3: Simulation result of Case 2.

5. Simulation Results

With AHV longitudinal motion model as the controlled object, the tracking simulation of velocity and altitude reference commands is performed. The initial velocity and altitude of AHV are taken as $V = 2500$ m/s and $h = 27000$ m. Velocity step and altitude step are chosen as $\Delta V = 200$ m/s and $\Delta h = 200$ m. Both the velocity and altitude reference inputs are given by a second-order reference model with a damping ratio of 0.9 and a natural frequency of 0.1 rad/s. The design parameters of the controller are chosen as $k_V = 5$, $k_{V1} = 8$, $k_{V2} = 0.01$, $k_{\gamma1} = 2$, $k_{\gamma2} = 0.01$, $k_h = 0.9$, $k_{h1} = 30$, and $\lambda = 7$. The design parameters of the adaptive law are taken as $\mu_V = \mu_h = 0.05$. The design parameters of FD are chosen as $R_1 = R_2 = 0.05$, $a_{11} = a_{13} = a_{21} = a_{23} = 0.5$, and $a_{12} = a_{14} = a_{22} = a_{24} = 0.1$. The node number of RBFNN is chosen as $N_1 = N_2 = 20$. In the velocity subsystem, the center c_1 of the Gaussian function is evenly spaced in [2500 m/s,

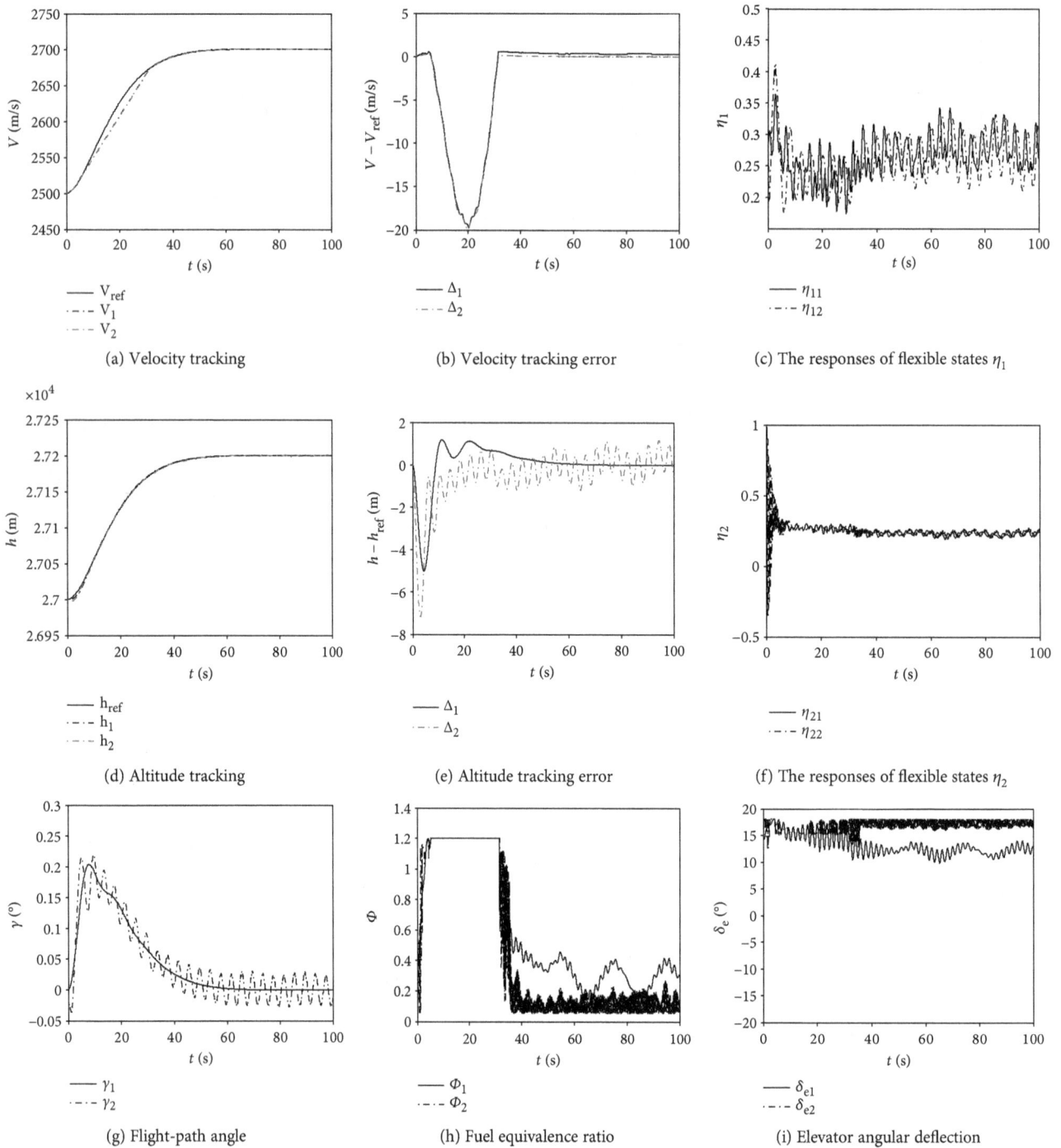

(a) Velocity tracking

(b) Velocity tracking error

(c) The responses of flexible states η_1

(d) Altitude tracking

(e) Altitude tracking error

(f) The responses of flexible states η_2

(g) Flight-path angle

(h) Fuel equivalence ratio

(i) Elevator angular deflection

FIGURE 4: Simulation result of Case 3.

3100 m/s] $\times$ [−0.1, 1], and the width $\mathbf{b}_1$ is selected as 6.56. In altitude subsystem, the center $\mathbf{c}_2$ of the Gaussian function is evenly spaced in [−1°, 1°] $\times$ [0°, 5°] $\times$ [−5°/s, 5°/s] $\times$ [0 rad, 0.35 rad]; the width $\mathbf{b}_2$ is selected as 0.01. Three cases are considered in the simulation study.

Remark 9. In the above stability analysis of the control law, in order to ensure the stability of the closed-loop control system, the range of values of related parameters (such as $k_{V1} > 1/4$, $k_{h1} > 1/4$) is given. At the same time, in order to

ensure tracking accuracy, the selection principle of related parameters is also analyzed (e.g., parameters k_{V1} and k_{h1} should be taken large enough, and parameters μ_V and μ_h should be taken small enough). The selection of parameters in the simulation is based on the above principles and is determined by repeated debugging.

Case 1. Considering Remark 1, we assume that actuators are constrained as $\Phi \in [0.05, 1.5]$ and $\delta_e \in [−20°, 20°]$. Simultaneously, to verify the robustness of the controller, it is assumed that there is a perturbation of ±40% for the aerodynamic coefficient of the AHV model, which is expressed as $C = C_0[1 + 0.4 \sin(0.1\pi t)]$, where C_0 represents the nominal value and C stands for the simulation value. And after 50 s, add external disturbances $d_1 = 2 \sin(0.1\pi t)$ and $d_2 = 0.02 \sin(0.1\pi t)$ to (1) and (5) in the AHV model, respectively. The subscript "1" in the figures shows the result obtained by the proposed method, and the subscript "2" shows the result obtained by the method of [27].

Case 2. Combining Remark 2, we further suppose that actuators are subject to stricter constraints, resulting in a smaller executable range than the theoretical value, which are taken as $\Phi \in [0.05, 1.2]$ and $\delta_e \in [-18°, 18°]$. Other simulation conditions are the same as Case 1.

Case 3. The control problem of nonlinear systems with actuator constraints has been considered extensively in the existing results, such as [28, 29]. In order to compare with the method in this paper, the adaptive dynamic surface control method considering actuator constraints proposed in [29] is applied to the AHV model. All conditions in the simulation are the same as in Case 2. The subscript "1" represents the result of the proposed method. The subscript "2" represents the result of the method proposed in [29].

When the control inputs are constrained as described in Case 1 (see Figures 2(h) and 2(i)), both the proposed method and method of [26] can guarantee that V and h track their reference commands V_{ref} and h_{ref} (see Figures 2(a) and 2(d)). However, the tracking accuracy and anti-interference ability of the proposed method in this paper are significantly improved than those of the method in [26] (see Figures 2(b) and 2(e)). At the same time, the changes in the flight-path angle and flexible states are also smoother than [26] (see Figures 2(c), 2(f), and 2(g)). When actuators of AHV are subject to stricter constraints (see Figures 3(h) and 3(i)), the stability of the closed-loop system cannot be guaranteed with the method of [26]. But through using the proposed method in this paper, the control objective can still be achieved (see Figures 3(a), 3(b), 3(d), and 3(e)). And the proposed method can effectively inhibit flexible states (see Figures 3(c) and 3(f)). In Case 3, the method proposed in [29] can guarantee that V and h track their reference commands when the actuators are constrained (see Figures 4(a) and 4(d)). However, since the method does not use the relevant means to estimate the uncertain dynamics of the model, the tracking error is significantly larger than the method proposed in this paper (see Figures 4(b) and 4(e)). At the same time, compared with the method in [29], the control curve obtained by the method is smoother (see Figures 4(h) and 4(i)), which is more suitable for engineering applications.

6. Conclusion

(1) An adaptive neural control law based on input saturation function is designed in this paper, which guarantees the stability of the closed-loop system and the boundedness of tracking error when the actuator reaches saturation.

(2) When the AHV model encounters perturbation of aerodynamic coefficient and external disturbances, the proposed method can still ensure good tracking accuracy, which proves that the algorithm has strong robustness.

(3) The controller designed in this paper can still effectively suppress the flexible states when the actuator is constrained. Simulation examples verify the effectiveness and superiority of the design method.

(4) In this paper, only the most common amplitude saturation problem of the actuator is considered, and the rate and bandwidth constraint problem is not discussed. These will serve as a further research direction.

Conflicts of Interest

The authors declare that they have no conflicts of interest.

Acknowledgments

This study was cosupported by the National Natural Science Foundation of China (Grant no. 61573374 and no. 61703421).

References

[1] E. A. Morelli, "Flight-test experiment design for characterizing stability and control of hypersonic vehicles," *Journal of Guidance, Control, and Dynamics*, vol. 32, no. 3, pp. 949–959, 2009.

[2] D. Preller and M. K. Smart, "Longitudinal control strategy for hypersonic accelerating vehicles," *Journal of Spacecraft and Rockets*, vol. 52, no. 3, pp. 993–999, 2015.

[3] L. G. Wu, H. An, J. X. Liu, and C. H. Wang, "Recent progress in control of air-breathing hypersonic vehicles," *Journal of Harbin Institute of Technology*, vol. 48, no. 10, pp. 1–16, 2016.

[4] H. B. Duan and P. Li, "Progress in control approaches for hypersonic vehicle," *Science China Technological Sciences*, vol. 55, no. 10, pp. 2965–2970, 2012.

[5] B. Xu and Z. K. Shi, "An overview on flight dynamics and control approaches for hypersonic vehicles," *Science China Information Sciences*, vol. 58, no. 7, pp. 1–19, 2015.

[6] F. Yang, R. Yuan, J. Yi, G. Fan, and X. Tan, "Direct adaptive type-2 fuzzy neural network control for a generic hypersonic flight vehicle," *Soft Computing*, vol. 17, no. 11, pp. 2053–2064, 2013.

[7] B. Xu, D. X. Gao, and S. X. Wang, "Adaptive neural control based on HGO for hypersonic flight vehicles," *Science China Information Sciences*, vol. 54, no. 3, pp. 511–520, 2011.

[8] B. Xu, D. Wang, F. Sun, and Z. Shi, "Direct neural discrete control of hypersonic flight vehicle," *Nonlinear Dynamics*, vol. 70, no. 1, pp. 269–278, 2012.

[9] X. Bu, X. Wu, Z. Ma, and R. Zhang, "Nonsingular direct neural control of air-breathing hypersonic vehicle via back-stepping," *Neurocomputing*, vol. 153, pp. 164–173, 2015.

[10] X. Bu, X. Wu, J. Huang, Z. Ma, and R. Zhang, "Minimal-learning-parameter based simplified adaptive neural back-stepping control of flexible air-breathing hypersonic vehicles without virtual controllers," *Neurocomputing*, vol. 175, Part A, pp. 816–825, 2016.

[11] X. Bu, X. Wu, R. Zhang, Z. Ma, and J. Huang, "A neural approximation-based novel back-stepping control scheme for air-breathing hypersonic vehicles with uncertain parameters," *Proceedings of the Institution of Mechanical Engineers, Part I: Journal of Systems and Control Engineering*, vol. 230, no. 3, pp. 231–243, 2016.

[12] C. Y. Sun, C. X. Mu, and Y. Yu, "Some control problems for near space hypersonic vehicles," *Acta Automatica Sinica*, vol. 39, no. 11, pp. 1901–1913, 2013.

[13] J. Li and B. Zuo, "Adaptive terminal sliding mode control for air-breathing hypersonic vehicles under control input constraints," *Acta Aeronautica et Astronautica Sinca*, vol. 33, no. 2, pp. 220–233, 2012.

[14] B. Xu, Z. Shi, C. Yang, and S. Wang, "Neural control of hypersonic flight vehicle model via time-scale decomposition with throttle setting constraint," *Nonlinear Dynamics*, vol. 73, no. 3, pp. 1849–1861, 2013.

[15] B. Xu, X. Huang, D. Wang, and F. Sun, "Dynamic surface control of constrained hypersonic flight models with parameter estimation and actuator compensation," *Asian Journal of Control*, vol. 16, no. 1, pp. 162–174, 2014.

[16] M. A. Bolender and D. B. Doman, "Nonlinear longitudinal dynamical model of an air-breathing hypersonic vehicle," *Journal of Spacecraft and Rockets*, vol. 44, no. 2, pp. 374–387, 2007.

[17] J. T. Parker, A. Serrani, S. Yurkovich, M. A. Bolender, and D. B. Doman, "Control-oriented modeling of an air-breathing hypersonic vehicle," *Journal of Guidance, Control, and Dynamics*, vol. 30, no. 3, pp. 856–869, 2007.

[18] X. Bu, X. Wu, Z. Ma, R. Zhang, and J. Huang, "Novel auxiliary error compensation design for the adaptive neural control of a constrained flexible air-breathing hypersonic vehicle," *Neurocomputing*, vol. 171, pp. 313–324, 2016.

[19] L. Fiorentini and A. Serrani, "Adaptive restricted trajectory tracking for a non-minimum phase hypersonic vehicle model," *Automatica*, vol. 48, no. 7, pp. 1248–1261, 2012.

[20] L. Fiorentini, *Nonlinear Adaptive Controller Design for Air-Breathing Hypersonic Vehicles*, The Ohio State University, Columbus, 2010.

[21] R. M. Sanner and J.-J. E. Slotine, "Gaussian networks for direct adaptive control," *IEEE Transactions on Neural Networks*, vol. 3, no. 6, pp. 837–863, 1992.

[22] A. J. Calise, N. Hovakimyan, and M. Idan, "Adaptive output feedback control of nonlinear systems using neural networks," *Automatica*, vol. 37, no. 8, pp. 1201–1211, 2001.

[23] J. H. Park, S. H. Huh, S. H. Kim, S. J. Seo, and G. T. Park, "Direct adaptive controller for nonaffine nonlinear systems using self-structuring neural networks," *IEEE Transactions on Neural Networks*, vol. 16, no. 2, pp. 414–422, 2005.

[24] S. S. Ge and C. Wang, "Adaptive NN control of uncertain nonlinear pure-feedback systems," *Automatica*, vol. 38, no. 4, pp. 671–682, 2002.

[25] X. Bu, "Guaranteeing prescribed output tracking performance for air-breathing hypersonic vehicles via non-affine back-stepping control design," *Nonlinear Dynamics*, vol. 91, no. 1, pp. 525–538, 2018.

[26] X. Wang, Z. Chen, and G. Yang, "Finite-time-convergent differentiator based on singular perturbation technique," *IEEE Transactions on Automatic Control*, vol. 52, no. 9, pp. 1731–1737, 2007.

[27] X. Bu, X. Wu, R. Zhang, Z. Ma, and J. Huang, "Tracking differentiator design for the robust backstepping control of a flexible air-breathing hypersonic vehicle," *Journal of the Franklin Institute*, vol. 352, no. 4, pp. 1739–1765, 2015.

[28] H. Wang, B. Chen, X. Liu, K. Liu, and C. Lin, "Robust adaptive fuzzy tracking control for pure-feedback stochastic nonlinear systems with input constraints," *IEEE Transactions on Cybernetics*, vol. 43, no. 6, pp. 2093–2104, 2013.

[29] L. Chen and Q. Wang, "Adaptive dynamic surface control for unknown pure feedback non-affine systems with multiple constraints," *Nonlinear Dynamics*, vol. 90, no. 2, pp. 1191–1207, 2017.

7

Modeling of Particle Trajectory and Erosion of Large Rotor Blades

Adel Ghenaiet

Faculty of Mechanical and Process Engineering, University of Sciences and Technology (USTHB), BP 32, El-Alia, Bab-Ezzouar, 16111 Algiers, Algeria

Correspondence should be addressed to Adel Ghenaiet; ag1964@yahoo.com

Academic Editor: Shaoping Wang

When operating in hostile environments, engines components are facing a serious problem of erosion, leading to a drastic drop in aerodynamic performance and life-cycle. This paper outlines the modeling and simulation of particle trajectory and erosion induced by sand particles. The governing equations of particle dynamics through the moving of large rotor blades are introduced and solved separately from the flow field by using our in-house particle tracking code based on the finite element method. As the locations of impacts are predicted, the erosion is assessed by semiempirical correlations in terms of impact conditions and particle and target surface characteristics. The results of these computations carried out for different concentrations of suspended dust (sand) cloud generated at takeoff conditions reveal the main areas of impacts with high rates of erosion seen over a large strip from the blade suction side, around the leading edge and the pressure side of blade. The assessment of the blade geometry deterioration reveals that the upper corner of blade suffers from an intense erosion wear.

1. Introduction

Aeroengines manipulating airflows laden by dust or volcanic ash suffer from extreme erosion wear, especially the front components. Inevitably, a great deal of dust, generated by a ground vortex that has the capability of picking up particles from the ground, enters the engine. In addition, the fine particles of dust cloud generated by rotor turning remain suspended for a considerable time, and when the blades move they are continually bombarded by the entrained particles [1]. In these circumstances, erosion is often seen on the leading edge as well as the aft of rotor blade body, thus leading to premature stall and producing a sudden change in torque and a rise in required power [2]. The role of erosion has been well recognized in turbomachinery applications, where the damage is evident in pitting on the blade leading edge and trailing edge and ensues in increased surface roughness [3]. Predominantly, in this context, erosion by solid particles and other particulates have shown detrimental effects on the aerodynamics of blades and life-cycle. One of the main effects of deterioration and fault is the modification of compressor and turbine performance maps [4] and subsequently the degradation of engine performance.

Sage and Tilly [5] were among the early researchers who attempted to quantify the erosion in turbomachinery, and since the past decades, this phenomenon has been the subject of many analytical and experimental investigations. Hussein and Tabakoff [6, 7] were the first to examine the role of particle material, density, and size on impacts locations and rebounds in 3D particle trajectory calculations through axial flow turbomachines. Later, Hamed and Fowler [8] demonstrated that, within twisted vanes, 3D particle trajectories are greatly influenced by the countering of hub and tip for different particle sizes. Hamed and Tabakoff [9] developed a methodology to predict blade surface erosion using the statistical impact data computed from the particle trajectory simulations and the correlations derived from erosion test results, which have been widely used in both axial and radial turbomachinery for automotive and gas turbine applications. Early studies on high-pressure compressors [10, 11] and fans [12] atmospheric erosive regimes indicated that rotor-blade erosion occurred over the outer 50% of blade span.

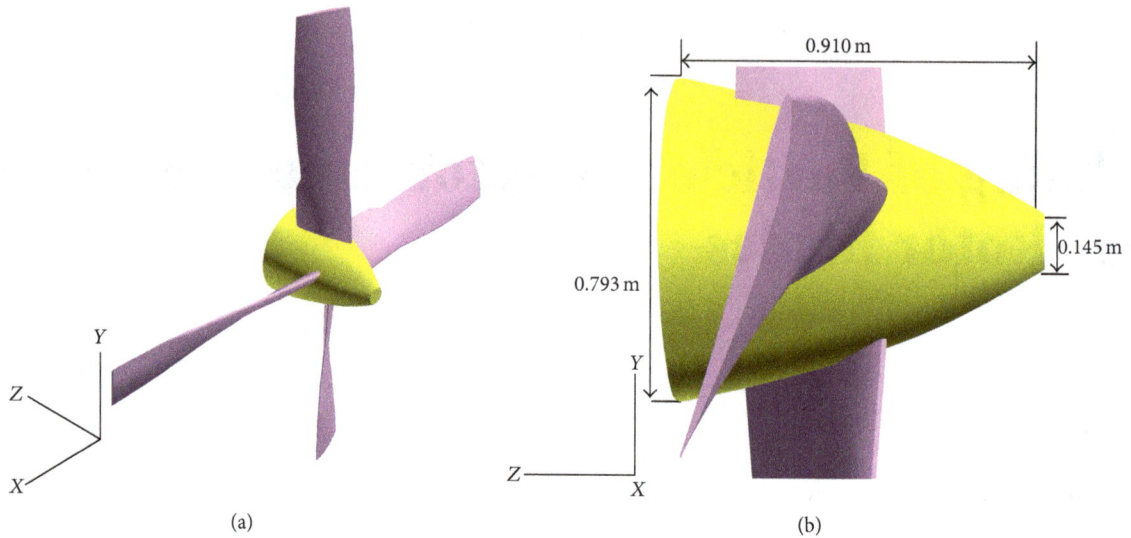

FIGURE 1: (a) Straight twisted four blades propeller. (b) Spinner and blades.

Balan and Tabakoff [11] studied experimentally the effects of sand ingestion on an axial flow compressor and found severe erosion of blade leading edge which increased the flow incidence that shifted the pressure rise coefficient and efficiency. Ghenaiet et al. [12, 13] found an increased tip to casing clearance and a reduction of blade chord in a high-speed axial fan following sand ingestion, and also noticed a 10% drop in efficiency and pressure rise coefficient due to blade leading edge blunting and erosive wear over the upper corner of blade. Hamed et al. [14] have reviewed in the detail the erosion and deposition in turbomachinery and the associated performance degradation of both aeronautical and ground based gas turbines. More recently, Ghenaiet [15] presented numerical results of erosion through the front compression stage of a turbofan, revealing the impact frequency, rates of erosion and critical areas of extreme erosion. Also, Schrade et al. [16] investigated numerically and experimentally the erosive change of a compressor blade shape based on different amounts of standardized Arizona test dust.

According to the literature, the study of erosion is still challenging due to numerous factors of complex interaction influencing the erosion process. These include flow conditions, geometry and material of blades, impact conditions, type of erodent, synergy of different erodents and exposure time. Moreover, most of the researches concentrated on compressors and turbines, but propellers practically received less interest. Therefore, further studies of particle dynamics and erosion facing open rotor blades (propellers), more susceptible to manipulate airflows laden by solid particles, are required to foresee the critical areas prone to erosion wear and subsequent blade shape deterioration. As noticed, hitherto there is no satisfactory protection for the leading edge even with stainless-steel coating [2].

This present study is a contribution to tackle the problem of erosion of large open rotor blades (propeller) at takeoff operation. For this purpose, our in-house particle tracking code [12, 13, 15] was adapted to this configuration of turbomachinery. The flow data were obtained separately and then followed the computations of particle trajectories and the determination of locations and conditions of impacts which served in evaluating the erosion rates and in the assessment of blade profile deterioration. The obtained results considering sand (quartz) particle (0–1000 mm) depict erosion wear that increases with particle concentration and reveal the main eroded areas along the fore of blade suction side and leading edge as well as the spinner.

2. Case Study

This study concerns a model of straight twisted propeller (Figure 1(a)) with large four blades, of a diameter equal to 4.117 m, made from aluminum alloy and operated by a single shaft gas turbine running at a constant speed of 13820 rpm via a gear box of a reduction ratio of 1/13.52. The overall length of spinner made from composite material (Figure 1(b)) is 0.91 m and its maximum diameter is 0.793 m. The 3D geometry was reconstituted by stacking different profiles from root to tip according to the measured blade twist. At the takeoff operation, the blade setting angle is equal to 25 deg measured between the blade chord and the tangential direction (at the reference radius $R75$), and the advance speed is equal to 74.6 m/s (145 knots).

3. Computational Domain

The computational domain (Figure 2) contains a rotating domain modeling one blade passage and a stationary domain consisting of extended inlet and outlet domains and a top domain over the propeller tip. At the pitchwise sides of these domains, the periodic boundary condition is used.

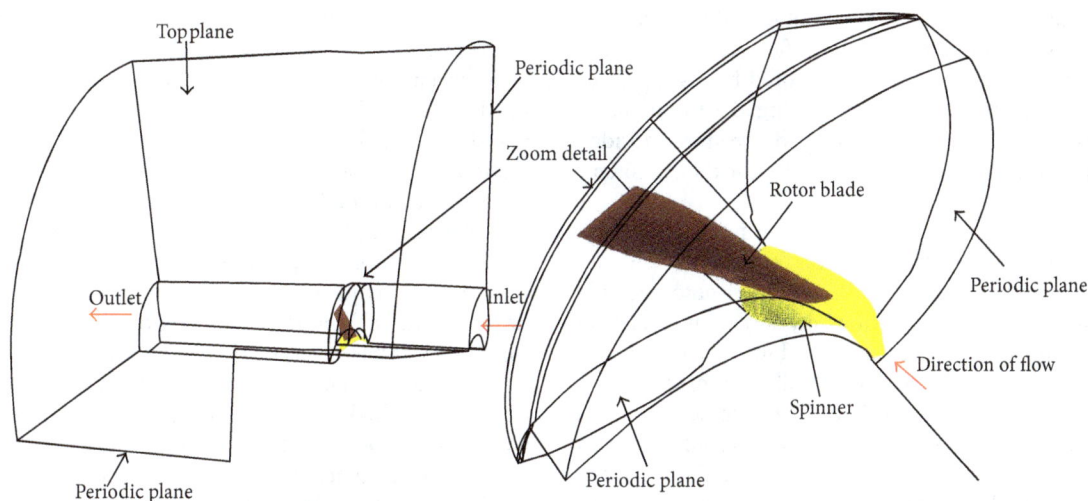

FIGURE 2: Full computational domain showing the rotor blade and the spinner and the boundary conditions.

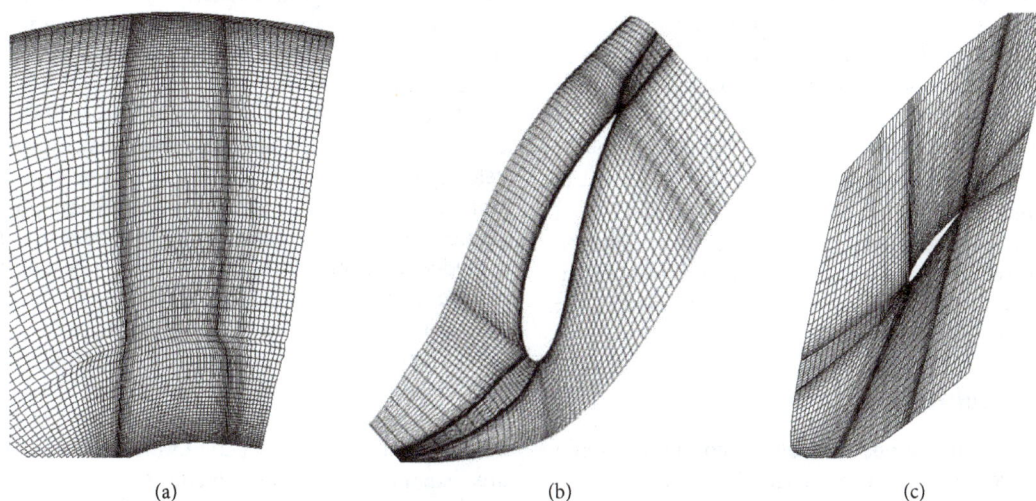

FIGURE 3: Grids of H type plotted from hub to tip meridional plane (a) and over hub and $R75$ planes (b, c).

Many of the difficulties encountered when applying CFD to an open rotor (a propeller) arise due to removal of casing existing in a conventional turbomachinery. The distances at upstream and downstream of the rotating domain are taken to sufficiently prevent any influence of the finite domain size on the aerodynamic performance of rotor. The mesh in the radial direction was sufficiently extended to capture the propeller stream-tube and to minimize the effects of the finite domain size. The locations of the rotating domain interfaces planes and the distances upstream and downstream of the spinner are critical and have a strong influence upon flow solution. Accordingly, the sizes of upstream, downstream, and top domains were examined to get a compromise with the particle trajectories computations that used the same grids. The distances upstream and downstream of the blade were taken as equal to $0.5c$ (c: axial mean chord), and the extended distances for inlet, outlet, and top were equal to $1.5D$, $2.5D$, and $3D$ (D: rotor diameter), respectively.

The rotating domain was meshed by CFX-TurboGrid, whereas the stationary domain used CFX-TASCgrid. Multi-block H type meshes were generated (Figure 3) with dense clustering around the blade in order to guarantee a better boundary layer resolution, in addition to the refinements near hub and tip. The transition from the rotating to stationary domain occurred with a smooth change of computation cells, to make sure that the conservative quantities are satisfied. As far as the quality of the meshes was concerned, some parameters have to be checked such as grid orthogonality characterized by a skew angle (between vertices) between 15 and 120 deg, minimum positive volumes, and cell aspect ratio not exceeding 100. The effect of computational domain mesh size was investigated for both the rotating domain and the stationary separately. To study the effect of propeller mesh on the propulsive efficiency, seven grids were used and the propulsive efficiency was assessed. The number of cells from one grid to another was scaled by a factor of 1.05 in all three directions. As a conclusion, about 450000 nodes were selected for the rotor domain. The influence of the mesh size of the stationary domain was explored by testing three different grids. Because of the computing time that might increase drastically for the third large grid size, it was decided to keep the first grid size of 350000 nodes for the stationary domain.

The adopted turbulence closure model for RANS equations is the k-ω model (Shear Stress Transport) with an automatic near-wall treatment [17]. This latter automatically switches from wall functions to a low-Reynolds near-wall formulation as the mesh is refined, but this cannot be guaranteed in most applications at all walls and some regions, for instance, blade junction to hub. This turbulence model was chosen for several reasons. First, the effects of free stream turbulence and surface roughness are easily included in the model [18]. Second, the transition can be calculated using the low-Reynolds number version of the model [19]. Third, Menter has shown that this model does well for flows with adverse pressure gradients [20, 21]. Finally, it should behave well numerically since it avoids using the distance to the wall and the complicated damping functions requiring a high computer power. Near walls, nodes were positioned in such a way that the value of $y^+ = \rho y_p u_t / \mu_f$, $u_t = V_f \sqrt{f/2}$ (V_f is flow velocity), and the friction factor is based on a flat plate $f = 0.025 \mathrm{Re}_x^{-0.1428}$. For an average value $y^+ = 40$, the last relation is simplified to $y_p = 357.77 c \mathrm{Re}_c^{-0.92}$, based on an average blade chord and used to position the near-wall first lines. In order to cover regions of high Reynolds number, the maximum flow velocity at the blade tip was used in the previous relation.

4. Flow Field Solution

Particle trajectory simulations through the components of a turbomachine are based on the numerical integration of particle equations of motion, which require the 3D flow solution of the RANS (Reynolds averaged Navier-Stokes) equations for turbulent flows by means of CFD (computational fluid dynamics) tools. In this study, the flow solution used the code CFX-TASCflow [17] which is a finite volume based solver. A sliding interface is used to connect two regions together (different frames of reference), in order to account for the change in the frame of reference and support steady-state predictions in the local frame. The interface sides must be a surface of revolution and sweep out the same surface of revolution and account internally for the pitch change by scaling up or down the local flows as they cross the interface. The boundary conditions used are periodic boundaries applied at one pitch of the blade, whereas symmetry is applied at the intersection planes joining on the axis of rotation. A rotating wall at the speed of 107 rd/s is selected for the blade and the spinner. A constant total pressure of 104541.4 Pa and temperature 288.15 K are applied at the inlet of the upstream domain, whereas at the outlet a static pressure of 101300 Pa is set at a single face near the top. A free stream velocity of 74.6 m/s is set over the top domain. At the inlet, the turbulence intensity was set at 5% and a value of turbulent viscosity μ_t was evaluated using Wilcox's model [18, 19].

The flow field within the blades is obtained in a relative frame and assumed to be steady. By adopting the approach of a frozen rotor, the stationary and rotating frames have fixed positions during the calculations. In the finite volume approach [17], all the diffusion terms are evaluated by summation of the derivatives of the shape function. The advection terms are computed by using a linear profile scheme (LPS) and a mass weighted skew upstream differencing scheme, which incorporate the physical advection correction. Initially the robust upwind scheme was used and then changed to a linear profile scheme for producing better results. The local time step was based on the tip blade chord and flow velocity. To ensure a good convergence, two criteria were considered: an RMS residual less than 10^{-5} and an imbalance in the conservation equations less than 10^{-2}, which are usually sufficient for an adequate convergence in most of the engineering applications. The RMS residual is the square root of the mean of all square residuals throughout the computational domain. The target imbalance for the conservation equations is used to ensure that the global balances are met, which is specified as the maximum fractional imbalance equal to the ratio of the subdomain imbalance and the maximum one over all subdomains.

As a result, Figure 4 plots the vectors of flow velocity near the blade hub, at $R75$ and blade tip, showing an increase in velocity with blade span. The relative flow velocity increases from the blade leading edge till a critical point over the suction side and afterwards tends to reduce toward the trailing edge. As noticed, the shape of spinner and its extended contour cause a flow blockage that increases the axial velocity and reduces the flow incidence onto the leading edge. At the takeoff operating regime, the relative Mach number is shown to increase along the blade span (Figure 5) to reach higher values beyond $R75$, but its value at blade tip does not exceed 0.712. Also, the relative Mach number contours depict a wake area expanding behind the blade. As the static pressure over the lower surface of blade is larger than at the upper one and the surrounding, the flow tends to curl as being forced from a high-pressure region just underneath the tip of blade to a low-pressure region. As a result, there is a spanwise component of the flow from the tip toward the root causing the flow streamlines at the upper surface to bend toward the root. Similarly, on the lower surface, there is a spanwise component of flow from root to tip forcing the streamlines to bend towards the tip. These two streams combine at trailing edge and the velocity difference causes the air to roll up into a vortex just inboard. This mechanism is observed in the formation of vortices (Figure 6) detaching from the blade tip to follow helicoidal flow paths in the same direction of blade rotation, which later dissipates with the downstream distance.

5. Particle Dynamics

The basis for particle trajectory simulations in turbomachinery continues to be Eulerian and Lagrangian with one-way coupling between the particles and flow. The Lagrangian approach considers the tracking of individual particles from different starting positions, thus giving an ability to handle in more detail some physical aspects, such as the interaction

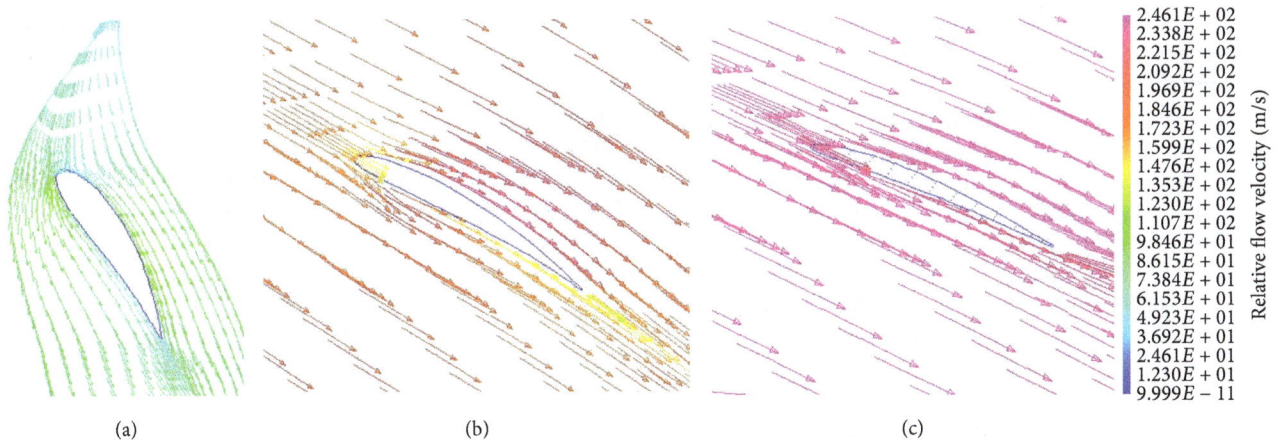

Figure 4: Relative flow velocity vectors: (a) near hub, (b) at $R75$, and (c) at blade tip.

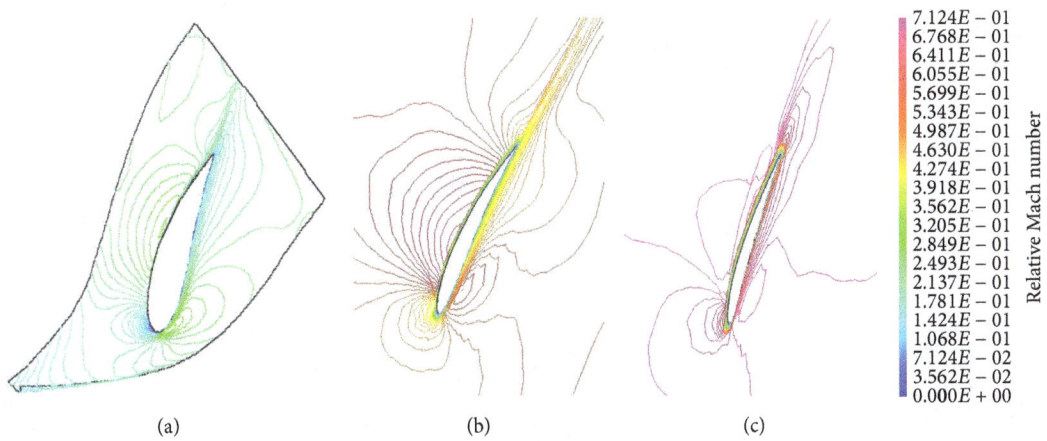

Figure 5: Relative Mach number (a) near hub, (b) at $R75$, and (c) at tip.

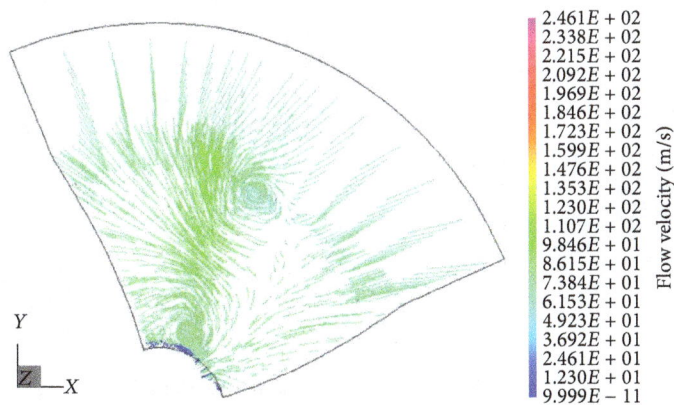

Figure 6: Counter-rotating vortex downstream of the rotor blade.

with walls. This approach is somewhat more economical since the effect of particle phase on the flow solution is neglected for very low volume fractions.

The particle trajectory equations are derived from the superposition of different involved forces, but in most particulate air flows the drag force is dominant. The general expression for the drag coefficient C_D is given (as below) according to Haider and Levenspiel [22], for Reynolds number from 0.01 to 2.6×10^5. For small Reynolds numbers (Re < 0.5), the viscous effect is dominating and this is referred to as the Stokes regime; $C_D = 24/\mathrm{Re}$. The constants A, B, C, and D depend on the particle shape:

$$\vec{F}_D = \frac{\pi}{8} d_p^2 \rho_f C_D \left(\overrightarrow{V_f} - \overrightarrow{V_p} \right) \left\| \vec{V}_f \right| - \left| \vec{V}_p \right\|,$$

$$C_D = \frac{24}{\mathrm{Re}_p} \left(1 + A\mathrm{Re}_p^B \right) + \frac{C}{\left(1 + D/\mathrm{Re}_p \right)}, \qquad (1)$$

$$\mathrm{Re}_p = \frac{\rho_f}{\mu_f} d_p \left| \vec{V}_f - \vec{V}_p \right|.$$

If a particle is sufficiently large and there is a large velocity gradient, there will be a lift force called Saffman force due to fluid shearing forces, which depends on the particle based Reynolds number and the shear flow Reynolds number [23] as follows:

$$\vec{F}_S = \frac{\pi \rho_f}{8} d_p^3 C_{LS} \left(\vec{V}_f - \vec{V}_p \right) \times \nabla \vec{V}_f,$$

$$C_{LS} = \frac{4.1126}{\sqrt{\mathrm{Re}_S}} f \left(\mathrm{Re}_p, \mathrm{Re}_S \right), \qquad (2)$$

$$\mathrm{Re}_S = \frac{\rho_f}{\mu_f} d_p^2 \left| \vec{\Omega}_f \right|.$$

The particle inertia forces, namely, the centrifugal and Coriolis forces, are derived from the second derivative of the vector position. By considering the drag and Saffman force as the main external forces, the following set of second-order nonlinear differential equations is derived:

$$\frac{\partial^2 r_p}{\partial t^2} = G\left(V_{fr} - V_{pr} \right) + r_p \left(\omega + \frac{V_{p\vartheta}}{r_p} \right)^2 + \frac{F_{Sr}}{m_p},$$

$$\frac{\partial^2 \theta_p}{\partial t^2} = G\frac{\left(V_{f\vartheta} - V_{p\vartheta} \right)}{r_p} - 2\frac{V_{pr}}{r_p} \left(\omega + \frac{V_{p\vartheta}}{r_p} \right) + \frac{F_{S\vartheta}}{r_p m_p},$$

$$\frac{\partial^2 z_p}{\partial t^2} = G\left(V_{fz} - V_{pz} \right) + \frac{F_{Sz}}{m_p}, \qquad (3)$$

$$G = \frac{3}{4 d_p} \frac{\rho_f}{\rho_p}$$

$$\cdot C_D \sqrt{\left| V_{fr} - V_{pr} \right|^2 + \left| V_{f\vartheta} - V_{p\vartheta} \right|^2 + \left| V_{fz} - V_{pz} \right|^2}.$$

Turbulence Effect. The fluctuating components corresponding to a particular eddy are obtained from the local turbulence properties. Turbulence effect is assumed to prevail as long as the particle-eddy interaction time (minimum between eddy lifetime and transit time, Brown and Hutchinson [24]) is less than the eddy lifetime. The eddy lifetime and dissipation length scale are estimated according to Shirolkar et al. [25]:

$$t_e = A\frac{k}{\varepsilon},$$

$$l_e = \frac{Bk^{3/2}}{\varepsilon}, \qquad (4)$$

$$\frac{A}{B} = \sqrt{1.5}.$$

The transit time scale is estimated from a linearized form of equation of motion given by Gosman and Ionnides [26]:

$$t_r = -\tau_p \mathrm{Ln} \left[1 - \frac{l_e}{\tau_p \left\| \overrightarrow{V_f} \right| - \left| \overrightarrow{V_p} \right\|} \right], \qquad (5)$$

$$\tau_p = \frac{24 \rho_p d_p^2}{18 \mu_f C_D \mathrm{Re}_p}.$$

Boundary Conditions. The boundary conditions implemented in the particle trajectory code include the interface between stationary frame and rotating frame, where the vector of particle velocity is decomposed into a tangential and a relative velocity. The tangential component is modified by adding the circumferential velocity, while the other components remain the same. At the periodical lateral sides, the velocity vector components are conserved while the tangential coordinate is modified by the angular shift. At the axis of revolution (condition of symmetry), for the velocity vector, a pure reflection is used. The impact physics is required to trace the particle trajectory after an impact with a surface, where the restitution coefficients (a measure of the kinetic energy exchange upon impact) are used to define variations in the magnitude and direction of a particle velocity. The impact angle β_1 is defined as the angle formed between the vector velocity of a particle at an impact point and its tangential component. In the course of particle trajectory computation, the restitution coefficients are estimated statistically based on the experimental values of mean and standard deviation (Tabakoff et al. [27]), given as polynomial regressions:

$$\frac{V_{P2}}{V_{P1}} = \sum_{i=0}^{4} a_i \beta_i^i,$$

$$\frac{\beta_{P2}}{\beta_{P1}} = \sum_{i=0}^{4} b_i \beta_i^i. \qquad (6)$$

Particle Distribution. The particle mass rate was obtained by multiplying the inlet air volume flow rate at the inlet by the mean value of particle concentration. The number and sizes of sand particles depend on the cumulative distribution curve 0–1000 μm (of mean diameter and variance equal to 237.4 μm and 164.5 μm, resp.) and the specified concentration profile. In the present simulations, particle concentration was varied in between 10 and 500 mg/m^3. As sand particles were released randomly with size distribution from 0 to 1000 μm, an iterative procedure, involving particle number and size in addition to the radial and circumferential seeding positions, was repeated until a convergence in the total mass of particles.

Solving Procedure. The flow data at the grids nodes were used to interpolate for the local flow properties, in the course of particle trajectory integration based on the Runge-Kutta-Fehlberg seventh-order technique. The integration time step depends on the computational cells sizes and the local flow velocities. However, this time step is reduced within the integration procedure to keep the leading truncation error within the tolerance. If a particle interacts with an eddy, the interaction time is considered as the effective time step. Near a boundary condition, a more accurate time step is reevaluated to get an impact within a half-diameter distance. After an impact, the restitution coefficients are used and the derived particle fragmentation factor (based on experiments by Tan and Elder [28]) is considered.

The particle tracking algorithm is based on the finite element method which requires transforming a particle position into its local coordinates by solving nonlinear equations (7). If these values do not exceed unity, that means the particle is still in the same cell; otherwise the algorithm starts to check all surrounding cells and the cell increment for a specified directrix is used to update the exact cell:

$$x_p = \sum_{i=1}^{8} N_i x_i,$$

$$y_p = \sum_{i=1}^{8} N_i y_i, \qquad (7)$$

$$z_p = \sum_{i=1}^{8} N_i z_i.$$

6. Erosion Assessment

This later depends on the physical properties of target surface and erodent, particle concentration and size, and the velocity and angle of impact. Finnie [29] attempted to develop the basic theoretical analysis of sand erosion based on Hertz's contact theory, but his model did not exactly predict the weight loss for high impact angles. Bitter suggested the mechanism of sand erosion which consists of deformation wear and cutting wear [30]. Bitter's model that accounts for erosion at all impact angles gives the sufficient prediction for both of ductile and brittle materials, but it is too complex. Neilson and Gilchrist [31] modified Bitter's model by assuming that the total erosion is an arithmetic combination of brittle and ductile contributions, and thus erosion loss can be predicted at intermediate impact angles. However, the resulting

equations are still complex as requiring experimentally determined parameters. The most successful erosion prediction model used in turbomachinery was developed by Grant and Tabakoff [32, 33], based on erosion measurements of 2024 aluminum alloy at particle speeds of 61–183 m/s and sand (quartz) as abrasive particles (20–200 micron). This latter is implementing two mechanisms, one predominant at low impact angle and another at high impact angle, as follows:

$$E = K_1 f\left(\beta_1\right) V_{P1}^2 \cos^2 \beta_1 \left(1 - R_\theta^2\right) + K_3 \left(V_{P1} \sin \beta_1\right)^4, \qquad (8)$$

$$f\left(\beta_1\right) = 1 + CK\left(K_2 \sin \frac{90}{\beta_o} \beta_1\right),$$

$$CK = \begin{cases} 1 & \beta_1 \le \beta_o \\ 0 & \beta_1 > \beta_o. \end{cases} \qquad (9)$$

Erosion rate is expressed as the amount (milligram) of material removed per unit of mass (gram) of impacting particles. R_θ is the tangential restitution factor derived from (6); the unit of velocity should be in ft/s. The material constants $K_1 = 3.67 \times 10^{-6}$, $K_2 = 0.585$, and $K_3 = 6 \times 10^{-12}$ and the angle of maximum erosion $\beta_o = 20$ deg are available for aluminum based alloy.

It was shown by many authors that the erosion damage increases with particle size up to some plateau value. Further, the influence of particle size on erosion is more pronounced at higher particle velocities [32]. The effect of particle size was included to correct the predicted erosion rate based on semiempirical correlation (8). It should be noted that, for the spinner made from a composite material, the specific restitution coefficient and erosion correlation were adopted from the experiments due to Drensky et al. [34].

The local values of mass erosion (milligram) are calculated from the local erosion rates and then cumulated over a given mesh element surface A_e in order to compute the equivalent erosion rate expressed in mg/g/cm^2. This latter was interpolated for the same node sharing the elements faces to get the nodal value distribution, which served to plot the contours of equivalent erosion rates:

$$E_{eq} = \frac{1}{A_e \sum_1^{N_p} m_{p_i}} \sum_1^{N_p} m_{p_i} E_i. \qquad (10)$$

The depth of penetration per a unit of time for an element face is calculated from the cumulative mass erosion of the given face. For the elements sharing the same node, the nodal values of penetration are estimated based on bilinear interpolation. As a result, the new coordinates are evaluated by knowing the normal vector and the depth of penetration at the given node. The assessment of blade geometry deterioration in terms of percentages of averaged reduction in blade

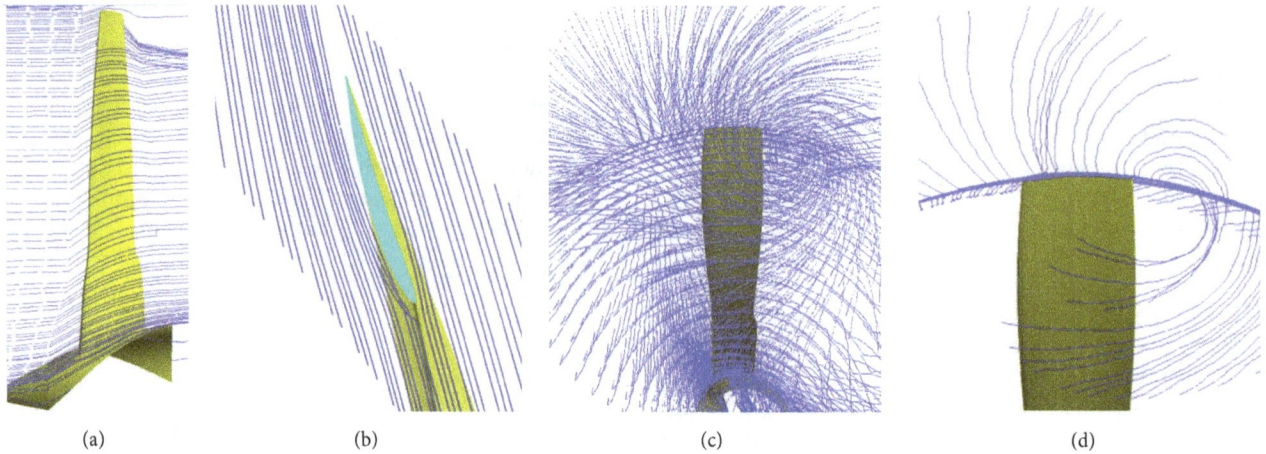

FIGURE 7: Sample of sand particle (10 μm) trajectories: (a) streamwise from hub to shroud, (b) near the tip, (c) downstream of rotor blade, and (d) downstream of rotor blade reflecting particles crossing over the blade tip.

chord and blade thickness is based on the whole blade length as follows:

(i) Average reduction in blade chord:

$$\Delta c_{av} (\%) = 100 \left[1 - \frac{1}{c_{av} h} \sum_{j=1}^{m} \delta c_j \cdot \delta h_j \right]. \qquad (11)$$

(ii) Average reduction in blade thickness:

$$\Delta e_{av} (\%) = 100 \left[1 - \frac{1}{e_{av} h} \sum_{j=1}^{m} \frac{1}{c_j} \sum_{i=1}^{n} \delta e \cdot \delta c_i \right]. \qquad (12)$$

7. Results and Discussion

The air and particles experience different degrees of turning through the rotor blade depending on their sizes. The deviations from air flow paths increase with particle inertia to provoke repeated impacts with the various surfaces of rotor blade. Figure 7 shows samples of simulated particle trajectories corresponding to small size particles, for instance, 10 microns, which tend to follow the flow paths closely (Figure 7(a)) along the blade passage from hub to shroud. Many of them are shown to collide with the spinner with repeated impacts. At outer radii, these particles are deviated downward owing to the rotational flow structure emanating from the tip and also because of moving from a rotating frame to a stationary one. The other view around the blade near tip (Figure 7(b)) depicts small particles circulating around the blade, and many of them impact around the leading edge and bounce closely to hit another time the suction side. Due to the formation of flow vortices (Figure 6) downstream of the blade, these tiny particles tend to follow a helicoidal trajectory (Figure 7(c)) and are strongly influenced by the flow turbulence. Detailed Figure 7(d) describes the trajectories of particles arriving toward the tip of blade, showing particles being entrained by the vortex, and many others are centrifuged outward from the blade tip.

When inertia of particles becomes more important, such as the case of 250 microns, they deviate considerably from the flow streamlines as shown in Figure 8. Many of these large size particles impact spinner wall (Figure 8(a)) and hence are deflected upward to follow ballistic trajectories. Also, owing to high centrifugal forces imparted, these particles are seen to deviate towards outer radii. At exit from rotor, and because of changing from a rotating frame to a stationary one, the large particles are heavily deviated tangentially. As seen in Figure 8(b), after hitting the fore part of suction side, these particles are deflected and follow bowed trajectories. Detailed Figure 8(c) depicts trajectories of large sand particles when arriving around the outer blade radius; they are shown to be entrained over the blade tip and receive high centrifugation projecting them upward from the rotor. Even for these heavy particles, some of them are being entrained by the vortex causing them to deviate downward (Figure 8(c)) after crossing the rotor.

Figure 9 shows that all particle trajectories related to size distribution (0–1000 μm) released upstream of the rotor combine all the features related to small and large size particles. Large particles are shown to deviate upwards due to high imparted centrifugal forces as compared to the drag force. These particles are accelerated in the axial direction and then impact the spinner to be largely deflected. Sand particles after crossing the rotor blade are deviated considerably towards the tangential direction and are entrained by the vortex structure behind the rotor blade.

Figure 10 shows impacts induced by a reduced sample (for the sake of plots clarity) of sand particles of a random size distribution (0–1000 μm). As revealed, the region of maximum impact frequency and erosion rate appears on the hub portion and the spinner, in addition to a large band at the front suction side and the back of pressure side. Figure 10(a) shows that a large number of particles tend to impact around the leading edge of propeller blade, and many others hit the suction side over a large strip from the leading edge and bounce to reach the pressure side towards the mid of trailing

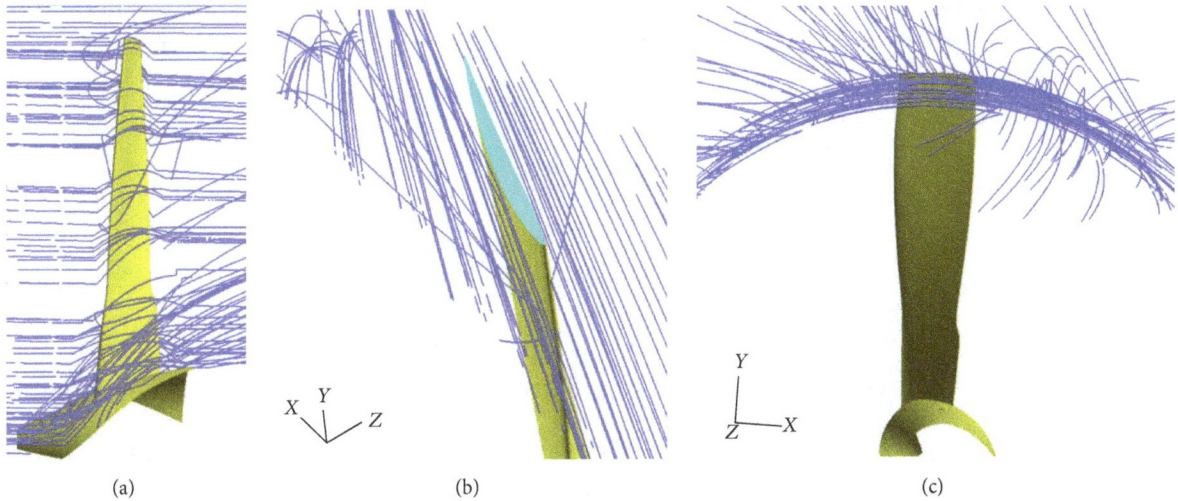

FIGURE 8: Sample of sand particle (250 μm) trajectories: (a) streamwise from hub to shroud, (b) near the tip, and (c) downstream of rotor blade reflecting particles crossing over the blade tip.

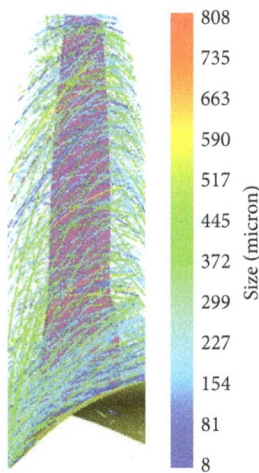

FIGURE 9: Sample of sand (0–1000 μm) particles trajectories crossing the rotor blade.

edge. Also, some particles are shown to impact around the trailing edge from suction side. Impacts on spinner are concentrated on the fore part, with a large crowd of particles seen along suction side at blade junction and around leading edge from both sides. Large impact velocities (Figures 10(b) and 11(b)) reaching a velocity of 223.5 m/s are predicted above the upper part and around the leading edge from both sides, owing to high flow velocities and rotation effect. As consequence, the high local rates of erosion are observed over the front part of suction side and around the leading edge from both sides (Figures 10(c) and 11(c)). The critical area of pronounced erosion is seen over the tip corner. Also there are regions of impacts towards the trailing edge from both sides.

The impacts frequency, the velocities, and angles of impacts and their distribution over the impacted surfaces

strongly affect the erosion patterns, estimated by semiempirical equation (8). The values of mass erosion (mg) at discrete points of an element face are calculated based on the local erosion rates in mg/g and cumulated to compute the equivalent erosion levels in mg/g/cm^2. Figures 12 and 13 depict the equivalent erosion rates corresponding to a low concentration (10 mg/m^3) and a high concentration (500 mg/m^3) of sand particles. All figures show noticeable erosion wear of the blade leading edge from the root to tip related to particles arriving at high velocity and also because of spanwise twist of the rotor blade that changed effectively the angle of particle incidence. The areas of erosion wear are seen to expand towards the tip in a triangular pattern with a spot area over the tip corner. On the blade pressure side, erosion is seen from the leading edge, in addition to a region of low erosion rates towards the trailing edge. On the spinner, there is a large spread of erosion but with low rates. Practically, similar patterns of erosion are obtained for each concentration, but the local mass erosion rates differ. As expected for this large rotor blade, the actual equivalent rates of erosion are lower in comparison with, for instance, the front stages of an axial compressor, according to the numerical results obtained by Ghenaiet [15], and this because of moderate flow velocity related to low rotational speed and the relatively large blade size.

The estimated mass erosion and blade geometry deterioration after one hour of sand ingestion are presented in Table 1. It is clear that the material removal and geometry changes are related to the size and mass (concentration) of particles impacting the blade surfaces and duration of exposure. The eroded mass from the rotor blade and spinner (Figure 14) and the subsequent geometry deteriorations estimated as percentages reductions of chord, thickness, and blade length are plotted in Figure 15, depicting an increase with concentration. The erosion wear is manifested by a drastic reduction in blade chord towards the tip and

TABLE 1: Erosion parameters and geometry deterioration of one blade after one hour.

| Erosion parameters | Different concentrations (mg/m^3) | | | | | | |
| | sand (0–1000 μm) | | | | | | |
	10	25	50	100	250	400	500
Sand particle rate (mg/s)	2.53	6.32	12.66	25.28	63.25	101.24	126.59
Ingested mass of sand (g)	9118.12	22751.19	45582.79	91002.2	227689.9	364483.2	455739.2
Erosion of blade (mg)	321.67	793.76	1568.49	3179.50	7820.65	12659.03	15734.07
Erosion of spinner (mg)	15.95	39.54	82.52	163.65	413.26	662.18	823.61
Reduction of chord at $R75$ (%)	$2.197E-03$	$6.34E-03$	$7.87E-03$	$1.62E-02$	$6.00E-02$	0.1029	0.1183
Reduction of chord at tip (%)	$4.449E-03$	$8.02E-03$	$1.28E-02$	$2.31E-02$	$7.62E-02$	0.1111	0.1503
Average reduction of $R75$ blade thickness (%)	$1.344E-02$	$4.774E-02$	$8.638E-02$	$1.636E-01$	$3.915E-01$	$6.485E-01$	$8.130E-01$
Average reduction of tip blade thickness (%)	$3.621E-02$	$7.169E-02$	$1.335E-01$	$3.766E-01$	1.365	1.954	2.321
Decrease of blade length (%)	$1.96E-05$	$2.37E-05$	$3.04E-05$	$5.57E-05$	$1.40E-04$	$2.071E-04$	$2.38E-04$

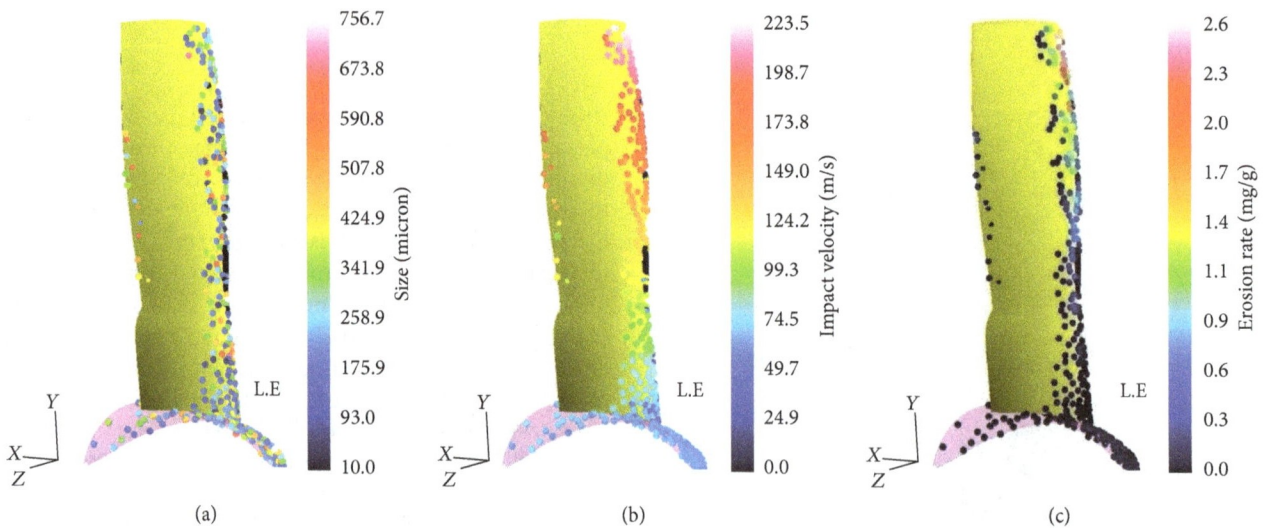

FIGURE 10: Sample of local impacts on suction side: (a) particles sizes (μm), (b) impact velocities (m/s), and (c) erosion rates (mg/g).

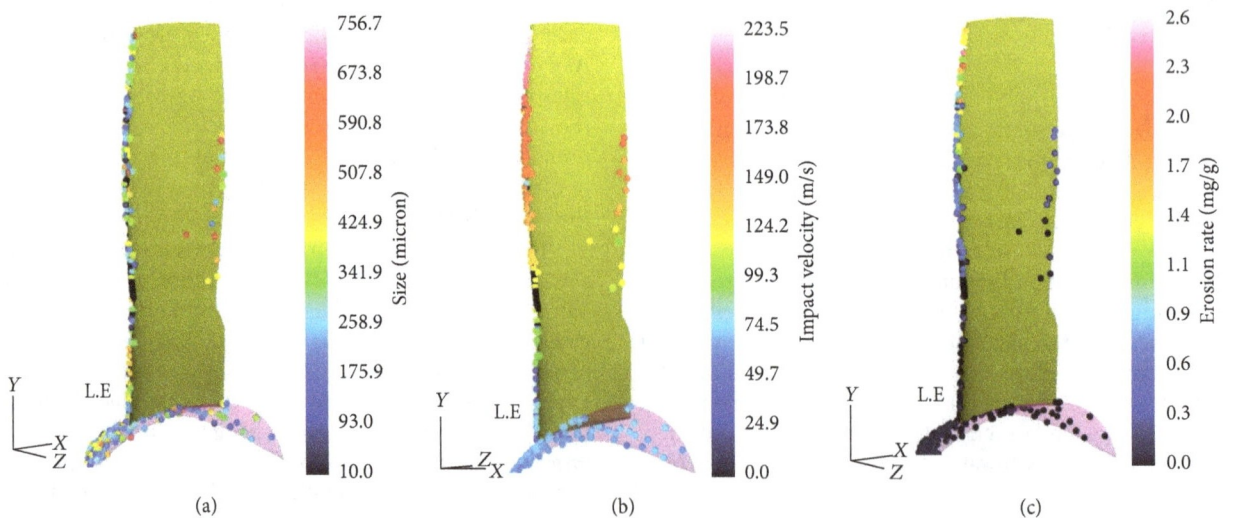

FIGURE 11: Sample of local impacts on pressure side: (a) particles sizes (μm), (b) impact velocities (m/s), and (c) erosion rates (mg/g).

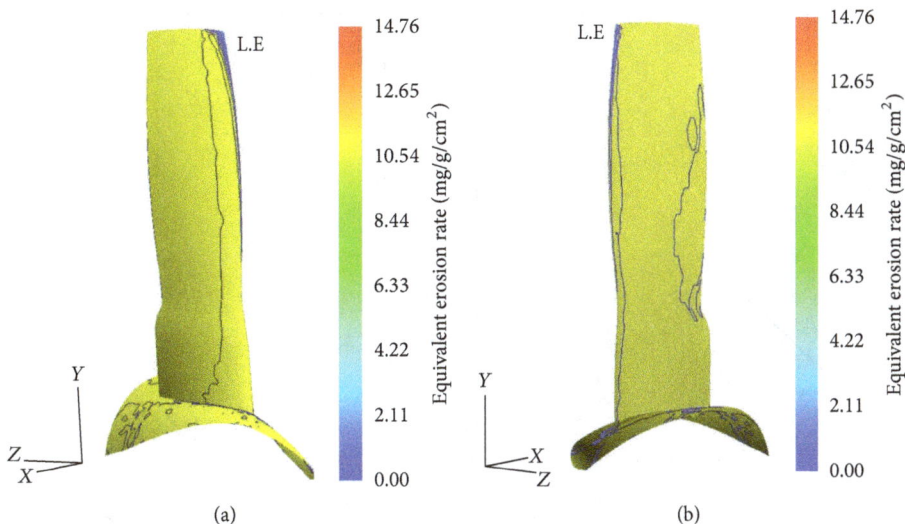

FIGURE 12: Equivalent erosion rates (mg/g/cm^2) at low concentration (10 mg/m^3): (a) Suction side and (b) pressure side.

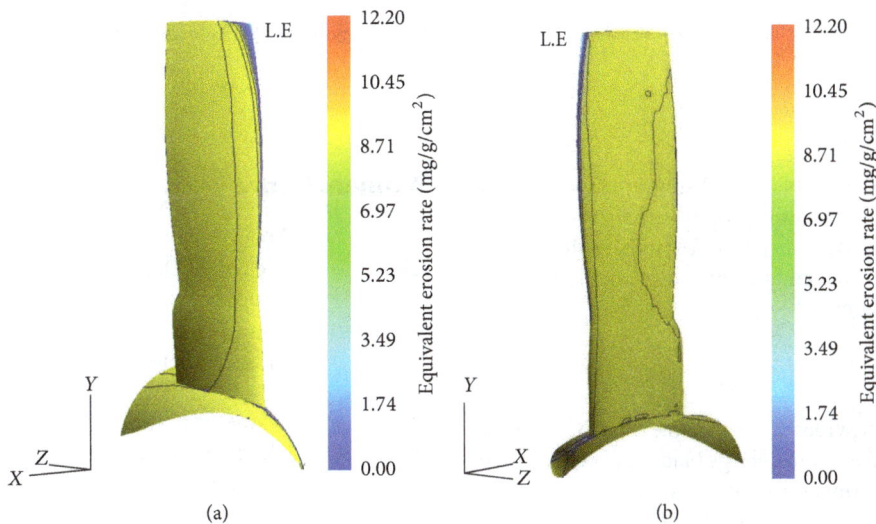

FIGURE 13: Equivalent erosion rates (mg/g/cm^2) at high concentration (500 mg/m^3): (a) Suction side and (b) pressure side.

a distortion (blunt) of leading edge, which are generally the main sources in the aerodynamic performance degradation. Accordingly, the mass erosion of one blade increases with particle concentration (between 10 and 500 mg/m^3) from 321.67 mg to 15734.07 mg, as illustrated in Figure 14(a). Also, the mass erosion of one quarter of spinner increases with particle concentration (between 10 and 500 mg/m^3) from 15.95 mg to 823.61 mg, but erosion is less important compared to rotor, which is well illustrated in Figure 14(b). The overall erosion rate is almost constant and has an average value of 3.47×10^{-2} mg/g. According to Table 1 and Figure 15(a), the reduction in blade length is from $1.96 \times 10^{-5}\%$ to $2.38 \times 10^{-4}\%$, after one hour of sand ingestion for concentration between 10 and 500 mg/m^3. As seen in Figure 15(b), the blade chord at radius R75 is reduced from $2.197 \times 10^{-3}\%$ to 0.1183%, whereas

that at tip is from $4.449 \times 10^{-3}\%$ to 0.1503%. Moreover, the thickness at radius R75 is reduced from $1.344 \times 10^{-2}\%$ to 0.8130%, whereas that at tip is from $3.621 \times 10^{-2}\%$ to 3.321%, as seen in Figure 15(c).

As exhibited in Figure 16, the erosion phenomenon is shown to distort the leading edge and reduce the chord and thickness of blade, progressively beyond 40% span till the tip. The process of erosion wear is clearer over the fore part of the suction side for all blade sections and the thickness is subsequently reduced and even more towards the tip section. Erosion damage is especially evident on the outboard sections of the rotor blades, where the tip approaches the sonic speed [1]. In addition, the distortion at leading edge affects the flow circulation and moves the stagnation point, which should alter the static pressure distribution and the aerodynamic

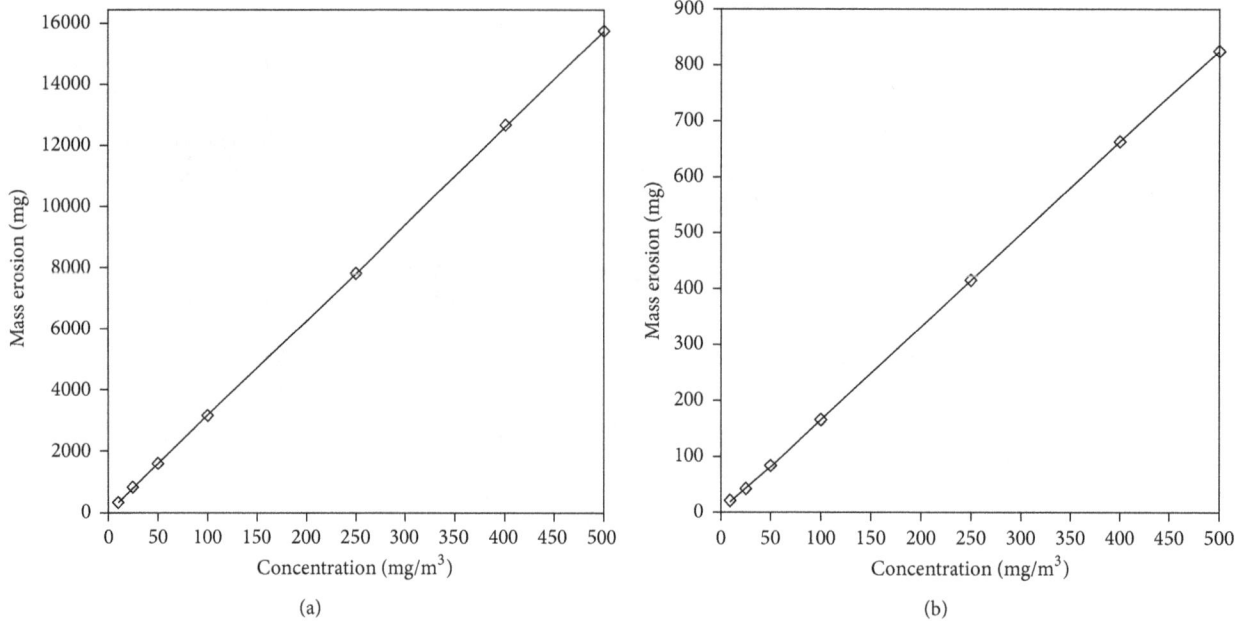

FIGURE 14: Mass erosion (mg) after one hour of sand ingestion: (a) blade and (b) spinner.

quality of the upper part of blade and subsequently the produced torque. As the maximum propeller thrust is obtained from 40% to 80% of blade span ($R40$–$R80$), there should be a subsequent drop in torque and performance coefficient to be evaluated in a next paper.

8. Conclusion

The present paper provides a reasonable insight into the particle dynamics and erosion of large blades propellers when exposed to different concentrations of sand particles. Tiny particles are shown to circulate around the rotor blade and stick to the flow path and also impact around the leading edge. Downstream of blade, these particles are deviated tangentially and follow helicoidal trajectories. When inertia of particles becomes more important, they deviate considerably from the flow path as a result of high drag and centrifugation. Many of these large particles impact the spinner and deflect to follow ballistic trajectories. At exit from rotor, they are deviated tangentially and projected upward, and many of them are entrained by the vortex and deviated downward. Multiple impacts are predicted around the blade leading edge due to its direct exposure to incoming particles and over suction side by particles arriving at high velocity and angle incidence. Extreme erosion area is revealed over the outer half of blade with blunting of the leading edge and thinning of the corner part of blade. As the maximum propeller thrust is expected to be from 40% to 80% of blade span, a subsequent drop in aerodynamic performance is related to this part. This information will in turn help to design propeller blades for minimum erosion and to envisage a coating for the critical regions with a hard material.

Nomenclature

A:	Area (m^2)
A_e:	Area of an element (m^2)
c:	Chord (m)
C:	Concentration (mg/m^3)
C_D:	Drag coefficient
C_L:	Lift coefficient
D:	Diameter of rotor (m)
d:	Diameter of particle (m)
e:	Thickness (m)
E:	Erosion rate (mg/g)
E_{eq}:	Equivalent erosion rate (mg/g/cm^2)
F:	Force (N)
h:	Length or height of blade (m)
k:	Turbulent kinetic energy (m^2/s^2)
K_1, K_2, K_3:	Material constants
l_e:	Dissipation length scale
m:	Mass (kg)
N_i:	Shape function
N_p:	Number of impacts on an element face
P:	Pressure (Pa)
r:	Radius, radial coordinate (m)
$R75$:	Radius at 75% of blade length
Re:	Reynolds number
t:	Time (s)
t_e:	Eddy lifetime (s)
U:	Peripheral velocity (m/s)
V:	Absolute velocity (m/s)
W:	Relative velocity (m/s)
z:	Axial coordinate (m)
x, y, z:	Cartesian coordinates.

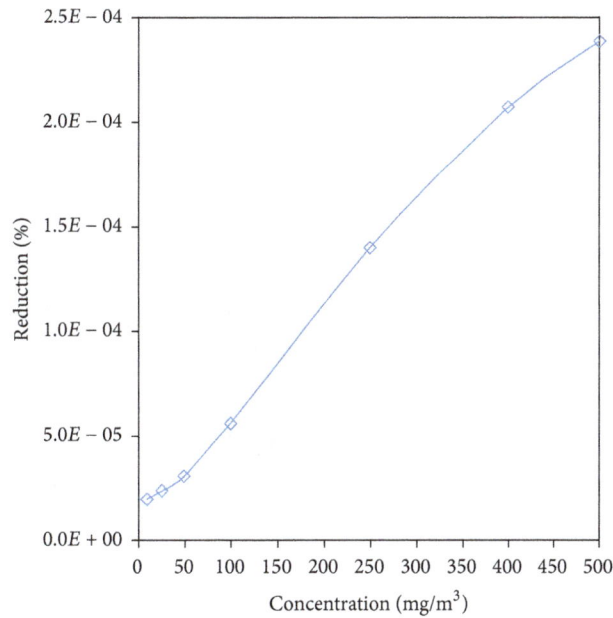

(a)

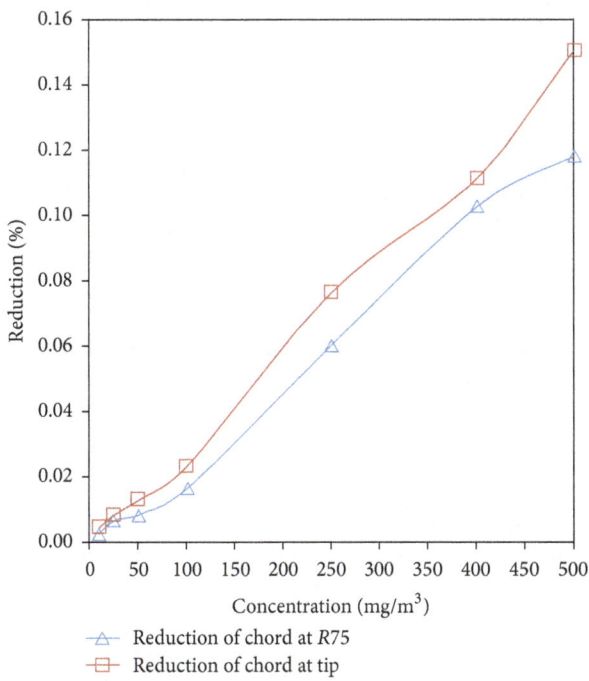

(b)

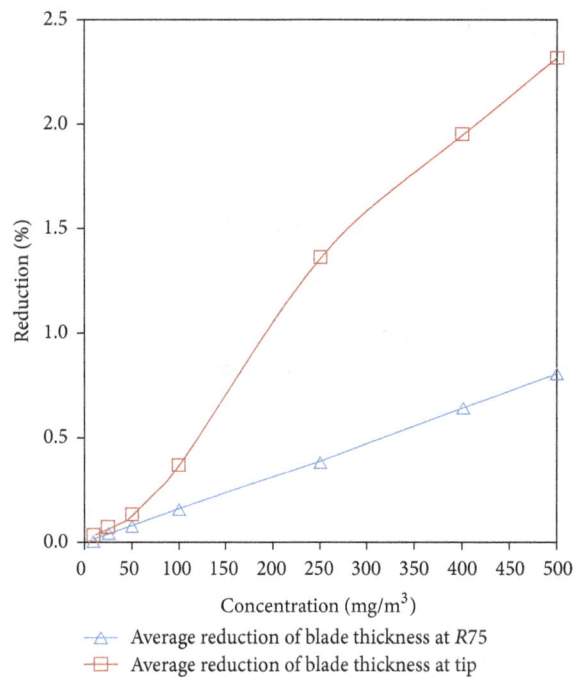

(c)

FIGURE 15: Blade geometry deteriorations with concentration after one hour of sand ingestion: (a) reduction of length, (b) reduction of chord, and (c) reduction of thickness.

Greek Letters

β: Impact angle (deg)
ε: Turbulence rate of dissipation (m^2/s^3)
ρ: Density (kg/m^3)
μ: Dynamic viscosity (kg/m·s)
ω: Speed of rotation (rad/s)
θ: Angular position (rad)
τ_p: Relaxation time (s).

Subscript

av: Average
D: Drag
f: Fluid
p: Particle
o: Related to maximum erosion
r: Radial
ϑ: Tangential component

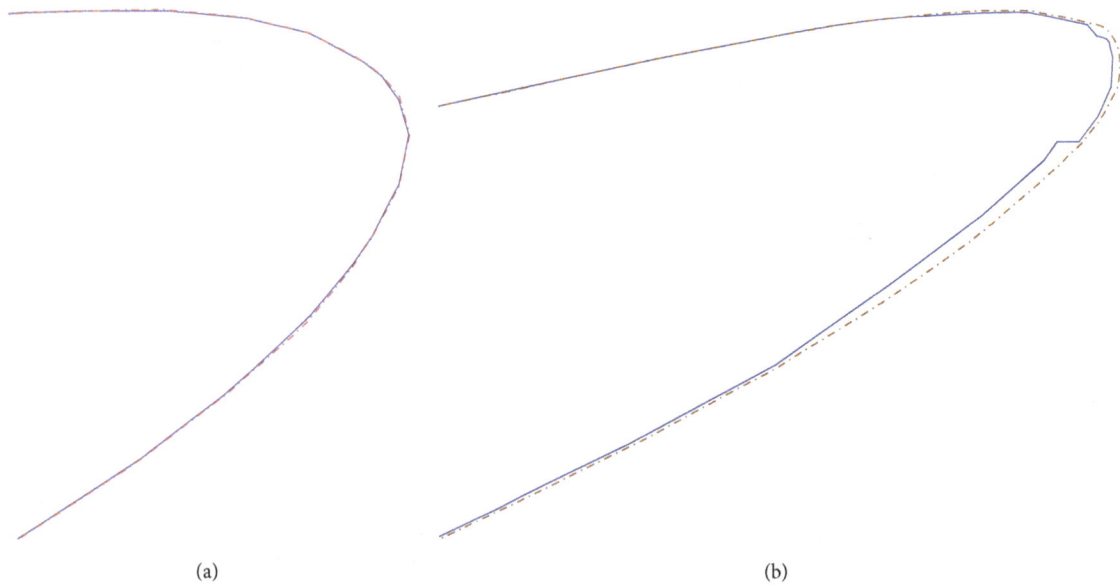

FIGURE 16: Eroded profiles after one hour of sand ingestion of a concentration (500 mg/m^3): (a) $R75$ profile; (b) tip profile (dashed line is the initial profile and solid line is the eroded profile).

S: Saffman
z: Axial direction
1, 2:   At impact and rebound from a surface.

Abbreviations

CFD: Computational fluid dynamics
RANS: Reynolds averaged Navier-Stokes
LPS: Linear profile scheme.

Conflict of Interests

The author declares no conflict of interests.

References

[1] T. Warren, S. C. Hong, C.-J. Yu, and E. L. Rosenzweig, "Enhanced erosion protection for rotor blades," in *Proceedings of the American Helicopter Society 65th Annual Forum*, pp. 2678–2686, Grapevine, Tex, USA, May 2009.

[2] P. Brotherhood and D. W. Brown, "Flight measurements of the effects of simulated leading-edge erosion on helicopter blade stall, torsional loads and performance," Reports and Memoranda 3809, R.A.E., Bedford, UK, 1976.

[3] W. Tabakoff, A. Hamed, and V. Shanov, "Blade deterioration in a gas turbine engine," *International Journal of Rotating Machinery*, vol. 4, no. 4, pp. 233–241, 1998.

[4] M. Morini, M. Pinelli, P. R. Spina, and M. Venturini, "Influence of blade deterioration on compressor and turbine performance," *ASME Journal of Engineering for Gas Turbines and Power*, vol. 132, no. 3, Article ID 032401, 11 pages, 2009.

[5] W. Sage and G. P. Tilly, "The significance of particles sizes in sand erosion of small gas turbines," *The Royal Aeronautical Society Journal*, vol. 73, pp. 427–428, 1969.

[6] M. F. Hussein and W. Tabakoff, "Dynamic behavior of solid particles suspended by polluted flow in a turbine stage," *Journal of Aircraft*, vol. 10, no. 7, pp. 434–440, 1973.

[7] M. F. Hussein and W. Tabakoff, "Computation and plotting of solid particle flow in rotating cascades," *Computers & Fluids*, vol. 2, no. 1, pp. 1–15, 1974.

[8] A. Hamed and S. Fowler, "Erosion pattern of twisted blades by particle laden flows," *Journal of Engineering for Power*, vol. 105, no. 4, pp. 839–843, 1983.

[9] A. Hamed and W. Tabakoff, "Experimental and numerical simulations of the effects of ingested particles in gas turbine engines," Tech. Rep. AGARD-CP-558, Erosion, Corrosion and Foreign Object Effects in Gas Turbines, 1994.

[10] G. P. Sallee, H. D. Kruckenburg, and E. H. Toomey, "Analysis of turbofan engine performance deterioration and proposed follow-on tests," Tech. Rep. NASA-CR-134769, 1978.

[11] C. Balan and W. Tabakoff, "Axial compressor performance deterioration," in *Proceedings of the 20th Joint Propulsion Conference*, vol. No of *AIAA Paper no. 84-1208*, Cincinnati, Ohio, USA, June 1984.

[12] A. Ghenaiet, R. L. Elder, and S. C. Tan, "Particles trajectories through an axial fan and performance degradation due to sand ingestion," in *Proceedings of the ASME Turbo Expo: Power for Land, Sea, and Air*, ASME Paper 2001-GT-0497, 12 pages, New Orleans, La, USA, June 2001.

[13] A. Ghenaiet, S. C. Tan, and R. L. Elder, "Numerical simulation of the axial fan performance degradation due to sand ingestion," in *Proceedings of the ASME Turbo Expo: Power for Land, Sea, and Air*, ASME Paper no. GT2002-30644, pp. 1181–1190, Amsterdam, The Netherlands, June 2002.

[14] A. Hamed, W. Tabakoff, and R. Wenglarz, "Erosion and deposition in turbomachinery," *AIAA Journal of Propulsion and Power*, vol. 22, no. 2, pp. 350–360, 2006.

[15] A. Ghenaiet, "Study of sand particle trajectories and erosion into the first compression stage of a turbofan," *Journal of Turbomachinery*, vol. 134, no. 5, Article ID 051025, 17 pages, 2012.

[16] M. Schrade, S. Staudacher, and M. Voigt, "Experimental and numerical investigation of erosive change of shape for high-pressure compressors," in *Proceedings of the ASME Turbo Expo: Turbine Technical Conference and Exposition*, ASME Paper GT2015-42061, 9 pages, Montreal, Canada, June 2015.

[17] CFX-TASCflow, Version 2.11, AEA Technology Engineering Software, Ltd. 2001.

[18] D. C. Wilcox, *Turbulence Modeling for CFD*, DCW Industries, La Canada, Calif, USA, 1994.

[19] D. C. Wilcox, "Simulation of transition with a two-equation turbulence model," *AIAA Journal*, vol. 32, no. 2, pp. 247–255, 1994.

[20] F. R. Menter, "Performance of popular turbulence models for attached and separated adverse pressure gradient flows," *AIAA Journal*, vol. 30, no. 8, pp. 2066–2072, 1992.

[21] F. R. Menter, "Improved two-equation k-ω turbulence model for aerodynamic flows," NASA TM-103975, 1992.

[22] A. Haider and O. Levenspiel, "Drag coefficient and terminal velocity of spherical and nonspherical particles," *Powder Technology*, vol. 58, no. 1, pp. 63–70, 1989.

[23] M. Sommerfeld, *Theoretical and Experimental Modeling of Particulate Flow—Overview and Fundamentals*, VKI Lecture Series, Von Karman Institute, Brussels, Belgium, 2000.

[24] D. J. Brown and P. Hutchinson, "The interaction of solid or liquid particles and turbulent fluid flow fields—a numerical simulation," *Journal of Fluids Engineering*, vol. 101, no. 2, pp. 265–269, 1979.

[25] J. S. Shirolkar, C. F. M. Coimbra, and M. Q. McQuay, "Fundamental aspects of modeling turbulent particle dispersion in dilute flows," *Progress in Energy and Combustion Science*, vol. 22, no. 4, pp. 363–399, 1996.

[26] A. D. Gosman and E. Ionnides, "Aspects of computer simulation of liquid fuelled combustors," in *Proceedings of the 19th Aerospace Science Meeting*, AIAA 81-0323, St. Louis, Mo, USA, January 1981.

[27] W. Tabakoff, A. Hamed, and D. M. Murugan, "Effect of target materials on the particle restitution characteristics for turbomachinery application," *Journal of Propulsion and Power*, vol. 12, no. 2, pp. 260–266, 1996.

[28] S. C. Tan and R. L. Elder, "Assessment of particle rebounds and fragmentation characteristics," Progress Report, Cranfield University, 1992.

[29] I. Finnie, "Erosion of surfaces by solid particles," *Wear*, vol. 3, no. 2, pp. 87–103, 1960.

[30] J. G. A. Bitter, "A study of erosion phenomena. Part II," *Wear*, vol. 6, no. 3, pp. 169–190, 1963.

[31] J. H. Neilson and A. Gilchrist, "Erosion by a stream of solid particles," *Wear*, vol. 11, no. 2, pp. 111–122, 1968.

[32] G. Grant and W. Tabakoff, "Erosion prediction in turbomachinery due to environmental solid particles," in *Proceedings of the 12th AIAA Aerospace Sciences Meeting*, AIAA Paper no. 74-16, Washington, DC, USA, January-February 1974.

[33] G. Grant and W. Tabakoff, "Erosion prediction in turbomachinery resulting from environmental solid particles," *AIAA Journal of Aircraft*, vol. 12, no. 5, pp. 471–478, 1975.

[34] G. Drensky, A. Hamed, W. Tabakoff, and J. Abot, "Experimental investigation of polymer matrix reinforced composite erosion characteristics," *Wear*, vol. 270, no. 3-4, pp. 146–151, 2011.

8

Numerical and Experimental Investigation of Computed Tomography of Chemiluminescence for Hydrogen-Air Premixed Laminar Flames

Liang Lv, Jianguo Tan, and Yue Hu

Science and Technology on Scramjet Laboratory, College of Aerospace Science and Engineering, National University of Defense Technology, Changsha, Hunan 410073, China

Correspondence should be addressed to Jianguo Tan; jianguotan@126.com

Academic Editor: Xuesong Li

Computed tomography of chemiluminescence (CTC) is a promising technique for combustion diagnostics, providing instantaneous 3D information of flame structures, especially in harsh circumstance. This work focuses on assessing the feasibility of CTC and investigating structures of hydrogen-air premixed laminar flames using CTC. A numerical phantom study was performed to assess the accuracy of the reconstruction algorithm. A well-designed burner was used to generate stable hydrogen-air premixed laminar flames. The OH^* chemiluminescence intensity field reconstructed from 37 views using CTC was compared to the OH^* chemiluminescence distributions recorded directly by a single ICCD camera from the side view. The flame structures in different flow velocities and equivalence ratios were analyzed using the reconstructions. The results show that the CTC technique can effectively indicate real distributions of the flame chemiluminescence. The height of the flame becomes larger with increasing flow velocities, whereas it decreases with increasing equivalence ratios (no larger than 1). The increasing flow velocities gradually lift the flame reaction zones. A critical cone angle of 4.76 degrees is obtained to avoid blow-off. These results set up a foundation for next studies and the methods can be further developed to reconstruct 3D structures of flames.

1. Introduction

The development of combustion theory and the improvement of advanced combustion equipment need detailed studies of flame structures. The application of laser-based diagnostic techniques, such as particle image velocimetry (PIV), planar laser induced fluorescence (PLIF), and planar laser induced Rayleigh scattering, can perform in situ, nonintrusive diagnostics of flame characteristics. These methods have advantages of obtaining planar information with high spatial and temporal resolution but at the cost of expensive and complex experimental devices, such as powerful lasers, ICCD cameras, and various optical components.

Compared to these laser-based diagnostic techniques, the chemiluminescence technique is relatively simple and cheap since there is no laser equipment. Chemiluminescence is a natural indicator of flame structures and combustion characteristics, because it intrinsically represents electronically excited radicals formed by exothermic reactions within reaction zones. The most common chemiluminescence is from OH (A-X) with wavelength of 310 nm, CH (A-X) of 430 nm, and Swan bands of C_2 (d-a) between 470 nm and 550 nm [1]. These three types of chemiluminescence can indicate flame structures [2, 3], combustion characteristics of heat release rate [4], local equivalence ratio [5], and stability [6]. Although the chemiluminescence measurement is an effective method to study flame structures and combustion characteristics, it only captures integral results in light paths. This shortcoming that is unable to resolve spatial distributions greatly limits the application of this diagnostic technique. An effective way to solve the problem and obtain 3D flame structures is the combination of the chemiluminescence measurement with the computed tomography (CT) technique, which leads to a prospective tool for flame studies named as computed tomography of chemiluminescence (CTC). In 1988, Hertz and Faris [7] first used CTC to conduct a simple 2D reconstruction.

Since then, studies on the CTC technique have been widely carried out.

The reconstruction algorithm and the experimental system are the foundations of the CTC technique. The optimization of them attracts worldwide interests. Denisova et al. [8] developed an algorithm based on local regularization and maximum entropy to meet the characteristics of the narrow emitting region and the great intensity gradient, which results in a better reconstruction even with few cameras and severe noises. Goyal et al. [9] further optimized the algorithm and obtained a new one using the maximization of entropy methodology, which significantly reduced errors as compared to 2D slice-by-slice reconstruction algorithms. Cai et al. [10] developed a hybrid algorithm to solve the measurement problem of 3D chemiluminescence intensity fields and validated this by both numerical and experimental methods. Despite the considerable progress, the reconstruction algorithm of CTC still has a great potential for further optimization in terms of computational speed, reconstruction precision, and efficiency by improving calculation models, access methods of the weight function, and the full use of the prior information. Gilabert et al. [11] and Hossain et al. [12] designed a series of experimental systems by using an optical transmission unit and optical fiber bundles, respectively, to increase the efficiency of CTC and reduce the cost. Fiber-based endoscopes (FBEs) were also applied for instantaneous 3D combustion measurements combined with tomography. Kang et al. [13] conducted a series of tests to quantify the accuracy of the view registration and the spatial resolution of the FBEs. These studies have further improved CTC by optimizing the reconstruction algorithm and designing the experimental system, which make CTC more suitable for combustion diagnostics. Numerical tests have been performed to evaluate the reconstruction algorithm. Most tests are performed on the assumption of parallel beams, which is uncertain in practical experiments as the depth of focus is limited, and thus more direct experimental verifications are needed to assess the feasibility and the accuracy of CTC. However, such convincing experiments are rarely reported.

Based on the development of the reconstruction algorithm and the experimental system, the CTC technique has been widely used in combustion diagnostics. Floyd et al. [14] reconstructed the 3D chemiluminescence field of a turbulent opposed jet flame with the CTC technique to obtain the flame surface density, wrinkling factor, flame normal direction, and heat release rate. They [15] also studied the instantaneous 3D information on flame geometry and excited species concentrations on a matrix burner using 10 simultaneous camera measurements. Anikin et al. [16] reconstructed 2D OH* chemiluminescence distributions with a spatial resolution of about 1 mm in turbulent diffusion flames with an exposure time of $200\,\mu s$. Li [17] first obtained instantaneous 3D combustion structures of turbulent jet flames imaging across a sizable measurement volume at kHz based on tomographic chemiluminescence technique, and phantoms simulating the experimental flames were also performed for comparison and validation purposes [18]. Many reconstructed results of different flames have been obtained, but most studies just focus on the reconstruction of chemiluminescence field

regardless of the relationship between the chemiluminescence information and the combustion characteristics.

The studies and applications of the CTC technique are still insufficient although considerable progress has been achieved in recent years. In this work, a phantom study is carried out to assess the ability of the ART algorithm. Then, a well-designed burner is employed to generate stable conical hydrogen-air premixed laminar flames, which are used to perform an experimental verification to further assess the feasibility of CTC. The OH* chemiluminescence intensity fields, which are important indicators of reaction zones of flame fronts, are reconstructed using the CTC technique. With the use of these reconstructions, the flame structures and effects of flow velocities and equivalence ratios are studied.

2. The CTC Method

2.1. The ART Algorithm. The CTC method contains two main processes: acquisition of chemiluminescence projections from different angles and reconstruction of the intensity field of the chemiluminescence from these projections using a suitable reconstruction algorithm. The algebraic reconstruction technique (ART) algorithm is used in this study.

The ART algorithm is suitable for the CTC among a series of robust reconstruction algorithms due to these following reasons: the ART algorithm is a kind of iterative reconstruction algorithm, which is quite appropriate for the reconstruction from few projections and with severe noises; the ART algorithm allows for more flexible projection geometries, which brings great potential of applications for different flames; the ART algorithm is easy to be improved with a priori information, which greatly enhances the efficiency of the reconstruction. However, the ART algorithm also has some disadvantages, and the most serious one is its high dependence on computing resources. With the development of the computing devices, this problem is gradually solved, especially in 2D reconstructions. In this work, we only focus on 2D reconstructions and the most serious disadvantage of the ART algorithm is not a problem. Based on the reasons above, the ART algorithm is employed in this work.

The reconstruction problem is reduced to solving an underdetermined system of linear equations. The system is obtained by approximating each projection measurement I_{qp} as a finite sum taken through the discrete object (the flame) domain as shown in the following [15]:

$$I_{qP} = \sum_{v=1}^{N_v} w_{qpv} f_v, \tag{1}$$

where I_{qp} is the intensity value of the pth projection in the qth view (an image contains a series of projections measured from a projection angle using a ICCD camera); f_v is the average scalar value of the pixel v, which is the goal; w_{qpv} is the weight coefficient representing the contribution of the pixel v to the pth (maximum value N_p) projection in the qth (maximum value N_q) view, which is approximated as the proportion of the intersection area between the projection beam and the pixel area to the pixel area. With both the necessity to obtain the correct weight coefficient for each projection and the importance of enhancing the computing efficiency considered, a simplified calculation approach is employed in this study, where the pixel is divided into 10 × 10 subgrids, and then the sum of area of these subpixels whose center passed through the beam is computed as the intersection area. In this work, the object domain is divided into pixels that are indexed by a single index v from 1 to N_v.

A relaxed additive version of the ART algorithm is used in this work as follows [15]:

$$F^{(i)} = F^{(i-1)} - \gamma \times \frac{\left(W_i \times F^{(i-1)} - I_i\right)}{W_i \times W_i^T} \times W_i^T. \qquad (2)$$

In this equation, F is a N_v-dimensional vector of f_v; I is a $(N_q \times N_p)$-dimensional vector of I_{qp}; W is a $[(N_q \times N_p) \times N_v]$-dimensional matrix of w_{qpv}; γ is a relaxation factor, which is used to improve convergence in the presence of noises, and it is 0.5 in this work as the noise is not so high; index i ($i = q \times p$) represents both the index of the projection in all views and the number of the iteration. Once all indexes have been addressed, the next round of iteration begins.

A simple criterion described by Mishra et al. [19] is employed to reduce the computational cost. Once the absolute difference of the sum of the chemiluminescence intensity (f_v) is below a threshold value $0.0001 \times |F^i|$, the reconstruction is considered to be converged.

In this work, the 2D reconstruction domain was divided into 131072 pixels ($N_v = 131072$). 37 projection angles ($N_q = 37$) were chosen, and the number of projections in each view was 255 ($N_p = 225$) as the amount of the reconstruction grids was small.

2.2. Phantom Study. Phantom studies are typically used to assess the feasibility and the accuracy of the reconstruction algorithm, which is the foundation of the CTC technique. A phantom object is designed with a size of 1 × 0.5 (relative length with unit of 1, as the length unit is meaningless in a phantom study) as shown in Figure 1, which is divided into 200 × 100 pixels and the detailed parameters are listed in Table 1. It includes four geometric shapes of one rectangle and three intersecting rings. This phantom object is similar to the shape of the flame structure, used in our experimental studies. The corners of the rectangle can be used to assess the abilities of the ART algorithm to reconstruct detailed information and the intersecting rings can help to assess the spatial resolution.

Projections were achieved in every 5 degrees from 0 degrees (the angle of the projection orientation to the base of the phantom object) to 180 degrees. Up to 37 views in total were collected, and each view included 225 projections. The criterion was followed strictly.

Through the analysis of the flow field and the experience of previous work, the relative position of the flame can be determined before the reconstruction. This will be greatly helpful to improve the accuracy of the reconstruction. The method is known as the introduction of prior knowledge. It was implemented in this work. Figure 2 shows an initial distribution of the reconstruction, which is determined by the prior knowledge. The detailed parameters are listed in Table 2.

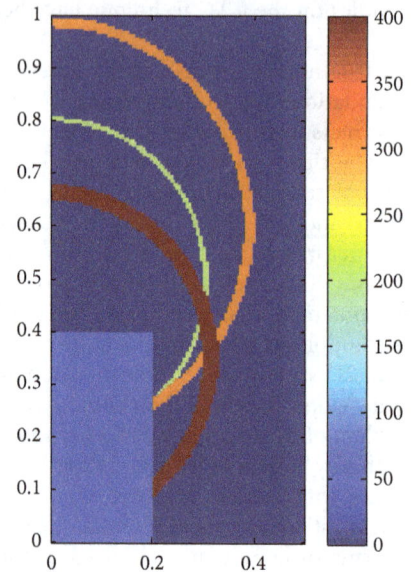

FIGURE 1: The phantom object for testing the reconstruction algorithm.

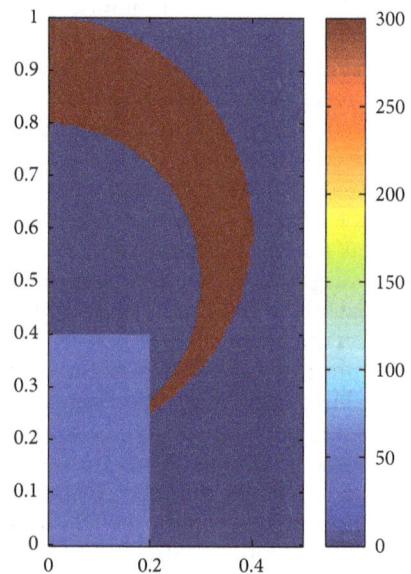

FIGURE 2: The initial distribution of the reconstruction.

TABLE 1: Geometric parameters of the phantom object.

Shape	Width (in-radius)	Length (out-radius)	Center coordinate	Intensity
Rectangular	0.20	0.40	(0.10, 0.20)	50
Ring	0.30	0.31	(0.00, 0.50)	200
Ring	0.38	0.40	(0.00, 0.60)	300
Ring	0.30	0.33	(0.00, 0.35)	400

The reconstructed results are shown in Figure 3, which shows a good agreement with the phantom object in general distribution information of the geometric shapes, but small

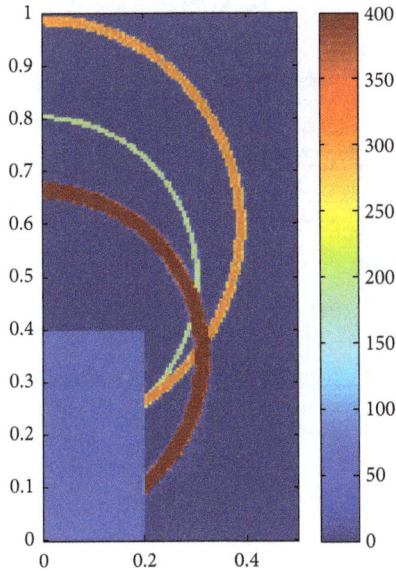

FIGURE 3: The reconstructed result of the phantom object.

TABLE 2: Geometric parameters of the initial distribution.

Shape	Width (in-radius)	Length (out-radius)	Center coordinate	Intensity
Rectangular	0.20	0.40	(0.10, 0.20)	50
Ring	0.30		(0.00, 0.50)	300
Ring		0.40	(0.00, 0.60)	300

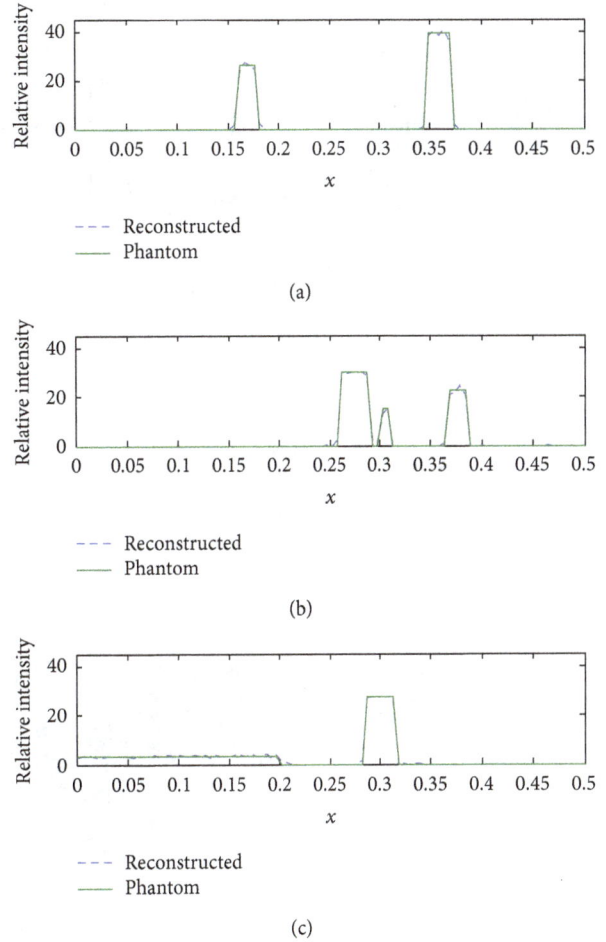

FIGURE 4: Comparison of relative intensity distributions of the phantom object and the reconstructed result at three different relative heights of 0.25, 0.50, and 0.75 from the bottom to the top (the axis "x" represents the relative distance to the left boundary of the phantom).

noises can be seen and the detailed information has some errors.

In order to further assess the accuracy of the ART algorithm, the maximum deviation coefficient E_m and the normalized root mean square error (NRMSE) E_n were calculated. They are defined as follows, respectively:

$$E_m = \frac{\max(F)}{f_{v\max}}, \tag{3}$$

$$E_n = \sqrt{\frac{\sum_{v=1}^{N_v}(F_v - f_v)^2}{\sum_{v=1}^{N_v}(f_v)^2}}, \tag{4}$$

where F is a N_v-dimensional vector of the reconstructed result; f_v is the average scalar value of the pixel v; $f_{v\max}$ is the average scalar value in the same pixel with the maximum element of F. E_m represents the maximum deviation between the phantom object and the reconstruction. It indicates the worst situation in the reconstruction. E_n represents a general deviation level. These two coefficients can effectively assess the accuracy of the reconstructions and verify the feasibility of the ART algorithm.

Relative intensity distributions (at different heights above the bottom) were compared as shown in Figure 4 at three different relative heights of 0.25, 0.50, and 0.75 from bottom to the top, respectively. The relative intensities, which

were achieved from the intensity values divided by the average values on each height, showed a good agreement with each other. In this phantom, E_n is calculated as 0.059, which is a very small one. It quantitatively represents a good performance of the ART algorithm. However, in the areas of large intensity gradient, errors can be found up to 9.3% (E_m is 0.093), which indicates a weakness of the ART algorithm in reconstructions of detailed information. In spite of this, the reconstructed result indicated main features of the structure of the phantom object, which verify the feasibility of the ART algorithm.

3. Experimental Arrangement

3.1. Combustion System. The burner and its gas supply system used in this study are illustrated in Figure 5(a). The burner had two blocks with the height of 15 mm, which were arranged on the perforated plate with a distance of 2 mm. It mainly consisted of the premix chamber, flame arrester device, perforated plate, and protective gas chamber. The

(a)

Gas source   MFC   Pressure regulating valve   Switching valve

Filter   Dryer   Back-fire relief valve   Check valve

(b)

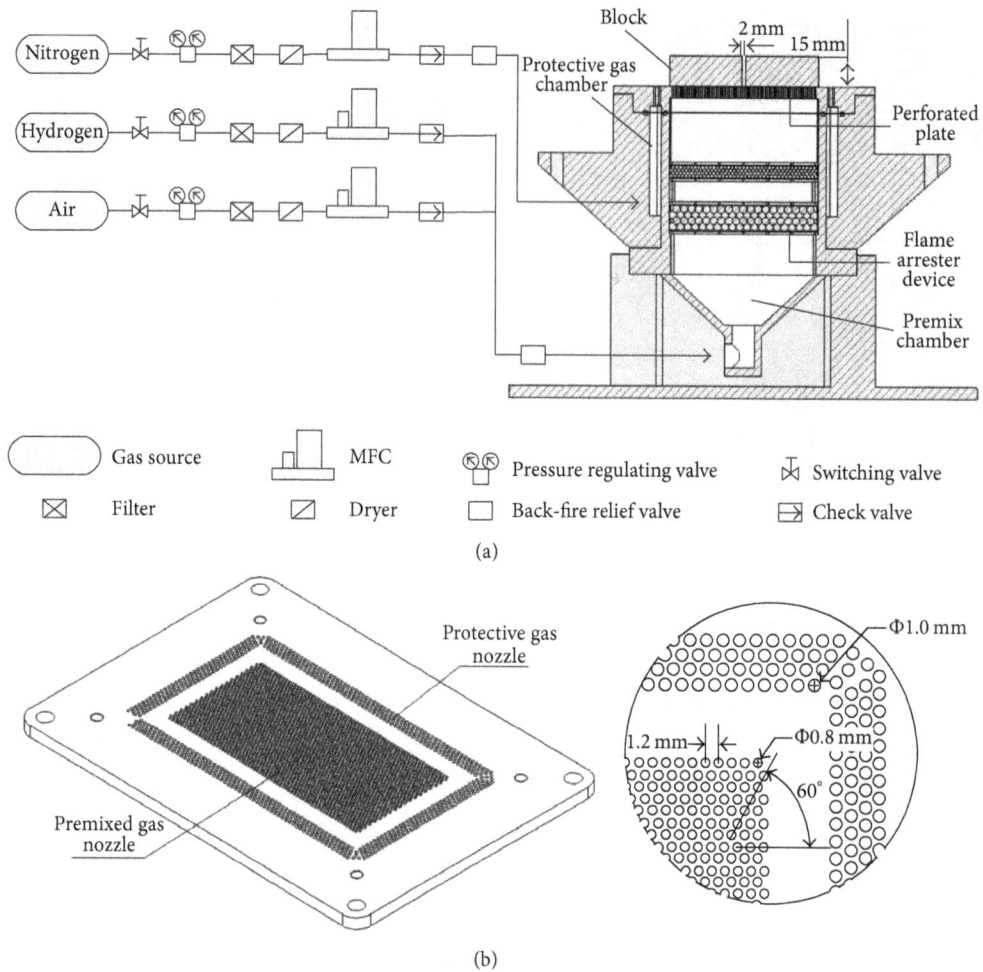

FIGURE 5: Schematics of the combustion system: (a) burner and its gas supply system; (b) arrangement of the circular nozzles.

arrangement of the circular nozzles on perforated plate was shown in Figure 5(b). Flame arrester device mounted in the premix chamber mainly consisted of steel balls with two different diameters of 2 mm and 1 mm, respectively, to get rid of backfire. Check valves and backfire relief valves were also employed in the pipelines for safety reason. Gases were adjusted by Alicat Scientific mass flow controllers (MFCs), which could provide adequate flow control with high accuracy. Premixed reactants (hydrogen-air) entered the burner and were further mixed in the premix chamber to ensure the accuracy of the equivalence ratio of the reactant stream above the blocks at the burner exit. N_2 was used as a protective gas to provide a relatively closed region eliminating the interference of the ambient air. Stable hydrogen-air premixed laminar flames, which were perfect for the verification experiment, were achieved due to the special structure of the burner.

In this work, 9 different flame conditions were designed to study the effects of the flow velocity and the equivalence ratio on the flame structure. Considering the stability of the flames, the flow velocities were limited from 11 m/s to 15 m/s and the equivalence ratios were changed from 0.35 to 0.55, as shown in Table 3.

TABLE 3: A list of 9 flame conditions with different flow velocities and equivalence ratios.

Identifier	Equivalence ratio	Flow velocity (m/s)
F1	0.45	14.36
F2	0.45	13.04
F3	0.45	11.73
F4	0.45	15.67
F5	0.45	16.95
F6	0.50	14.36
F7	0.55	14.36
F8	0.40	14.36
F9	0.35	14.36

3.2. Optical System. Figure 6(a) illustrates two orientations of the projection view and the side view. The projection view orientation demonstrated a plane, in which projections were obtained, while the side view orientation demonstrated an orientation perpendicular to the projection plane in which integral images were recorded directly. Integral images of the OH^* chemiluminescence from the side view were used as a

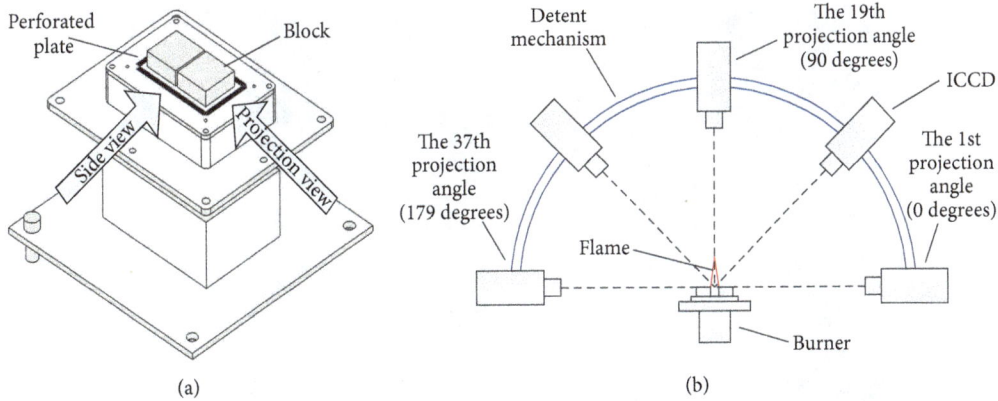

FIGURE 6: Schematics of the optical system: (a) two typical view orientations; (b) a single camera mounted on a guideway and rotated around a fixed center.

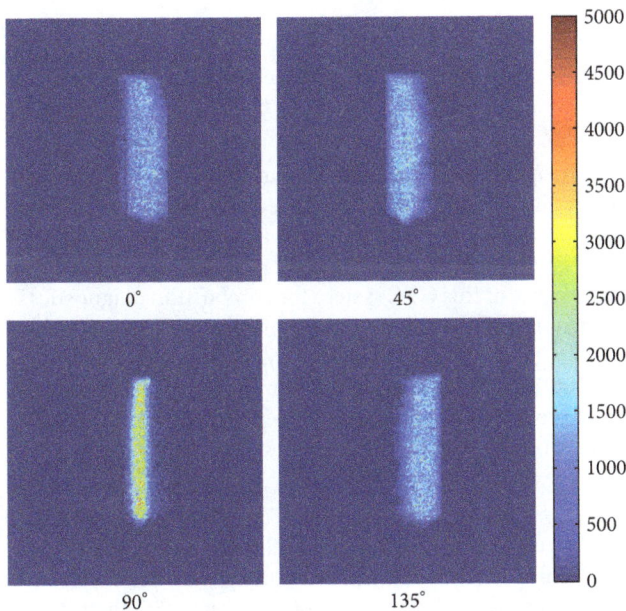

FIGURE 7: Projection images from four projection angles of 0, 45, 90, and 135 degrees.

real 2D distribution of the OH* chemiluminescence for comparison with the reconstructed results. An annular guideway working as a detent mechanism was mounted vertically to the burner exit located at the center, as shown in Figure 6(b). A single camera was mounted on the guideway and rotated around the center, while maintaining the alignment and the distance of 750 mm from the center. Every 5 degrees from 0 degrees to 180 degrees, added up to a total of 37 angles, were chosen as projection angles. Due to symmetric view, the projections from 0 degrees and 180 degrees were identical, of which the angle of 179 degrees was chosen to be the 37th projection angle as a result. Views were measured from 37 projection angles as mentioned above using a PI-MAX I 512 × 512-pixel ICCD camera with a 50 ms exposure time and a gain of 20, which were appropriate for getting a high signal-to-noise ratio (SNR). An interference band-pass filter of 307 nm

with 17% transmission at peak frequency and a full width at half maximum of 12 nm was added in front of the UV lens for acquiring OH* chemiluminescence.

Calibration was carried out to ensure the precise location of the center, along which the camera rotated. The boundaries of the chemiluminescence images were easily recognized as the shape of the flame in this study was simple. The midline of the boundaries was taken as the reconstruction plane (which here was a line in an image) in each view, which was also helpful to get the projection geometries.

Figure 7 shows 4 images (views) from the projection angles of 0, 45, 90, and 135 degrees. The midline of the boundaries of the chemiluminescence distribution in each of the images was used as the projections for reconstruction as mentioned above.

4. Results and Discussion

4.1. Experimental Verification for CTC. The reconstructed result obtained using the CTC technique is presented in Figure 8(a), along with the side image (the integral image achieved from the side view as mentioned above) shown in Figure 8(b). The side image was processed by position transformation to make it easier to compare. It is clear that the reconstructed result can reflect the distributions of the OH* chemiluminescence of the flame despite the slight differences in detail.

In order to further evaluate the feasibility of CTC, a relative intensity distribution is compared between two images in Figure 8. Figure 9 shows the comparison at three different heights above the exit plane of 2 mm, 6 mm, and 10 mm, respectively. As shown in Figure 8, the relative intensities agree well with each other, and E_n is calculated as 0.185. Differences can be found especially on the bottom part with the maximum deviation up to 26.7%, which shows a weakness of the ART algorithm to reconstruct the field with a cavity surrounded by the valuable information (areas of large intensity gradient). In spite of this, the reconstructed result can indicate main features of the flame structures, which verify the feasibility of CTC.

(a)   (b)

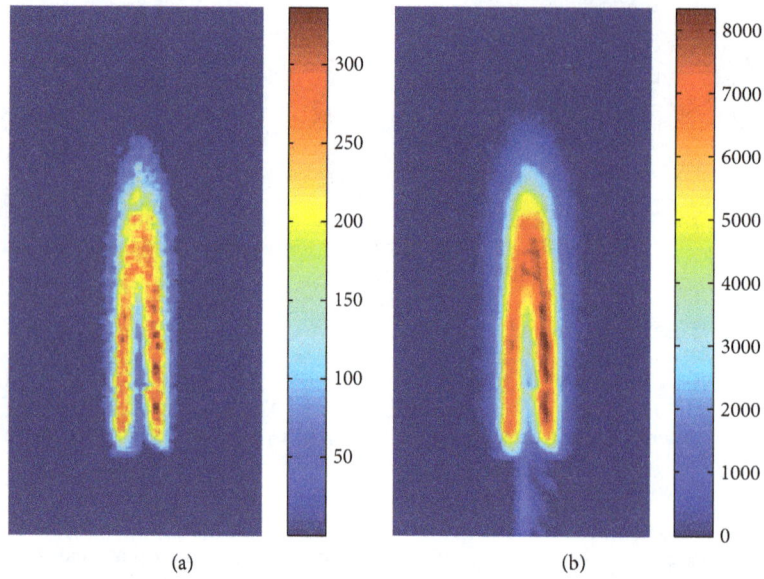

FIGURE 8: Comparison of the reconstructed result (a) and the side image (b) of the OH* chemiluminescence.

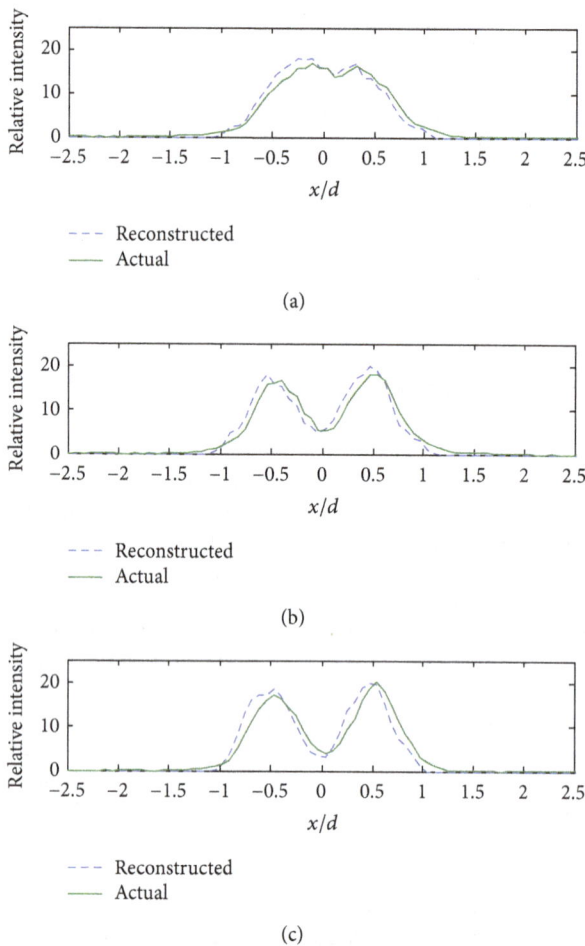

(a)

(b)

(c)

FIGURE 9: Comparison of relative intensity distributions of the reconstructed results and the side image at three different heights of 2, 6, and 10 mm from the bottom to the top (the axis "x/d" represents the ratio of the distance to the center of the burner and the width of the exit).

The coefficients (E_n and E_m) in this reconstruction are bigger than those in the phantom study by about 3.1 times and 2.9 times, respectively. This is mainly caused by the experiment devices and the errors in projecting system. The uncertainty of this CTC system for combustion diagnostics is determined by the maximum deviation (26.7% in this work). It is a big one for quantitative analysis. However, the relative distribution of the OH* chemiluminescence in flames are fairly precise. It is good enough to perform the study of the flame structures by the CTC technique.

4.2. Flame Structures Based on CTC. Figure 10 shows the reconstructions of the OH* chemiluminescence for 8 flame conditions. A blow-off phenomenon appeared in F9 and the reconstructed result was not displayed as no OH* chemiluminescence was obtained in this condition. These reconstructed results of the OH* chemiluminescence intensity fields can represent the real OH* radiation distributions. OH* distribution is an important indicator of the reaction zones of the flame fronts and it can characterize the flame structure. The intensity values are increasing with larger flow velocities as shown in Figure 10, due to the increasing reactant mixtures. Figure 10 also shows the same trend with the increasing equivalence ratios.

In order to characterize the flame structure information, the height of the flame (HF) and the ratio of relative distributions of the flame chemiluminescence (RDF) were employed in the OH* data analysis. The height of the flame (HF) is defined as

$$\text{HF} = Y_d - Y_e, \qquad (5)$$

where Y_d is the vertical coordinate of the dividing line between the flame and the surroundings determined by the threshold value, which is 1/3 peak value in each image, and Y_e is the vertical coordinate of the burner exit. The ratio of

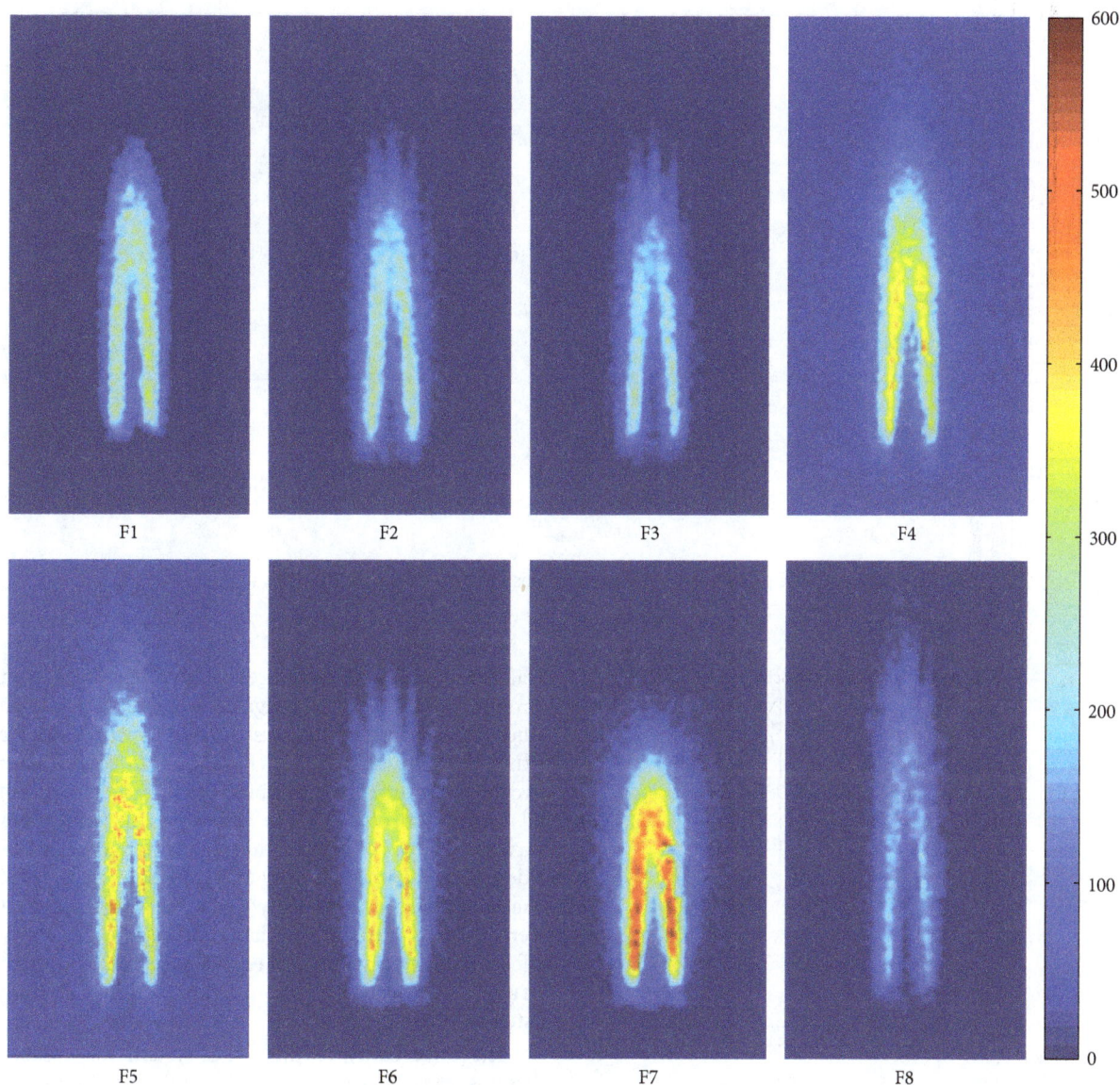

FIGURE 10: Reconstructions of the OH* chemiluminescence in different flame conditions.

relative distributions of the flame chemiluminescence (RDF) is defined as

$$\text{RDF} = \frac{I_{\text{ts}}}{I_{\text{ws}}}, \qquad (6)$$

where I_{ts} is the sum of the chemiluminescence intensities of the top half of the flame; I_{ws} is the sum of the chemiluminescence intensities of the whole of the flame. Flame structures can be preliminarily characterized by these two parameters.

HFs and RDFs with different flow velocities and equivalence ratios are presented in Figure 11. HF becomes larger with increasing flow velocities as shown in Figure 11(a), whereas it decreases with increasing equivalence ratios (no larger than 1) as shown in Figure 11(b). HF relies on the relative relations between the flow velocity and the flame propagation velocity, which is connected with the equivalence ratio. The largest flame propagation velocity is usually obtained near

the equivalence ratio of 1 with other parameters unchanged. That means a larger flame propagation velocity with a larger equivalence ratio (no larger than 1) in this study. Figure 12 shows a geometric relationship of the flame structure and the velocities. It is clear that HF is proportional to $\tan \theta$, which is determined by the flow velocity and the flame propagation velocity. It shows the same regularities with the experimental results. The 9th flame condition shows a blow-off phenomenon, as the cone angle is too sharp to hold the flame. Therefore, the cone angle of 4.76 degrees, which is calculated by HF and the exit width, in 8th flame condition can be considered a critical cone angle, which will be a useful criterion to avoid the blow-off phenomenon, although the actual one should be sharper.

Figure 11(a) shows a positive effect of flow velocities on RDFs, indicating that the reaction zone was gradually lifted with increasing flow velocities. The equivalence ratio shows

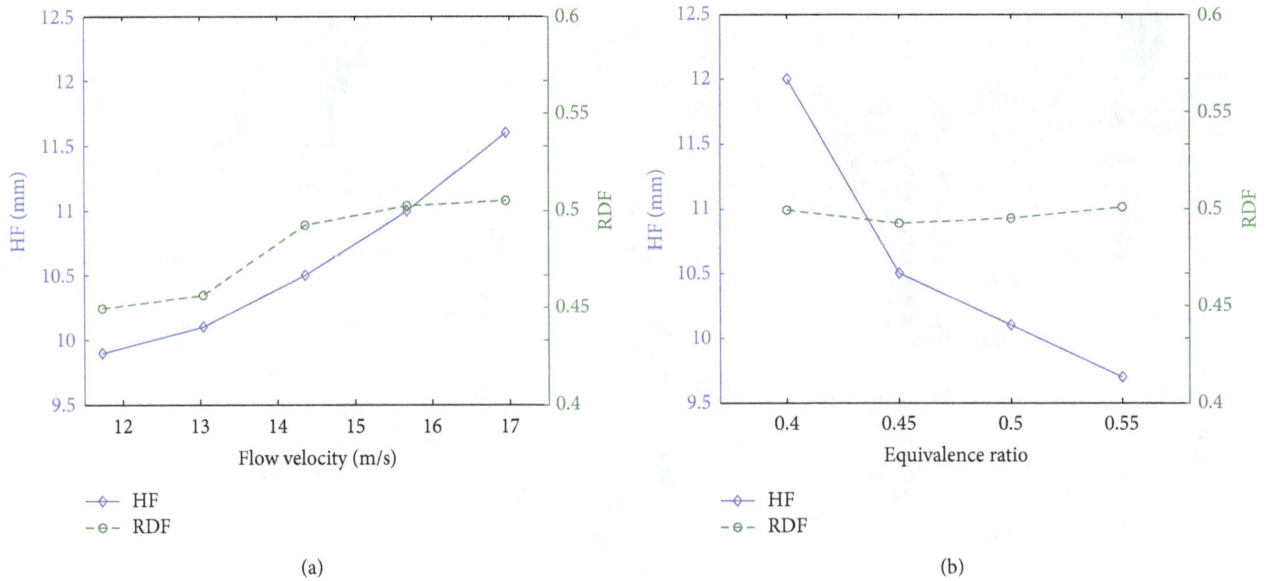

FIGURE 11: HFs and RDFs in different flame conditions: (a) flow velocities; (b) equivalence ratios.

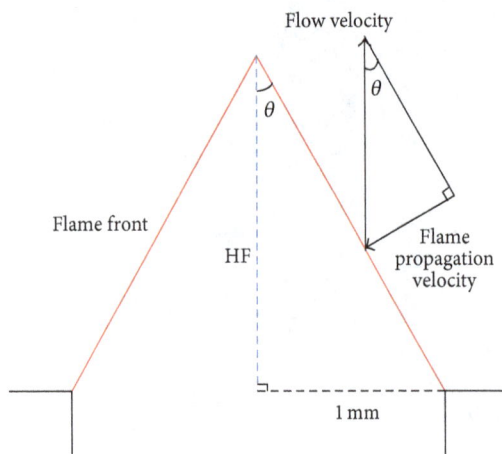

FIGURE 12: Geometric relationships of the flame structure and the velocities.

a minor effect as seen in Figure 11(b). Therefore, RDF is independent of HF and is just related to the flow velocity, suggesting RDF is a useful parameter of the flame structure. Figure 11(a) also reveals a critical flow velocity of about 15 m/s, which leads RDF to 0.5. It is an indicator of the OH* chemiluminescence distributions. It suggests that the reactions relatively gather to the top of the flame and the flame gradually separates from the exit of the burner, while the flow velocity increases to 15 m/s. It should be noted that this is just a rough estimate. Some other factors may also contribute to RDF, which calls for more experimental data and further studies.

5. Conclusions

In conclusions, the numerical and experimental verifications of computed tomography of chemiluminescence (CTC) were

conducted and the structures of hydrogen-air premixed laminar flames were investigated using CTC. A phantom study was performed to assess the ability of the ART algorithm. A well-designed burner was built to generate a stable hydrogen-air premixed laminar flame for the experimental verification of CTC. The experimental verification was carried out by comparing the reconstructed result to the side image recorded directly from the side view. The OH* chemiluminescence intensity fields reconstructed by CTC were used to indicate the flame structures and study the effects of flow velocities and equivalence ratios on the flame structure. Detailed conclusions obtained from this study are listed as follows.

The uncertainty of this CTC system for combustion diagnostics is determined as 26.7% in this work, and it is not suitable for quantitative analysis. However, the reconstructed results can still effectively indicate the relative distributions of the flame chemiluminescence. It is good enough to perform the study of the flame structures by the CTC technique.

The height of the flame is affected by the flow velocity and the equivalence ratio. With an increase of HF, the cone angle of the flame structure decreases until a blow-off phenomenon appears at a critical cone angle of 4.76 degrees.

The increasing flow velocities gradually lift the reaction zone, while the equivalence ratio has a minor effect on it.

It should be noted that the CTC technique still fails to show the flame structure with high accuracy in areas with a large intensity gradient, which calls for further optimizations. In addition, the extension of the CTC method to reconstruct complex 3D turbulent flames is also required in the next step.

Competing Interests

There are no competing interests related to this paper.

Acknowledgments

The authors would like to express their thanks for the support from the National Natural Science Foundation of China (Grants nos. 11272351 and 91441121).

References

[1] A. G. Gaydon, H. G. Wolfhard, and R. M. Fristrom, *Flames: Their Structure, Radiation and Temperature*, Chapman and Hall, London, UK, 1978.

[2] K. Kuwahara and H. Ando, "Analysis of barrel-stratified lean-burn flame structure by two-dimensional chemiluminescence measurement," *JSME International Journal B*, vol. 37, no. 3, pp. 650–658, 1994.

[3] J. Kojima, Y. Ikeda, T. Nakajima et al., "Chemiluminescence-based diagnostics for the flame-front structure of premixed flames," *Transactions of the Japan Society of Mechanical Engineers Part B*, vol. 69, no. 678, pp. 482–489, 2003.

[4] M. Röder, T. Dreier, and C. Schulz, "Simultaneous measurement of localized heat-release with OH/CH$_2$O-LIF imaging and spatially integrated OH*-chemiluminescence in turbulent swirl flames," *Proceedings of the Combustion Institute*, vol. 34, no. 2, pp. 3549–3556, 2013.

[5] Y. K. Jeong, C. H. Jeon, and Y. J. Chang, "Evaluation of the equivalence ratio of the reacting mixture using intensity ratio of chemiluminescence in laminar partially premixed CH4-air flames," *Experimental Thermal and Fluid Science*, vol. 30, no. 7, pp. 663–673, 2006.

[6] B. Taupin, G. Cabot, G. Martins, D. Vauchelles, and A. Boukhalfa, "Experimental study of stability, structure and CH* chemiluminescence in a pressurized lean premixed methane turbulent flame," *Combustion Science and Technology*, vol. 179, no. 1-2, pp. 117–136, 2007.

[7] H. M. Hertz and G. W. Faris, "Emission tomography of flame radicals," *Optics Letters*, vol. 13, no. 5, pp. 351–362, 1988.

[8] N. Denisova, P. Tretyakov, and A. Tupikin, "Emission tomography in flame diagnostics," *Combustion and Flame*, vol. 160, no. 3, pp. 577–588, 2013.

[9] A. Goyal, S. Chaudhry, and P. M. V. Subbarao, "Direct three dimensional tomography of flames using maximization of entropy technique," *Combustion and Flame*, vol. 161, no. 1, pp. 173–183, 2014.

[10] W. Cai, X. Li, F. Li, and L. Ma, "Numerical and experimental validation of a three-dimensional combustion diagnostic based on tomographic chemiluminescence," *Optics Express*, vol. 21, no. 6, pp. 7050–7064, 2013.

[11] G. Gilabert, G. Lu, and Y. Yan, "Three-dimensional tomographic reconstruction of the luminosity distribution of a combustion flame," *IEEE Transactions on Instrumentation and Measurement*, vol. 56, no. 4, pp. 1300–1306, 2007.

[12] M. M. Hossain, G. Lu, and Y. Yan, "Three-dimensional reconstruction of combustion flames through optical fiber sensing and CCD imaging," in *Proceedings of the 2011 IEEE International Instrumentation and Measurement Technology Conference (I2MTC '11)*, pp. 79–83, Binjiang, China, May 2011.

[13] M. W. Kang, X. Li, and L. Ma, "Three-dimensional flame measurements using fiber-based endoscopes," *Proceedings of the Combustion Institute*, vol. 35, no. 3, pp. 3821–3828, 2015.

[14] J. Floyd, P. Geipel, and A. M. Kempf, "Computed Tomography of Chemiluminescence (CTC): instantaneous 3D measurements and Phantom studies of a turbulent opposed jet flame," *Combustion and Flame*, vol. 158, no. 2, pp. 376–391, 2011.

[15] J. Floyd and A. M. Kempf, "Computed Tomography of Chemiluminescence (CTC): high resolution and instantaneous 3-D measurements of a matrix burner," *Proceedings of the Combustion Institute*, vol. 33, no. 1, pp. 751–758, 2011.

[16] N. B. Anikin, R. Suntz, and H. Bockhorn, "Tomographic reconstruction of 2D-OH*-chemiluminescence distributions in turbulent diffusion flames," *Applied Physics B*, vol. 107, no. 3, pp. 591–602, 2012.

[17] X. Li, "Three-dimensional measurements of turbulent jet flames at KHz rate based on tomographic chemiluminescence," in *Proceedings of the Aerospace Sciences Meeting*, Reno, Nev, USA, January 2000.

[18] X. Li and L. Ma, "Volumetric imaging of turbulent reactive flows at kHz based on computed tomography," *Optics Express*, vol. 22, no. 4, pp. 4768–4778, 2014.

[19] D. Mishra, J. P. Longtin, R. P. Singh, and V. Prasad, "Performance evaluation of iterative tomography algorithms for incomplete projection data," *Applied Optics*, vol. 43, no. 7, pp. 1522–1532, 2004.

9

Numerical-Experimental Assessment of a Hybrid FE-MB Model of an Aircraft Seat Sled Test

F. Caputo ⓘ,[1] **A. De Luca** ⓘ,[1] **F. Marulo,**[2] **M. Guida** ⓘ,[2] and **B. Vitolo**[3]

[1]*Department of Engineering, University of Campania Luigi Vanvitelli, Aversa, Italy*
[2]*Department of Industrial Engineering, University of Naples "Federico II", Napoli, Italy*
[3]*Geven, Zona Asi Boscofangone, Nola, Napoli, Italy*

Correspondence should be addressed to F. Caputo; francesco.caputo@unicampania.it

Academic Editor: Roberto G. Citarella

This paper deals with the development of an established hybrid finite element multibody (FE-MB) model for the simulation of an experimental sled test of a single row of a double passenger seat placed in front of a fuselage bulkhead, by considering a single anthropomorphic Hybrid II 50th dummy arranged on one of the seat places. The numerical investigation has been carried out by focusing on the passenger passive safety. Specifically, the occupant injury assessment has been quantitatively monitored by means of the head injury criterion (HIC), which, based on the average value of the dummy head acceleration during a crash event, should not exceed, according to the standards, the value of 1000. Numerical results provided by the hybrid model have been compared with the experimental ones provided by the Geven S.p.A. company and with the results carried out by a full FE model. The hybrid model simulates with a good level of accuracy the experimental test and allows reducing significantly the computing time with respect to the full FE one.

1. Introduction

Nowadays, the importance of passive safety is becoming even more important for the transport field to the point of influencing the design practice [1–3]. In fact, several experimental and numerical studies are addressed to investigate and improve the crashworthiness of vehicles by paying attention to the passengers' safety. Compliance with the passive safety specifications leads to a little increase of the vehicle weight, which can be tolerated in view of an increasing of the chance of survival in case of an accident.

In the aerospace field, several components, such as the seat system, have been completely redesigned in order to improve the capability of the aircraft to protect passengers during crash events. As crash event, it must be intended, for example, an emergency landing.

The vehicle crashworthiness can be measured also in terms of the occupant injuries [4–12]. Among the several parameters that can be monitored, the head injury criterion (HIC) one plays a key role, denoting possible head injuries [13–18]. HIC parameter, based on the average value of the

passenger's head acceleration acting during a crash event, should be lower than 1000 in order to guarantee the passengers' safety. Another parameter is the maximum compressive load measured between the pelvis and the lumbar column of the anthropomorphic test dummy (ATD), which should not exceed 1500 lb (6.67 kN), or, if torso restraints are used, tension loads in individual straps should not exceed 1750 lb (7.78 kN) [1, 2].

Several experimental and numerical studies dealing with such matter have been proposed in literature: the former are characterized usually by destructive tests performed on such large instrumented full-scale structures such as aircraft, fuselage section, and seats, even equipped with ATDs, which cannot be easily repeated due to the high costs and the complexity of the tests, requiring often advanced laboratories, the use of complex acquisition sensor networks, complex and long result analysis, and so on. Concerning the numerical investigations, an established predictive numerical model can give a significant contribution in the assessment of passengers' safety, allowing to overcome all aforementioned issues related to experimental test. In particular, a numerical tool gives also the possibility, under a certification by analysis (CBA) approach [19–22], to perform optimization analyses aimed to virtually achieve the optimal structural solution, decreasing the number of experimental tests and the costs required for the developmental phase. Concerning the disadvantages, numerical simulation of crash phenomena involving ATDs needs a high computational power. Moreover, the establishment of the model requires facing with the assessment of the assumed assumptions as well as hypotheses, the improvement of the level of accuracy, and consequently, the availability of high computational power.

This paper investigates on the passive safety addressed to the aircraft seat system.

A well-designed seat should allow passengers to not entrap themselves independently and escape the aircraft, by leading to good chance to survive, after a crash. Standards that investigate on the realistic dynamic performance of aircraft seats can be found in literature in order to emphasize occupant impact protection and to analyse the full-scale aircraft impact tests.

This paper deals with an improved hybrid finite element multibody (FE-MB) model for the simulation of an experimental sled test of a single row of a double passenger seat placed in front of a fuselage bulkhead by considering a single anthropomorphic dummy arranged on one of the seat places. Tests have been developed at the laboratory of Geven S.p.A., which is equipped with a sled decelerator testing system compliant with certification requirements from the FAR25 for TSO C127a regulations [1]. The development of the proposed numerical model started from a preliminary hybrid FE-MB model presented by authors in [23]. Specifically, the new modelling has been carried out in order to improve the level of accuracy of the predicted passenger kinematics. The hybrid modelling strategy has been carried out in order to exploit both FE method level of accuracy and the lower MB computational costs [23]. Whilst in the MB approach, which requires less computational costs, the dummy is modelled by rigid bodies, defined by both mass and moments of inertia (connected by suitable characteristics joints); in the FE approach, the dummy is modelled by means of finite elements containing more details than the former, which lead to several difficulties in terms of model management and higher computational costs.

In a seat sled test, all components, such as dummy, seat, and restraint system, may be modelled by means of both MB and FE approaches, leading to a less or more accurate simulation, respectively.

FE codes allow a very detailed modelling of all components, such as safety belts, dummies, and structural parts, by leading to very complex models, which are usually characterized by several million of degrees of freedom with negative feedback on the computing time and on the model versatility. In fact, the management of small design changes implies strong efforts in terms of modelling.

On the contrary, the MB method, to the detriment of a less level of accuracy related to the nondeformability of the modelled components, can be helpful for designers to simulate quickly the structural response of a structure under several configurations. More properly, MB methods lend mainly themselves to the prediction of the kinematics of assembly components under complex loading conditions more than to the investigation of the stress-strain field.

As a matter of fact, this modelling method is widely used in a preliminary design stage where it is still interesting to investigate more structural solutions.

So in order to enjoy the accuracy of the FE method, as well as the low computational time provided by the MB method, the hybrid approach can be used, allowing improvement of the modelling where needed by means of the FE method as well as lowering the computational time. The lowering of the computational time can be achieved by considering the MB approach for the parts of the analysed system, whose deformations do not influence the dynamic system responses and for which only kinematic aspects must be taken into account.

In order to assess the prediction capability of the developed hybrid FE-MB model, the simulated biomechanical parameters, such as the acceleration of the head of the passenger, with the relative calculation of the head injury criterion (HIC) and the loads transmitted to his lower limbs have been compared with the experimental ones. Moreover, the predicted ATD trajectory has been compared with the trajectory simulated by a full FE model in previous papers [22, 23].

2. Experimental Dynamic Testing of Airplane Seats

The experimental test, provided by Geven S.p.A., is aimed to demonstrate the compliance of the seat passenger system with FAR 25.562 [1]. A Hybrid II 50th passenger dummy has been arranged on a double seat positioned in front of a relatively stiff bulkhead (Figure 1). The main parameter monitored during the test is the acceleration of the head, with the relative calculated HIC. The experimental test provided value of HIC higher than the limit one expressed in the standard. However, this aspect does not affect the purpose of determining a methodology for a numerical-experimental correlation.

The sled and passenger seat systems are launched at the prescribed speed against a steel bar deceleration system in order to reproduce the required simulation pulse. The resultant longitudinal deceleration over time is shown in Figure 2.

A dedicated test rig has been set up to reproduce the overall seat installation within the aircraft cabin. Prototype seat has been installed on the test sled with effective seat track. Proper seat belt installation required a test rig able to guarantee the correct position of aircraft/belt interface points with respect to the seat. Additionally, the test rig has been oversized in order to minimize the effects of any deformation occurring during the test.

2.1. Hybrid FE-MB Model. In a hybrid FE-MB model, the user diversifies the modelling, where possible, by integrating within the same solver rigid bodies with deformable finite element components, with the advantage of computing time reduction. The paper [23] reported an alternative strategy of simulation, named coupling, that as well as the hybrid one

FIGURE 1: Experimental test.

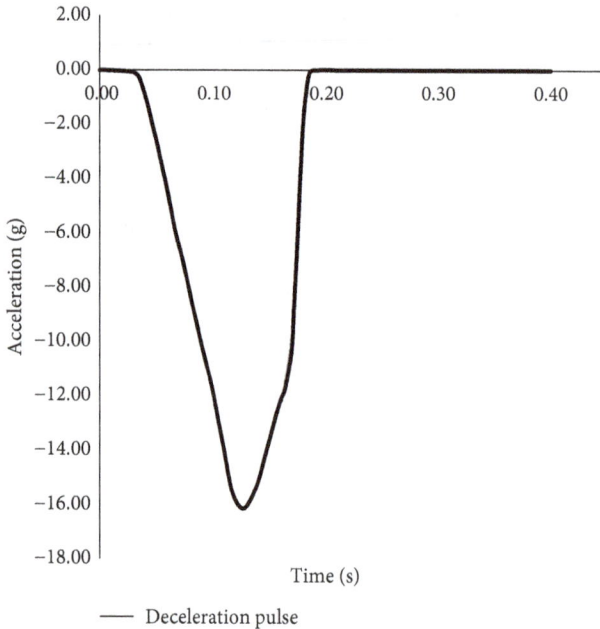

FIGURE 2: Resultant deceleration versus time curve characterizing the seat sled test.

gathers features and advantages of both FE and MB methods; according to this strategy, two separate solvers (one FEM and the other MB) work in parallel, exchanging information, and the connection between the software is represented by contact and constraint forces acting on the elements of the models. Generally, the effectiveness of these techniques lies in the combination of multibody versatility with finite element level of accuracy; these simulations are more flexible than full FEM. The hybrid method, contrary to the coupling approach, does not use an external FE code; for this reason, the computational costs are strongly reduced. Specifically, it has been assessed in a preliminary study [22, 23] that the simulation of the deformability of both seat and restraint system, which cannot be accomplished by means of the MB method, influences significantly the kinematics of the ATD during the seat sled test. This aspect can be empathised by comparing the ATD trajectory simulated by a full FE model (Figure 3(a)) with the trajectory simulated by a full MB one (Figure 3(b)). In fact, according to Figure 3(b), the ATD of the full MB model does not hit the bulkhead, unlike that of the full FEM (Figure 3(a)).

As a result, the modelling of the whole seat system by FEM, in the hybrid FE-MB model, appears to be the most efficient strategy to simulate the seat sled test. Consequently, the other parts, such as the ATD and the bulkhead, are modelled by means of rigid bodies according to the MB approach to reduce the computing time.

The improvement of this hybrid FE-MB model with respect to the one presented by authors in [23] lies in a different setup of the kinematic joints of the MB components. The interconnection structure of a multibody system depends strictly on the definition of the kinematic joints. The equations of motion (Newton-Euler) (1 and 2) of a rigid body, referred to its centre of gravity, are

$$m_i \ddot{r}_i = F_i, \tag{1}$$

$$J_i \cdot \dot{\omega}_i + \omega_i \times J_i \cdot \omega_i = T_i, \tag{2}$$

where m_i is the mass, J_i is the inertia tensor with respect to the centre of gravity, ω_i is the angular velocity vector, F_i is the resultant force vector, and T_i is the resultant torque vector relative to the centre of gravity. For each body, F_i and T_i include the constraint forces and torques due to joints which cannot be determined until the acceleration of the system is known, in contrast with all other forces and torques which depend only on position and velocity quantities. Equations (2) and (3) are multiplied by a variation of the position vector, δr_i, and a variation of the orientation, $\delta \pi_i$, and the resulting equations are summed for all bodies of the system.

$$\sum \delta r_i \cdot \{m_i \ddot{r}_i - F_i\} + \delta \pi_i \cdot \{J_i \cdot \dot{\omega}_i + \omega_i \times J_i \cdot \omega_i - T_i\} = 0. \tag{3}$$

When the variations δr_i and $\delta \pi_i$ of connected bodies are such that the constraints caused by the joint are not violated, the constraint forces and torques in joints will cancel.

The model has been developed within the TNO Madymo® software [24, 25] environment (Figure 4), which contains both MB and FE solvers. In the finite element module, solid hexa, penta, 1D beam, and shell elements can be chosen. However, since the Madymo is a native MB code, the FE module is not characterized by the same accuracy of other native FE codes, especially for 3D finite elements. Hence, in order to improve the precision of the simulations, it has been preferred to model the whole seat by means of shell element type for a total of 105,226 elements and 151,219 nodes.

The modelling of inertial properties of the seat system has been guaranteed by the definition of both materials' density and thickness for each shell element.

The adopted hybrid approach allows the use of different integration methods for the equations of motion for both FE and MB modules. For short-duration crash analyses, explicit integration methods are preferred.

The hybrid approach is in any case based on the assumptions that the parts considered rigid do not influence the behaviour of deformable parts [7]. The deceleration pulse (Figure 2) has been applied to the seat fixed to the slide, along the seat sled test longitudinal direction as shown in Figure 4. Gravity and initial velocity have been applied to all parts of the model. The test case selected for the experimental test consists of a metallic double seat, fabricated from aluminium 2024-T351 and 7075-T651 components. Cushions are made of foam material [26], which constitutive law has been shown in Figure 5 in true stress-strain. For each material, an elastic-plastic model has been selected.

(a)

(b)

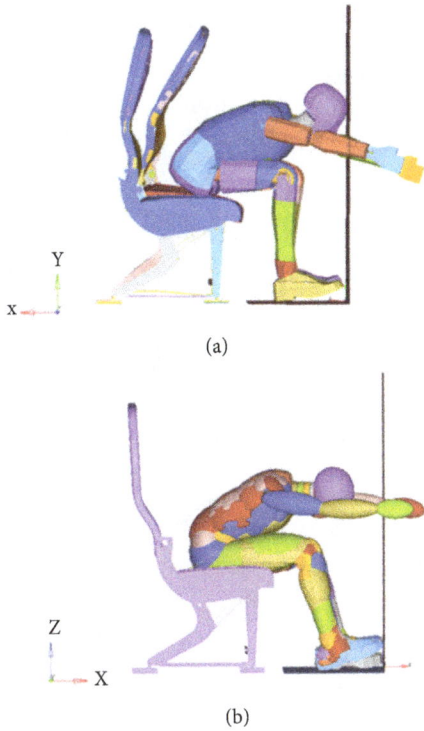

FIGURE 3: Full FE (a) and full MB (b) models.

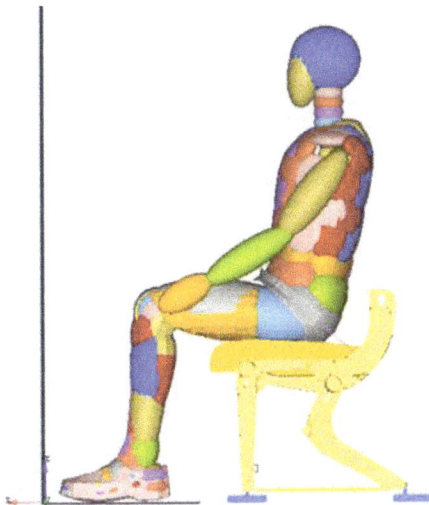

FIGURE 4: Hybrid model.

FIGURE 5: Foam constitutive curve.

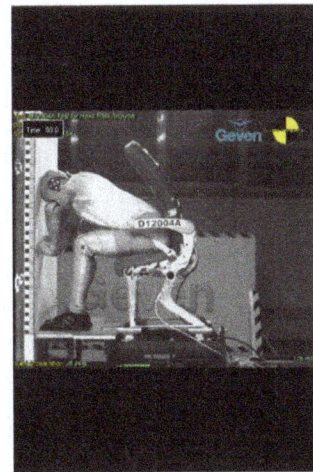

(a)

Hybrid FEM/MB
Loadcase 1 : time = 0.175000
Frame 36

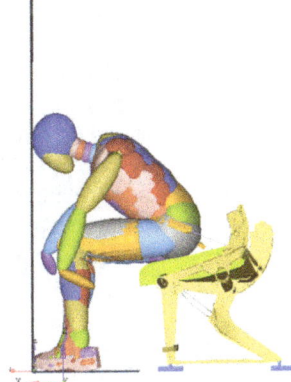

(b)

FIGURE 6: Experimental test (a) and hybrid model (b).

3. Results and Discussion

To assess the reliability of the proposed hybrid model, the head resultant acceleration has been numerically and experimentally monitored, allowing consequently the calculation of the experimental and predicted HIC values.

The numerical and experimental frames corresponding to the instant of time in which the ATD head hits the bulkhead are shown in Figure 6. According to Figure 6, it is possible to observe that the seat deformation (Figure 6(a)) is in good agreement with the predicted one (Figure 6(b)).

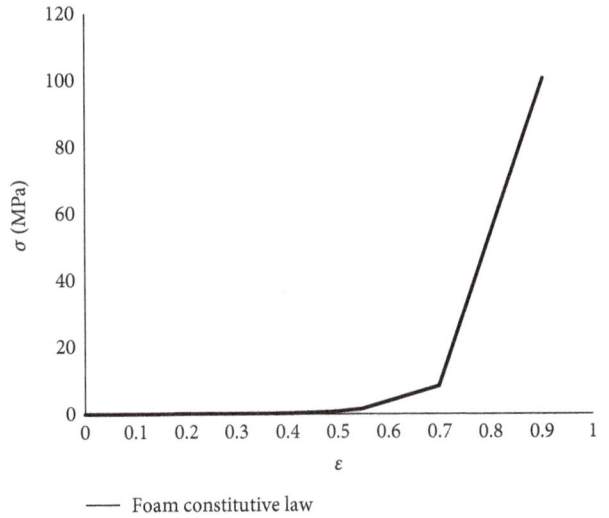

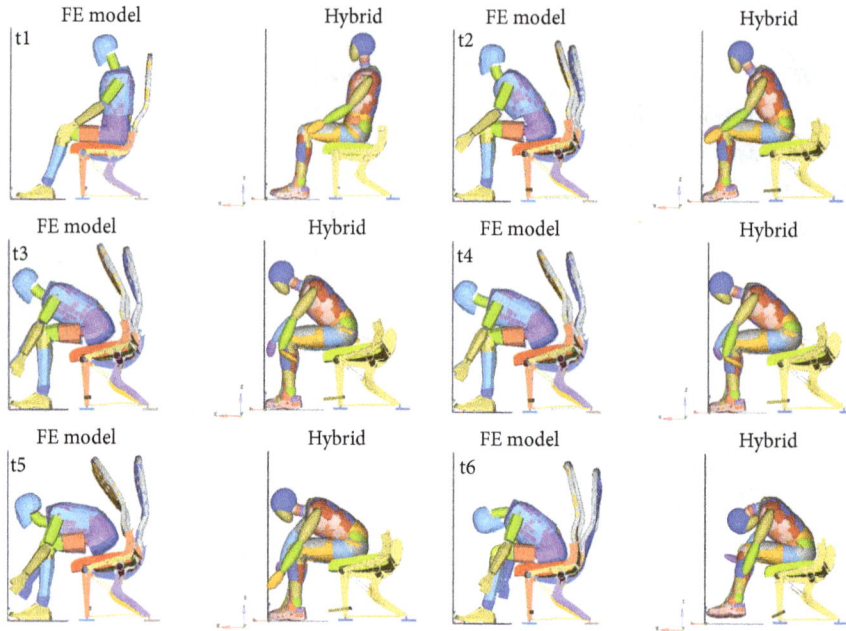

FIGURE 7: Full FE and hybrid model results.

Moreover, results of the hybrid model have been compared to those predicted by the full FE model presented by the authors in [22, 23]. Some numerical frames extracted by both full FE and hybrid models have been compared in Figure 7.

Figure 7 shows a good correlation in terms of seat system and ATD kinematics.

Moreover, the predicted head path, induced mainly by the effects of the deformation on the seat frame, has been compared with the experimental one in Figure 8.

According to Figure 8, a good level of accuracy can be noticed.

Concerning the head resultant accelerations, Figure 9 compares results provided by the experimental and numerical investigations. Acceleration versus time curves have been filtered with SAE filter 1000 [27]. For a better comparison, the numerical curves have been shifted a few milliseconds to make sure that their maximum occurs at the same instant as that of the experimental one.

From Figure 9, it can be noticed that even if all predicted acceleration peaks are well-predicted, the same cannot be said for HIC values. Specifically, HIC value carried out by the full FE simulation is significantly higher than the ones provided by the experimental test and hybrid model. The full FE overestimation can be attributed to the fact that HIC value is calculated by (4), which considers mainly the area under the full FE acceleration versus time curve larger than the other ones.

$$\text{HIC} = \max\left\{ (t_2 - t_1)\left[\frac{1}{(t_2 - t_1)} \int_{t_1}^{t_2} a(t)dt \right]^{2.5} \right\}, \quad (4)$$

where $a(t)$ is the resultant head acceleration measured in g and t_1 and t_2 are the extremes of the integration interval

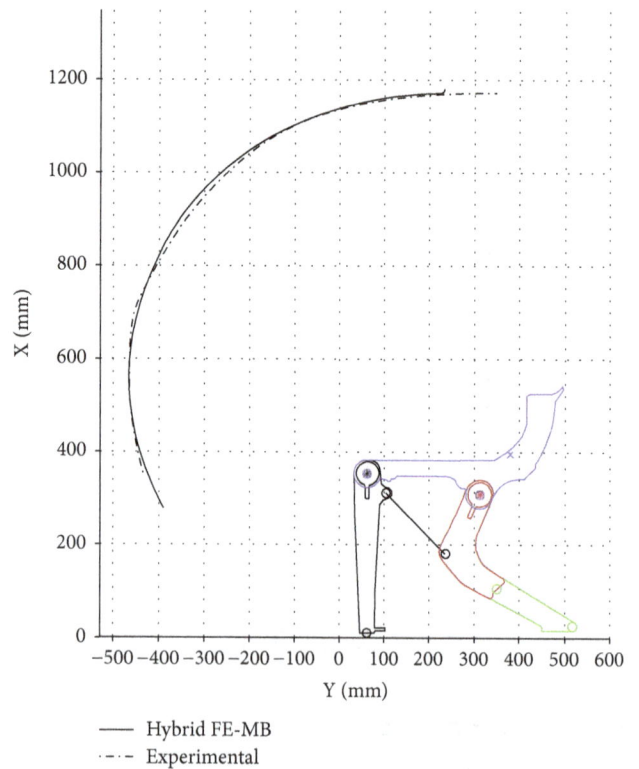

FIGURE 8: Head path correlation.

containing the head acceleration peak measured in seconds for the HIC calculation.

It is very important to emphasize that the computational time of the numerical simulations are about 20 hours for the full FE simulation and about 2 hours for the hybrid one.

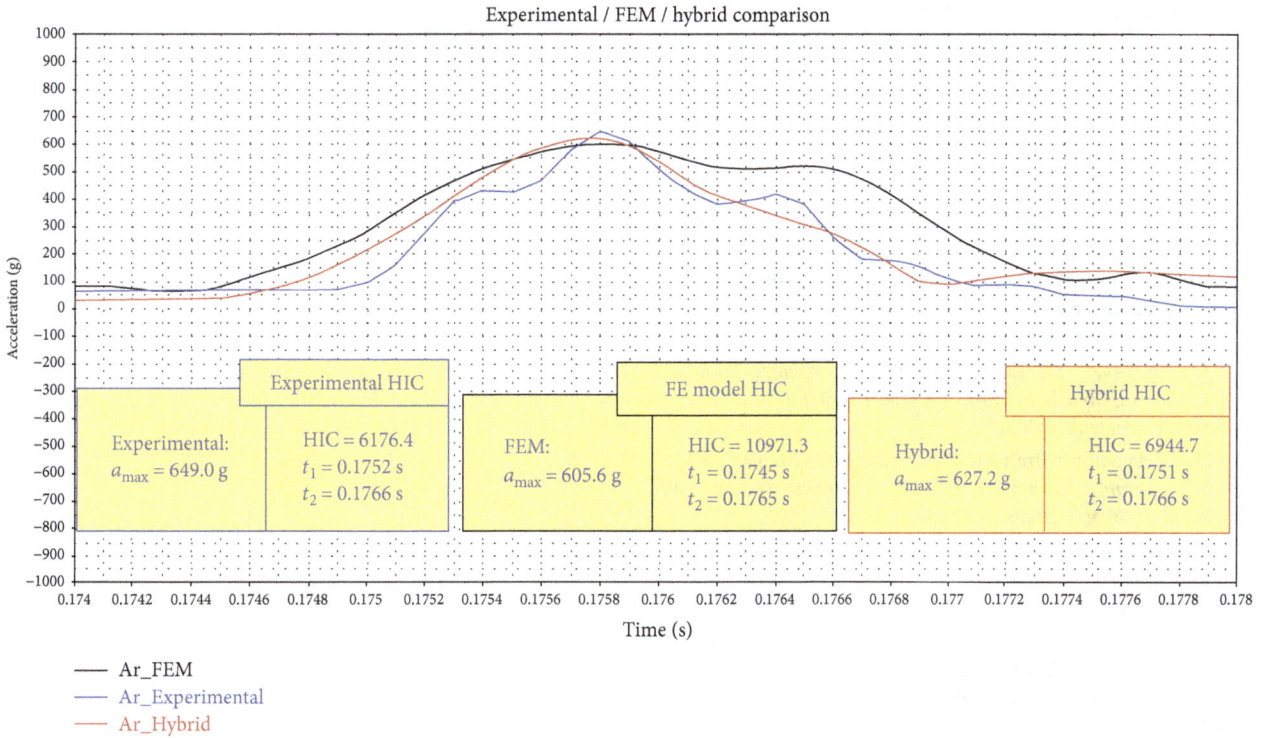

FIGURE 9: Head resultant accelerations.

4. Conclusions

In this paper, a hybrid FE-MB model has been developed, and its reliability has been assessed against the experimental test results provided by the Geven S.p.A. company and the numerical results carried out by a full FE simulation, presented by the authors in previous papers [22, 23]. The hybrid model allowed simulating the seat sled test, reducing significantly the computational costs with respect to those requested by a full FE strategy and at the same time improving the level of accuracy that can be achieved by a full MB model which does not permit the modelling of the deformability of the seat system. As a result, the hybrid approach is a good solution to exploit both FE accuracy and MB lower computing time.

The performed numerical-experimental result correlation demonstrates the efficiency of the proposed hybrid model in simulating the phenomenon. Moreover, the correlation of the numerical results achieved by the hybrid model with the results carried out by the full FE simulation showed that the former is also able to simulate the kinematics of both seat and dummy during the crash event.

Conflicts of Interest

The authors declare that there are no conflicts of interest regarding the publication of this paper.

Acknowledgments

The research activity has been developed within IMM (Interiors with Multifunctional Materials) Research Project funded by Regione Campania, Italy.

References

[1] Federal Aviation Regulation FAR 25.562, "Emergency landing dynamic conditions," in *Federal Aviation Regulation*, Federal Aviation Administration FAA, 2006.

[2] National Aerospace Standards - NAS 809, *Specification - Aircraft Seats and Berths*.

[3] Federal Aviation Regulation, *Dynamic Evaluation of Seat Restraint Systems and Occupant Protection on Transport Airplanes*, Federal Aviation Administration, 2006.

[4] N. Dhole, V. Yadav, and G. Olivares, *Certification by Analysis of a Typical Aircraft Seat*, National Institute for Aviation Research, Wichita, Kansas, 2012.

[5] P. Bhonge, *A Methodology for Aircraft Seat Certification by Dynamic Finite Element Analysis*, Department of Mechanical Engineering, Wichita State University, 2008.

[6] T. M. Hermann, M. W. Holmes, and D. Hartley, "Development of a dynamic aircraft seat analysis capability," in *SAE Technical Paper Series*, September 2001.

[7] J. S. Ruan, C. Zhou, T. B. Khalil, and A. I. King, *Techniques and Applications of Finite Element Analysis of the Biomechanical Response of the Human Head to Impact*, CRC Press LLC, 2001.

[8] R. Happee, A. Janssen, E. Fraterman, and J. Monster, "Application of MADYMO occupant models in DYNA/MADYMO coupling," in *4th European LS-DYNA Users Conference*, Ulm, Germany, 2003.

[9] H. Boucher and C. D. Waagmeester, "Enhanced FAA-HYBRID III numerical dummy model in MADYMO for aircraft occupant safety assessment," in *SAE Technical Paper Series*, September 2003.

[10] H. J. Mertz, P. Prasad, and G. Nusholtz, "Head injury risk assessment for forehead impacts," in *SAE Technical Paper Series*, February 1996.

[11] J. S. H. M. Wismans, *Injury Biomechanics*, Eindhoven University of Technology, Division of Fundamentals (WFW), 1994.

[12] D. H. Laananen, "Crashworthiness analysis of commuter aircraft seats and restraint systems," *Journal of Safety Research*, vol. 22, no. 2, pp. 83–95, 1991.

[13] SAE Aerospace Standard AS8049B, *Performance Standard for Seats in Civil Rotorcraft, Transport Aircraft, and General Aviation Aircraft*, 2005.

[14] Federal Aviation Administration, "Advisory Circular 25.562-1B," *Dynamic Evaluation of Seat, Restraint Systems and Occupant Protection on Transport Airplanes*, 2006.

[15] Federal Aviation Regulation, *Methodology for Dynamic Seat Certification by Analysis for Use in Parts 23, 25, 27, and 29 Airplanes and Rotorcraft*, U.S. Department of Transportation, 2003.

[16] P. Prasad and H. J. Mertz, "The position of the United States Delegation to the ISO Working Group 6 on the use of HIC in the automotive environment," in *SAE Technical Paper Series*, June 1985.

[17] C. D. Waagmeester, E. Hassel, and C. G. Huijskens, "Design optimization and evaluation of a three point harness seat design for airbus A320 aircraft family," in *The Fourth Triennial International Aircraft Fire and Cabin Safety Research Conference*, November 2004.

[18] C. Olschinka and A. Schumacher, "Dynamic simulation of flight passenger seats," in *Proceeding of the 5th LS-DYNA Anwenderforum*, Ulm, October 2006.

[19] H. Lankarani and P. Bhone, "Finite element modelling strategies for dynamic aircraft seats," in *SAE Technical Paper Series*, August 2008.

[20] H. Gaber, D. Browen, and C. Molnar, "Modelling of commuter category aircraft seat," in *U.S. Department of Transportation*, 2004.

[21] J. Christl, S. Kunz, P. Bayrasy, I. Kalmykov, and J. Kleinert, "FEA-MBS-coupling approach for vehicle dynamics," in *Proceedings of 2nd European Conference on Coupled MBS-FE Applications*, pp. 29–32, Pembroke, Ontario, October 2015.

[22] M. Guida, A. Manzoni, A. Zuppardi, F. Caputo, F. Marulo, and A. De Luca, "Development of a multibody system for crashworthiness certification of aircraft seat," *Multibody System Dynamics*, 2018.

[23] F. di Napoli, A. De Luca, F. Caputo, F. Marulo, M. Guida, and B. Vitolo, "Mixed FE-MB methodology for the evaluation of passive safety performances of aeronautical seats," *International Journal of Crashworthiness*, pp. 1–12, 2018.

[24] TASS BV, *MADYMO Reference Manual Version 7.5*, 2013.

[25] TASS BV, *MADYMO Coupling Manual Version 7.5*, 2013.

[26] "Development of a seat cushion replacement component test method for dynamically certified seat," 2001, AGATE report, C-GEN-3433B-1 (Rev N/C).

[27] Society of Automotive Engineers, *Recommended Practice: Instrumentation for Impact Test - Part 1*, Electronic Instrumentation, SAE J211/1, 1995.

10

Structural Damage Detection using Nonlinear Vibrations

M. Carminati⑩ **and S. Ricci**⑩

Department of Aerospace Science and Technology, Politecnico di Milano, Via La Masa 34, 20156 Milano, Italy

Correspondence should be addressed to S. Ricci; sergio.ricci@polimi.it

Academic Editor: Linda L. Vahala

Nonlinear vibrations emerging from damaged structures are suitable indicators for detecting defects. When a crack arises, its behavior could be approximated like a bilinear stiffness. According to this scheme, typical nonlinear phenomena as the presence of superharmonics in the dynamic response and the variation of the oscillation frequency in time emerge. These physical consequences give the opportunity to study damage detection procedures with relevant improvements with respect to the typical strategies based on linear vibrations, such as high sensitivity to small damages, no need for an accurate comparison model, and behavior not influenced by environmental conditions. This paper presents a methodology, which aims at finding suitable nonlinear phenomena for the damage detection of three contact-type damages in a panel representing a typical aeronautical structural component. At first, structural simulations are executed using MSC Nastran models and reduced dynamic models in MATLAB in order to highlight relevant nonlinear behaviors. Then, proper experimental tests are developed in order to look for the nonlinear phenomena identified: presence of superharmonics in the dynamic response and nonlinear behavior of the lower frequency of vibration, computed using the CWT (continuous wavelet transform). The proposed approach exhibits the possibility to detect and localize contact-type damages present in a realistic assembled structure.

1. Introduction

In the actual aeronautical field, the damage detection methodologies are achieving a significant importance due to their great influence on both safety and cost/time optimization. Traditional maintenance schemes based on statistical predictions of times between controls are not optimal for cost savings. Moreover, maintenance implies two kinds of costs for operators [1]: direct costs (repairs) and indirect costs (stop imposed to the airplanes). Both of them are becoming excessive due to the augmentation of the operational life of several airplanes. Indeed, the probability for airplanes to suffer from unpredicted flight loads increases with that, forcing the trend of maintenance costs to grow. These aspects justify the interest in new inspection methodologies able to detect cracks and defects but at the same time guaranteeing time and cost reductions.

The traditional approaches to damage detection are represented by the NDT (nondestructive testing) techniques. They are off-line and localized inspections, developed since the early to mid of 1960s, mainly related to visual inspections,

ultrasonic inspections, eddy currents, acoustic emissions, radiography, thermography, and shearography [2, 3]. For the goals before highlighted, the interest has recently switched to structural health monitoring (SHM) schemes. They are online and global procedures, based on the idea of directly applying a system of detection giving the following benefits: absence of accessibility requirements for operating the tests, absence of necessity to predict possible collocations of damages before executing the trials, and a single time session for tests. SHM methods are classified as follows [4]: "modal-based methods," characterised by the comparison between modal parameters of the current structure with those of the undamaged one; "response-based methods," characterised by the analyses of the results to proper dynamic tests; and "model-based methods," characterised by the comparisons of results from experimental tests with those ones from numerical simulations.

The typical SHM methods are the linear vibrational schemes [5, 6], developed from the hypotheses of linear, time-independent structures. Among them, there has been a wide research on modal techniques since the 1970s [7].

These ones are based on the idea that a damage causes a reduction of the local physical stiffness, causing the modification of modal parameters (eigenfrequencies, damping ratios, and eigenvectors). Therefore, the comparison between these physical elements in the current state and in the undamaged one is suitable for detecting damages. These methodologies exhibit some weaknesses. At first, the traditional linear eigenfrequencies analyses are weakly sensitive to small damages. According to [8], a crack owning an area 15–20% of the cross section of a beam structure determines a reduction of the natural frequency of the element from 1% to 5%. This aspect is the consequence of the global character of modal methods, and thus it can be overpassed by augmenting the amount of data to analyze. In this scenario, the actual research is exhibiting interests in data mining (DM) techniques, such as the artificial neural networks (ANN), principal component analysis (PCA) [9], and genetic algorithms (GA), able to reduce the amount of data to be processed for SHM. Reference [10] represents a very recent review paper that includes a large and complete bibliography of papers about DM in structural damage detection. Then, traditional linear analyses are influenced by external conditions as temperature, which drive modal parameters to change. Finally, traditional linear analyses require accurate pieces of information on the undamaged structure as they focus on a comparison. This aspect implies the necessity for proper undamaged models or test results on the undamaged structures.

Nonlinear vibrational methods in SHM context are emerging in order to overcome the previous limitations. The basic idea around them is that a damage causes a nonlinear behaviour in the structure where it grows. For this reason, it is sufficient to detect the nonlinearities in order to state elements on the defect presence. According to [11], a simple but realistic way to represent the mechanical behaviour of a damage is the usage of a bilinear stiffness: when the crack is open, no material can locally sustain loads and thus a stiffness reduction happens; when it is closed, the originally undamaged stiffness is restored. This model justifies several nonlinear phenomena in vibrations that have been used in some research studies available in literature: absence of homogeneity [12], variation of frequency with amplitude of vibration [11], superharmonic generation [8, 11], and sum/difference of the input harmonics in the dynamic response [13]. The potential of nonlinear vibrational methods is related to their benefits: sensitivity to small defects, making them appealing in the assessment of damages at the very early moments of the growing process; absence of comparing models, influencing the time for the analyses; and insensitivity to environmental conditions, remarkable in the aeronautical context. According to these benefits, the research has concentrated on analytical/numerical modeled structures, with breathing cracks inside. The current paper presents an extension of these methods in a structure reproducing an aeronautical stiffened panel, in which three different damaged scenarios have been taken into account. Each defect is represented by the absence of connections between a portion of a stiffener and a base plate. The procedure aims at executing the first two steps of the detection procedure scheduled by Rytter in [14]: detection and localization, leaving the damage quantification step for the next research phase.

2. Theoretical Background

In order to justify the phenomena emerging from nonlinearities suitable for the damage detection procedure here discussed, a short theoretical background on nonlinear systems is mandatory.

2.1. Nonlinear Phenomena. A nonlinear dynamical system owns particular characteristics compared to those of the linear processes. Firstly, the dynamical response to an input excitation depends both on the frequency of the excitation and its amplitude. For a linear system, instead, the only dependence is that from the excitation frequency. Then, nonlinear systems exhibit sensitivity to initial conditions. In the case of linear time-invariant systems, the impulse response is the linear combination of specific shapes, called eigenvectors, vibrating at specific eigenfrequencies. These invariant sets do not depend on initial conditions. Moreover, given a certain group of harmonics in the input signal, the output of a nonlinear system may present subharmonics or superharmonics. Finally, due to the high sensitivity to initial conditions, nonlinear systems can exhibit a chaotic behavior also with deterministic nonlinear models.

The mathematician Volterra proposed an extension of the conventional input-output relation used to describe dynamical systems in the time domain [15]. For a linear time-invariant system, the output $\mathbf{y}(t)$ is related to the input $\mathbf{u}(t)$ by the following convolution:

$$\mathbf{y}(t) = \int_{-\infty}^{+\infty} \mathbf{h}(\tau)\mathbf{u}(t - \tau)\,d\tau, \tag{1}$$

where $\mathbf{h}(t)$ is the impulse response of the system. The form of the relation for nonlinear systems becomes the Volterra series:

$$\mathbf{y}(t) = \sum_{i=1}^{n} \mathbf{y}_i(t), \tag{2}$$

where

$$\mathbf{y}_n(t) = \int_{-\infty}^{+\infty}\int_{-\infty}^{+\infty} \cdots \int_{-\infty}^{+\infty} \mathbf{h}_n(\tau_1, \tau_2, \ldots, \tau_n)\mathbf{u}(t - \tau_1)\mathbf{u} \tag{3}$$
$$\cdot (t - \tau_2) \ldots \mathbf{u}(t - \tau_n)d\tau_1\,d\tau_2 \ldots d\tau_n.$$

The terms $\mathbf{h}_i(t)$ are called "kernels." This relation is correct under specific hypothesis [16], always included in "weakly" nonlinear systems. Each kernel can be transformed into the frequency domain by using the multidimensional Fourier transform (MFT):

$$\mathbf{H}_n(\omega_1, \omega_2, \ldots, \omega_n) = \int_{-\infty}^{+\infty}\int_{-\infty}^{+\infty} \cdots \int_{-\infty}^{+\infty} \mathbf{h}_n$$
$$\cdot (\tau_1, \tau_2, \ldots, \tau_n)e^{-i(\omega_1\tau_1 + \omega_2\tau_2 + \cdots + \omega_n\tau_n)}d\tau_1 d\tau_2 \ldots d\tau_n. \tag{4}$$

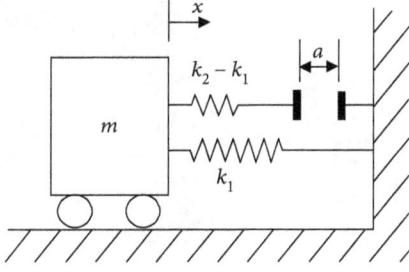

FIGURE 1: 1 dof system with bilinear spring.

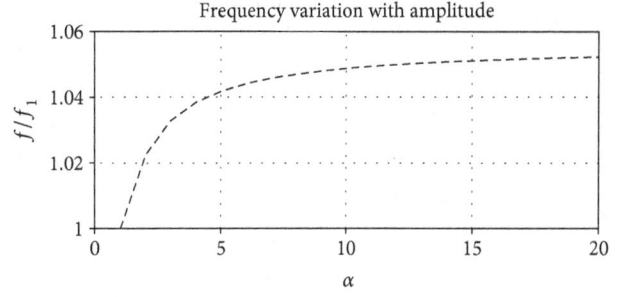

FIGURE 2: Normalized frequency with different amplitudes.

The terms $\mathbf{H}_n$ are the higher-order frequency response functions (HOFRF). It can be easily noticed how the Volterra series including only its first term corresponds to (1). Thus, the input-output relation for linear systems is a particular case of (2). The MFT of the single kernel is the Fourier transform (FT) of the impulse response, which represents the frequency response function (frf) for asymptotically stable linear systems.

The Volterra series can be used in order to justify the superharmonic presence in the output of nonlinear systems. Indeed, in the case of a periodic excitation composed by the harmonic $\omega(u(t) = e^{i\omega t})$, the expression of the response $y(t)$ is the following one:

$$y(t) = H_1(\omega)e^{i\omega t} + H_2(\omega, \omega)e^{i2\omega t} + \cdots + H_n(\omega, \omega, \ldots, \omega)e^{in\omega t}. \tag{5}$$

The response exhibits multiple integers of the harmonic ω through the effect of the HOFRF. It is possible to match the values of each HOFRF to the Fourier transform of the input and the output [17].

Another relevant aspect described by the Volterra series is the absence of homogeneity. While each single convolution of the series is homogeneous, the response is the combination of homogeneous convolutions with different numbers of input functions. Consequently, differences in the amplitude of the input cause a nonhomogeneous system response.

2.2. A Simple 1 Degree of Freedom Example. In order to highlight the two effects underlined at the end of the previous paragraph for nonlinear systems and relate them to a bilinear spring mechanism representing a damage, it is here proposed a simple 1 degree of freedom (dof) example coming from the work of Neild [11]. Figure 1 represents a mass m linked to the wall by a bilinear spring: the spring constant is k_1 until the displacement x is less than a; then, it becomes k_2. Distance a in the model can be both considered as a crack opening size or a contact gap. The first representation is that of a "breathing crack," as indicated in [11]. The second one suits with the contact-type damages analyzed in the work here presented.

The equations of motion for the two different situations are here proposed:

$$m\ddot{x} + k_1 x = 0, \quad \text{if } x \leq a,$$
$$m\ddot{x} + k_2 x + (k_1 - k_2)a = 0, \quad \text{if } x > a. \tag{6}$$

The solutions of the equations for the two cases can be easily found once the initial conditions are defined. Considering the motion evolving from an initial position $x(0) = -\alpha a$, with an initial speed equal to zero, the following equation expresses the motion behaviour:

$$x = C \cos\left(\sqrt{\frac{k}{m}}\,t + \varphi\right) - \frac{K}{k}t, \tag{7}$$

where $k = k_1$ and $K = 0$ when $x \leq a$ and $k = k_2$ and $K = (k_1 - k_2)a$ when $x > a$. Constants C equals $-\alpha a$ and constant φ equals 0 in the first case. When $x > a$, their values are the following ones:

$$C = a\frac{k_1}{k_2}\sqrt{1 + \frac{k_2}{k_1}\left(\alpha^2 - 1\right)},$$
$$\varphi = \arctan\left(-\sqrt{\frac{k_2}{k_1}\left(\alpha^2 - 1\right)}\right). \tag{8}$$

Through the analytic calculation of [11], the nondimensional $f_s = f/(\omega_1/2\pi)$ can be computed, where $\omega_1 = \sqrt{k_1/m}$ and f equals the frequency of oscillation. Once the mathematical model is solved, it is possible to determine the nondimensional frequency of vibration for different values of α (Figure 2). Moreover, it is possible to highlight the second and third normalized superharmonics by applying the fast Fourier transform (FFT) to the normalized displacement stories. They are equal to their value divided by $f_1 = \omega_1/2\pi$ (Figure 3).

The results clearly underline the consequences of the bilinear spring in the system response. Figure 2 shows how the vibration frequency depends on the amplitude of the oscillation; Figure 3 underlines the presence of the superharmonics in the dynamic response.

3. Typical Aeronautical Panel: Preliminary Numerical Studies

The panel analyzed in the current work reproduces a typical aeronautical structure. It is the one used in [9, 18]. It consists in a flat rectangular plate, having 1000 mm for width and 1200 mm for length, with a thickness of 1.3 mm. It is edged by two transversal supports with rectangular section

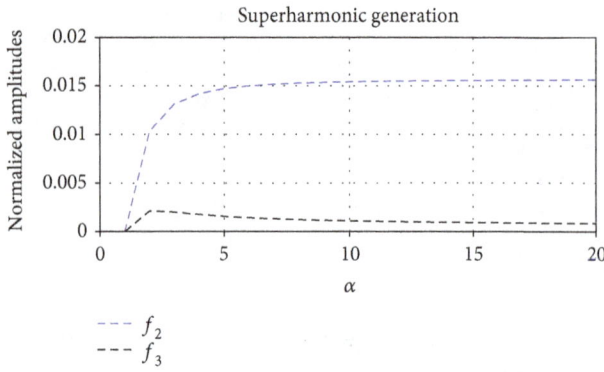

FIGURE 3: Normalized superharmonics with different amplitudes.

FIGURE 4: Aluminum panel used for the analyses.

TABLE 1: Damaged scenario characteristics.

Case	Description
D2	Separation, 80 mm long; ending stiffener position
D3	Separation, 120 mm long; central stiffener position
D5	Separation, 40 mm long; ending stiffener position

FIGURE 5: Detail of D2 damage.

(40 mm × 4 mm), running along the short sides. Moreover, five stiffeners equally spaced reinforce the longitudinal direction: they are all T-shaped, with a length of 1120 mm, a flange and a web of 40 mm, and a constant thickness of 4 mm. The whole panel is built using aluminum alloy 2024. Figure 4 shows the panel.

The three damaged scenarios are represented by the absence of connections between one stiffener and the base plate (two lines of screws are used to connect each stiffener to the base plate). They are called D2, D3, and D5. Table 1 summarizes the characteristics of each one of them.

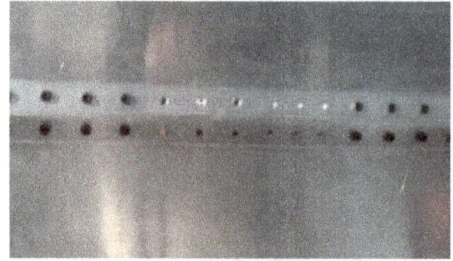

FIGURE 6: Detail of D3 damage.

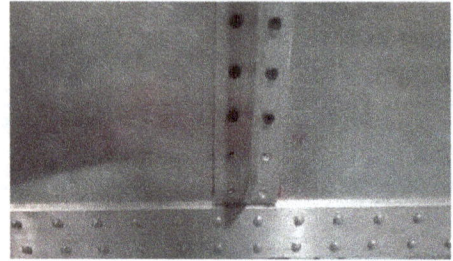

FIGURE 7: Detail of D5 damage.

TABLE 2: First 20 eigenfrequencies for the undamaged panel and for the three damages with their % variations.

Mode	f (Hz)	D2, f (Hz)	D2, %	D3, f (Hz)	D3, %	D5, f (Hz)	D5, %
1	7.95	7.79	−2.01	7.95	0.00	7.83	−1.51
2	11.38	11.37	−0.09	11.38	0.00	11.38	0.00
3	23.67	23.44	−0.97	23.67	0.00	23.51	−0.68
4	27.90	26.67	−4.41	27.90	0.00	27.60	−1.08
5	47.72	43.15	−9.58	47.72	0.00	46.56	−2.43
6	49.82	48.93	−1.79	49.82	0.00	49.11	−1.43
7	72.67	71.01	−2.28	72.58	−0.12	71.47	−1.65
8	83.43	73.63	−11.75	83.43	0.00	80.68	−3.30
9	87.70	85.73	−2.25	87.57	−0.15	86.44	−1.44
—	—	93.42	—	—	—	—	—
10	99.22	99.41	0.19	98.89	−0.33	98.20	−1.03
11	101.11	99.88	−1.22	101.10	−0.01	99.76	−1.34
12	101.22	109.68	8.36	100.97	−0.25	100.20	−1.01
13	114.46	114.63	0.15	114.45	−0.01	112.51	−1.70
14	117.42	117.48	0.05	115.53	−1.61	116.92	−0.43
15	118.05	117.86	−0.16	117.70	−0.30	117.86	−0.16
16	121.78	121.75	−0.02	121.77	−0.01	121.75	−0.02
17	121.82	126.57	3.90	121.65	−0.14	119.86	−1.61
18	128.90	131.10	1.71	128.88	−0.02	128.15	−0.58
19	137.98	137.16	−0.59	137.68	−0.22	137.36	−0.45
20	139.43	139.27	−0.11	139.34	−0.06	139.33	−0.07

Figures 5–7 show the different cases.

Table 2 shows the first 20 eigenfrequencies of the undamaged free-free panel and exhibits the eigenfrequencies and their percentage variations for the three damaged cases considered. All the results have been obtained by means of numerical FEM mode analyses in MSC Nastran, by

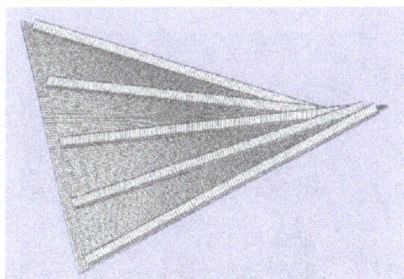

FIGURE 8: Mode 1.

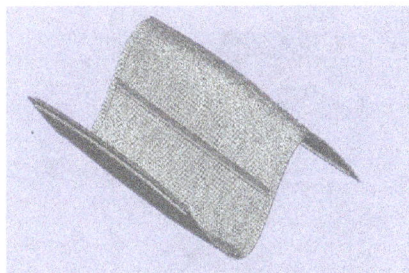

FIGURE 9: Mode 2.

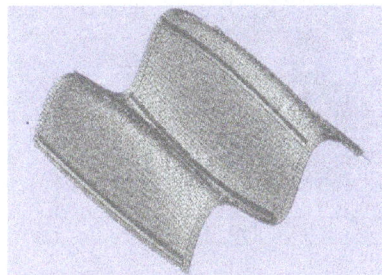

FIGURE 10: Mode 3.

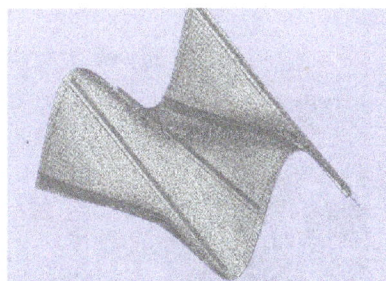

FIGURE 11: Mode 4.

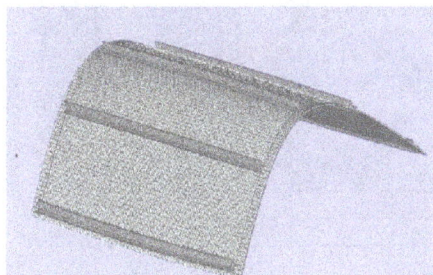

FIGURE 12: Mode 5.

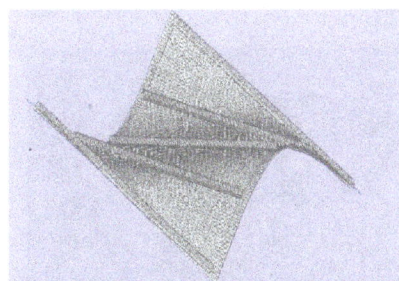

FIGURE 13: Mode 6.

representing each damaged area using not merged nodes between the damaged stiffener and the panel. In the case of D2, an uncorrelated mode at frequency 93.42 Hz has been identified.

Figures 8–13 show the numerical shapes for the first six undamaged modes.

Figures 14–16 show the MAC (modal assurance criterion) matrixes obtained comparing the first undamaged 20 modes with the corresponding damaged ones.

Each MAC matrix has been obtained by comparing the transversal (out of plane) displacements of a set of 120 nodes equally spaced along the lateral and longitudinal directions of the panel.

Figure 14 clearly exhibits the not correlated mode at 93.42 Hz (mode number 10) emerging for D2, which interrupts the dominant behavior of the main diagonal of the matrix.

In order to highlight the presence of nonlinear phenomena, preliminary numerical tests have been run both using complete FEM (finite element method) models with the software MSC Nastran and reduced models with the

software MATLAB. The panel has been analyzed as a free-free structure.

3.1. FEM Analyses on Complete Models. The original undamaged panel model has been developed in Nastran by using 12800 CQUAD4 elements (12000 for the base plate mesh, 800 for the transversal supports) and 560 CBEAM elements for the stiffeners. Each damaged scenario has been obtained from the original model by using a certain amount of GAP elements, connecting pairs of nodes in the vertical direction: they are nonlinear tools used to implement a bilinear stiffness. In particular, the penalty approach has been used in order to simulate the contact mechanism in each damaged area: when there is a contact between the two nodes connected by the gap, a stiffness of 10^9 N/m acts in order to avoid interpositions. The cards used to implement these gaps are CGAP and PGAP for the property definition [19].

For each scenario, a limited number of gap elements have been used only for the central pairs of nodes involved in

Undamaged panel:

Mode	f(Hz)	1	2	3	4	5	6	7	8	9	10	11	12	13	14	15	16	17	18	19	20
1	7.95	1	0.01	0.01	0.02	0.06	0.07	0	0.03	0.01	0.02	0	0.01	0.01	0.01	0.01	0	0.10	0.04	0.04	0
2	11.38	0.01	1	0.01	0	0	0.01	0.01	0	0.05	0	0.03	0.02	0.06	0.04	0.06	0	0	0.05	0.08	0.23
3	23.67	0.02	0.01	1	0.06	0.03	0	0.01	0	0	0.02	0.01	0.01	0.05	0.08	0	0.01	0	0.04	0.09	0
4	27.9	0	0.01	0.04	0.99	0.18	0.02	0.04	0.12	0.01	0.01	0.15	0.14	0.2	0.18	0.1	0.04	0.09	0.12	0.04	0
5	47.72	0	0.01	0.02	0.07	0.79	0.5	0.02	0.3	0.36	0.02	0.03	0.03	0.04	0.07	0.09	0.05	0.01	0.02	0.04	0.07
6	49.82	0.07	0	0.02	0.07	0.5	0.86	0	0.26	0.02	0.1	0.04	0.05	0.07	0.03	0.01	0	0.15	0.09	0.08	0
7	72.67	0.01	0.01	0.01	0.04	0.02	0	0.99	0.01	0.12	0.01	0.02	0.06	0.02	0.07	0.04	0	0.02	0.02	0.01	0.02
8	83.43	0	0	0.04	0.03	0.1	0.02	0.02	0.68	0.12	0.49	0.03	0.04	0.33	0.1	0.07	0.03	0.01	0.03	0.12	0
9	87.7	0	0.05	0.01	0.01	0.27	0.2	0.08	0.29	0.92	0.3	0.12	0.05	0.09	0.01	0.05	0	0.01	0.02	0.07	0
10	99.22	0.01	0	0.02	0.24	0.03	0.01	0.04	0.16	0.04	0.13	0.6	0.59	0.21	0.12	0.05	0.01	0.01	0.06	0.08	0
11	101.1	0.06	0.01	0	0.12	0.01	0.11	0.35	0.03	0.04	0.02	0.25	0.35	0	0.01	0.05	0.23	0.05	0	0.04	0.03
12	101.2	0	0.08	0	0.04	0.09	0.03	0.01	0.29	0.08	0.59	0.07	0.29	0.5	0.2	0.06	0.01	0	0.12	0.13	0.01
13	114.5	0	0	0.08	0.01	0.01	0	0.04	0.12	0.14	0.2	0.08	0.01	0.6	0.79	0.08	0.1	0.03	0.14	0.01	0.06
14	117.4	0	0.04	0	0	0.08	0.07	0.01	0.08	0	0.08	0.04	0.01	0.25	0.19	0.89	0.3	0.02	0.11	0.07	0.06
15	118.1	0	0	0	0.06	0.02	0.01	0.03	0.04	0.03	0.03	0.05	0.01	0.15	0	0.25	0.95	0.03	0.08	0.04	0.04
16	121.8	0.11	0	0	0.02	0.12	0.14	0	0.04	0	0.07	0.04	0.06	0.12	0.15	0.13	0.03	0.95	0.2	0.11	0
17	121.8	0	0.01	0.01	0.29	0.1	0.01	0.06	0.14	0.01	0.13	0.01	0.04	0.28	0.39	0.3	0.09	0.3	0.55	0.35	0.01
18	128.9	0	0	0.10	0.01	0.05	0.01	0.01	0.13	0.01	0.06	0.02	0	0.11	0.12	0.03	0.01	0.02	0.69	0.71	0.05
19	138	0	0.25	0	0.01	0.06	0.02	0.01	0.02	0.02	0.03	0.03	0.01	0.05	0.02	0.03	0.03	0.01	0.05	0	0.97
20	139.4	0	0	0	0	0.01	0.02	0.01	0.01	0.01	0	0.01	0	0.01	0.03	0.02	0.01	0	0.02	0.06	0.19
f(Hz)		7.79	11.4	23.4	26.7	43.2	48.9	71	73.6	85.7	93.4	99.4	99.9	110	115	117	117	122	127	131	137
Mode		1	2	3	4	5	6	7	8	9	10	11	12	13	14	15	16	17	18	19	20

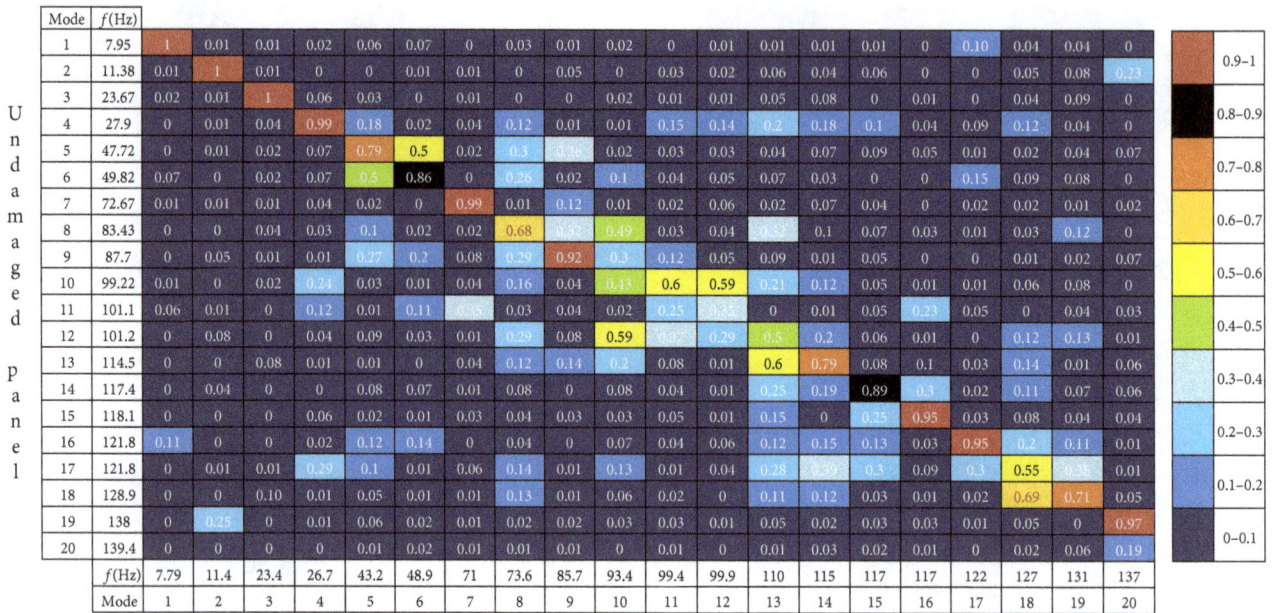

D2 damaged panel

FIGURE 14: Undamaged panel–D2 damaged panel MAC.

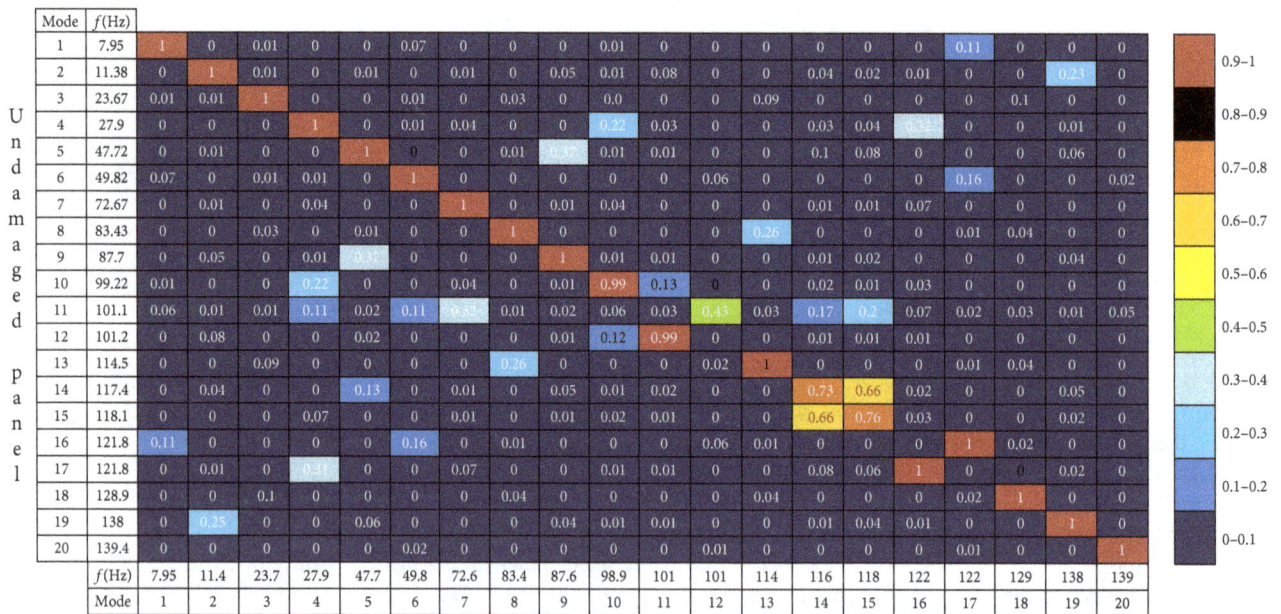

Undamaged panel:

Mode	f(Hz)	1	2	3	4	5	6	7	8	9	10	11	12	13	14	15	16	17	18	19	20
1	7.95	1	0	0.01	0	0	0.07	0	0	0	0.01	0	0	0	0	0	0	0.11	0	0	0
2	11.38	0	1	0.01	0	0.01	0	0.01	0	0.05	0.01	0.08	0	0	0.04	0.02	0.01	0	0	0.23	0
3	23.67	0.01	0.01	1	0	0	0.01	0	0.03	0	0.0	0	0	0.09	0	0	0	0	0.1	0	0
4	27.9	0	0	0	1	0	0.01	0.04	0	0	0.22	0.03	0	0	0.03	0.04	0.31	0	0	0.01	0
5	47.72	0	0.01	0	0	1	0	0	0.01	0.37	0.01	0.01	0	0	0.1	0.08	0	0	0	0.06	0
6	49.82	0.07	0	0.01	0.01	0	1	0	0	0	0	0	0.06	0	0	0	0	0.16	0	0	0.02
7	72.67	0	0.01	0	0.04	0	0	1	0	0.01	0.04	0	0	0	0.01	0.01	0.07	0	0	0	0
8	83.43	0	0	0.03	0	0.01	0	0	1	0	0	0	0	0.26	0	0	0.01	0.04	0	0	0
9	87.7	0	0.05	0	0.01	0.37	0	0	0	1	0.01	0.01	0	0	0.01	0.02	0	0	0	0.04	0
10	99.22	0.01	0	0	0.22	0	0	0.04	0	0.01	0.99	0.13	0	0	0.02	0.01	0.03	0	0	0	0
11	101.1	0.06	0.01	0.01	0.11	0.02	0.11	0.37	0.01	0.02	0.06	0.03	0.43	0.03	0.17	0.2	0.07	0.02	0.03	0.01	0.05
12	101.2	0	0.08	0	0	0.02	0	0	0.01	0	0.12	0.99	0	0	0.01	0.01	0.01	0	0	0	0
13	114.5	0	0	0.09	0	0	0	0	0	0	0.26	0	0.02	1	0	0	0	0.01	0.04	0	0
14	117.4	0	0.04	0	0	0.13	0	0.01	0	0.05	0.01	0.02	0	0	0.73	0.66	0.02	0	0	0.05	0
15	118.1	0	0	0	0.07	0	0	0.01	0	0.01	0.02	0.01	0	0	0.66	0.76	0.03	0	0	0.02	0
16	121.8	0.11	0	0	0	0.16	0	0.01	0	0	0	0	0.06	0.01	0	0	1	0.02	0	0	0
17	121.8	0	0.01	0	0.31	0	0	0.07	0	0	0.01	0.01	0	0	0.08	0.06	0	1	0	0.02	0
18	128.9	0	0	0.1	0	0	0	0	0.04	0	0	0	0	0	0.04	0	0	0	1	0.02	0
19	138	0	0.25	0	0	0.06	0	0	0	0.04	0.01	0.01	0	0	0.01	0.04	0.01	0	0	1	0
20	139.4	0	0	0	0	0	0.02	0	0	0	0	0.01	0	0	0	0	0	0.01	0	0	1
f(Hz)		7.95	11.4	23.7	27.9	47.7	49.8	72.6	83.4	87.6	98.9	101	101	114	116	118	122	122	129	138	139
Mode		1	2	3	4	5	6	7	8	9	10	11	12	13	14	15	16	17	18	19	20

D3 damaged panel

FIGURE 15: Undamaged panel–D3 damaged panel MAC.

order to reduce the time required for the analyses with the SOL 129 ("nonlinear transient solution"): damages D2 and D3 present three gaps and damage D5 presents two gaps.

The campaign of FEM analyses with complete models has regarded the following tests:

(i) Double sine excitation tests: a 5 Hz vertical (out of plane) sinusoidal excitation and a 15 Hz one have been located on the opposite sides of the damaged stiffener, with amplitudes of 50 N and duration of 10 s. Acceleration stories in the excited nodes and in the other three nodes along the damaged stiffener have been windowed with the Hanning window and finally analyzed by using the FFT (all the elaborations have been executed in MATLAB). Figures 17–19 show the locations of the exciting nodes (node 12 for 5 Hz excitation and node 93 for 15 Hz excitation) and the

Mode	f(Hz)																				
1	7.95	1	0	0.01	0.01	0.03	0.07	0	0.01	0.01	0.01	0	0	0.01	0	0	0.03	0.1	0.01	0.03	0.01
2	11.38	0	1	0.01	0	0.01	0.01	0.01	0	0.05	0.03	0.04	0.04	0.01	0.03	0.01	0.03	0	0.02	0	0.25
3	23.67	0.02	0.01	1	0.01	0.01	0	0.01	0.02	0	0	0.01	0.01	0.09	0.02	0.01	0.01	0	0.08	0.03	0.01
4	27.9	0	0.01	0.01	1	0.04	0.01	0.04	0.03	0.01	0.14	0.11	0.1	0.06	0.07	0.07	0.3	0.1	0.03	0.02	0.01
5	47.72	0	0.01	0.01	0.02	0.93	0.33	0.01	0.09	0.37	0.01	0.02	0.03	0.01	0.11	0.05	0	0	0	0.02	0.08
6	49.82	0.07	0	0.01	0.03	0.36	0.94	0	0.08	0.01	0.04	0.03	0.06	0.03	0.01	0	0.02	0.15	0.03	0.07	0.03
7	72.67	0.01	0.01	0.01	0.04	0.01	0	1	0.02	0.09	0.03	0.02	0.04	0.02	0.01	0.01	0.07	0.02	0.01	0.02	0.02
8	83.43	0	0	0.04	0.01	0.04	0.02	0.02	0.96	0.12	0.14	0.05	0.05	0.29	0.05	0.02	0.05	0.01	0.02	0.09	0.03
9	87.7	0	0.05	0.01	0	0.34	0.13	0.05	0.13	0.98	0.07	0.1	0.02	0.07	0.05	0	0.01	0	0.01	0.01	0.06
10	99.22	0.01	0	0	0.23	0.01	0.01	0.04	0.07	0.02	0.81	0.4	0.41	0.04	0.03	0.02	0.07	0.01	0.02	0.1	0.04
11	101.1	0.06	0.01	0	0.11	0.01	0.11	0.34	0.02	0.04	0.03	0.27	0.34	0.01	0.05	0.22	0.09	0.05	0.03	0.03	0.01
12	101.2	0	0.08	0	0.01	0.02	0.02	0.01	0.12	0.04	0.54	0.54	0.53	0.12	0.06	0.01	0.1	0	0.04	0.12	0.04
13	114.5	0	0	0.08	0	0.01	0	0.03	0.22	0.07	0.09	0.08	0.01	0.93	0.3	0.09	0.08	0.02	0.1	0.05	0.03
14	117.4	0	0.04	0	0	0.11	0.04	0.01	0.02	0.01	0.03	0.04	0.01	0.2	0.87	0.34	0.17	0.01	0.05	0.07	0.04
15	118.1	0	0	0	0.07	0.01	0	0.03	0.01	0.02	0.01	0.05	0	0.17	0.34	0.92	0.16	0.03	0.04	0.08	0
16	121.8	0.11	0	0	0.01	0.07	0.15	0	0.01	0	0.03	0.04	0.06	0.04	0.03	0	0.36	0.95	0.07	0.05	0.03
17	121.8	0	0.01	0	0.31	0.03	0.01	0.06	0.05	0	0.04	0.02	0.04	0.07	0.1	0.01	0.86	0.31	0.15	0.25	0.1
18	128.9	0	0	0.1	0	0.01	0.01	0.01	0.08	0.01	0.03	0.02	0.01	0.03	0.06	0.02	0.1	0.02	0.97	0.18	0.02
19	138	0	0.25	0	0	0.06	0.01	0.01	0.01	0.03	0.02	0.02	0.01	0.04	0.02	0.03	0.02	0.01	0.05	0.38	0.9
20	139.4	0	0	0	0	0.01	0.02	0.01	0	0.01	0	0	0	0.02	0	0	0.02	0	0	0.14	0.26
f(Hz)		7.83	11.4	23.5	27.6	46.6	49.1	71.5	80.7	86.4	98.2	99.8	100	113	117	118	122	120	128	137	139
Mode		1	2	3	4	5	6	7	8	9	10	11	12	13	14	15	16	17	18	19	20

Color scale legend:
- 0.9–1
- 0.8–0.9
- 0.7–0.8
- 0.6–0.7
- 0.5–0.6
- 0.4–0.5
- 0.3–0.4
- 0.2–0.3
- 0.1–0.2
- 0–0.1

(Row label: Undamaged panel)

D5 damaged panel

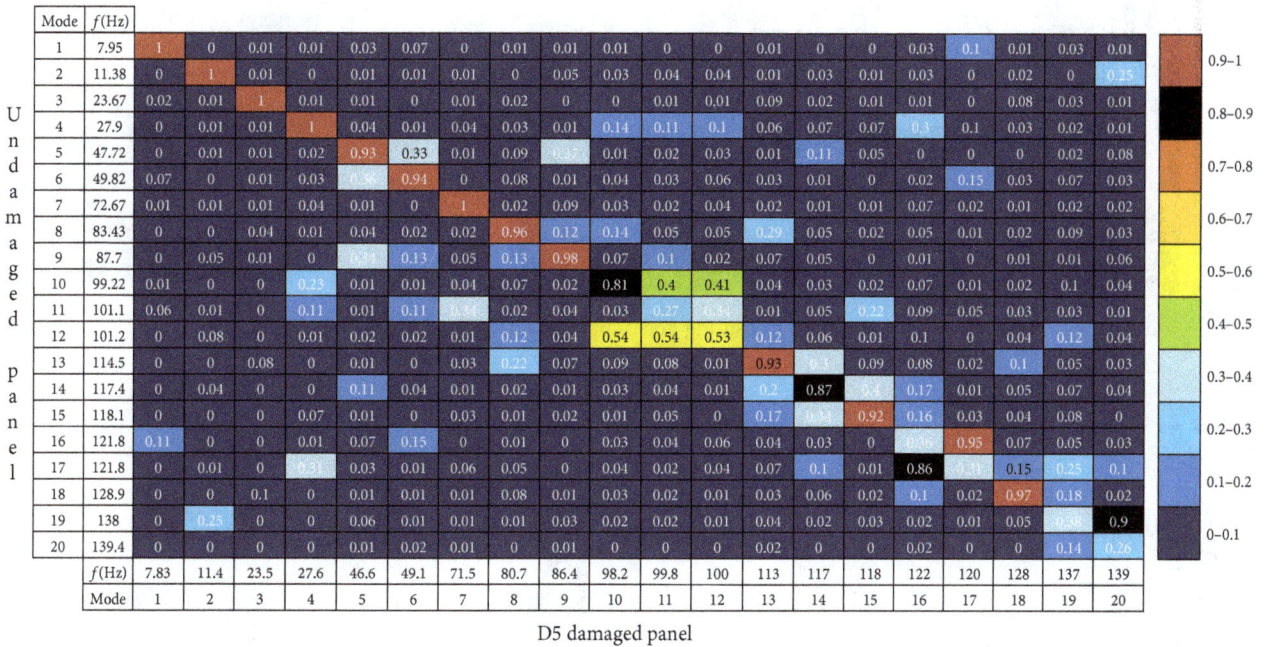

FIGURE 16: Undamaged panel–D5 damaged panel MAC.

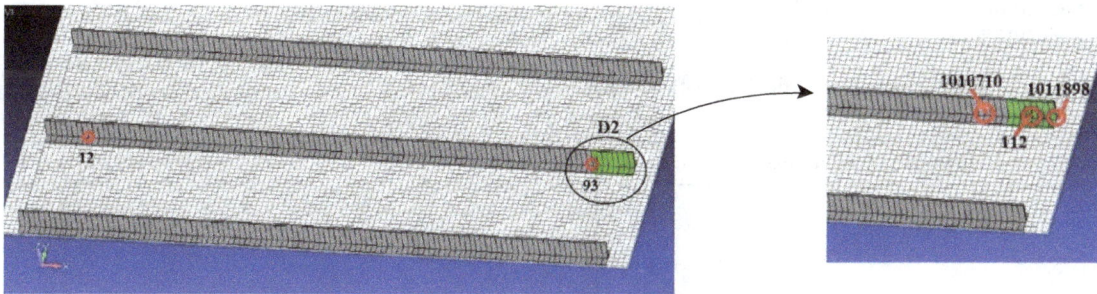

FIGURE 17: Excitation and measuring nodes for D2 damage along the second stiffener.

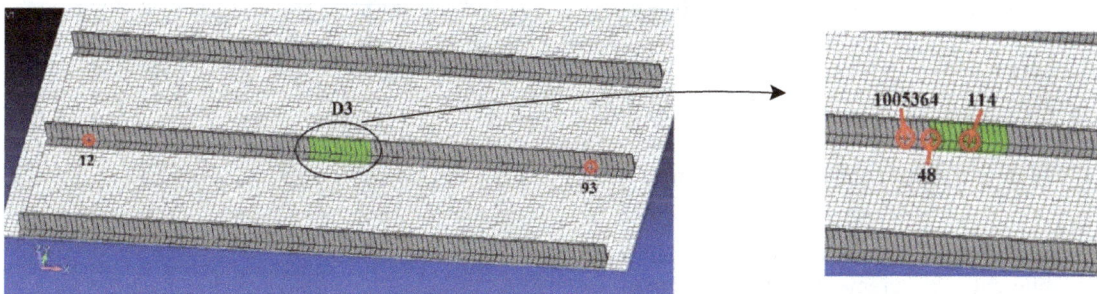

FIGURE 18: Excitation and measuring nodes for D3 damage along the second stiffener.

positions of the other three measuring nodes for each damage, along the second stiffener.

The double sine excitation at the indicated frequencies has been chosen in order to look for possible linear combinations, sum, or difference of the excitation frequencies in the outputs. The use of a difference between the two frequencies matching a structural eigenfrequency determines a relevant response, as indicated in the literature [13]. As a benefit of nonlinear methods is to provide detection without having preliminary pieces of information on the structure, it has been decided to consider 5 Hz and 15 H. They are low frequencies having a difference which does not meet a panel eigenfrequency.

In the case of the damaged scenario D2, a 30 Hz response in the results has been obtained. Figure 20

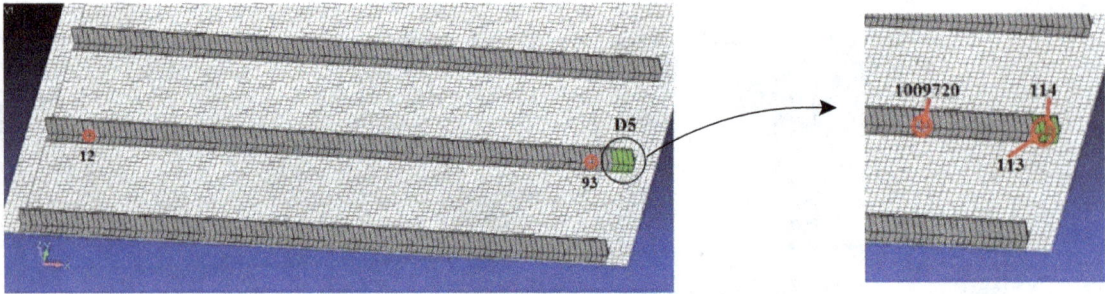

FIGURE 19: Excitation and measuring nodes for D5 damage along the second stiffener.

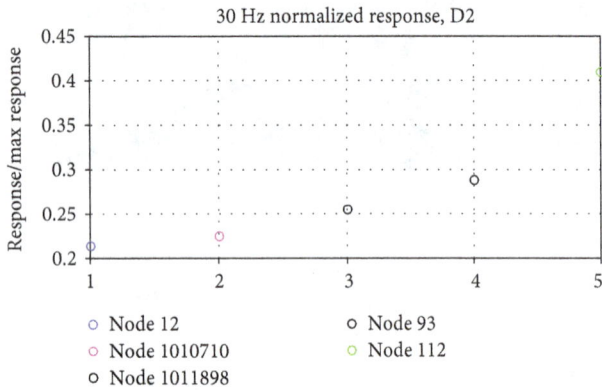

FIGURE 20: D2, 30 Hz normalized responses, double excitation tests.

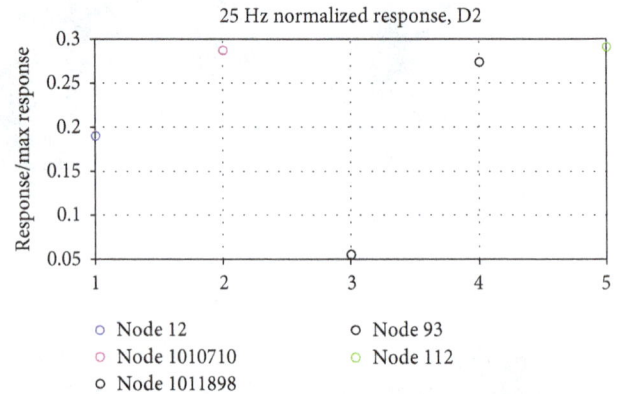

FIGURE 22: D2, 25 Hz normalized responses, double excitation tests.

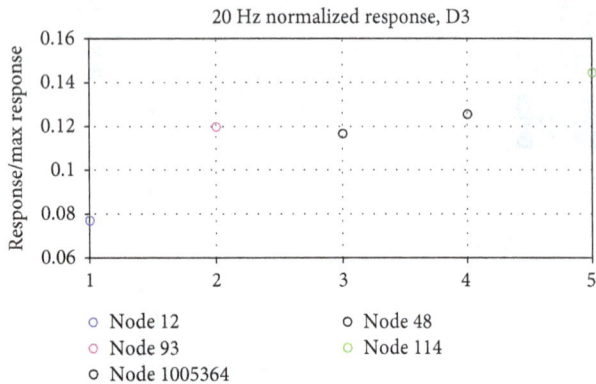

FIGURE 21: D3, 20 Hz normalized responses, double excitation tests.

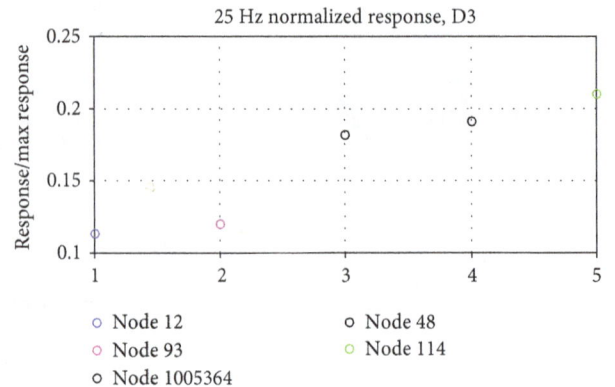

FIGURE 23: D3, 25 Hz normalized responses, double excitation tests.

reports the behavior of the normalized response for the different measurement points, obtained by dividing its value by the maximum one in the spectrum. In the figure, the nodes in the abscissa go closer to the damaged area from the left to the right. Thus, the amount of superharmonic contribution increases with the proximity to the damage. Similar results have been obtained for the 20 Hz contribution in the damaged scenario D3 (Figure 21) and for the 25 Hz contribution in each damaged scenario (Figures 22–24). These representations highlight the possibility to use the amount of the nonlinear response as an indicator for the localization.

(ii) Tests for superharmonics: for each damaged scenario, two different numerical tests have been executed applying a 50 N amplitude vertical sinusoidal force next to each damaged area. It has been decided to use the same frequencies of the previous tests, in order to highlight possible similarities in the results. Thus, in the first case, the frequency is 5 Hz and in the second case, it is 15 Hz for a duration of 12 s. In the case of the 15 Hz excitation, scenario D2 has exhibited the superharmonic 30 Hz highlighted in the tests done before (Figure 25).

(iii) Decay tests: for each damaged scenario, an impulsive vertical load has been applied next to each damaged

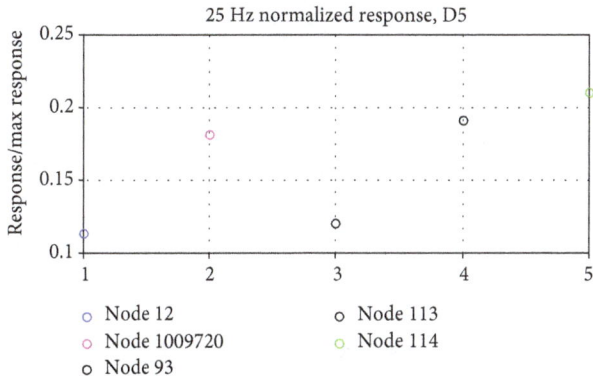

FIGURE 24: D5, 25 Hz normalized responses, double excitation tests.

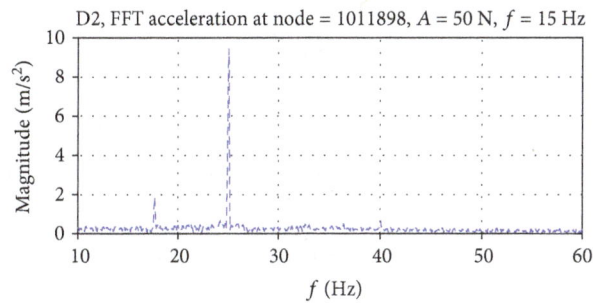

FIGURE 25: D2, direct response FFT, 15 Hz excitation test.

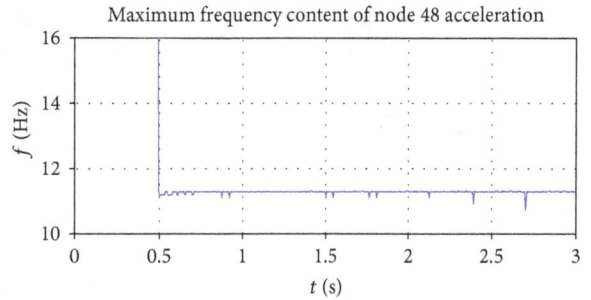

FIGURE 26: D3, decay test CWT, acceleration in the excitation node.

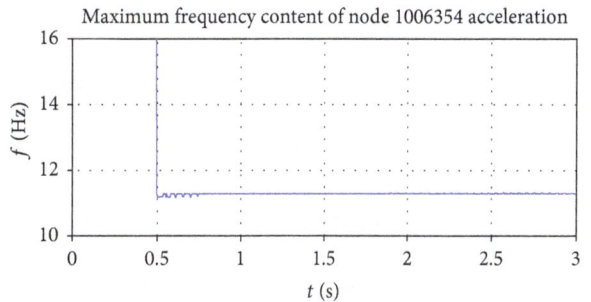

FIGURE 27: D3, decay test CWT, acceleration in the damaged area.

area, represented as a triangular excitation of duration 0.01 s and amplitude 20 N. A damping factor of 1% has been set for a frequency about 7 Hz, which is close to the first eigenfrequency of the undamaged plate. Each simulation with SOL 129 has lasted 5 s, with a time step suitable as to follow the very rapid load story (an initial step of 0.001 s has been used). Accelerations in time have been computed in each excitation node and in a node located in the damaged area. Then, they have been processed by using the continuous wavelet transform (CWT) based on the Morlet wavelets [20–22]. This tool guarantees the decomposition of a time signal in components well localized in time and frequency called "wavelets." Therefore, the CWT allows looking for changes in the behavior of a signal during its evolution. Before its application, the analytic signal [11] has been obtained and then the curves exhibiting the maximum frequency content for different times have been presented in output. The purpose of the analyses has been the search for possible variations of the frequency of vibration in time, due to the nonlinear behavior of each damage. Figures 26 and 27 show the output curves for the D3 case: they exhibit slow fluctuations around the second eigenfrequency of the undamaged plate, about 11 Hz. These little waves can be related to the variation of the local stiffness due to the contact mechanism, but models considering the

entire contact are required in order to focus this phenomenon in a better way.

3.2. FEM Analyses on Reduced Models. The analyses described before have the limitation of long times required, and this does not give the opportunity to model the entire damaged areas with the gaps. For this reason, further analyses have been executed by using reduced-order models, following the Craig-Bampton approach [23] in conjunction with the concentrated nonlinearities [24]. In particular, the boundary set of each model includes those dof connecting the bilinear springs simulating the contact mechanism (all the dof in each contact area), the excitation dof, and the measurement dof. By using MSC Nastran SOL 103, the Craig-Bampton mass and stiffness matrixes have been obtained in output (it has been chosen to consider a number of fixed interface modes equal to 14, in order to cover a wide range of frequencies). From the results with the complete models, it has been decided to focus on the following nonlinear phenomena: superharmonics and variations of vibration frequencies in time. Thus, the following tests have been done with the software MATLAB using the matrixes obtained:

(i) Sine excitation tests: a vertical sinusoidal excitation (amplitude = 50 N) at different frequencies (5, 10, 15, 20, 25, and 30 Hz) has been placed in a node located next to each damaged area. The direct acceleration responses, obtained by using the forward finite difference method on the velocities, have been windowed with the Hanning window and then elaborated through the FFT. The first simulations done on undamped models have shown unclear

TABLE 3: D2, superharmonic contributions, sine excitation tests.

	1st superharmonic	2nd superharmonic	3rd superharmonic
5 Hz	Yes	No	Yes
10 Hz	Yes	No	Yes
15 Hz	Yes	Yes	Yes
20 Hz	Yes	Yes	Yes
25 Hz	Yes	Yes	No

TABLE 4: D5, superharmonic contributions, sine excitation tests.

	1st superharmonic	2nd superharmonic	3rd superharmonic
20 Hz	Yes	No	Yes
25 Hz	Yes	No	No

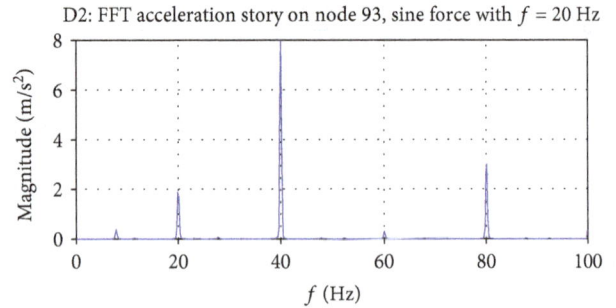

FIGURE 28: D2, direct response FFT, 20 Hz sine excitation.

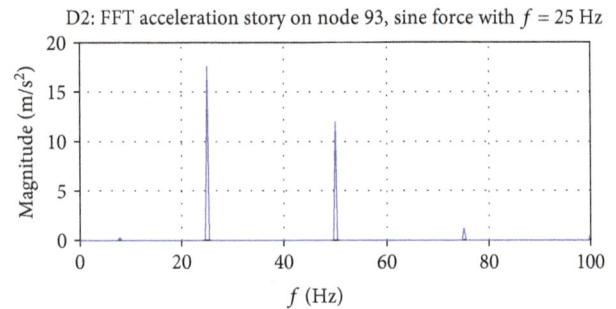

FIGURE 29: D2, direct response FFT, 25 Hz sine excitation.

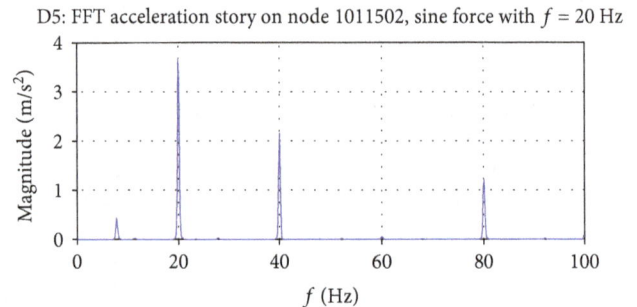

FIGURE 30: D5, direct response FFT, 20 Hz sine excitation.

responses in those cases exhibiting nonlinear behaviors: indeed, the absence of damping causes a limited dissipation of those contributions at frequencies higher than the half of the sampling frequency (time step has been taken equal to 0.005 s). For this reason, tests have been repeated on damped models: a damping matrix proportional to the stiffness matrix has been used, with the proportional constant taken as to give a damping of 1% for a frequency of 7 Hz. The function used to solve the system of nonlinear ordinary differential equation is "ode15s," a stiff solver. In this case, tests have been done for the damaged scenarios D2 and D5, as D3 has not exhibited superharmonics in the previous simulations. Tables 3 and 4 summarize the results obtained in terms of superharmonic contributions found for the two scenarios.

Figures 28 and 29 report the FFT of the 5 s acceleration stories for the damaged scenario D2 with 20 Hz and 25 Hz excitations; the same results for D5 are shown in Figures 30 and 31. These figures highlight the evidence of the superharmonic presence for the 20 Hz and 25 Hz cases.

A theoretical interpretation of this aspect comes from several works in which it is stated that, in the case of crack breathing into beam structures, the higher first superharmonic behavior is excited when the excitation frequency matches one-half of the eigenfrequency [17, 25]. Therefore, the possibility for superharmonics increases when the excitation frequency has multiples close to the eigenfrequencies of the structure. As the panel analyzed exhibits two linear modes in the range between 40 Hz and 50 Hz (see Table 2), the excitations 20 Hz and 25 Hz satisfy this condition.

(ii) Sine excitation tests with multiple acceleration measurements: once the previous tests have revealed the presence of superharmonics, the 20 Hz excitation

simulations have been repeated by computing the accelerations in 10 nodes along the damaged stiffeners, at a distance one to each other of about 106.6 mm (see Figure 32 for the node location). This procedure has aimed at finding possible correlations between the superharmonic contributions and each damaged area (for both D2 and D5, the damage is located at one end of the stiffener). In the case of D2 scenario, the first and third superharmonic contributions have resulted sensitive to the localization. Figure 33 reports the normalized trend for the first superharmonic amplitudes. For D5, the same output is shown in Figure 34, exhibiting the localization trend. Damages D2 and D5 are located close to node 93 (see Figures 17 and 18).

(iii) Sine excitation tests repeated with a different force position. The multiple acceleration measurement tests have been executed exciting in node 12, which

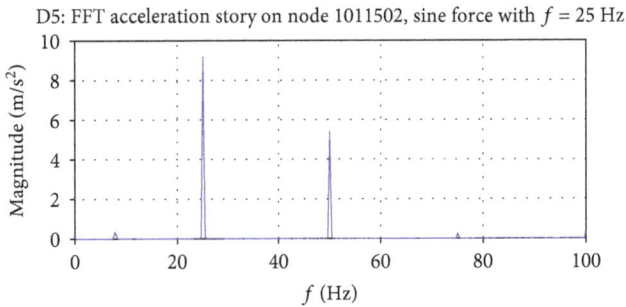

FIGURE 31: D5, direct response FFT, 25 Hz sine excitation.

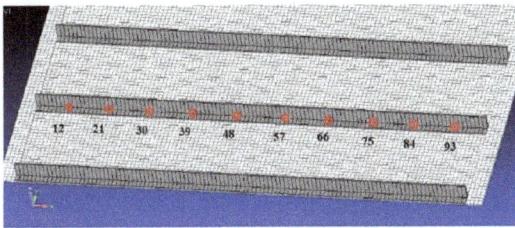

FIGURE 32: Positions of the measuring nodes for experimental superharmonic tests.

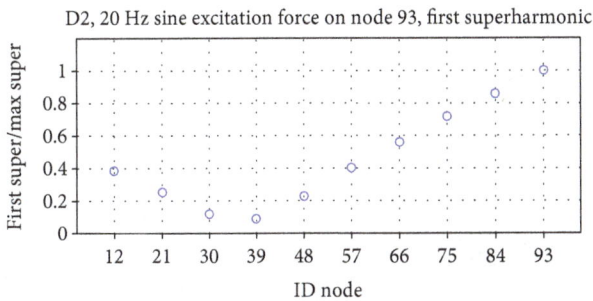

FIGURE 33: D2, 20 Hz excitation, normalized 1st superharmonic.

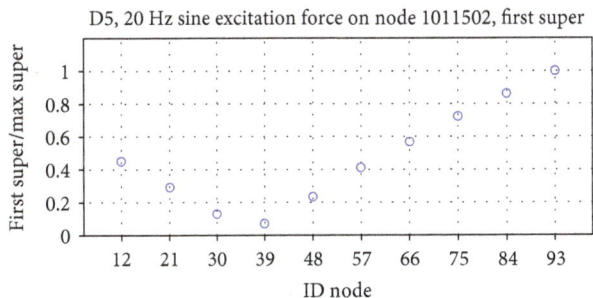

FIGURE 34: D5, 20 Hz excitation, normalized 1st superharmonic.

is located on the opposite side of the damages D2 and D5 (see Figures 17 and 18). In this case, the relevant trend is that obtained for the first superharmonic contributions of the D2 scenario (Figure 35). Even if the localization trend has been obtained also for D5, the absolute values are not significant in this case (the 40 Hz peak in node 93 equals 0.0361 m/s^2, while the 20 Hz one is 3.995 m/s^2).

(iv) Decay tests: as done for the tests with complete models, an impulsive vertical load has been applied next to each damaged area, with the same characteristics of the one used before. A damping matrix proportional to the stiffness one has been introduced in the model. 4 s acceleration signals with a resolution of 0.005 s have been recorded in the excitation node and in a node close to each damaged area. Each signal has been transformed into its analytic version and processed by using the CWT with Morlet wavelets. Results have highlighted relevant fluctuations of the main vibration frequency around an eigenfrequency after a transition time, for the accelerations in the damaged area nodes (Figures 36–38).

(v) As observed for the D3 results with the complete models, the fluctuations in this case are around the second eigenfrequency (11 Hz). On the other hand, they are around the first eigenfrequency (8 Hz) for D2 and D5 cases. This fact is due to the relation between the position of the excitation node and the mode shapes: the first mode, with a frequency about 8 Hz, is a torsion (see Figure 8); the second mode, with a frequency about 11 Hz, is a flexion around the longitudinal direction (see Figure 9) and thus forces the central part of the structure to vibrate vertically. Thus, the central excitation used for D3 excites the second mode more than the first one.

3.3. *Comments on Numerical Results.* The numerical tests described have been executed with the aim of finding suitable nonlinear phenomena for an experimental damage detection technique. From the results discussed, it has been decided to focus on two phenomena highlighted by the use of the reduced-order simulations: the presence of superharmonics with sine excitation tests and the fluctuations of the main vibration frequency for decay tests. Both these aspects are theoretically justified by the analyses described in Section 2. Indeed, the superharmonic presence is a consequence of the nonlinear response described by the Volterra series, while the variation of frequency of vibration results from the activation of a local bilinear stiffness behavior. Double sine excitation tests executed with the complete models have not been considered in the experimental activity as single excitations are more practical to be managed than double excitations, with the consequent impact on time required for the analyses and ease of testing.

In order to exhibit a comparison between the numerical results obtained in the cases of complete and reduced models, Figures 39 and 40 report the first superharmonic contributions found for the 20 Hz excitation on node 93 along the acceleration measuring nodes used for sine excitation tests with multiple acceleration measurements. The damaged case considered is D2. These contributions have been obtained for both the simulations, with correlated orders of magnitudes. Moreover, Figures 41 and 42 show the comparison between the FFT of the direct response (that obtained on node 93).

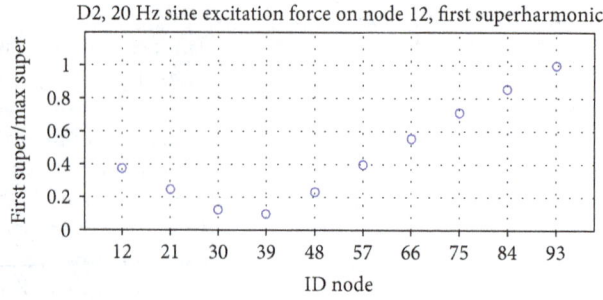

FIGURE 35: D2, 20 Hz excitation in node 12, normalized 1st superharmonic.

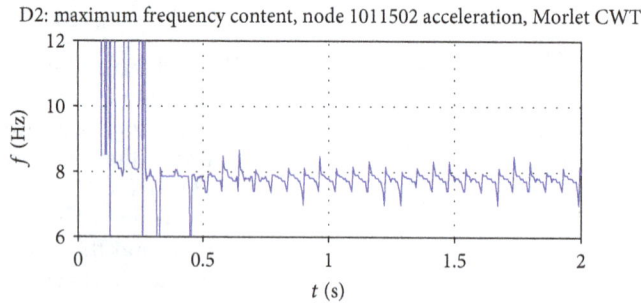

FIGURE 36: D2, decay test CWT, acceleration in the damaged node.

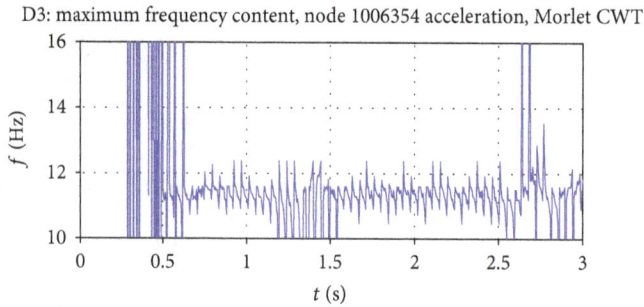

FIGURE 37: D3, decay test CWT, acceleration in the damaged node.

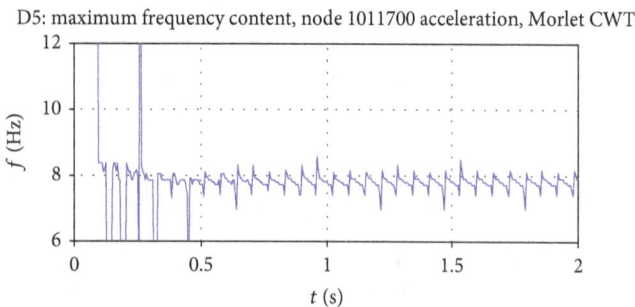

FIGURE 38: D5, decay test CWT, acceleration in the damaged node.

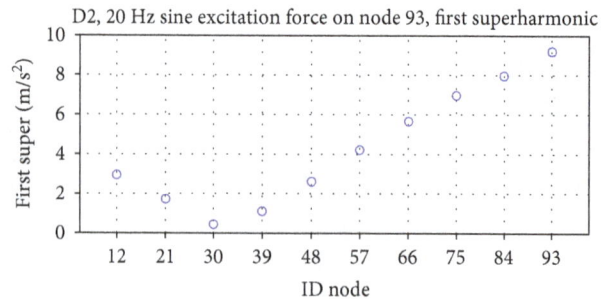

FIGURE 39: D2, 20 Hz excitation 1st superharmonic, full model.

The comparisons shown must be completed highlighting that full models consider a limited area for the gap mechanism implementation in each damaged place. Moreover, reduced model simulations have been executed introducing a damping has written in Section 3.2. These relevant aspects, together with the simplification introduced by the Craig-Bampton approach, justify the difference in the FFT contributions, the detection of relevant contributions at higher superharmonics only in the case of reduced models, and the evident difference in the fluctuations resulting from decay tests.

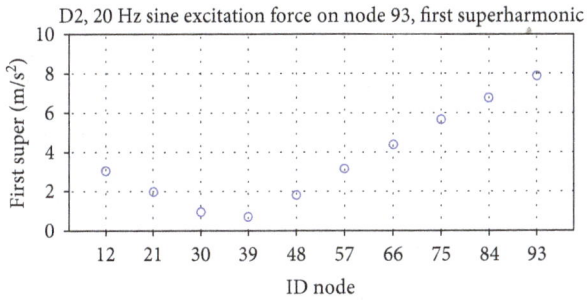

FIGURE 40: D2, 20 Hz excitation 1st superharmonic, reduced model.

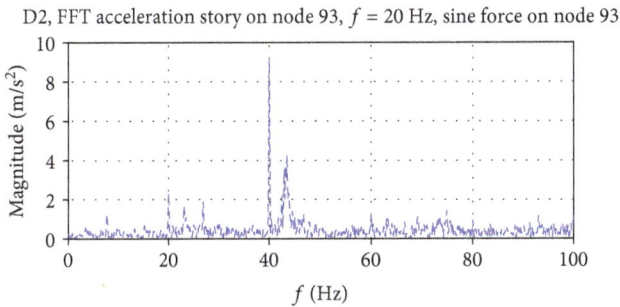

FIGURE 41: D2, node 93 acceleration, FFT, full model.

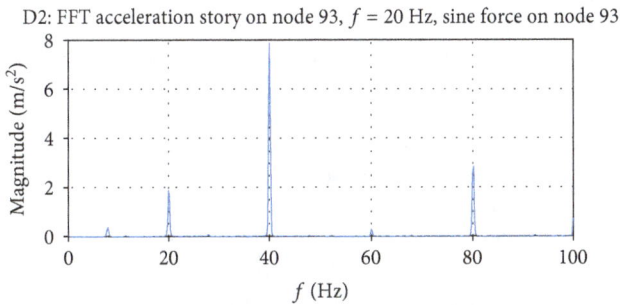

FIGURE 42: D2, node 93 acceleration, FFT, reduced model.

FIGURE 43: Detail of the hanging system.

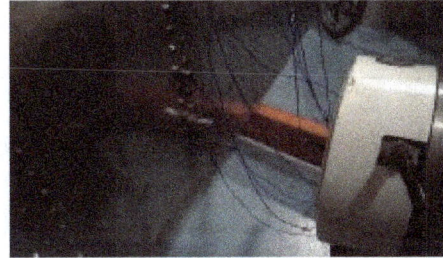

FIGURE 44: Detail of the rod for panel-shaker connection.

TABLE 5: Load levels used for the sine excitation tests.

Load level	Voltage range (V)	Load range (N)
1	0.05	12.4
2	0.1	23.6
3	0.15	34.1
4	0.2	44.3
5	0.25	54
6	0.3	62.7

It is important to underline that the numerical simulations have been based on model assumptions. In particular, the intuitive idea of the local bilinear stiffness behavior of each contact has been followed without an energetic formalization of the contact mechanism. Moreover, the contact has been modeled only in each damaged area, so where it has been supposed that the nonlinear phenomena show their main effects. Finally, each link between the stiffeners and the baseplate has been realized merging the common nodes.

The detail of time integration parameters and model setup for the numerical analyses is present in [18].

4. Experimental Tests

The experimental tests have been conducted with the panel suspended by the use of an elastic cable, which supports the structure at the two extremities. Figure 43 shows the detail of the hanging system. The interference of the low frequency of oscillation due to the cord is limited as to consider the structure in the free-free condition. Both the sine excitation tests and the decay tests have dealt with the measure of accelerations by using the "PCB Piezotronics" accelerometers. The data acquisition and postprocessing are carried out using the Test-Lab software by Siemens.

All the accelerometers are located on the side of the baseplate without the stiffeners.

4.1. Sine Excitation Tests for Superharmonics. Tests for superharmonics have been conducted by forcing the reference structure and the damaged scenarios in two different positions. The first one corresponds to that of node 12 and the second one is the place of node 93. They are located on the opposite sides of the damaged stiffeners, as shown in Figures 17–19. For each single excitation, six load levels have been imposed. Figure 44 shows the detail of the stinger for the panel-shaker connection, and Table 5 reports the six load levels (the voltage range imposed to the signal generator and the corresponding load level measured by the load cell).

For each excitation, ten accelerations have been recorded along the second stiffener in those points corresponding to the measuring nodes of the numerical simulations. Figure 32 shows the location of the measuring and excitation nodes along the second stiffener. Each single test has lasted 10 s. Each signal has been filtered with an imposed band of 1024 Hz and then amplified before its digitalization.

TABLE 6: Number of superharmonic peaks (N) and maximum superharmonic frequency content (f).

Case	N, node 12	f, node 12 (Hz)	N, node 93	f, node 93 (Hz)
Reference	7	160	2	60
D2	2	80	24	980
D3	2	60	26	820
D5	19	700	9	820

The final numerical data have been processed by using the software MATLAB (windowing with the Hanning window and FFT). For each condition (reference structure, D2, D3, and D5), the following 24 tests have been executed:

(i) 20 Hz sinusoidal excitation in node 12, with the six load levels

(ii) 20 Hz sinusoidal excitation in node 93, with the six load levels

(iii) 25 Hz sinusoidal excitation in node 12, with the six load levels

(iv) 25 Hz sinusoidal excitation in node 93, with the six load levels

The choice for the excitation frequencies result from the numerical analyses.

By analyzing the direct responses, the presence of superharmonics has been found also for the reference condition. Moreover, several superharmonics have been detected for each damaged scenario analyzed [18]. This aspect is probably due to the discrete nature of each connection and marks a fundamental consequence: the detection alone by finding the nonlinear phenomena is not possible. Table 6 indicates, for each direct response with the 20 Hz excitation, the number of superharmonic peaks (N) identified and the corresponding frequency range (f): the counted peaks are those ones having a contribution higher than the 10% of the highest response in each spectrum.

The analyses of the direct responses for the 25 Hz excitation have revealed the presence of superharmonics with more limited contributions. For this reason, it has been decided to focus on the 20 Hz excitation cases. As done for the numerical tests, the first superharmonic contributions (40 Hz) in all the measuring nodes have been compared. In particular, for all the cases, these contributions and their normalized values (obtained dividing the data by the highest value in each test) have been reported into figures representing the first superharmonics along the measuring nodes for all the load levels.

4.1.1. Reference Structure. Figures 45 and 46 report the behavior of the first superharmonics along the measuring nodes for all the load levels, exciting in node 12 and node 93. The trends are mainly symmetrical, and the nonlinear phenomena is approximately proportional changing the load levels, as visible from the normalized trends.

4.1.2. D2 Scenario. Figures 47 and 48 report the same figures for the D2 scenario. In the case of excitation in node 93, the force is located in that side of the stiffener closer to the damaged area. As a consequence, the contributions grow towards that direction, with the exception of lower load levels (0.05 V and 0.1 V). In the case of excitation in node 12, even if the maximum contributions are those in the excitation point, the increasing of the load level determines a nonlinear augmentation of the contribution in the damaged area, as visible from the normalized trend. The trigger is reached with load level 0.2 V.

4.1.3. D3 Scenario. Figures 49 and 50 report the same figures for the D3 case. As for the reference structure, the trends are mainly symmetrical.

4.1.4. D5 Scenario. Figures 51 and 52 report the same figures for the D5 scenario. In this case, both the trends own the contributions growing towards node 93, the point closer to the damaged area.

4.1.5. Comments on Results. The analyses of the first superharmonics have revealed the possibility to localize the end damages (D2 and D5) in a qualitative way by comparing the responses to the different load levels and exciting at the opposite sides. Only for those cases of end damages, growing trend oriented towards a predominant direction has been observed. Moreover, for the case of excitation in node 12 in the D2 scenario, the nonhomogeneity explained by the Volterra series has resulted in the increasing of the amount of the contributions of those nodes located in the damaged area with the augmentation of the load level. Therefore, the load level acts like a trigger able to activate the nonlinear contact mechanism. Figure 53 reports the ratios between the first superharmonic magnitudes in load level 0.3 V and load level 0.05 V for all the measuring points for this case: the highest value is that of the damaged area.

Similar conclusions for the qualitative localization are possible analyzing other superharmonics. In particular, [18] reports the results obtained with the second (60 Hz) and the third (80 Hz) ones.

4.2. Decay Tests. Decay tests have been executed by using the PCB instrumented hammer shown in Figure 54.

For each damaged scenario, two impulsive excitations have been applied: the first one located next to the damaged region and the second one located at one end of the damaged stiffener. For each test, three accelerations have been recorded: one in point A, located near the damaged area; one in point B, located in the damaged area; and one in the point corresponding to node 12, at one end of the stiffener. Figures 55–57 show the location of the three measuring points for each damaged scenario. Each 10 s signal has been firstly filtered with an imposed band of 1024 Hz and then amplified before its digitalization. The final numerical data have been processed by using the software MATLAB (transformation through the Hilbert transform as to obtain the analytic version of each signal and CWT).

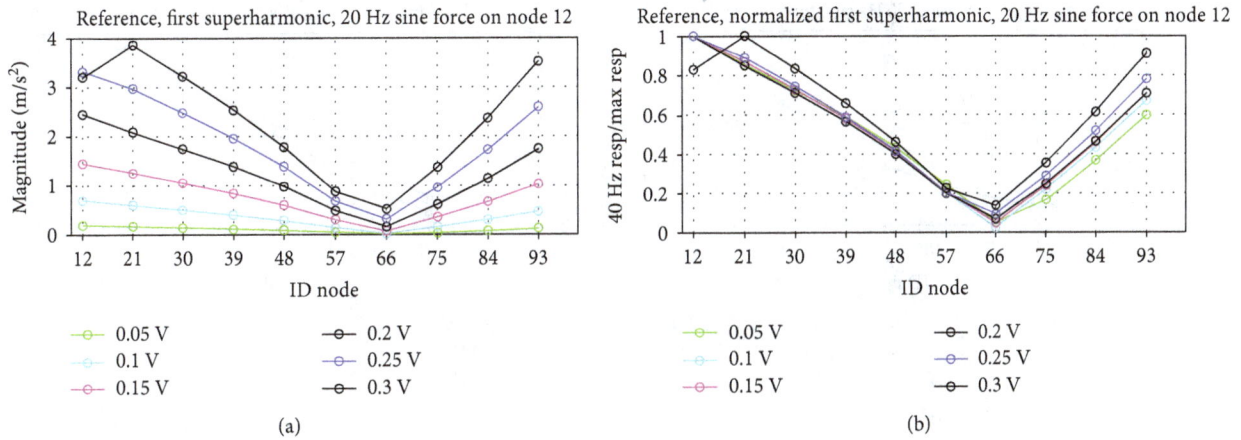

FIGURE 45: 40 Hz FFT magnitudes, 20 Hz sine excitation force applied to node 12, reference structure.

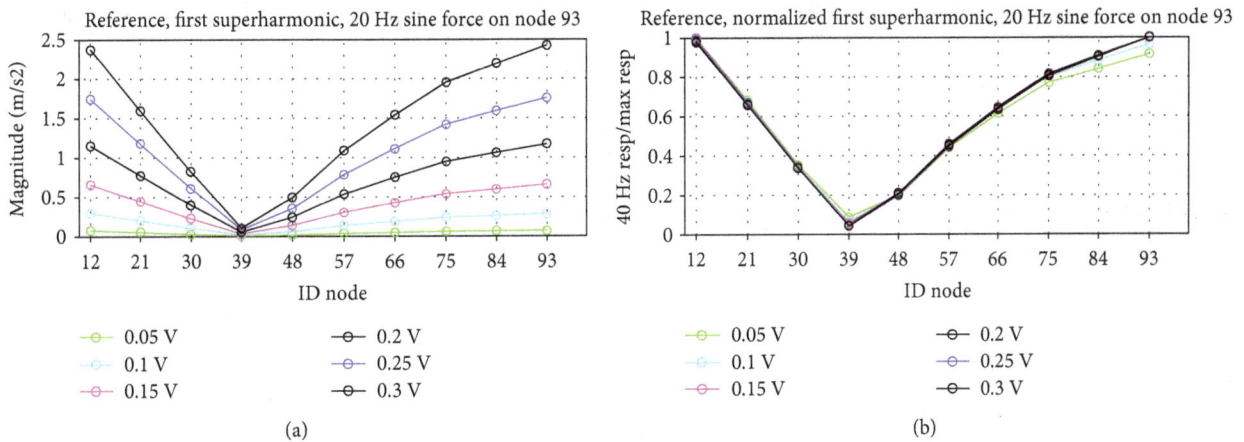

FIGURE 46: 40 Hz FFT magnitudes, 20 Hz sine excitation force applied to node 93, reference structure.

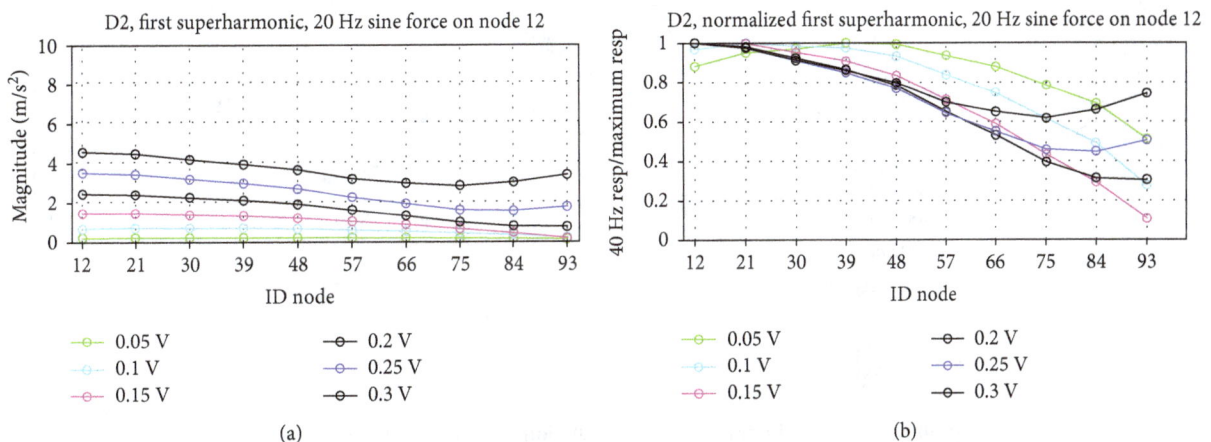

FIGURE 47: 40 Hz FFT magnitudes, 20 Hz sine excitation force applied to node 12, D2 scenario.

4.2.1. D2 Scenario. The CWT has been firstly applied considering a frequency range between 5 Hz and 30 Hz. This allows the elimination of the low-frequency contribution related to the elastic cable used to suspend the structure. By the data processing, the main frequency content in time has been obtained. Two contributions have been observed in the acceleration stories: one at 8 Hz, so close to the first eigenfrequency of the undamaged structure, and one at 24 Hz [18]. In order to focus on the first eigenfrequency behavior as done in the numerical tests, the analysis has been repeated with a frequency range between 5 Hz and 10 Hz. Figures 58 and 59 report the comparisons of the maximum frequency content

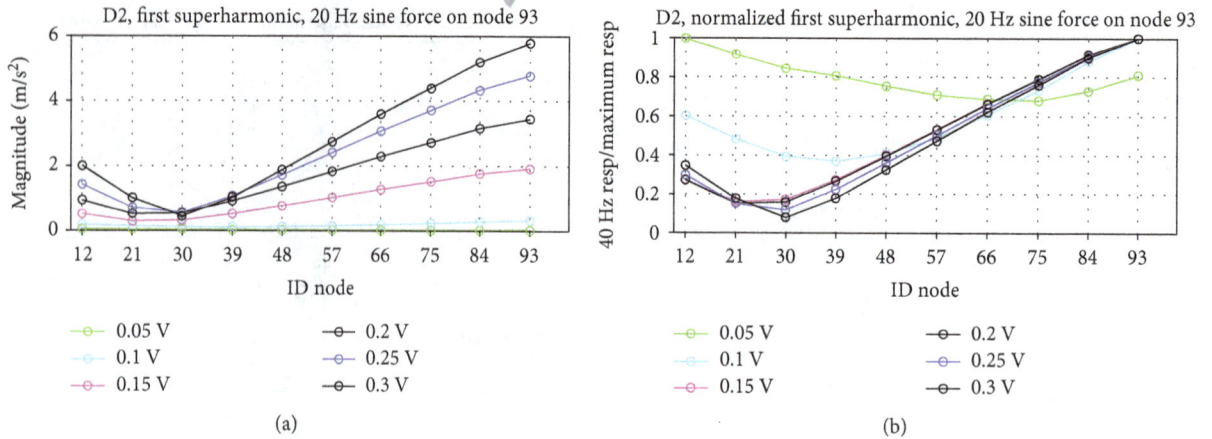

FIGURE 48: 40 Hz FFT magnitudes, 20 Hz sine excitation force applied to node 93, D2 scenario.

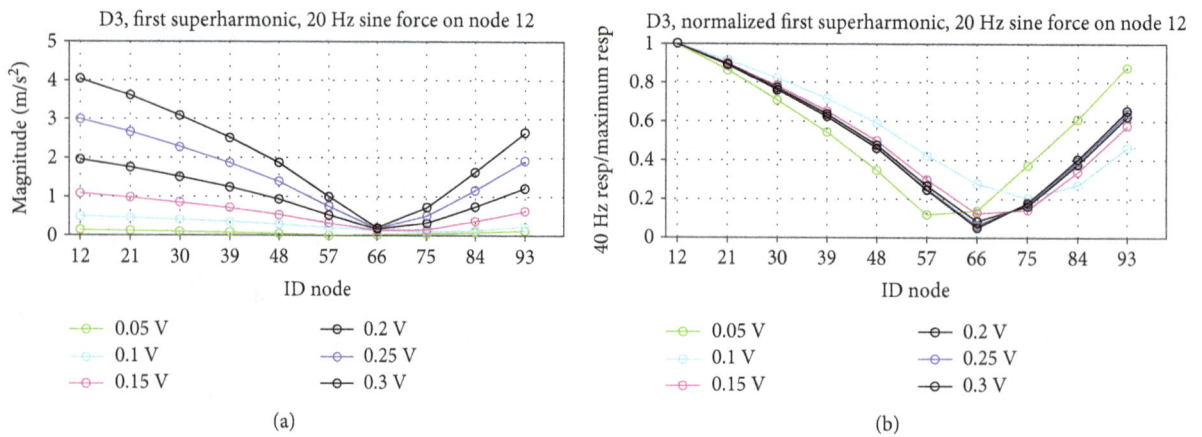

FIGURE 49: 40 Hz FFT magnitudes, 20 Hz sine excitation force applied to node 12, D3 scenario.

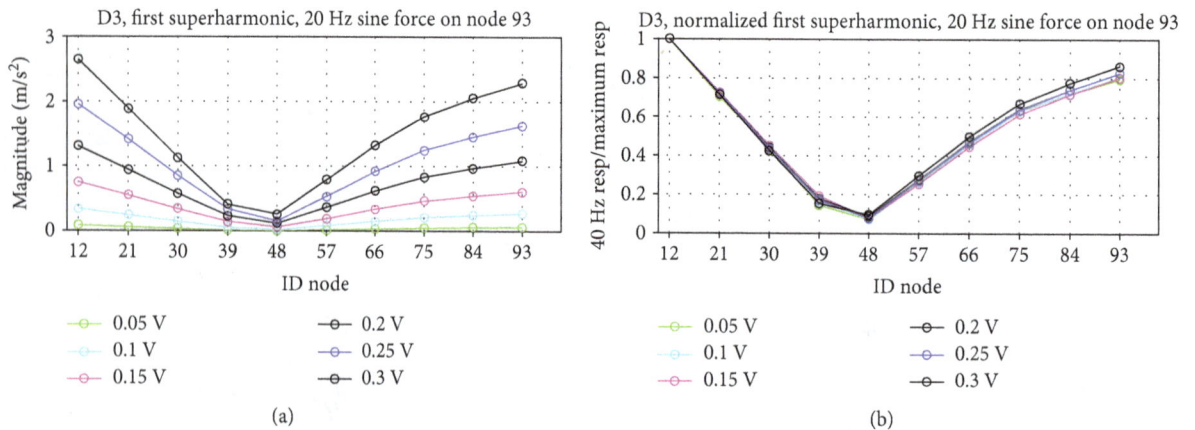

FIGURE 50: 40 Hz FFT magnitudes, 20 Hz sine excitation force applied to node 93, D3 scenario.

in this range for the two excitation tests done. The three curves included represent each acceleration measured.

Fluctuations in the contribution are evident. The oscillations for point B (black curves) are higher than the other ones, for both cases. Moreover, the oscillations in point A (blue curve) are higher than the ones in node 12 (red curves). As point B is the one in the damaged area, an explanation for the phenomenon is that this region has the highest variation of stiffness in time. From the results, it is also possible to state that all the structure exhibits main frequency oscillations. In order to compare this situation with the undamaged one, the same tests have been repeated with the original structure. Results are reported in Figures 60 and 61.

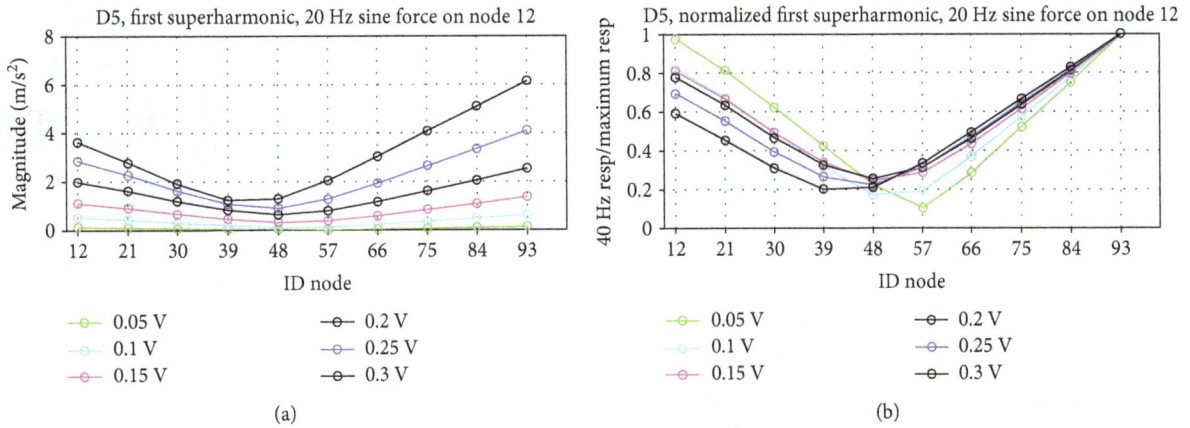

FIGURE 51: 40 Hz FFT magnitudes, 20 Hz sine excitation force applied to node 12, D5 scenario.

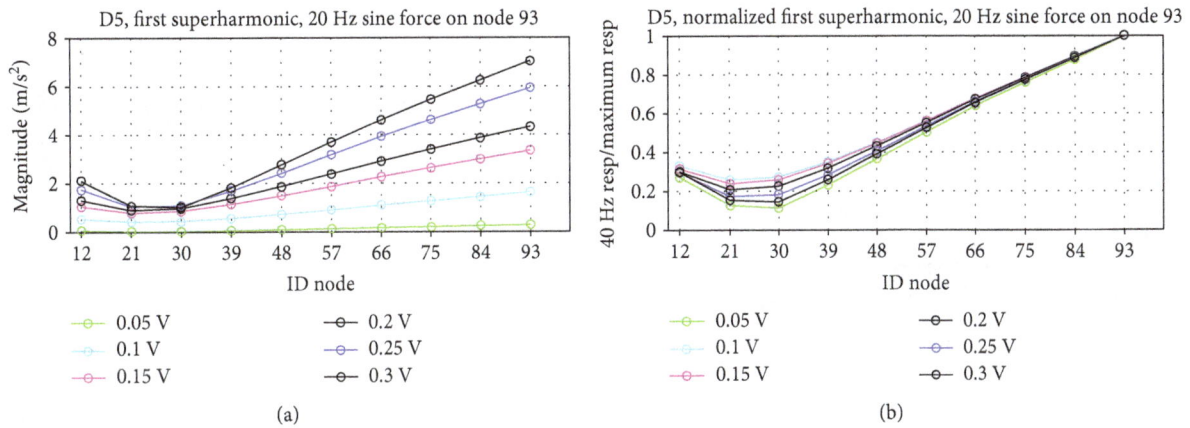

FIGURE 52: 40 Hz FFT magnitudes, 20 Hz sine excitation force applied to node 93, D5 scenario.

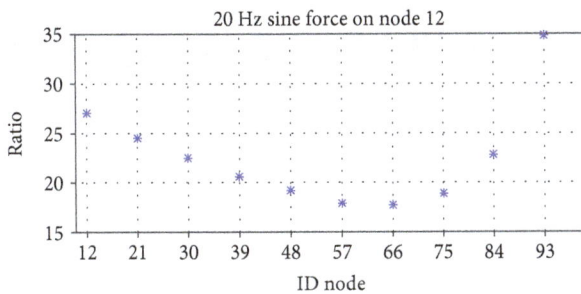

FIGURE 53: D2, 20 Hz sine excitation force in node 12, ratios between load levels.

In this case, the oscillation ranges are the same for all the points. The visible abrupt ends are due to the absence of further more acceleration time data after the final time indicated: the test recordings last 14-15 s.

4.2.2. D3 Scenario. The elaborations done for the D3 case are the same as those for the D2 one, with the exception that the range of frequencies used for the second analyses goes from 5 Hz to 13 Hz. In the test done exciting near the damage, the acceleration analysis for point *B* has revealed a very

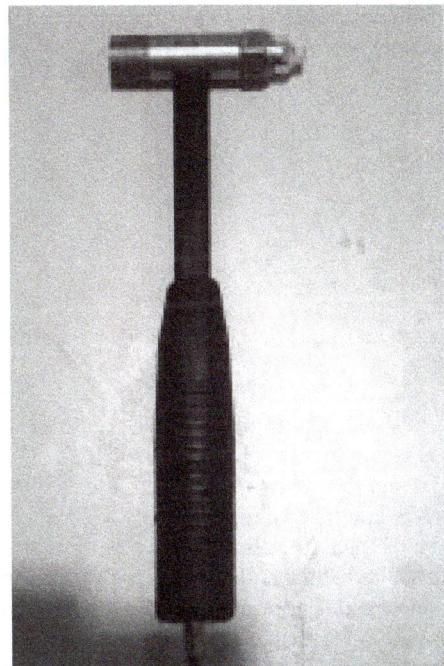

FIGURE 54: Instrumented hammer used for decay tests.

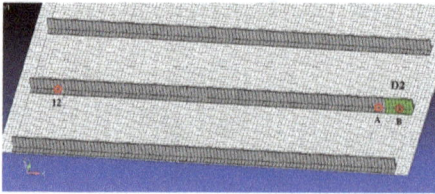

FIGURE 55: Positions of the measuring nodes for decay tests, damage D2.

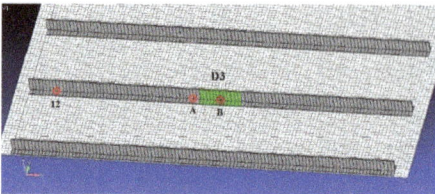

FIGURE 56: Positions of the measuring nodes for decay tests, damage D3.

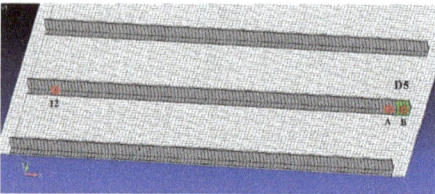

FIGURE 57: Positions of the measuring nodes for decay tests, damage D5.

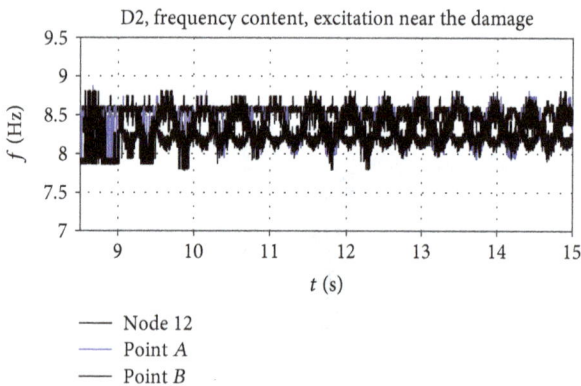

FIGURE 58: D2, maximum frequency, excitation near the damage.

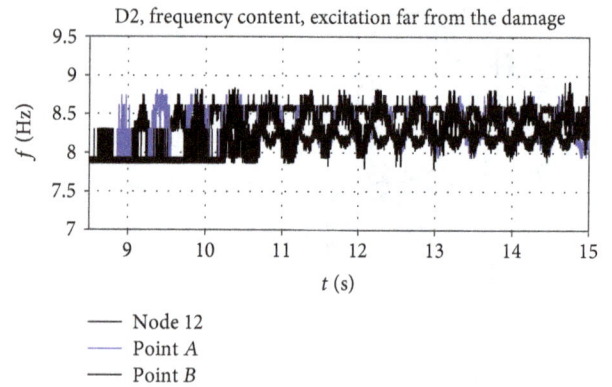

FIGURE 59: D2, maximum frequency, excitation far from the damage.

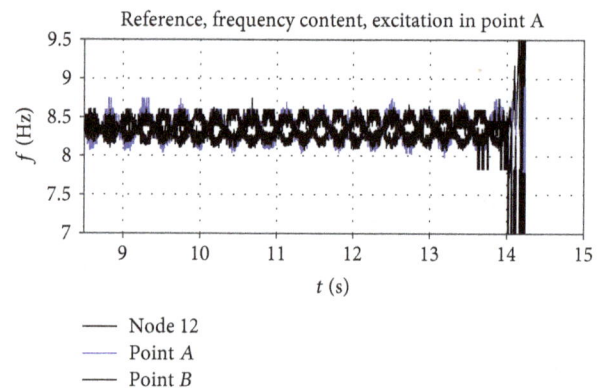

FIGURE 60: Undamaged case, excitation near the damage.

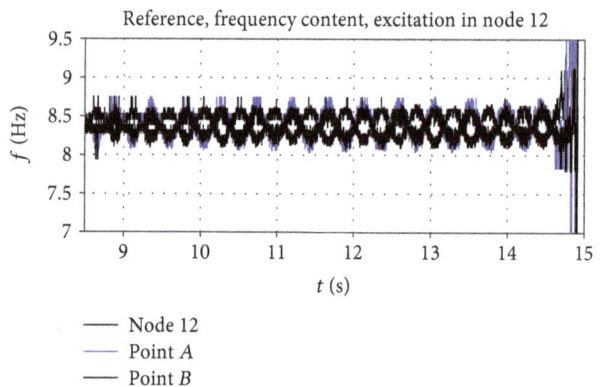

FIGURE 61: Undamaged case, excitation far from the damage.

strong fluctuation around the second eigenfrequency of the undamaged structure (11 Hz), as reported in Figure 62. Moreover, Figure 63 shows the comparisons of the main frequency oscillations for the accelerations recorded exciting far from the damage (the abrupt ends are present according to what was stated at the end of the previous section).

As visible from Figure 63, the acceleration in node 12 owns a dominant frequency content represented by the first eigenfrequency. On the other hand, as known from the numerical analyses, point A and point B exhibit fluctuations around the second eigenfrequency. The amplitude of the

oscillations in point B located in the damaged area is higher than the one for point A.

4.2.3. D5 Scenario. The elaborations done for the D5 case are the same as those for the previous cases. The frequency ranges for the results reported go from 5 Hz to 10 Hz. Figures 64 and 65 report the comparisons of the maximum frequency content in this range for the two excitation tests done (abrupt ends are present according to what was stated at the end of Section 4.2.1).

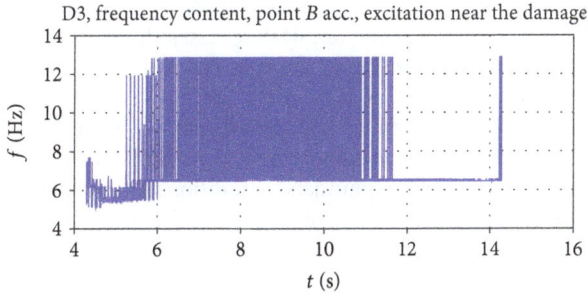

FIGURE 62: D3, maximum frequency, point B, excitation near the damage.

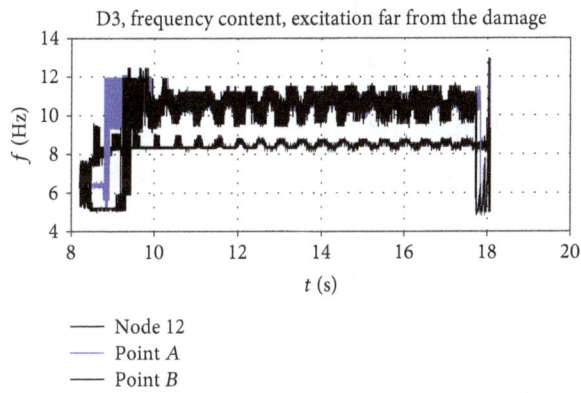

—— Node 12
—— Point A
—— Point B

FIGURE 63: D3, maximum frequency, excitation far from the damage.

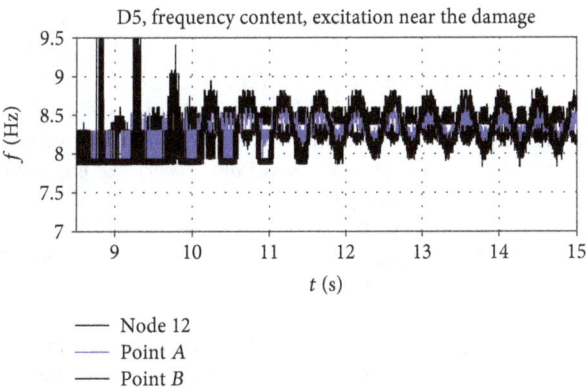

—— Node 12
—— Point A
—— Point B

FIGURE 64: D5, maximum frequency, excitation near the damage.

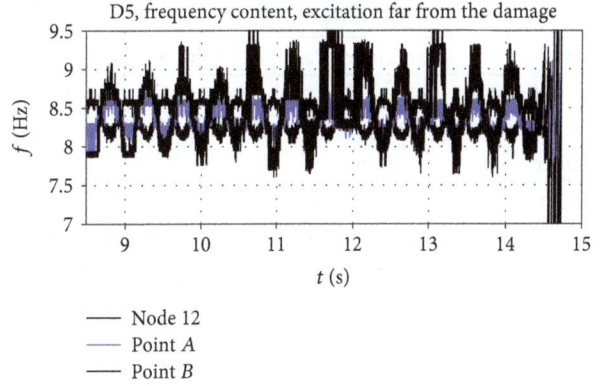

—— Node 12
—— Point A
—— Point B

FIGURE 65: D5, maximum frequency, excitation far from the damage.

TABLE 7: D2, excitation near the damage, localization indexes.

Index	Node 12	Point A	Point B
f, min (Hz)	7.93	7.86	7.78
f, max (Hz)	8.74	8.82	8.82
f, range (Hz)	0.81	0.96	1.04

TABLE 8: D2, excitation far from the damage, localization indexes.

Index	Node 12	Point A	Point B
f, min (Hz)	7.87	7.86	7.79
f, max (Hz)	8.75	8.81	8.93
f, range (Hz)	0.88	0.95	1.14

TABLE 9: D3, excitation far from the damage, localization indexes.

Index	Point A	Point B
f, min (Hz)	9.66	9.38
f, max (Hz)	11.67	11.67
f, range (Hz)	2.01	2.29

Again, the trend of the amplitudes increasing moving from node 12 to point A and point B is evident.

4.2.4. *Comments on Results.* The analyses from decay tests have underlined some relevant points. At first, the main frequency fluctuation expected from numerical tests has been found for all the measurement points. As understood by the superharmonic tests, this element is the consequence of the nonlinear behavior of the structure itself. Then, it clearly emerges that the oscillations found for the lower dominant frequencies have higher amplitudes when related to the damaged area. Thus, comparing these amplitudes among different acceleration elaborations, it is possible to detect and localize the damages. Finally, the nonlinear phenomenon observed, justified by the variation of the local stiffness due to the contact, has shown some remarkable elements. Among them, all the fluctuations identified by the CWT elaborations have exhibited a 2 Hz oscillation behavior, with higher contributions at the top and the bottom of each one of them. Also, the local effect of damages has resulted stronger than the global nonlinearity of the structure. In this context, the CWT has resulted a powerful tool for the identification of a nonlinear and nonstationary phenomenon.

In order to define a proper quantitative index for the localization, Tables 7–11 report the maxima and minima frequency values obtained for each elaboration, with the resulting range.

In order to obtain the value reported, proper time intervals have been fixed for the data available in order to include only the dominant frequency oscillations. For all

TABLE 10: D5, excitation near the damage, localization indexes.

Index	Node 12	Point A	Point B
f, min (Hz)	7.93	7.87	7.83
f, max (Hz)	8.62	8.62	8.95
f, range (Hz)	0.69	0.75	1.12

TABLE 11: D5, excitation far from the damage, localization indexes.

Index	Node 12	Point A	Point B
f, min (Hz)	7.93	7.87	7.60
f, max (Hz)	8.75	8.75	9.60
f, range (Hz)	0.82	0.88	2.00

the cases, the frequency range increases from node 12 to point B, so the damage approaches.

5. Conclusions

The paper has presented an extension of the nonlinear vibrational methods in a structure reproducing an aeronautical panel, in which three different contact-type damages have been taken into account.

The results presented based on the search for two nonlinear behaviours (superharmonics, variation of the lower frequency of vibration in time) have highlighted the following points in relation to the damage detection procedure:

(i) The nonlinear phenomena here investigated have also emerged with the undamaged structure. Thus, the detection itself by finding these characteristics has not resulted a possibility.

(ii) The localization of the contact damages has been possible through the comparisons of the responses measured. In particular, the variations of the lower frequency of vibration in time have resulted usefulness to detect all the damages analyzed, without requiring the amount of elaboration done for the end defect detection using the superharmonics. Indeed, the amplitude of these variations has resulted sensitivity to the damage location. Moreover, the CWT used to analyze the decay tests data has been identified as a powerful tool for underlining such a nonlinear and nonstationary behaviour.

The numerical procedure executed in order to highlight the presence of suitable nonlinear phenomena has been able to underline effects also found in the experimental tests. Therefore, the contact mechanism model implemented using the local penalty approach has resulted validity for understanding possible nonlinear evidences but needs to be furtherly elaborated in order to ensure true predictions of the analyzed structure. Despite the model assumptions reported in Section 3.3, the nonlinear behavior found in the undamaged structure reveals the requirement of simulations including all the contacts with their whole interferences.

Finally, in order to better explore the nonlinear vibrational methods for applicative cases, further analyses are mandatory: extension of the methods used (step and sine for superharmonics, CWT used for the monitoring of superharmonics in time, and decay tests with more damages of different extensions), new conditions (composites structures, different boundary conditions), and different kinds of damages.

Conflicts of Interest

The authors declare that there are no conflicts of interest regarding the publication of this paper.

Acknowledgments

Special thanks are due to Mauro Terraneo from Vicoter for his support during the experimental activity and to Dr. Kamal Rezvani for his previous activity on the design, manufacturing, and testing of the reference panel used in this work.

References

[1] W. J. Staszewski, C. Boller, and G. R. Tomlinson, Eds., *Health Monitoring of Aerospace Structures: Smart Sensor Technologies and Signal Processing*, Wiley, 2004.

[2] Y. Bar-Cohen, *Emerging NDE Technologies and Challenges at the Beginning of the 3rd Millennium—Part I*, NDT, 2000.

[3] Y. Bar-Cohen, *Emerging NDE Technologies and Challenges at the Beginning of the 3rd Millennium—Part II*, NDT, 2000.

[4] T. Stepinski, T. Uhl, and W. Staszewski, Eds., *Advanced Structural Damage Detection: from Theory to Engineering Applications*, Wiley, 2013.

[5] S. W. Doebling, C. R. Farrar, and M. B. Prime, "A summary review of vibration-based damage identification methods," *The Shock and Vibration Digest*, vol. 30, no. 2, pp. 91–105, 1998.

[6] J. J. Sinou, "A review of damage detection and health monitoring of mechanical systems from changes in measurements of linear and non-linear vibrations," in *Mechanical Vibrations: Measurement, Effects and Control*, pp. 643–702, Nova Science Publishers, Inc, 2009.

[7] P. Cawley and R. D. Adams, "The location of defects in structures from measurements of natural frequencies," *The Journal of Strain Analysis for Engineering Design*, vol. 14, no. 2, pp. 49–57, 1979.

[8] S. L. Tsyfansky and V. I. Beresnevich, "Non-linear vibration method for detection of fatigue cracks in aircraft wings," *Journal of Sound and Vibration*, vol. 236, no. 1, pp. 49–60, 2000.

[9] K. Rezvani, *Vibration-based damage identification techniques*, [Ph.D. thesis], Politecnico di Milano, 2015.

[10] M. Gordan, H. A. Razak, Z. Ismail, and K. Ghaedi, "Recent developments in damage identification of structures using data mining," *Latin American Journal of Solids and Structures*, vol. 14, no. 13, pp. 2373–2401, 2017.

[11] S. A. Neild, *Using non-linear vibration techniques to detect damage in concrete bridges*, [Ph.D. thesis], University of Oxford, 2001.

[12] S. O. Vismara, *Non-linear spacecraft component parameters identification based on experimental results and finite element model updating*, [M.S. thesis], Politecnico di Milano, 2015.

[13] J. N. Sundermeyer and R. L. Weaver, "On crack identification and characterization in a beam by non-linear vibration analysis," *Journal of Sound and Vibration*, vol. 183, no. 5, pp. 857–871, 1995.

[14] A. Rytter, *Vibrational based inspection of civil engineering structures*, [Ph.D. thesis], Department of Building Technology and Structural Engineering, Aalborg University, 1993.

[15] V. Volterra, *Theory of Functionals and of Integral and Integro-Differential Equations*, Dover Publications, 1959.

[16] D. M. Storer, *Dynamic analysis of non-linear structures using higher order frequency response functions*, [Ph.D. thesis], University of Manchester, 1991.

[17] R. Ruotolo, C. Surace, P. Crespo, and D. Storer, "Harmonic analysis of the vibrations of a cantilevered beam with a closing crack," *Computers & Structures*, vol. 61, no. 6, pp. 1057–1074, 1996.

[18] M. Carminati, *Damage detection using nonlinear vibrations*, [M.S. thesis], Politecnico di Milano, 2016.

[19] "MSC nastran 2013 dynamic analysis user's guide, MSC software," https://simcompanion.mscsoftware.com/infocenter/index?page=content&id=DOC10354&actp=RSS.

[20] C. K. Chui, *Wavelet Analysis and Its Applications-I: an Introduction to Wavelets*, Academic Press, 1992.

[21] C. K. Chui, *Wavelet Analysis and Its Applications-II: a Tutorial in Theory and Applications*, Academic Press, 1992.

[22] M. Rucka and K. Wilde, "Application of continuous wavelet transform in vibration based damage detection method for beams and plates," *Journal of Sound and Vibration*, vol. 297, no. 3–5, pp. 536–550, 2006.

[23] M. C. C. Bampton and R. R. Craig Jr., "Coupling of substructures for dynamic analyses," *AIAA Journal*, vol. 6, no. 7, pp. 1313–1319, 1968.

[24] K. Roughen, J. Huang, D. Hammerand, D. Stuewe, and S. Hertz, "Efficient simulation of structural dynamic systems with discrete nonlinearities," in *52nd AIAA/ASME/ASCE/AHS/ASC Structures, Structural Dynamics and Materials Conference*, pp. 1–7, Denver, CO, USA, 2011.

[25] S. Loutridis, E. Douka, and L. J. Hadjileontiadis, "Forced vibration behaviour and crack detection of cracked beams using instantaneous frequency," *NDT & E International*, vol. 38, no. 5, pp. 411–419, 2005.

11

Drilling Load Model of an Inchworm Boring Robot for Lunar Subsurface Exploration

Weiwei Zhang, Shengyuan Jiang, Dewei Tang, Huazhi Chen, and Jieneng Liang

State Key Laboratory of Robotics and System, Harbin Institute of Technology, Harbin 150001, China

Correspondence should be addressed to Shengyuan Jiang; jiangshy@hit.edu.cn

Academic Editor: Paul Williams

In the past decade, the wireline robot has received increasing attention due to the advantages of light weight, low cost, and flexibility compared to the traditional drilling instruments in space missions. For the lunar subsurface in situ exploration mission, we proposed a type of wireline robot named IBR (Inchworm Boring Robot) drawing inspiration from the inchworm. Two auger tools are utilized to remove chips for IBR, which directly interacted with the lunar regolith in the drilling process. Therefore, for obtaining the tools drilling characteristics, the chips removal principle of IBR is analyzed and its drilling load model is further established based on the soil mechanical theory in this paper. And then the proposed theoretical drilling load model is experimentally validated. In addition, according to the theoretical drilling load model, this paper discusses the effect of the drilling parameters on the tools drilling moments and power consumption. These results imply a possible energy-efficient control strategy for IBR.

1. Introduction

Scientific targets such as the physical and mechanical properties and heat flux of regolith are critical evidences for human studying the planetary origin and evolution [1]. Drilling is an effective method and is widely utilized in the planetary subsurface exploration missions [2, 3]. Though these missions are capable of sampling the regolith or rack by using of the drill with its inbuilt coring tube, these drills are generally designed less than two meters because of the constraints of power, payload, and volume. For obtaining more valuable and reliable scientific data, the depth of implementing investigation should be more than three meters which is beyond the significant thermal cycles [4]. If long drills such as MARTE [5] and SPECES [6] which coupled with multiple drill sections are utilized to achieve greater depth, they will unavoidably increase the mass and complexity of the system.

To sidestep the drawbacks of traditional drill, a type of wireline robot has been proposed for deep subsurface exploration. The wireline robot has the capability of accessing the target position in the planet and implementing scientific investigation by internal instruments. Compared with the traditional drill, the wireline robot does not have rigid connection to the surface but has a tether which provides

the wireline robot with power and data communication [7]. According to the boring method of penetrating or excavating, the wireline robot can be classified as two types of penetrator and excavating robot. Serval penetrators such as MUPUS [8] (Rosetta mission), Insight [9] (InSight mission), and KRET [10] (future lunar robotic mission) were proposed, and they can penetrate into the planetary subsurface by using of the impact driven by penetrator's internal hammer mechanism. Besides that, a bioinspired penetrator based on the working mechanism of wood wasp ovipositors was proposed for avoiding the needed external force in the space missions [11]. Although these penetrators have advantages of light weight and small dimensions, it is difficult for them to penetrate hard regolith or rocks and thus they are just suitable for subsurface exploration of the shallow depth in relatively loose regolith. Drawing inspiration from nature, several bioinspired excavating robots including IDDS [12], Auto-Gopher [13], and Earthworm-type robot [14] were proposed. The excavating robot can break and remove the regolith for making space and advancement, and it is generally equipped with anchor mechanism used to supply the enough down force when applied on the borehole wall. Although the excavating robot potentially has stronger drilling ability than the traditional drill and the penetrator, it also has the disadvantages in

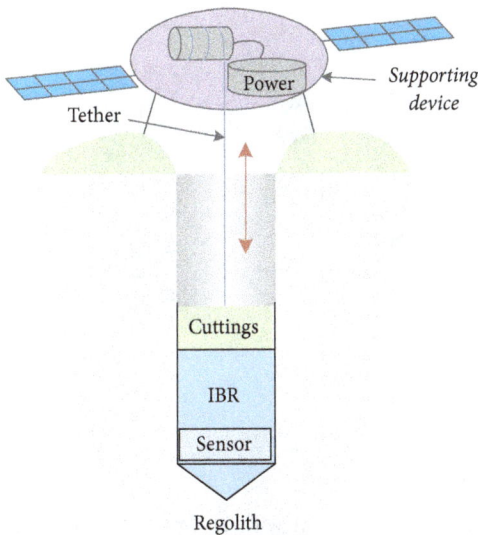

FIGURE 1: Subsurface exploration system.

complex operation and structure especially for a robotic mission.

Therefore, an Inchworm Boring Robot was proposed for China's lunar robotic subsurface exploration mission [15]. The IBR has two critical tools of drill bit used for cutting the regolith or rocks and the auger was used to remove the chips. In the case of energy limit, it is necessary to reasonably select operating parameters to reduce working resistance and energy consumption of IBR. Therefore, it is of great significance to study the interaction between tools and lunar regolith for tools design and motion control strategy. The interaction between drill bit and lunar soil or rock had already been analyzed [16, 17], so that this paper mainly focuses on the establishment of auger drilling load model based on the soil mechanics theory. According to the drilling load model, this paper takes the drilling moment and power consumption as the evaluation index and studies the drilling parameters matching which lays the foundation for the motion control strategy of IBR.

The remainder of the paper is organized as follows. The IBR and its working principle are introduced in Section 2. In Section 3, chips removal in the excavating mode and discharging mode are analyzed. Section 4 gives the establishment of the drilling load model of IBR. In Section 5, the drilling load model was experimentally validated and the effect of drilling parameters on the drilling load is analyzed for the future motion control strategy. Section 6 concludes this paper.

2. Working Principle of IBR

Subsurface exploration system shown in Figure 1 was proposed to implement scientific investigations of scientific targets such as the physical and mechanical properties and heat flux of lunar regolith. As depicted in Figure 1, the subsurface exploration system is mainly made up of IBR and supporting device. IBR automatically drills into planetary

subsurface with the power supply of supporting device according to one tether. In the drilling process, the hard original regolith was broken into loose cuttings, and the cuttings were simultaneously conveyed to chamber at the end of IBR. Once the volume of cuttings accumulated in the chamber reaches the set value, IBR will return back to the surface and discharge the cuttings. In the investigating process, IBR can collect the scientific data by its inner sensors.

The IBR consists of three modules, excavating module, discharging module, and propulsion module, as shown in Figure 2(a). Excavating module was equipped with drill head and excavating auger: the drill head was used to break the original regolith or rocks into cuttings and the excavating auger was used to convey the cuttings to the storage room. The discharging module was just equipped with a discharging auger used to convey the cutting to the back of the IBR from storage room, while excavating module and discharging module can be fixed on the borehole wall by excavating anchor and discharging anchor, respectively. Propulsion module can provide linear motion for excavating module and discharging module.

One drilling process is composed of four steps as shown in Figure 2(b).

Step 1. Discharging anchor engages the borehole wall.

Step 2. Excavating module drills forward with rotating and penetrating motion. At the same time, the excavating auger rotates in the opposite direction.

Step 3. Excavating anchor engages the borehole and discharging anchor disengages the borehole wall.

Step 4. Discharging module drills forward with rotating and penetrating motion.

Repeat the steps; IBR realizes the function of continued drilling into the regolith. In Steps 2 and 4, the IBR is mainly focused on excavating forward and discharging the cuttings. And thus the two steps can be defined as excavating mode and discharging mode for IBR.

3. Chips Removal of IBR

According to the components of IBR, two augers, EA (excavating auger) and DA (discharging auger), were used to remove the cuttings produced by the breaking of regolith or rocks by the DH (drill head). The structure parameters of the EA and DA are shown in Figure 3.

According to the working principle of IBR, two working modes of excavating mode and discharging mode can be defined in the whole drilling process as shown in Figure 4. In order to analyze the principle of chip removal for the two working modes, the following three assumptions need to be made:

(i) The flow channels (auger groove) of EA and DA are fully filled with the cuttings in the whole drilling process.

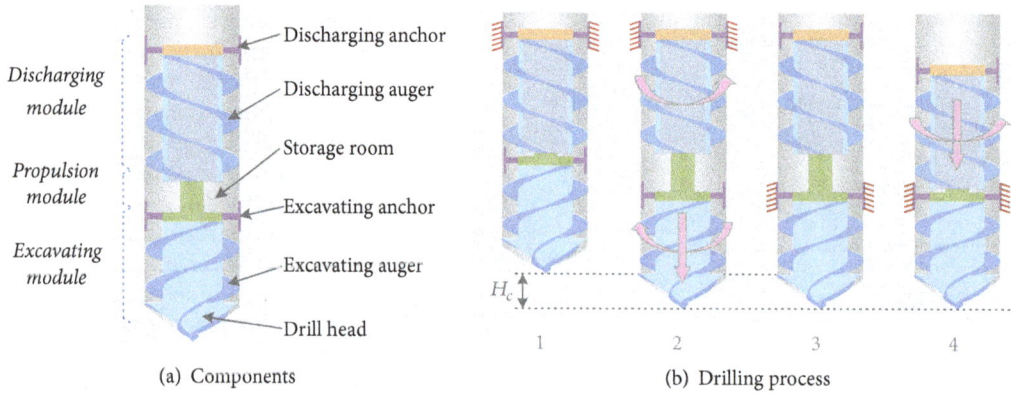

(a) Components   (b) Drilling process

FIGURE 2: Components and drilling process of IBR.

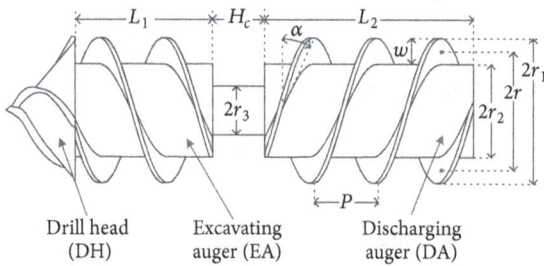

FIGURE 3: Structure parameters of EA and DA (r_1, r_2, and r are the inner radius, outer radius, and middle radius of the screw flight, resp. α is the screw angle. N is the number of the screw flights. L_1 and L_2 are the lengths of EA and DA, resp. r_3 is the inner radius of the storage room; H_c is the maximal height of the storage room. P is the screw pitch. w is the width of the screw flight).

(ii) The regolith is relatively compact compared with the cuttings; its bulk density is ρ_1. The compression among the cuttings on the screw flight is not taken into account, so that the cuttings have a constant bulk density ρ_2 in the analysis.

(iii) The cuttings in the storage room are probably compressed when the cuttings are not timely removed by DA, so that the bulk density of the compact cuttings increases and it is noted as ρ_3.

3.1. Discharging Mode. In the discharging mode, EA stops working and the DA conveys the compact cuttings to the end of IBR from the storage room. DA rotates and penetrates relative to the borehole; the rotary speed and penetrating velocity are ω_2 and v_2, respectively. For one certain instant, the motion of cuttings on the screw flight is equivalent to a planer motion and the instantaneous velocities are given in Figure 5.

In Figure 5, v_n represents the peripheral velocity of the auger, and $v_n = \omega_2 r$. v_r is the sliding velocity of the cuttings relative to the auger. v_a is resultant velocity of cuttings relative to the borehole, which has an incline angle β_2 and is composed of the velocities v_n, v_2, and v_r. Therefore, from the whole process of conveying, the cuttings are lifted and

do upward spiral motion with an opposite direction to the screw flight. As shown in Figure 5, v_a can be resolved into the useful lifting velocity v_{f2} and the peripheral velocity v_{h2} of the cuttings. Hence, the velocity relations of cuttings on the screw flight can be illustrated as follows:

$$v_n - v_r \cos \alpha = v_{h2}$$

$$v_r \sin \alpha - v_2 = v_{f2} \qquad (1)$$

$$v_{f2} = v_{h2} \tan \beta_2.$$

Substituting $v_n = \omega_2 r$ into (1) and eliminating the velocity v_r, the peripheral velocity v_{h2} and the lifting velocity v_{f2} of the cuttings can be solved:

$$v_{h2} = \frac{\omega_2 r \tan \alpha - v_2}{\tan \alpha + \tan \beta_2} \qquad (2)$$

$$v_{f2} = \frac{(\omega_2 r \tan \alpha - v_2) \tan \beta_2}{\tan \alpha + \tan \beta_2}. \qquad (3)$$

In Figure 4(b), Φ_2 is the cuttings mass flow rate from the storage room to the flow channel of DA. Φ_2 can be given as the reduced cuttings mass flow rate $\rho_3 A_3 v_2$ in the storage room and the cuttings mass flow rate $\rho_2 A_2 (v_{f2} + v_2)$ in the flow channel of DA. Therefore, the incline angle β_2 can be deduced based on the equation $\rho_3 A_3 v_2 = \rho_2 A_2 (v_{f2} + v_2)$.

$$\beta_2 = \arctan \frac{(\rho_3 A_3 - \rho_2 A_2) v_2}{\rho_2 A_2 \omega_2 r - \rho_3 A_3 v_2 \cot \alpha}, \qquad (4)$$

where A_2 represents the cross-sectional area of cuttings flow channel in DA or EA and $A_2 = \pi(r_1^2 - r_2^2)$. A_3 represents the cross-sectional area of cuttings flow channel in the storage room and $A_3 = \pi(r_1^2 - r_3^2)$.

3.2. Excavating Mode. In the excavating mode, DH breaks the original regolith into cuttings and these cuttings are then conveyed to the storage room by EA with the rotary and penetrating motion. The produced cuttings are filled and compressed in the storage room, and some of them are removed with the rotary movement of DA. As shown in

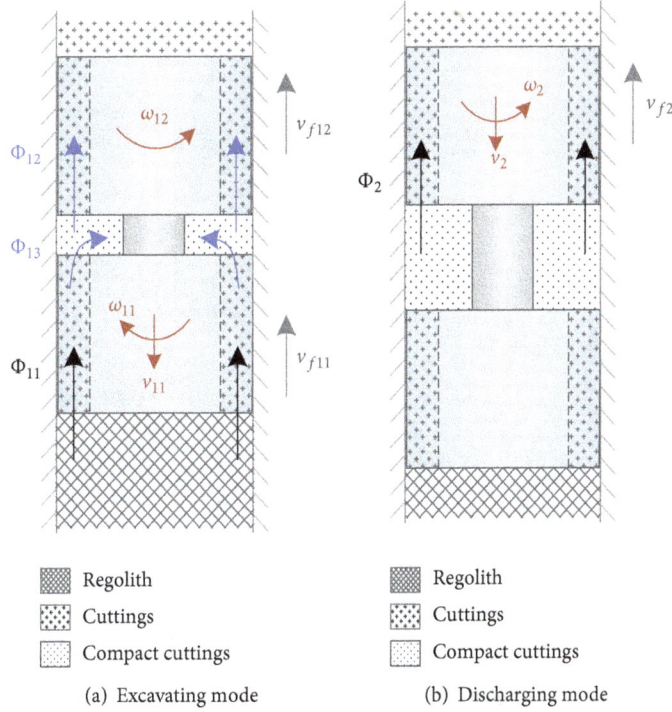

(a) Excavating mode

(b) Discharging mode

FIGURE 4: IBR working mode.

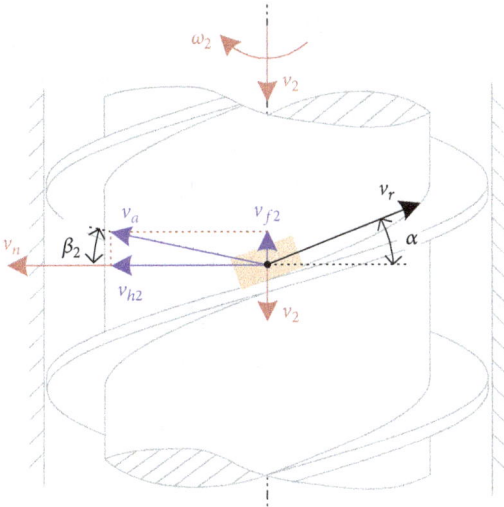

FIGURE 5: Velocities of cuttings on the screw flight.

Figure 4(a), v_{f11} and v_{f12} are the vertical upward velocities of cuttings in the EA and DA flow channel. Based on (3), the two velocities can be calculated as follows:

$$v_{f11} = \frac{(\omega_{11} r \tan \alpha - v_{11}) \tan \beta_{11}}{\tan \alpha + \tan \beta_{11}}$$

$$v_{f12} = \frac{\omega_{12} r \tan \alpha \tan \beta_{12}}{\tan \alpha + \tan \beta_{12}},$$

(5)

where ω_{11} and v_{11} represent the rotary speed and penetrating velocity of EA, respectively. ω_{12} represents the rotary speed

of DA. β_{11} and β_{12} are, respectively, the incline angles of cuttings resultant velocities in the EA and DA. As can be seen in Figure 4(a), Φ_{11} is the mass flow rate of the chips from the original regolith to the flow channel in EA. Φ_{11} can be given as the produced cuttings mass flow rate $\rho_1 A_1 v_{11}$ from the original regolith and the cuttings mass flow rate $\rho_2 A_2 (v_{f11} + v_{11})$ in the flow channel in EA. In the excavating mode, the mass flow rate Φ_{11} was divided into two parts: the mass flow rate Φ_{12} of the flow channel in the DA and the mass flow rate Φ_{13} in the storage room. Φ_{12} and Φ_{13} can be expressed as $\rho_2 A_2 v_{f12}$ and $\rho_3 A_3 v_{11}$, respectively. Therefore, two equations can be obtained as $\rho_1 A_1 v_{11} = \rho_2 A_2 (v_{f11} + v_{11})$ and $\Phi_{11} = \Phi_{12} + \Phi_{13}$, and the incline angles β_{11} and β_{12} are deduced as shown in the following equation:

$$\beta_{11} = \arctan \frac{(\rho_1 A_1 - \rho_2 A_2) v_{11}}{\rho_2 A_2 \omega_{11} r - \rho_1 A_1 v_{11} \cot \alpha}$$

$$\beta_{12} = \arctan \frac{(\rho_1 A_1 - \rho_3 A_3) v_{11}}{\rho_2 A_2 \omega_{12} r - (\rho_1 A_1 - \rho_3 A_3) v_{11} \cot \alpha},$$

(6)

where A_1 represents the cross-sectional area of the borehole and $A_1 = \pi r_1^2$.

4. Drilling Load Model of IBR

4.1. Soil-Auger Interaction.
For obtaining the soil-auger interaction model, a soil element with a small angle $d\theta$ was taken as the analyzed object on the screw flight as shown in Figure 6(a). In Figure 6(b), the distance from the center of soil element to the central axis of the auger is r. w and H are the height and width of the soil element, respectively. The left

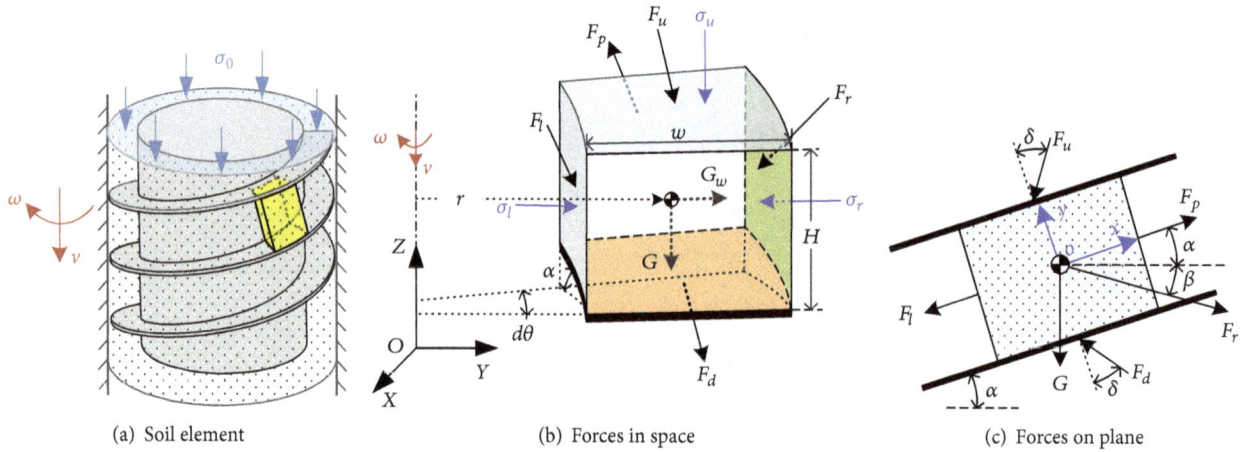

(a) Soil element   (b) Forces in space   (c) Forces on plane

FIGURE 6: Forces analysis of chips element.

lateral surface of the soil element represents the auger stem surface, its area was noted by A_L, and $A_L = Hr_2 \sec \alpha d\theta$. The right lateral surface of the soil element represents the borehole wall surface, its area was noted by A_R, and $A_R = Hr_1 \sec \alpha d\theta$. The top and bottom surfaces of the soil element represent the lower and upper surfaces of the screw flight, and the area can be defined as A_F: $A_F = wr \sec \alpha d\theta$. The front and back surfaces representing the contact surface between the soil element and neighboring soil on the screw flight were noted by A_C, and $A_C = Hw$. For the soil element, the pressure on the left, top, and right surfaces is, respectively, defined as σ_l, σ_u, and σ_r; the difference pressure acting on the front and back surfaces is $d\sigma$.

These forces shown in Figure 6(b) applied on the soil element can be expressed on the plane as shown in Figure 6(c), and then the static equilibrium equation (7) can be carried out in the x direction (the tangential direction of the screw flight) and the y axial direction (the normal direction of the screw flight).

$$F_r \cos (\alpha + \beta) + F_p - F_l - (F_d + F_u) \sin \delta - G \sin \alpha$$

$$= 0 \tag{7}$$

$$(F_d - F_u) \cos \delta - F_r \sin (\alpha + \beta) - G \cos \alpha = 0,$$

where F_d is the force of the soil element against the upper surface of the screw flight. F_l is the frictional force of the soil element against the lateral surfaces of auger stem, and $F_l = \sigma_l A_F \tan \delta$. F_u is the force of the soil element against the lower surface of the screw flight, and $F_u = \sigma_u A_F \sec \delta$. F_r is the frictional force of the soil element against the hole wall, and $F_r = \sigma_r A_R \tan \varphi$. F_p is the resultant force of neighboring soil applied on the soil element, and $F_p = d\sigma A_C$. G is the element gravity, and $G = \rho_2 H A_F g$. G_w is the centrifugal force of the element, and $G_w = \rho_2 H A_F v_h^2 / r$. By substituting

above forces into (7) and simplifying the equation, (8) can be obtained:

$$\frac{d\sigma}{d\theta} = K_l \sigma_l + K_u \sigma_u + K_r \sigma_r + K_c$$

$$K_l = \frac{r_2 \tan \delta}{w \cos \alpha},$$

$$K_u = \frac{2r \tan \delta}{H \cos \alpha}, \tag{8}$$

$$K_r = \frac{r_1 \cos (\alpha + \beta + \delta) \tan \varphi}{-w \cos \delta \cos \alpha},$$

$$K_c = \rho_2 g r (\tan \alpha + \tan \delta),$$

where δ is the friction angle between soil and metal and φ is the internal friction angle of the soil.

The average stress σ_p caused by the soil gravity on the four lateral surfaces of the soil element can be calculated as $\sigma_p = \rho_2 g H K_0 \cos \alpha / 2$ based on the soil mechanical theory, where K_0 is the stress coefficient and $K_0 = 1 - \sin \varphi$ [18]. The stress σ_a caused by the centrifugal force on the left and right lateral surfaces can be calculated as $\sigma_a = \rho_2 v_h^2 w / r$. According to the compression properties of soil, the stress σ will produce additional compressive stress on the surface of soil element. And the additional compressive stresses on the left, top, and right lateral surfaces are $K\sigma$, where K is the stress transfer coefficient and $K = (1 - \sin \varphi)/(1 + \sin \varphi)$ [18]. Therefore, under the action of its own gravity, centrifugal force, and the additional stress, the stresses applied on the left, top, and right surfaces are deduced as follows:

$$\sigma_l = \sigma_p - \sigma_a + K\sigma$$

$$\sigma_u = K\sigma \tag{9}$$

$$\sigma_r = \sigma_p + \sigma_a + K\sigma.$$

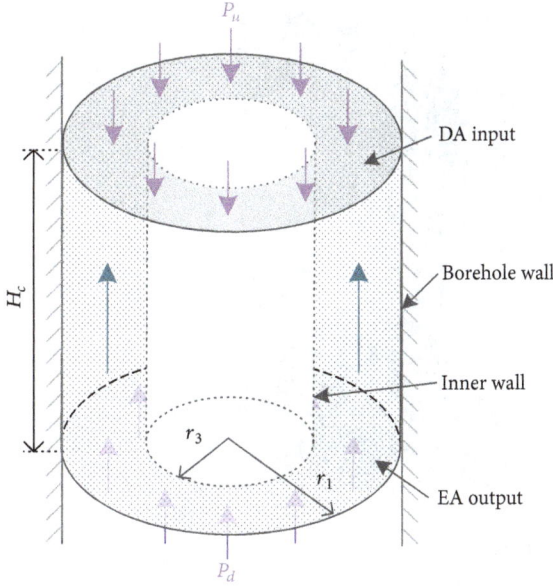

FIGURE 7: Stress of regolith in the storage room.

Substituting (9) into (8) and solving the differential equation, the explicit expression of the stress σ can be obtained:

$$\sigma = (\sigma_0 + B_s) \exp (A_s \theta) - B_s$$
$$A_s = K (K_l + K_r + K_u),$$
$$B_s = \frac{K_l (\sigma_P - \sigma_a) + K_r (\sigma_P + \sigma_a) + K_c}{K (K_l + K_r + K_u)}, \tag{10}$$

where σ_0 represents the preload acting on top of the auger flow channel.

4.2. Drilling Moments of Tools.

From (12), the stress state of the cuttings on the screw flight is affected by σ_0 which is the preload actnig on top of the auger flow channel. Therefore, it is necessary to analyze the stress of the regolith in the storage room and then give the obstructive stress acting on the top of EA flow channel. As shown in Figure 7, in the excavating mode, the cuttings discharged from EA were deposited in the storage and compressed, resulting in compressive stress P_u inside the cuttings. At the same time, the cuttings in the storage room were pushed up by the newly generated cuttings discharged from EA. As a result of the preload P_u and the fact that the frictions between the regolith and walls included borehole wall and the inner wall, the extrusion stress P_d was generated on the bottom of the storage room.

Based on the hyperbolic relation of soil strain and stress assumed in the Duncan-Chang model [19] and the relation of soil strain and void ratio in the condition of confined consolidation test, an e-p compression mode is established as shown in the following equation:

$$e_3 = e_2 - \frac{(1 + e_2) P_u}{A_p + B_p P_u}, \tag{11}$$

where e_2 and e_3 are the void ratios of the cuttings and the compressed cuttings in the storage room and $e_2 = \rho_s/\rho_2 - 1$ and $e_3 = \rho_s/\rho_3 - 1$. ρ_s is the granular density of regolith. A_p and B_p are the unknown coefficients that need to be determined by the confined compression test. By substituting e_2 and e_3 into (11), P_u is calculated as follows:

$$P_u = \frac{(\rho_3 - \rho_2) A_p}{\rho_3 - (\rho_3 - \rho_2) B_p}. \tag{12}$$

Based on Janssen's method [20] and considering the inner wall friction, the stress P_d is calculated by the following equation:

$$P_d = \frac{\rho_3 g}{E} \left(\exp (H_c E) - 1 \right) + P_u \exp (H_c E)$$
$$E = \frac{2\pi K (r_3 \tan \delta + r_1 \tan \varphi)}{A_3}. \tag{13}$$

In the conveying process of cuttings, the forces of F_l, F_u, and F_d acting on auger will produce the resisting moments which can be, respectively, expressed as $F_l r_2 \cos \alpha$, $F_u r \sin(\delta - \alpha)$, and $F_d r \sin(\delta + \alpha)$. By integrating the sum of above three resisting moments in the longitudinal direction, the total resisting moment M of auger can be deduced as follows:

$$M = N \int_0^{\theta_m} (K_{ml}\sigma_l + K_{mu}\sigma_u + K_{mr}\sigma_r + K_{mc}) d\theta$$
$$K_{ml} = Hr_2^2 \tan \delta,$$
$$K_{mu} = 2wr^2 \tan \delta \tag{14}$$
$$K_{mr} = Hr_1 r \sin (\alpha + \beta) \sin (\alpha + \delta) \tan \varphi \sec \alpha \sec \delta$$
$$K_{mc} = \rho_3 g w H r^2 \sec \delta \sin (\alpha + \delta),$$

where θ_m represents the angle of the screw flight: $\theta_m = 2\pi L/P$. From the soil-auger interaction analysis, M is the function of the parameters including σ_0, ω, v, and β, which can be noted as $M(\sigma_0, \omega, v, \beta)$. In the excavating mode, the drilling moment M_1 of EA can be obtained by substituting the parameters P_d, ω_{11}, v_{11}, and β_{11} into the function of M. In the discharging mode, there is no preload acting on the cuttings output port of the DA, and the drilling moment M_2 of EA can also be obtained by substituting the parameters ω_2, v_2, and β_2 into the function of M. Therefore, the drilling moments M_1 and M_2 can be calculated as follows:

$$M_1 = M (P_d, \omega_{11}, v_{11}, \beta_{11})$$
$$M_2 = M (0, \omega_2, v_2, \beta_2). \tag{15}$$

5. Experiments and Model Validation

5.1. Drilling Test Bed

5.1.1. Mechanical System.
For testing the drilling load of EA and DA, a drilling test bed shown in Figure 8 was developed. The drilling test bed is composed of the stander, two rotary

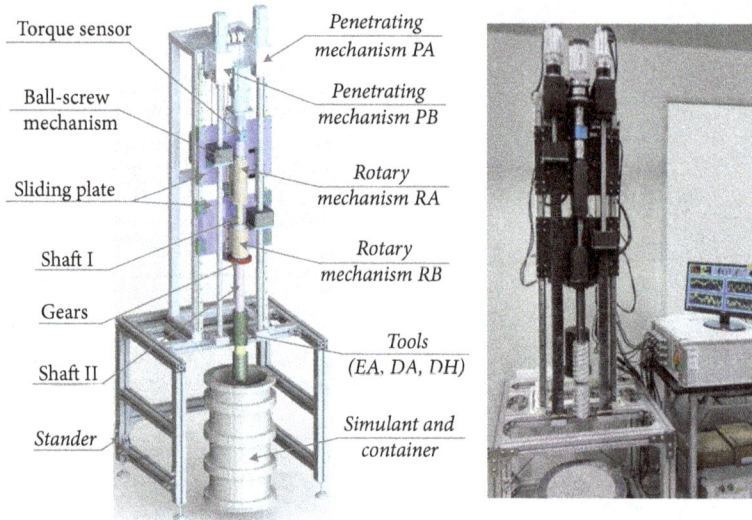

FIGURE 8: Drilling test bed.

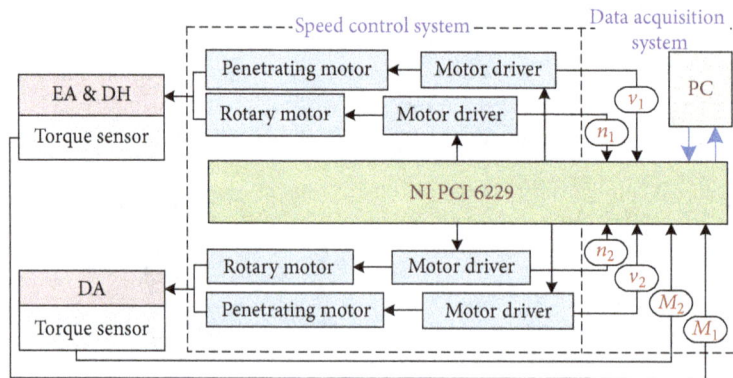

FIGURE 9: Speed control and data acquisition system.

mechanisms, RA and RB, and two penetrating mechanisms, PA and PB.

The two rotary mechanisms RA and RB are, respectively, fixed on the two sliding plates. RA consists of rotary motor, torque sensor, and shaft I. Shaft I and the EA are directly driven by the rotary motor for the rotary motion with nominal rotary speed of 300 rpm and nominal moment of 23 Nm. The rotary motor coupled with a reducer with the ratio of 1 : 5 is AC servo motor (MHMJ082) with 750 W power. For moment measurement of shaft I and EA, a torque sensor (HLT-20) which has the range of 0~20 Nm was installed between the rotary motor and shaft I. The rotary mechanism RB consists of rotary motor, torque sensor, gear transmission, and shaft II. Shaft II and the DA are driven by the paralleled rotary motor and a gear transmission mechanism for the rotary motion with nominal rotary speed of 200 rpm and nominal moment of 12 Nm. The rotary motor coupled with a reducer with the ratio of 1 : 10 is AC servo motor (MHMJ042) with 400 W power. For moment measurement of shaft II and DA, a torque sensor (HLT-10) which has the range of 0~10 Nm was installed between the rotary motor and shaft II.

Two penetrating mechanisms are arranged in parallel and they are used to provide penetrating motion for EA and DA, respectively. The two penetrating mechanisms have the same features; they consist of penetrating motor, ball-screw mechanism, and sliding plate. The ball-screw mechanism is driven by the penetrating motor and thus the two sliding plates can move linearly on the guide way for penetrating motion with a nominal velocity of 300 mm/min and a nominal force of 500 N. The penetrating motor coupled with a reducer with the ratio of 1 : 5 is also an AC servo motor (MHMJ042) with 100 W power.

5.1.2. Speed Control and Data Acquisition System. The test bed speed control and data acquisition system shown in Figure 9 has the function of motion control and signals and data acquisition. The two rotary motors and the two penetrating motors are, respectively, driven by the rotary drivers and penetrating drivers with the closed-loop speed control mode. In this control system, a data acquisition card (PCI 6229) produced by NI company was used to send control instructions to the motor drivers and also collect the signals of torque sensors M_1 and M_2, rotary speeds n_1 and n_2, and

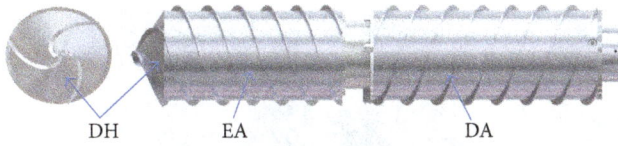

FIGURE 10: Tools: DH, EA, and DA.

penetrating velocities v_1 and v_2, and then these signals can be sent to the PC.

5.1.3. Tools and Simulant. The tools including DH, EA, and DA are shown in Figure 10. In the auger EA and DA, the number of screw flights $N = 3$; helix angel $\alpha = 15.2°$. The radii of the augers are $r_1 = 37$ mm, $r_2 = 33$ mm, and $r_3 = 25$ mm, and the lengths of EA and DA are $L_1 = 120$ mm and $L_2 = 150$ mm.

The raw material of the simulant is the Cenozoic alkaline olivine basalt which is dehumidified and crushed into particles in size of 0.1~1 mm. In the process of preparation, the raw material was steeply poured into the container and vibrated for the regolith simulant with consistent mechanical property along the height direction. The bulk densities of the regolith simulant and the cuttings are 2.18 g/cm³ and 1.54 g/cm³, respectively. The internal friction angle of simulant is 26°, the angle between simulant and tools is 17.6°, and the cohesion of the simulant is 0.05 kPa. The coefficients of A_p and B_p are, respectively, 2.2×10^5 and 5 determined by the confined compression test.

5.2. Experiments Results. In the drilling experiments, H_c is set to 30 mm, rotary speed n_{11} is 60 rpm, and penetrating velocities v_{11} and v_{12} both are 30 mm/min. In order to investigate the effect of rotary speeds n_{11} and n_{12} on the drilling moments, n_{11} was set to 20 rpm, 40 rpm, and 60 rpm and n_{12} was set to 5 rpm, 10 rpm, 20 rpm, 30 rpm, and 40 rpm for drilling experiments and each drilling experiment will be repeated three times.

5.2.1. Drilling Experiments. In the drilling test, IBR takes 120 s for one drilling process; the excavating mode and the discharging mode both take 60 s. As shown in Figure 11, DH is in contact with the simulant surface at $t = 0$ s, EA is just fully penetrated into the simulant at $t = 540$ s, and DA is just fully penetrated into the simulant at $t = 1080$ s.

The drilling moments of the EA and DA are recorded by two torque sensors with 100 Hz sample rate. Figure 12 shows the drilling moments when $n_{11} = 60$ rpm and $n_{12} = 40$ rpm, where the red curve represents the moment M_{1e} of EA in the excavating mode and the blue curve represents the moment M_{2e} of DA in the discharging mode.

From Figure 12, the whole drilling test can be divided into three stages: stages I, II, and III. In stage I, drill head breaks the simulant to cuttings; its drilling moment increases as the depth increases and reaches the maximum value M_{11} at $t = 60$ s. In stage II, drilling moment of the EA is slowly increasing, while the drilling moment of DA does not

increase because it has not interacted with the simulant yet. In stage III, the EA has already drilled into the simulant; thus the storage room is unavoidably filled with cuttings, which will result in a sharp increase in drilling moment of EA. In whole stage III, the drilling moment of EA has reached the maximum value for four times M_{12}, M_{13}, M_{14}, and M_{15}, and this paper selects the maximum value of them as the contrast of theory results. At the same time, the drilling moment of DA is slowly increasing with the drilling depth gradually increasing and reaches the maximum M_{21} at $t = 1080$ s. According to the above analysis, the experimental drilling moment M_1 of EA and the experimental drilling moment M_2 of DA can be noted as follows:

$$M_1 = \max(M_{12}, M_{13}, M_{14}, M_{15}) - M_{11}$$
$$M_2 = M_{21}.$$
(16)

5.2.2. Model Validation. Drilling moment M_1 of EA in excavating mode and the drilling moment M_2 of DA in discharging mode were calculated based on the drilling load model. Figure 13 gives the comparison between the theory curve and experimental results of the EA drilling moments. From Figure 13, the theoretical curve and the experimental results have the same tendency that the drilling moment of EA decreases with the increase of n_{11} or n_{12}. There was good consistency between the theoretical model and experimental results with the RMS error of 0.148 Nm, 0.141 Nm, and 0.121 Nm when n_{11} is 20 rpm, 40 rpm, and 60 rpm, respectively. Figure 14 gives the comparison between the theory and experimental results of the DA drilling moments. From Figure 14, the theoretical curve and the experimental results have the same tendency that the drilling moment of DA decreases with the increase of n_2. The theoretical model still has a good agreement with the experimental results with the RMS error of 0.089 Nm.

5.3. Discussion of K_{11}, K_{12}, and K_2. As can be seen from the experimental results, the drilling moments of tools are directly related to the drilling parameters including rotary speeds and penetrating velocities, and reasonable selection of these motion parameters is critical to the subsurface exploration mission. In the theory model, the drilling parameters mainly exist in the form of ratio of rotary speed and penetrating velocity. Therefore, three speed ratios $K_{11}(n_{11}/v_{11})$, $K_{12}(n_{12}/v_{11})$, and $K_2(n_2/v_2)$ are defined for discussing the effect of drilling parameters on the drilling moment requirements and power consumption of EA and DA.

According to the theoretical model, the surfaces of drilling moment requirement M_1 and power consumption P_1 of EA with K_{11} and K_{12} in the excavating mode are drawn as shown in Figure 15. As can be seen from Figure 15, when $1 < K_{11} < 2$, the driving moment decreases slowly with the increase of K_{11}, while the driving power consumption increases. And when $K_{11} > 1$, the power consumption decreases while the drilling moment sharply increases with decreasing of K_{11}. Therefore, considering the drilling moment requirement and power consumption of EA, the value of K_{11} is recommended in the range

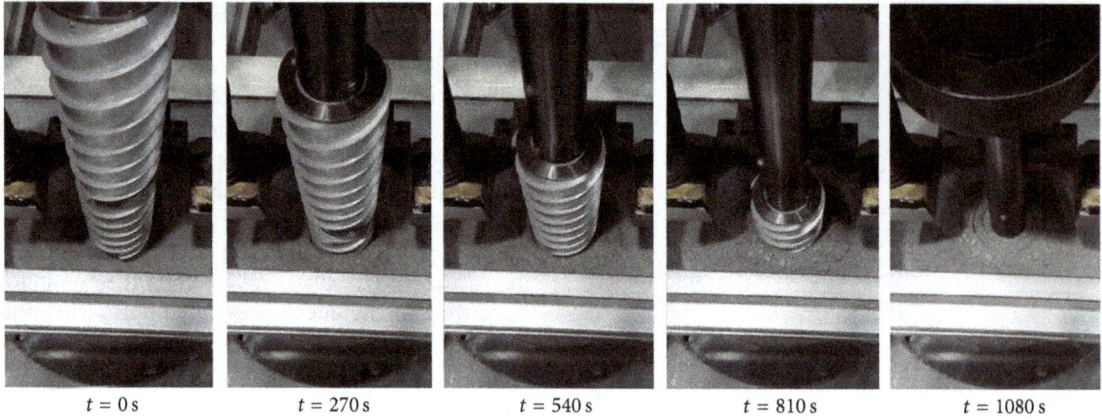

| $t = 0\,s$ | $t = 270\,s$ | $t = 540\,s$ | $t = 810\,s$ | $t = 1080\,s$ |

FIGURE 11: Inchworm drilling experiments process.

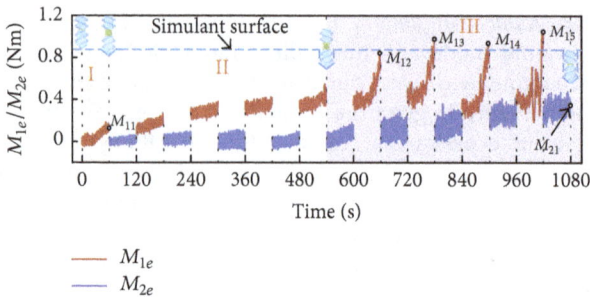

FIGURE 12: Drilling moments of EA and DA.

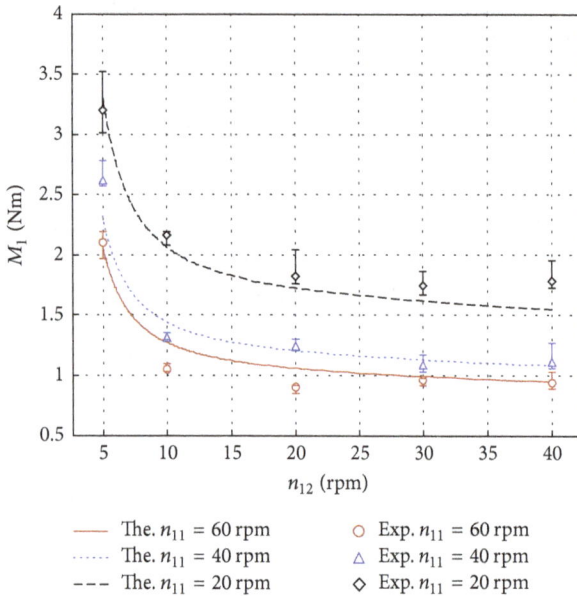

FIGURE 13: M_1 versus n_{11} and n_{12}.

The. $n_{11} = 60\,rpm$ ○ Exp. $n_{11} = 60\,rpm$
The. $n_{11} = 40\,rpm$ △ Exp. $n_{11} = 40\,rpm$
The. $n_{11} = 20\,rpm$ ◇ Exp. $n_{11} = 20\,rpm$

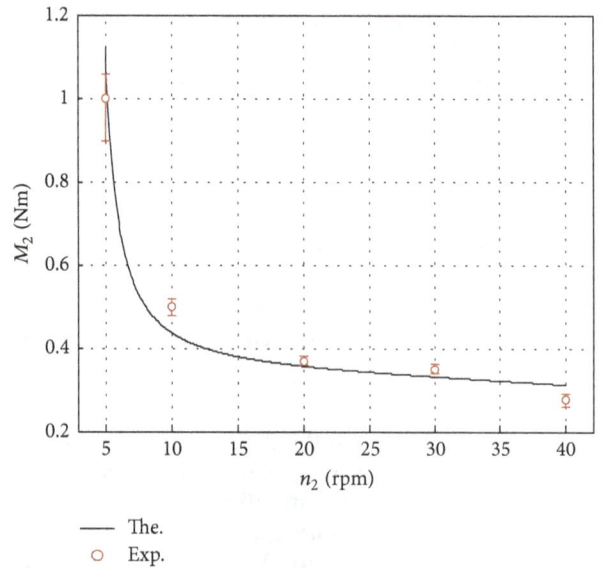

——— The.
○ Exp.

FIGURE 14: M_2 versus n_2.

mode are drawn as shown in Figure 16. As can be seen in Figure 16, when $K_2 < 1/3$, the drilling moment requirement M_2 increases drastically with the decrease of K_2. And when $K_2 < 1/3$, the power consumption P_2 increases linearly with the increase of K_2. Therefore, considering the drilling moment requirement and power consumption of DA, the value of K_2 in the discharging mode is recommended in the range of $1/3 < K_2 < 1/2$.

6. Conclusions

In this paper, an Inchworm Boring Robot was proposed for lunar subsurface exploration. Two auger tools of EA and DA are utilized to remove chips for IBR. For obtaining drilling characteristics of the two tools, the chips removal principles of IBR in excavating mode and discharging mode are analyzed, and then the relation of tools drilling parameters and the soil flow state in the flow channel of EA and DA are given. Subsequently, the drilling load model of tools

of $1 < K_{11} < 4/3$. For K_{12}, when $K_{12} < 1/2$, drilling moment requirement and power consumption of EA increase drastically as K_{12} decreases. Therefore, K_{12} should satisfy the condition $K_{12} < 1/2$ as far as possible.

The theoretical surfaces of drilling moment M_2 and power consumption P_2 of DA with K_2 in the discharging

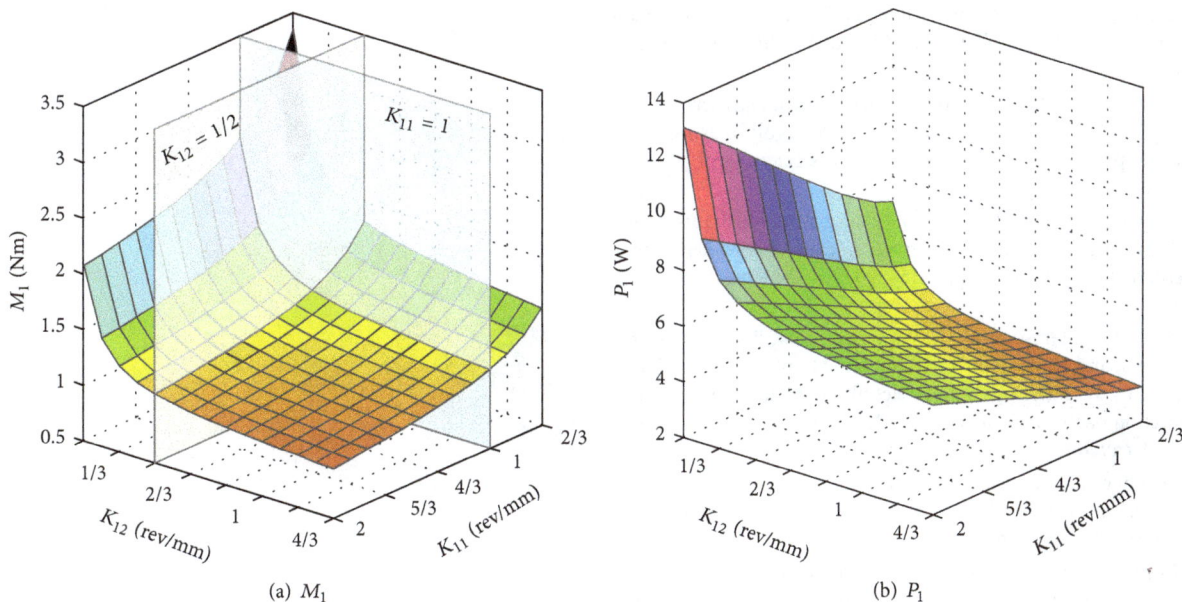

FIGURE 15: M_1 and P_1 versus K_{11} and K_{12}.

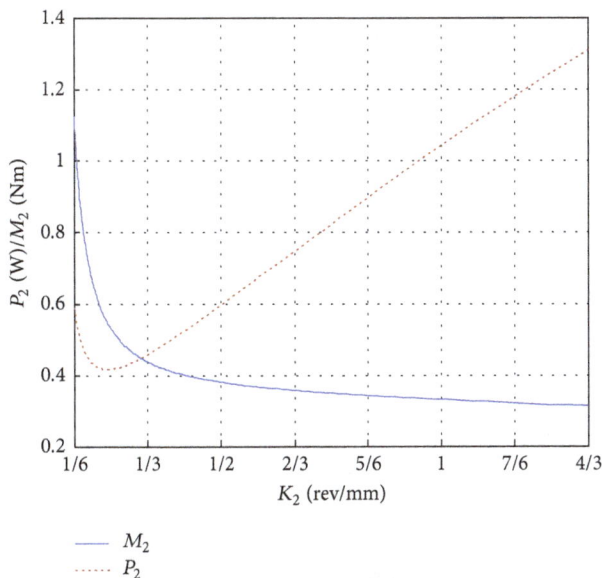

FIGURE 16: M_2 and P_2 versus K_2.

is further established based on the soil mechanical theory. Drilling experiments with different drilling parameters had been conducted to test the tools drilling moments, and the proposed theoretical drilling load models agree well with the experimental results within the RMS errors of 0.148 Nm. In addition, according to the theoretical model analysis, the effect of drilling parameters on the tools drilling moments and power consumption was discussed and the recommend value of speed ratios can be given as $1 < K_{11} < 4/3$, $K_{12} < 1/2$, and $1/3 < K_2 < 1/2$. These results can be used for the energy-efficient control strategy for IBR.

Conflicts of Interest

The authors declare that they have no conflicts of interest.

Acknowledgments

This work is supported by the National Natural Science Foundation of China (Grant no. 51575122) and the Self-Planned Task of State Key Laboratory of Robotics and System (HIT) (no. SKLRS201616B).

References

[1] K. Zacny, Y. Bar-Cohen, M. Brennan et al., "Drilling systems for extraterrestrial subsurface exploration," *Astrobiology*, vol. 8, no. 3, pp. 665–706, 2008.

[2] I. I. Stakheev, A. V. Ivanov, and E. K. Vulfson, "Comparative characteristics of the lunar-soil particle-size distribution at the landing sites of the automatic lunar stations Luna 16 and Luna 20," *Cosmic Research*, vol. 14, no. 3, pp. 381–386, 1976.

[3] A. Davé, S. J. Thompson, C. P. Mckay et al., "The sample handling system for the mars icebreaker life mission: From dirt to data," *Astrobiology*, vol. 13, no. 4, pp. 354–369, 2013.

[4] K. Zacny, S. Nagihara, M. Hedlund et al., "Pneumatic and percussive penetration approaches for heat flow probe emplacement on robotic lunar missions," *Earth, Moon and Planets*, vol. 111, no. 1-2, pp. 47–77, 2013.

[5] C. R. Stoker, H. N. Cannon, S. E. Dunagan et al., "The 2005 MARTE Robotic Drilling Experiment in Río Tinto, Spain: objectives, approach, and results of a simulated mission to search for life in the Martian subsurface," *Astrobiology*, vol. 8, no. 5, pp. 921–945, 2008.

[6] J. W. Reiter, J. L. Guerrero, D. Wu et al., "Advanced planetary drill technology and applications to future space missions," *Lpi Contributions*, p. 1287, 2005.

[7] Y. Bar-Cohen and K. Zacny, *Drilling in Extreme Environments*, Wiley-VCH Verlag GmbH & Co. KGaA, Weinheim, Germany, 2009.

[8] K. Seweryn, K. Skocki, M. Banaszkiewicz et al., "Determining the geotechnical properties of planetary regolith using low velocity penetrometers," *Planetary & Space Science*, vol. 99, no. 1, pp. 70–83, 2014.

[9] J. Grygorczuk and Ł. Wiśniewski, "Hammering Mechanism for HP3 Experiment (InSight)," in *Proceedings of the 43rd Aerospace Mechanisms Symposium*, pp. 415–428, NASA Ames Research, 2016.

[10] K. Seweryn, J. Grygorczuk, R. Wawrzaszek, M. Banaszkiewicz, T. Rybus, and Ł. Wiśniewski, "Low velocity penetrators (LVP) driven by hammering action—definition of the principle of operation based on numerical models and experimental tests," *Acta Astronautica*, vol. 99, no. 1, pp. 303–317, 2014.

[11] C. Pitcher and Y. Gao, "Analysis of drill head designs for dual-reciprocating drilling technique in planetary regoliths," *Advances in Space Research*, vol. 56, no. 8, pp. 1765–1776, 2015.

[12] T. Myrick, S. Frader-Thompson, J. Wilson et al., "Development of an inchworm deep subsurface platform for in situ investigation of Europa's icy shell," in *Proceedings of the Workshop on Europa's Icy Shell: Past, Present, and Future*, p. 7041, 2004.

[13] K. Zacny, G. Paulsen, Y. Bar-Cohen et al., "Wireline deep drill for exploration of Mars, Europa, and Enceladus," in *Proceedings of 2013 IEEE Aerospace Conference, AERO 2013*, usa, March 2013.

[14] H. Omori, T. Murakami, H. Nagai, T. Nakamura, and T. Kubota, "Development of a novel bio-inspired planetary subsurface explorer: initial experimental study by prototype excavator with propulsion and excavation units," *IEEE/ASME Transactions on Mechatronics*, vol. 18, no. 2, pp. 459–470, 2013.

[15] T. Dewei, Z. Weiwei, J. Shengyuan, S. Yi, and C. Huazhi, "Development of an Inchworm Boring Robot(IBR) for planetary subsurface exploration," in *Proceedings of IEEE International Conference on Robotics and Biomimetics, IEEE-ROBIO 2015*, pp. 2109–2114, chn, December 2015.

[16] W. Zhang, S. Jiang et al., "Design of a screw-cone drill for lunar regolith drilling mission," *Journal of Astronautics*, vol. 37, no. 12, pp. 61–69, 2016.

[17] P. Li, S. Jiang, D. Tang et al., "Design and testing of coring bits on drilling lunar rock simulant," *Advances in Space Research*, pp. 1–20, 2016.

[18] K. Terzaghi, *Theoretical Soil Mechanics*, John Wiley & Sons, New York, NY, USA, 1965.

[19] J. M. Duncan and C. Y. Chang, "Nonlinear analysis of stress and strain in soils," *Asce Soil Mechanics & Foundation Division Journal*, vol. 96, no. 5, pp. 1629–1653, 1970.

[20] R. M. Nedderman, *Statics and Kinematics of Granular Materials*, Cambridge University Press, 1992.

12

A Terminal Guidance Law based on Motion Camouflage Strategy of Air-to-Ground Missiles

Chang-sheng Gao, Jian-qing Li, and Wu-xing Jing

Department of Aerospace Engineering, Harbin Institute of Technology, Harbin, China

Correspondence should be addressed to Jian-qing Li; ljq18@hit.edu.cn

Academic Editor: Christopher J. Damaren

A guidance law for attacking ground target based on motion camouflage strategy is proposed in this paper. According to the relative position between missile and target, the dual second-order dynamics model is derived. The missile guidance condition is given by analyzing the characteristic of motion camouflage strategy. Then, the terminal guidance law is derived by using the relative motion of missile and target and the guidance condition. In the process of derivation, the three-dimensional guidance law could be designed in a two-dimensional plane and the difficulty of guidance law design is reduced. A two-dimensional guidance law for three-dimensional space is derived by bringing the estimation for target maneuver. Finally, simulation for the proposed guidance law is taken and compared with pure proportional navigation. The simulation results demonstrate that the proposed guidance law can be applied to air-to-ground missiles.

1. Introduction

Proportional navigation (PN), which has a simple form and is implemented easily, is widely used in the missile interception field. Through decades of development, proportional navigation law has been improved to different forms, including true proportional navigation (TPN), pure proportional navigation (PPN), augmented proportional navigation (APN), and bias-proportional navigation (BPN) [1–3]. The ultimate goal of these guidance laws is to make the line of sight (LOS) angle rate converge to zero as much as possible. However, the LOS angle rate convergence to zero is difficult for high maneuvering target. This comes with some of the inherent problems of PN guidance, such as lateral acceleration singularity at the end time when range-to-go or time-to-go approaches zero [4]. Traditional proportional navigation law requires the normal acceleration and the LOS rate is proportional to the ratio; the bias guidance law is to make the normal line of sight angular rate and acceleration give a small deviation term. The modified bias-proportional navigation deals with angle constraint by increasing two time-varying terms, but it requires a time-to-go estimation and the velocity of the missile to be constant [5, 6].

In recent years, the optimal guidance law is investigated intensively based on optimal control theory [7]. The different forms of guidance law can be achieved by different performance indexes, such as the minimum miss distance, the minimum consumption, and the minimum time. In [8, 9], optimal control laws, for a missile with arbitrary order dynamics trying to attack a stationary target, were proposed with a similar cost function and an LOS fixed the coordinate system. The proposed law was implemented for lag-free and first-order lag missile systems. In the optimal guidance law, the time-to-go has significant effects on the guidance commands and even performance index. Therefore, the key issue is how to accurately estimate the remaining time so that we can improve the performance of guidance law. Hexner et al. [10] derived an optimal guidance law by analyzing an intercept scenario in the framework of a linear quadratic Gaussian terminal control problem with bounded acceleration command. Ratnoo and Ghose [11] introduced a tracking filter to estimate the relative motion for obtaining estimates of the time-to-go. In the ideal case, the optimal guidance law can get a good trajectory, but the ballistic performance maybe gets poor in uncertainty [12].

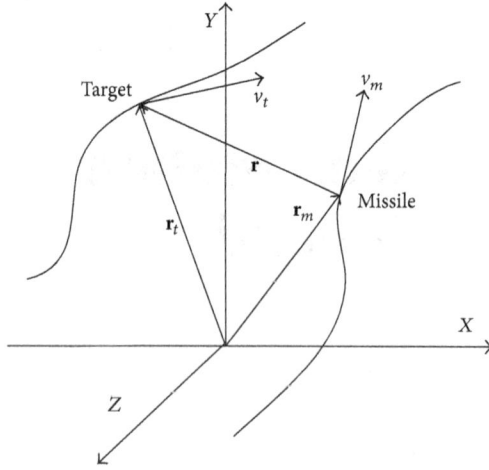

FIGURE 1: Missile-target engagement geometry.

Hence, the variable structure control theory is widely applied to guidance law design, due to its advantages of inherent robustness and simple algorithm. The robustness nature of sliding mode control can accommodate the target maneuvering and other disturbances. Shima [13] gives a deviated velocity pursuit guidance law, which is formulated using sliding mode control theory. Harl and Balakrishnan [14] present a guidance law based on a sliding manifold and develop a robust second-order sliding mode control law by using a backstepping concept. Shtessel and Tournes [15] develop an integrated autopilot and guidance algorithm by using higher-order sliding mode control for interceptors. This law which is robust to target maneuvers generates flight-path trajectory angular rates and attitude rate commands. Although the sliding mode can be robust to target maneuvers and missile's model uncertainties, this guidance law has a disadvantage that it needs the second derivative of LOS or other information of the target [16].

Motion camouflage (MC) theory was first proposed in 1995, and Srinivasan and Davey [17] explain the predatory strategy of insects with MC theory. This strategy can be simply described as in the process of the predator pursuing the target: the predator camouflages itself against a fixed background object so that the prey observes no relative motion between the predator and the fixed object. Because this strategy has some military value of the application, it has been used in spacecraft rendezvous, unmanned aerial vehicle (UAV) flight-path planning, and so forth [18–20]. Many scholars also have studied interception guidance. Mischiati and Krishnaprasad [21] studied the dynamics of motion camouflage interception model and convergence issues. Bakolas and Tsiotras [22] studied the robustness issues of motion camouflage guidance law under a two-dimensional flow field and compared the performance of different guidance laws. Justh and Krishnaprasad [23] established a missile and target motion model of Frenet frame, given a feedback guidance law based on the motion camouflage theory, and proved that the guidance law can make the line of sight angular rate convergence in finite time.

This paper proposes a dimension-reduction guidance law for attacking ground target based on motion camouflage strategy, which has compensation for target maneuvering. First, the interception condition of the missile is derived from motion camouflage characteristics which is obtained by the theory of motion camouflage. Then, the two-dimensional guidance law based on the condition is designed to the three-dimensional space model. This dimension reduction method not only simplifies the design steps of guidance law but also reduces the design difficulty. Finally, some simulations are carried out. The simulation results show the effectiveness of the proposed guidance law.

2. Dynamics Model

The relative relationships of the missile and the target are shown in Figure 1.

The relative displacement vector from the missile to the target is given by

$$\mathbf{r} = \mathbf{r}_t - \mathbf{r}_m = r\mathbf{e}_r, \tag{1}$$

where r is the relative distance between missile and target. $\mathbf{e}_r$ is a fixed unit vector along the line of sight. Differentiating $\mathbf{e}_r$ with respect to time yields

$$\dot{\mathbf{e}}_r = \boldsymbol{\omega} \times \mathbf{e}_r = \omega\mathbf{e}_\omega \times \mathbf{e}_r. \tag{2}$$

The vector $\mathbf{e}_\theta$ is defined as

$$\mathbf{e}_\theta = \mathbf{e}_\omega \times \mathbf{e}_r. \tag{3}$$

The set of unit vectors $(\mathbf{e}_r, \mathbf{e}_\theta, \mathbf{e}_\omega)$ constitutes a reference frame. This frame is a rotating coordinate system and the origin is the mass center of the missile. Differentiating (1) yields

$$\dot{\mathbf{r}} = \dot{r}\mathbf{e}_r + r\dot{\mathbf{e}}_r = \dot{r}\mathbf{e}_r + r\omega\mathbf{e}_\theta. \tag{4}$$

Obviously, the relative velocity vector is constituted by the radial velocity and the normal velocity. Let $\mathbf{a}_m$ and $\mathbf{a}_t$ be the maneuvering acceleration of missile and target; they are expressed in the rotating coordinate system as

$$\begin{aligned}\mathbf{a}_m &= a_{mr}\mathbf{e}_r + a_{m\theta}\mathbf{e}_\theta + a_{m\omega}\mathbf{e}_\omega \\ \mathbf{a}_t &= a_{tr}\mathbf{e}_r + a_{t\theta}\mathbf{e}_\theta + a_{t\omega}\mathbf{e}_\omega.\end{aligned} \tag{5}$$

Therefore, the relative acceleration of the missile and the target can be expressed as

$$\begin{aligned}\ddot{\mathbf{r}} &= \mathbf{a}_t - \mathbf{a}_m \\ &= (a_{tr} - a_{mr})\mathbf{e}_r + (a_{t\theta} - a_{m\theta})\mathbf{e}_\theta + (a_{t\omega} - a_{m\omega})\mathbf{e}_\omega.\end{aligned} \tag{6}$$

From the above equation, we can derive a second-order dynamic equation of relative movement as

$$\begin{aligned}\ddot{r} &= r\dot{\mathbf{e}}_r^2 - a_{mr} + a_{tr} \\ \ddot{\mathbf{e}}_r &= (a_{t\theta} - a_{m\theta})\frac{\mathbf{e}_\theta}{r} + (a_{t\omega} - a_{m\omega})\frac{\mathbf{e}_\omega}{r} - \dot{\mathbf{e}}_r^2\mathbf{e}_r - \frac{2\dot{r}}{r}\dot{\mathbf{e}}_r.\end{aligned} \tag{7}$$

Therefore, the guidance problem can be described as finding the acceleration of the missile a_{mr}, $a_{m\theta}$, and $a_{m\omega}$ to let r converge to zero in finite time.

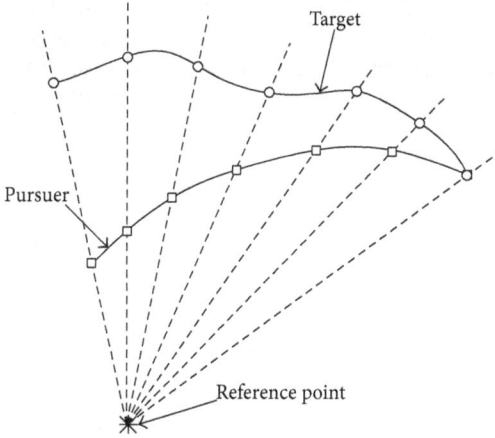

FIGURE 2: Motion camouflage scenario.

3. Guidance Law Implementation

3.1. Motion Camouflage Theory. Motion camouflage strategy is a new form of stealth strategy which describes the relative motion relationship of the pursuer, target, and reference point: the movement of them as shown in Figure 2.

The pursuer's path is controlled by the path control parameter (PCP) $c(t)$ as

$$\mathbf{x}_p = \mathbf{x}_r + c(t)\,\mathbf{x}_{er}, \tag{8}$$

where $\mathbf{x}_{er} = \mathbf{x}_e - \mathbf{x}_r$ are the relative distance vector from the reference point to the target. The selected PCP and reference point determine the speed and curvature of the trajectory in the constructed subspace.

If the position of the reference point is a fixed camouflage background, motion camouflage strategy is similar to the three-point guidance law. And if the reference point is chosen at the infinity, it is similar to the constant-bearing navigation. Therefore, motion camouflage strategy has both features of the three-point guidance law and constant-bearing navigation.

3.2. Guidance Law Based on Motion Camouflage. Let the pursuer and target be the missile and the ground target, respectively. And setting the reference point as infinity yields

$$\mathbf{r} = \mathbf{r}_t - \mathbf{r}_m = c(t)\,\mathbf{e}_r. \tag{9}$$

The component of the missile velocity transverse to the baseline is

$$\dot{\mathbf{r}}_{m\perp} = \dot{\mathbf{r}}_m - (\mathbf{e}_r \cdot \dot{\mathbf{r}}_m)\,\mathbf{e}_r, \tag{10}$$

and, similarly, that of the target is

$$\dot{\mathbf{r}}_{t\perp} = \dot{\mathbf{r}}_t - (\mathbf{e}_r \cdot \dot{\mathbf{r}}_t)\,\mathbf{e}_r. \tag{11}$$

The relative transverse component is

$$\boldsymbol{\lambda} = (\dot{\mathbf{r}}_t - \dot{\mathbf{r}}_m) - (\mathbf{e}_r \cdot (\dot{\mathbf{r}}_t - \dot{\mathbf{r}}_m))\,\mathbf{e}_r = \dot{\mathbf{r}} - (\mathbf{e}_r \cdot \dot{\mathbf{r}})\,\mathbf{e}_r. \tag{12}$$

The missile-target system is in a state of motion camouflage without collision on an interval iff $\boldsymbol{\lambda} = 0$ on that interval.

According to the fact that the final goal of guidance problem is such that the relative distance converges to zero, we consider the ratio as follows:

$$Z = \frac{\dot{r}}{|\dot{\mathbf{r}}|} \tag{13}$$

which compares the rate of change of the baseline length to the absolute rate of change of the baseline vector. If the baseline experiences pure lengthening, then the ratio assumes its maximum value, $Z = +1$. If the baseline experiences pure shortening, then the ratio assumes its minimum value, $Z = -1$.

Equation (13) can be written as

$$Z = \frac{\mathbf{r}}{r} \cdot \frac{\dot{\mathbf{r}}}{|\dot{\mathbf{r}}|}. \tag{14}$$

Thus, Z is the dot product of two unit vectors: one in the direction of $\mathbf{r}$ and the other in the direction of $\dot{\mathbf{r}}$. According to (12), the magnitude squared of $\boldsymbol{\lambda}$ is

$$|\boldsymbol{\lambda}|^2 = |\dot{\mathbf{r}}|^2 - 2(\mathbf{e}_r \cdot \dot{\mathbf{r}})^2 + (\mathbf{e}_r \cdot \dot{\mathbf{r}})^2 = |\dot{\mathbf{r}}|^2(1 - Z^2). \tag{15}$$

Obviously, the requirement of the component of the missile velocity is equal to that of the target: it could be transferred to $Z = -1$. Thus, our objective is to design a guidance law to guarantee $Z = -1$.

Differentiating Z gives

$$\dot{Z} = \frac{(\dot{\mathbf{r}} \cdot \dot{\mathbf{r}} + \mathbf{r} \cdot \ddot{\mathbf{r}}) \cdot r\,|\dot{\mathbf{r}}|}{(r\,|\dot{\mathbf{r}}|)^2} - \frac{\mathbf{r} \cdot \dot{\mathbf{r}} \cdot \dot{r}\,|\dot{\mathbf{r}}|}{(r\,|\dot{\mathbf{r}}|)^2} - \frac{\mathbf{r} \cdot \dot{\mathbf{r}} \cdot r \cdot \mathbf{r} \cdot \ddot{\mathbf{r}}}{(r\,|\dot{\mathbf{r}}|)^2\,|\dot{\mathbf{r}}|}$$

$$= \left(\frac{\dot{\mathbf{r}} \cdot \dot{\mathbf{r}} + \mathbf{r} \cdot \ddot{\mathbf{r}}}{r\,|\dot{\mathbf{r}}|}\right) - \left(\frac{\mathbf{r} \cdot \dot{\mathbf{r}}}{|\dot{\mathbf{r}}|}\right)\left(\frac{\mathbf{r} \cdot \dot{\mathbf{r}}}{r^3}\right)$$

$$\quad - \left(\frac{\mathbf{r} \cdot \dot{\mathbf{r}}}{r}\right)\left(\frac{\mathbf{r} \cdot \ddot{\mathbf{r}}}{|\dot{\mathbf{r}}|^3}\right) \tag{16}$$

$$= \frac{|\dot{\mathbf{r}}|}{r}\left[1 - \left(\frac{\mathbf{r}}{r} \cdot \frac{\dot{\mathbf{r}}}{|\dot{\mathbf{r}}|}\right)^2\right] + \frac{1}{|\dot{\mathbf{r}}|}\left[\frac{\mathbf{r}}{r} - \left(\frac{\mathbf{r}}{r} \cdot \frac{\dot{\mathbf{r}}}{|\dot{\mathbf{r}}|}\right)\frac{\dot{\mathbf{r}}}{|\dot{\mathbf{r}}|}\right]$$

$$\cdot \ddot{\mathbf{r}}.$$

We define

$$\boldsymbol{\xi} = \frac{1}{|\dot{\mathbf{r}}|}\left[\frac{\mathbf{r}}{r} - \left(\frac{\mathbf{r}}{r} \cdot \frac{\dot{\mathbf{r}}}{|\dot{\mathbf{r}}|}\right)\frac{\dot{\mathbf{r}}}{|\dot{\mathbf{r}}|}\right]. \tag{17}$$

Using the formula $\mathbf{a} \times (\mathbf{b} \times \mathbf{c}) = \mathbf{b}(\mathbf{a} \cdot \mathbf{c}) - \mathbf{c}(\mathbf{a} \cdot \mathbf{b})$ and (4), we compute

$$\boldsymbol{\xi} = -\frac{1}{|\dot{\mathbf{r}}|^3}\left[\dot{\mathbf{r}} \times \left(\dot{\mathbf{r}} \times \frac{\mathbf{r}}{r}\right)\right]$$

$$= -\frac{1}{|\dot{\mathbf{r}}|^3}\left[\dot{\mathbf{r}}\left(\dot{\mathbf{r}} \cdot \frac{\mathbf{r}}{r}\right) - \frac{\mathbf{r}}{r}(\dot{\mathbf{r}} \cdot \dot{\mathbf{r}})\right]$$

$$= -\frac{1}{|\dot{\mathbf{r}}|^3}\left[\dot{r}(\dot{r}\mathbf{e}_r + r\omega\mathbf{e}_\theta) - \mathbf{e}_r(\dot{r}\mathbf{e}_r + r\omega\mathbf{e}_\theta)^2\right] \tag{18}$$

$$= -\frac{1}{|\dot{\mathbf{r}}|^3}\left[(\dot{r}^2\mathbf{e}_r + \dot{r}r\omega\mathbf{e}_\theta) - \mathbf{e}_r(\dot{r}^2 + r^2\omega^2)\right]$$

$$= \frac{1}{|\dot{\mathbf{r}}|^3}\left(r^2\omega^2\mathbf{e}_r - \dot{r}r\omega\mathbf{e}_\theta\right).$$

Then

$$\boldsymbol{\xi} \cdot \ddot{\mathbf{r}} = \frac{1}{|\dot{\mathbf{r}}|^3} \left(r^2 \omega^2 \mathbf{e}_r - \dot{r} r \omega \mathbf{e}_\theta \right) \cdot \left(P\mathbf{e}_\theta + Q\mathbf{e}_\omega + R\mathbf{e}_r \right)$$

$$= \frac{1}{|\dot{\mathbf{r}}|^3} \left(r^2 \omega^2 R - \dot{r} r \omega P \right). \tag{19}$$

Substituting (19) into (16) yields

$$\dot{Z} = \frac{|\dot{\mathbf{r}}|}{r} \left(1 - Z^2 \right) - \frac{\dot{r} r \omega}{|\dot{\mathbf{r}}|^3} \left(a_{t\theta} - a_{m\theta} \right)$$

$$+ \frac{\omega^2 r^2}{|\dot{\mathbf{r}}|^3} \left(a_{tr} - a_{mr} \right). \tag{20}$$

As can be seen from the above results, the acceleration term $a_{t\omega}$ has been eliminated. Thus, we only design the tangential acceleration and normal acceleration of missile to make $\dot{Z} < 0$. According to the literature [3], the relative acceleration $(a_{tr} - a_{mr})$ is high-order small quantity relative to the other direction of the acceleration and can be neglected. Thus, the final task of the designed three-dimensional guidance law is to give the analytical expression of the acceleration $a_{m\theta}$.

We give the guidance law as

$$a_{m\theta} = \mu \frac{v_m}{\omega r} \left(\dot{r} - \frac{|\dot{\mathbf{r}}|^2}{\dot{r}} \right) + a_{t\theta} \tag{21}$$

and substitute into (20)

$$\dot{Z} = \left(1 - Z^2 \right) \left(\frac{|\dot{\mathbf{r}}|}{r} - \frac{\mu v_m}{|\dot{\mathbf{r}}|} \right). \tag{22}$$

We assume that the upper and lower bounds $[v_m^-, v_m^+]$ and $[v_t^-, v_t^+]$ exist such that

$$\frac{v_t}{v_m} \leq K < 1. \tag{23}$$

For the interception process, the relative velocity of the missile and the target should satisfy the following relationship:

$$v_m^- (1 - K) \leq |\dot{\mathbf{r}}| \leq v_m^+ (1 + K). \tag{24}$$

We define

$$\mu = \frac{v_m^+ (K + 1)}{v_m^-} \left(\frac{v_m^+ (K + 1)}{r_o} + \sigma \right) \tag{25}$$

and hence

$$\mu \geq \frac{v_m^+ (K + 1)}{v_m^-} \left(\frac{v_m^+ (K + 1)}{r} + \sigma \right), \tag{26}$$

where $r_o > 0, \sigma > 0$. Thus, for $r > r_o$, (22) becomes

$$\dot{Z} \leq \left(1 - Z^2 \right) \left[\frac{v_m^+ (1 + K)}{r} \right.$$

$$\left. - \frac{v_m^-}{v_m^+ (K + 1)} \frac{v_m^+ (K + 1)}{v_m^-} \left(\frac{v_m^+ (K + 1)}{r} + \sigma \right) \right] \tag{27}$$

$$= - \left(1 - Z^2 \right) \sigma.$$

Obviously, $\dot{Z} < 0$ can be held for $\sigma > 0$. Therefore, (21) can guarantee interception. However, the target acceleration information of guidance law is not measurable and is only estimated approximately. We assume that a constant W exists such that

$$|a_{t\theta}| \leq W. \tag{28}$$

The target acceleration $a_{t\theta}$ is replaced by $W \operatorname{sgn}(\omega)$ and (20) is given by

$$a_{m\theta} = \mu \frac{v_m}{\omega r} \left(\dot{r} - \frac{|\dot{\mathbf{r}}|^2}{\dot{r}} \right) + W \operatorname{sgn}(\omega). \tag{29}$$

The switching term would lead to the appearance of the chattering effect on acceleration. To remove the chattering, the signum function can be smoothened, usually replacing $\operatorname{sgn}(x)$ with a saturation function expressed as

$$\operatorname{sat}(x, \delta) = \begin{cases} 1 & x > \delta \\ \dfrac{x}{\delta} & |x| \leq \delta \\ -1 & x < -\delta. \end{cases} \tag{30}$$

The final three-dimensional guidance law can be designed as

$$\mathbf{a}_m = \left(\mu \frac{v_m}{\omega r} \left(\dot{r} - \frac{|\dot{\mathbf{r}}|^2}{\dot{r}} \right) + W \cdot \operatorname{sat}(\omega, \delta) \right) \cdot \mathbf{e}_\theta. \tag{31}$$

The guidance law only has a normal component of LOS, so the three-dimensional guidance law based on motion camouflage strategy can be converted directly into the rotation plane of LOS. If the target does not maneuver, the designed guidance law only requires the LOS rate, the relative distance, and the velocity of the missile. The proposed guidance law reduces the difficulty of detection (without obtaining pre-angle information) compared with the constant-bearing method. Also, it can ensure a smaller overload at the terminal stage than proportional navigation method, because the guidance law contains the relative motion information, although it needs more measurement information.

4. Simulation

4.1. Comparison of Different Gains. In order to verify the validity of the designed guidance law, the different coefficients μ will be given for comparison simulation. The initial position and the initial velocity of the missile are $\mathbf{r}_m = [131\,\text{km}, 5\,\text{km}, 0\,\text{km}]$ and $v_m = 700\,\text{m/s}$. The initial position and the initial velocity of the target are $\mathbf{r}_t = [142\,\text{km}, 0\,\text{km}, 2.54\,\text{km}]$ and $v_t = 60\,\text{m/s}$. Firstly, the target moves in a straight line and the guidance coefficients are specified as 0.5 and 2. The simulation results are shown in Figures 3–5.

As can be seen from the figures, the acceleration amplitudes of motion camouflage guidance law are closely related to the guidance coefficient. The trajectory and overload are also different when different guidance coefficients are taken.

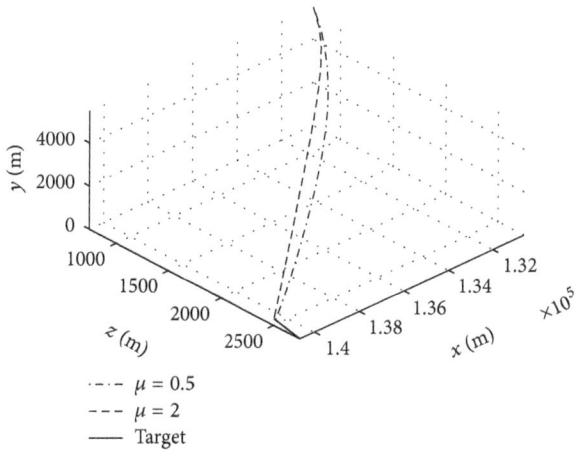

FIGURE 3: The interception trajectories of different ratios.

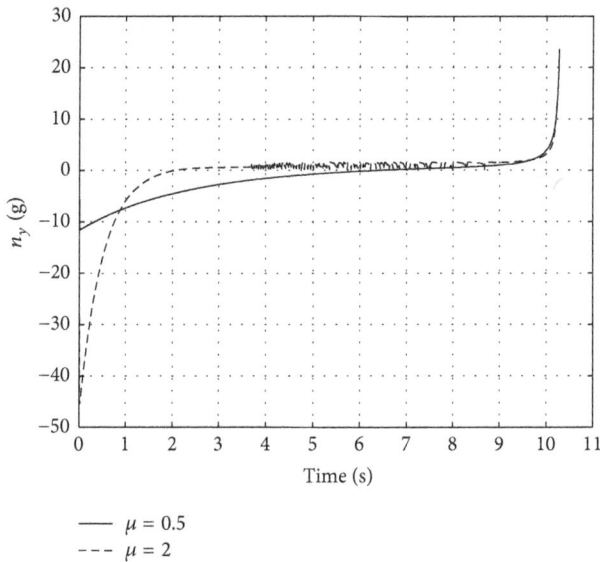

FIGURE 5: Lateral acceleration of different ratios.

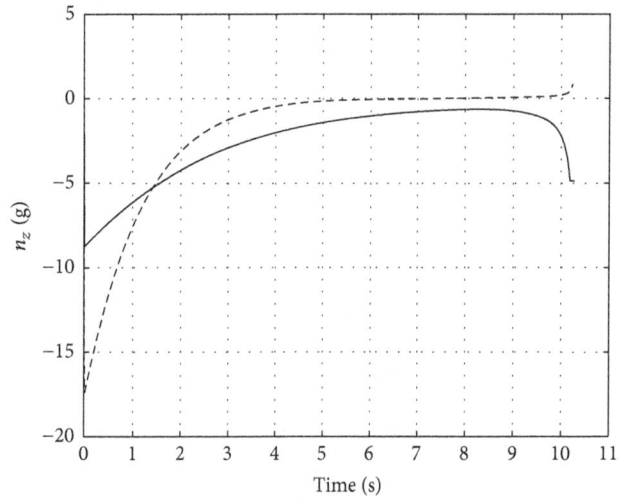

FIGURE 4: Longitudinal acceleration of different ratios.

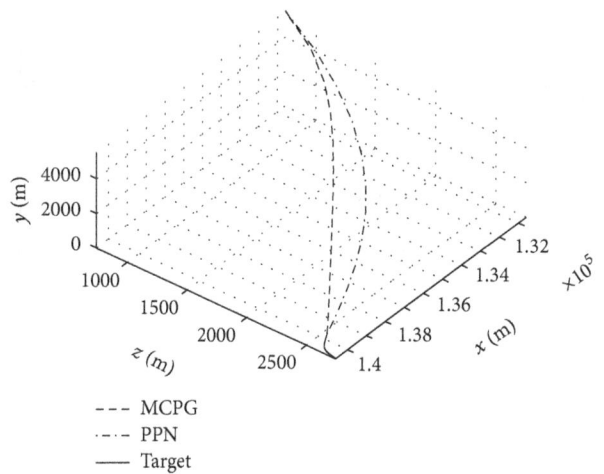

FIGURE 6: Interception trajectories.

Because the coefficient $\mu = 0.5$ is small, the overload of the missile is small in the initial stage and thus cannot keep up with the target. Whereas when $\mu = 2$ the missile overload is larger in the terminal stage, the miss distance is smaller than the other. Therefore, the guidance coefficient should be selected reasonably and it can guarantee that the proper overload of missile is smooth and can also achieve a smaller miss distance.

4.2. Comparison of Different Guidance Laws. The pure proportional navigation (PPN) is chosen for comparing with the proposed guidance law. The guidance coefficients of MCPG and PPN are 0.5 and 3, respectively. The simulation conditions are not changed. The target's maneuvering acceleration is $1g$. The simulation results are shown in Figures 6–10.

The miss distances of MCPG and PPN are 0.3254 m and 0.5851 m, respectively. In the initial stage of interception, the acceleration of MCPG is larger than that of PPN. However,

the acceleration of MCPG has a faster response so that the missile can track the maneuvering of the target better. Figure 9 presents the rate of rotation of LOS and illustrates that the MCPG can restrain the rate rotation before hitting the target.

Figure 10 shows the values of Z. Note that the proposed guidance law trends Z to -1 during the process of interception. The value of Z always fluctuates around -1. Thus, the relative motion satisfies the status of motion camouflage.

According to the above analysis, the MCPG has a large overload at the initial stage, but the proposed guidance law can satisfy the demand for rapid response and maneuvering.

5. Conclusion

This paper presents a three-dimensional guidance law for intercepting the ground target, which is based on the motion camouflage theory and the proposed dynamic equations. To

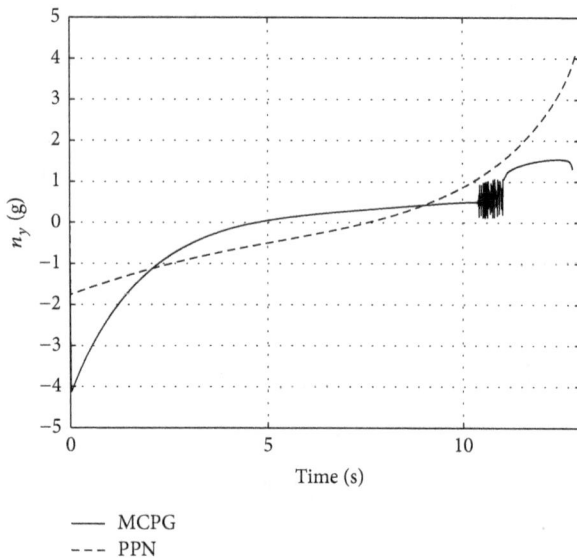

FIGURE 7: Longitudinal acceleration.

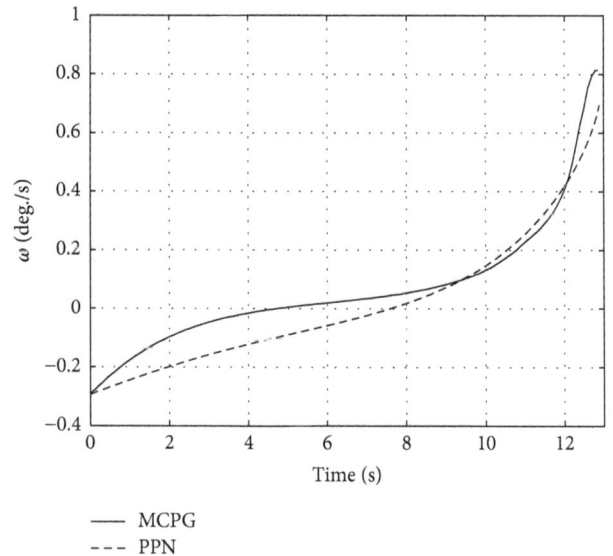

FIGURE 9: LOS angular velocity of missile.

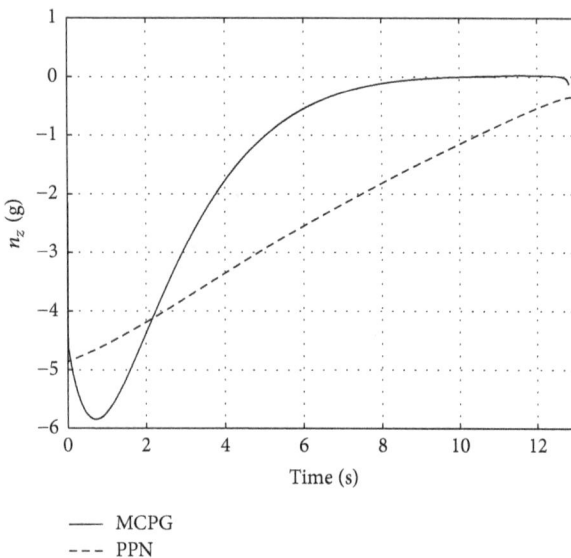

FIGURE 8: Lateral acceleration.

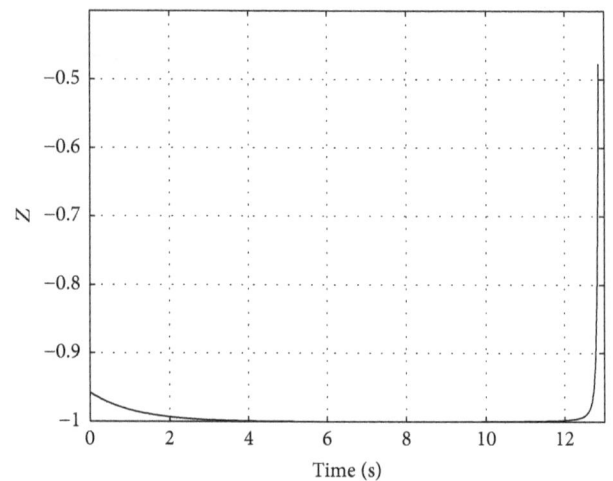

FIGURE 10: Variation of Z.

improve the robustness of the guidance law, a compensation is given for the target maneuver. The designed guidance law does not require too much measurement information so that it is implemented easily. Also, its expression only has a normal component of acceleration, and it can reduce the difficulty of the design process. According to the simulation and comparison of MCPG and PPN, the results show that the three-dimensional guidance law based on the motion camouflage theory can destroy ground targets effectively. The proposed guidance law has a faster response of acceleration and a smaller acceleration at the last moment.

Competing Interests

The authors declare that they have no competing interests.

Acknowledgments

This work is supported by Aerospace Science and Technology Innovation Foundation of China under Grant no. CASC-HIT13-1C03 and National Natural Science Foundation of China under Grant no. 11572097.

References

[1] P. Zarchan, *Tactical and Strategic Missile Guidance*, American Institute of Aeronautics and Astronautics, Reston, Va, USA, 5th edition, 2007.

[2] S. Ghosh, D. Ghose, and S. Raha, "Capturability analysis of a 3-D retro-PN guidance law for higher speed nonmaneuvering targets," *IEEE Transactions on Control Systems Technology*, vol. 22, no. 5, pp. 1864–1874, 2014.

[3] C.-D. Yang and C.-C. Yang, "Analytical solution of three-dimensional realistic true proportional navigation," *Journal of Guidance, Control, and Dynamics*, vol. 19, no. 3, pp. 569–577, 1996.

[4] C.-K. Ryoo, H. Cho, and M.-J. Tahk, "Optimal guidance laws with terminal impact angle constraint," *Journal of Guidance, Control, and Dynamics*, vol. 28, no. 4, pp. 724–732, 2005.

[5] B. S. Kim, J. G. Lee, and H. S. Han, "Biased PNG law for impact with angular constraint," *IEEE Transactions on Aerospace and Electronic Systems*, vol. 34, no. 1, pp. 277–288, 1998.

[6] Z. Yang, H. Wang, and D. Lin, "Time-varying biased proportional guidance with seeker's field-of-view limit," *International Journal of Aerospace Engineering*, vol. 2016, Article ID 9272019, 11 pages, 2016.

[7] J. Zhu, L. Liu, G. Tang, and W. Bao, "Optimal diving maneuver strategy considering guidance accuracy for hypersonic vehicle," *Acta Astronautica*, vol. 104, no. 1, pp. 231–242, 2014.

[8] C.-K. Ryoo, H. Cho, and M.-J. Tahk, "Closed-form solutions of optimal guidance with terminal impact angle constraint," in *Proceedings of the IEEE Conference on Control Applications*, pp. 504–509, Istanbul, Turkey, June 2003.

[9] B. S. Kim, J. G. Lee, H. S. Han, and C. G. Park, "Homing guidance with terminal angular constraint against nonmaneuvering and maneuvering targets," in *Proceedings of the AIAA Guidance, Navigation, and Control Conference*, pp. 189–199, New Orleans, La, USA, August 1997.

[10] G. Hexner, T. Shima, and H. Weiss, "LQC guidance law with bounded acceleration command," *IEEE Transactions on Aerospace & Electronic Systems*, vol. 44, no. 1, pp. 77–86, 2008.

[11] A. Ratnoo and D. Ghose, "Impact angle constrained interception of stationary targets," *Journal of Guidance, Control, and Dynamics*, vol. 31, no. 6, pp. 1816–1821, 2008.

[12] V. Shaferman and T. Shima, "Linear quadratic guidance laws for imposing a terminal intercept angle," *Journal of Guidance, Control, and Dynamics*, vol. 31, no. 5, pp. 1400–1412, 2008.

[13] T. Shima, "Deviated velocity pursuit," in *Proceedings of the AIAA Guidance, Navigation, and Control Conference*, pp. 4364–4379, Hilton Head, SC, USA, August 2007.

[14] N. Harl and S. N. Balakrishnan, "Impact time and angle guidance with sliding mode control," *IEEE Transactions on Control Systems Technology*, vol. 20, no. 6, pp. 1436–1449, 2012.

[15] Y. B. Shtessel and C. H. Tournes, "Integrated higher-order sliding mode guidance and autopilot for dual-control missiles," *Journal of Guidance, Control, and Dynamics*, vol. 32, no. 1, pp. 79–94, 2009.

[16] R. Su, Q. Zong, B. Tian, and M. You, "Comprehensive design of disturbance observer and non-singular terminal sliding mode control for reusable launch vehicles," *IET Control Theory and Applications*, vol. 9, no. 12, pp. 1821–1830, 2015.

[17] M. V. Srinivasan and M. Davey, "Strategies for active camouflage of motion," *Proceedings of the Royal Society of London B: Biological Sciences*, vol. 259, no. 1354, pp. 19–25, 1995.

[18] Y. Xu and G. Basset, "Sequential virtual motion camouflage method for nonlinear constrained optimal trajectory control," *Automatica*, vol. 48, no. 7, pp. 1273–1285, 2012.

[19] K. S. Erer and O. Merttopçuoğlu, "Indirect impact-angle-control against stationary targets using biased pure proportional navigation," *Journal of Guidance, Control, and Dynamics*, vol. 35, no. 2, pp. 700–704, 2012.

[20] G. Basset, Y. Xu, and K. Pham, "Bio-inspired rendezvous strate-gies and respondent detections," *Journal of Guidance, Control, and Dynamics*, vol. 36, no. 1, pp. 64–73, 2013.

[21] M. Mischiati and P. S. Krishnaprasad, "The dynamics of mutual motion camouflage," *Systems & Control Letters*, vol. 61, no. 9, pp. 894–903, 2012.

[22] E. Bakolas and P. Tsiotras, "Feedback navigation in an uncertain flowfield and connections with pursuit strategies," *Journal of Guidance, Control, and Dynamics*, vol. 35, no. 4, pp. 1268–1279, 2012.

[23] E. W. Justh and P. S. Krishnaprasad, "Steering laws for motion camouflage," *Proceedings of the Royal Society of London, Series A: Mathematical, Physical and Engineering Sciences*, vol. 462, no. 2076, pp. 3629–3643, 2006.

13

Analysis on Aircraft Brake Squeal Problem based on Finite Element Method

Ming Zhang,[1] **Ran Xu,**[1] **and Hong Nie**[1,2]

[1]*Key Laboratory of Fundamental Science for National Defense-Advanced Design Technology of Flight Vehicle,*
Nanjing University of Aeronautics and Astronautics, Nanjing 210016, China
[2]*State Key Laboratory of Mechanics and Control of Mechanical Structures, Nanjing University of Aeronautics and Astronautics,*
Nanjing 210016, China

Correspondence should be addressed to Ming Zhang; zhm6196@nuaa.edu.cn

Academic Editor: Hikmat Asadov

Brake squeal phenomenon is a problem that has been long studied using multiple methods and theories. Finite Element Method (FEM) has been applied to the study of brake squeal problem. First, a disc brake model has been established. Complex mode theory has been applied to the mode analysis and unstable vibration modes can be extracted subsequently. The form of unstable vibration mode has been studied. Then, transient dynamic simulation using explicit dynamic method has been performed. Response in both time and frequency domain has been analyzed. Two methods have been compared, considering accuracy and calculation consumption. Then, the effect of different parameters such as coefficient of friction, stiffness, and brake force fluctuation frequency on squeal phenomenon has been analyzed. It can be found that coefficient of friction and the brake stiffness have a positive correlation with the extent of brake squeal phenomenon, while the frequency of brake force fluctuation should remain as low as possible. Afterwards, a ring-shaped layer of viscoelastic damping material is constrained to outer margin of the stator to restrain the unstable modal. This method can change the vibration nature and improve the brake squeal problem.

1. Introduction

Along with the rising requirement of comfort in the field of civil aviation, brake noise problem has gained great attention because it considerably affects passengers' convenience [1–3]. Aircraft brake system generates vibration during its normal working process, followed by noise inevitably. It has been over 60 years since foreign researchers began working on brake squeal problem, while domestic researchers merely initiated their study in 1980s. In terms of the classification of squeal problem, a universal standard has not yet been put forth. However, researchers agree on naming different types of brake noise according to their respective vibration frequency. Most of the researchers tend to divide squeal phenomenon into three kinds. First type of brake noise possesses a frequency range from 100 Hz to 1 kHz, which is relatively low. This type of brake noise comes from the stick-slip motion between friction surfaces [4]. Applying other materials whose coefficient of friction is stable about slip speed could ease this type of brake noise effectively. Second type of brake noise is generated due to the occurrence of modal coupling among brake parts, with a frequency range between 1 kHz and 3 kHz. The third type of brake noise is generally caused by modal response of brake discs, whose frequency is between 3 kHz and 15 kHz. The last two types of brake noise are often named as "squeal," which has always been an important issue in the field of brake noise research [5–8].

Brake squeal phenomenon brings tremendous oscillation load to both brake mechanism and landing gear structure, which greatly interferes with pilot's regular flight control and passengers' comfort. Serious brake squeal may even damage the entire brake system [9]. As a result, it is quite necessary to study aircraft brake squeal problem thoroughly.

In the field of experiment, many researchers have found that it is very hard to replicate brake squeal in the lab [10]. Moreover, noise generated in the lab does not have a stable frequency and sound pressure level [11–14]. During

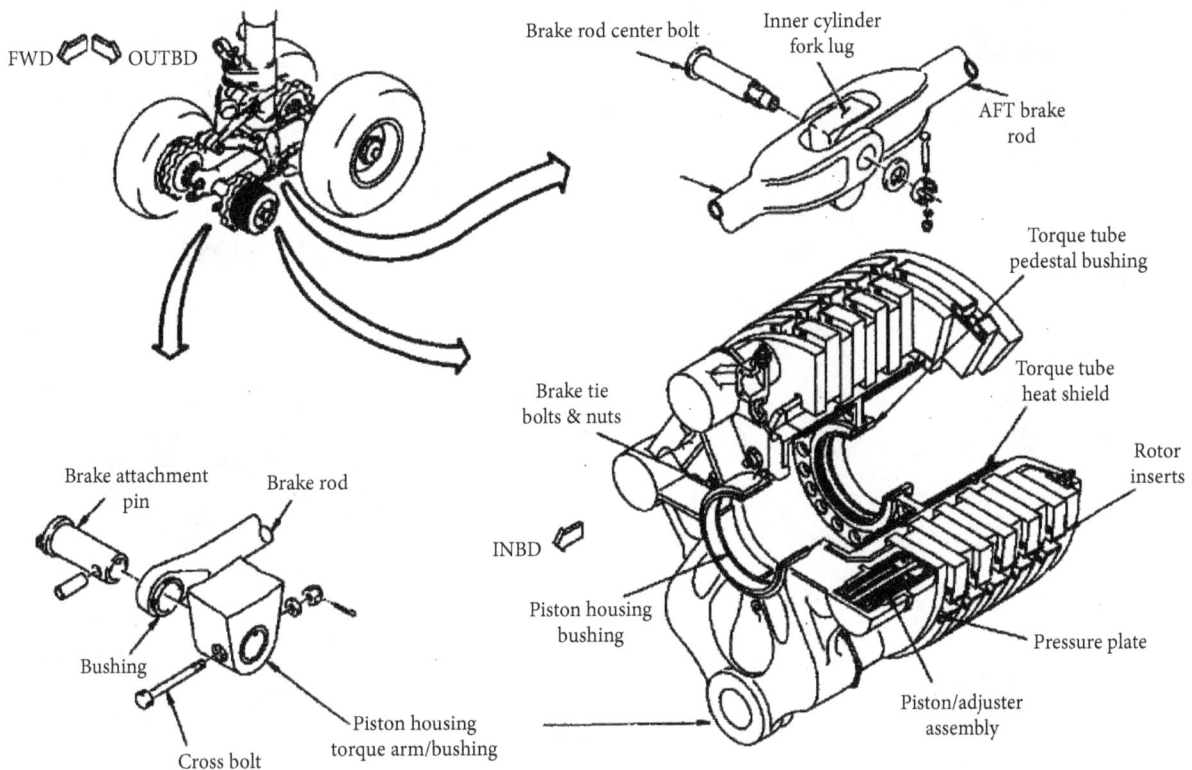

FIGURE 1: Schematic figure of aircraft brake system.

experiment, Oberst and Lai discovered that even brake force, speed, and temperature are stable; the sound pressure level of brake force still has a considerable straggling tendency [15]. Beloiu and Ibrahim discovered that even during one single experiment progress, there can still be an undeniable straggling tendency in sound pressure level, contact force of friction surface, and coefficient of friction [16]. So far, no clear result can be concluded from the experiments.

As the theoretical research proceeds over time, four main theories dealing with brake squeal problem have been put forward by researchers. They are stick-slip theory, lock-slip theory, negative slope of friction force theory, and modal coupling theory. The previous two theories emphasize stick-slip property; the latter two theories concentrate on structural instability [17]. In all, researchers have been studying the cause and influence factor of brake squeal problem for a long time. But unfortunately, a complete and universal theory to handle brake squeal problem has not been invented yet [18]. However, there is no doubt that modal coupling and modal response method show promising prospect in coping with brake squeal problem.

Combining with previous researchers' experience, an FEM model of an actual aircraft brake system has been constructed. Complex mode analysis is thoroughly introduced and applied to the aircraft brake model. Complex eigenvalues are extracted and analyzed. The vibration property of certain typical unstable frequencies has been analyzed. With the help of explicit dynamic analysis, transient dynamic analysis has been performed on the aircraft brake model. The displacement response in both time domain and frequency

domain has been acquired. Several influence factors are studied by applying complex mode method.

2. Aircraft Disc Brake Modeling

Schematic figure of a typical aircraft disc brake is shown in Figure 1. It mainly consists of pistons, torque tubes, brake rods, and a set of brake discs that stack alternatively. Hydraulic system provides pressure for the brake procedure. Pistons empowered by hydraulic system push a set of ganged stators against a set of ganged rotors, so brake force can be generated by friction between brake discs. Brake force is transferred to the entire landing gears structure through wheel axis and piston houses. Among the primary brake mechanisms, carbon brake discs carry out the most crucial functions of the brake system, which is friction. Carbon brake discs enable the aircraft to completely stop moving while the plane is taking off and landing; this makes the discs key components in the brake system.

A simplified brake model is built according to the FEM modeling principles. Subsequent improvements are conducted by setting up contact, friction, mesh, and element properties. This reliable model is the foundation stone for further analysis.

2.1. Physical Modeling. Brake discs model are currently applied in actual type of civil aircraft. The detailed CAD models are given in CATIA software. All the parts are made of carbon composite material, whose Young's modulus is 80 GPa, density is 1800 kg/m^3, and coefficient of friction on

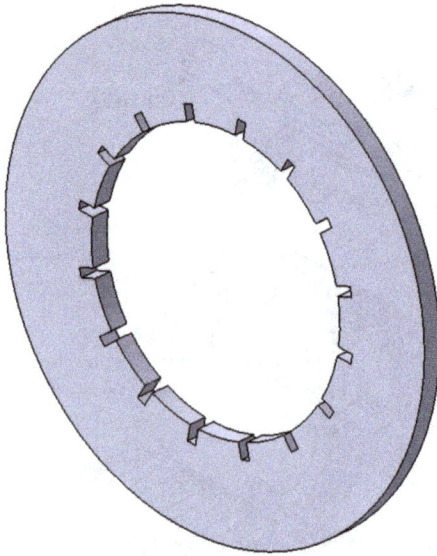

FIGURE 2: Simplified model of stator.

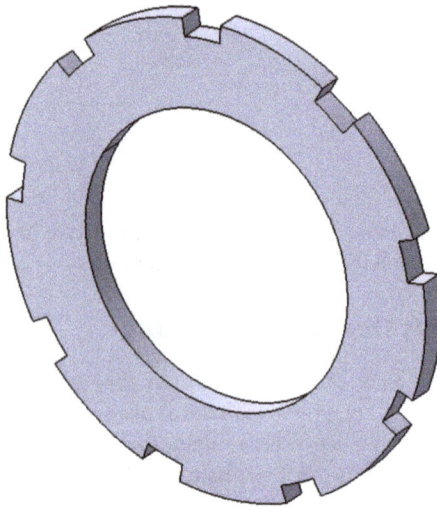

FIGURE 3: Simplified model of rotor.

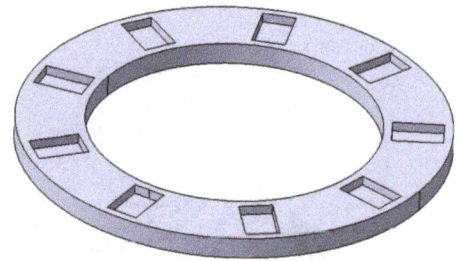

FIGURE 4: Simplified model of baseplate.

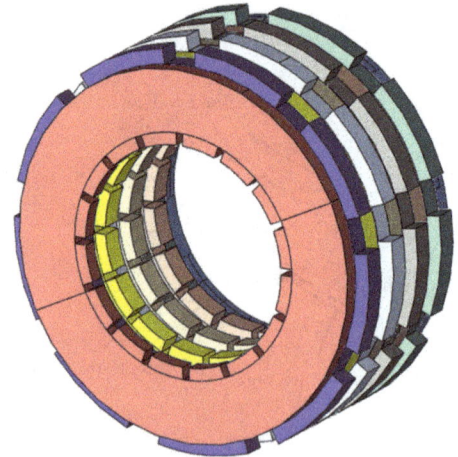

FIGURE 5: FEM model of assembly.

the contact surface is 0.3. Because straggling brake discs are the key objects in the analysis, several tiny holes, fillets, and chamfers are simplified or removed to ensure a more qualified mesh. Finished discs models are shown in Figures 2–4.

Import these simplified disc models into ABAQUS software and construct the assembly model according to their actual position. The completed assembly model is shown in Figure 5.

2.2. Improvements of the Model. During brake procedure, contact behavior exists in multiple positions. Simulating contact behavior both conveniently and accurately is the key to analyzing brake squeal problem. Construct surface-surface contact pair on every contact surface of the brake disc, so friction effect can be built in the software. Contact property includes penalty function, so coefficient of friction can be

introduced. There are two kinds of contact algorithms in ABAQUS software, which are small sliding and finite sliding. Small sliding algorithm is applied, because subsequent analysis does not contain large scale of displacement compared with the size of element.

In order to satisfy accuracy and time consumption requirements of FEM simulation, partitions of the entire model are divided based on geometry appearance of the model. Afterwards, an exquisite structural mesh can be established. Comparison between directly generated mesh and partition based mesh is shown in Figure 6. It is obvious that, by setting up partitions, mesh becomes neater and more concise, which provides great accuracy and efficiency for further analysis. There are 11440 elements on rotor model, 11165 elements on stator model, and 8136 elements on baseplate model.

Finally, verify the entire mesh condition for excessive distortion for excessive errors. At last, FEM model of aircraft brake is finished.

3. Complex Mode Analysis

After finishing constructing of the FEM model of the brake system, subsequent squeal analysis can be performed. This chapter concentrates on complex mode theory, unstable vibration modes are extracted, and unstable vibration modes are analyzed.

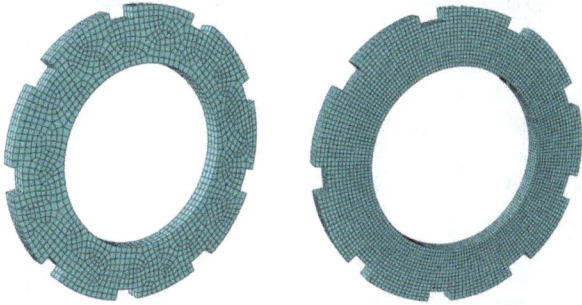

FIGURE 6: Comparison between different mesh strategies.

3.1. Theory of Complex Mode Analysis. In order to research brake squeal problem, modal analysis must be conducted to discover intrinsic vibration property of the entire model. Modal analysis means decoupling the differential equations which demonstrate vibration property of the system and transform physical coordinate into modal coordinate so relative modal parameters can be deduced.

As for ordinary real mode vibration analysis, mass matrix and stiffness matrix are both symmetric and positive definite. Damping matrix is a linear representation of mass and stiffness matrix, which can be decoupled. Moreover, natural vibration modes are real vectors which are orthogonal to mass, stiffness, and damping matrices. Therefore, modal coordinate is capable of simplifying the differential equations. However, in the aircraft disc brake model, damping matrix cannot be linearly represented by mass and stiffness matrices. As a result, modal parameters turn to be complex instead of real.

For an ordinary damped vibration system, damping force is in direct proportion to velocity. The vibration differential equation is

$$M\ddot{X} + C\dot{X} + KX = F, \tag{1}$$

where M, C, K, X, and F are all matrices, representing mass, damping, stiffness, displacement, and force matrices. This equation can also be illustrated in matrix form.

$$\begin{bmatrix} C & M \\ M & 0 \end{bmatrix} \begin{Bmatrix} \dot{X} \\ \ddot{X} \end{Bmatrix} + \begin{bmatrix} K & 0 \\ 0 & -M \end{bmatrix} \begin{Bmatrix} X \\ \dot{X} \end{Bmatrix} = \begin{Bmatrix} F \\ 0 \end{Bmatrix}. \tag{2}$$

Assume $2N \times 1$ state vector $\overline{X}$ is

$$\overline{X} = \begin{Bmatrix} X \\ \dot{X} \end{Bmatrix}. \tag{3}$$

So (2) can be simplified as

$$A\dot{\overline{X}} + B\overline{X} = F,$$

$$A = \begin{bmatrix} C & M \\ M & 0 \end{bmatrix}, \tag{4}$$

$$B = \begin{bmatrix} K & 0 \\ 0 & -M \end{bmatrix}.$$

Equation (4) is the state equation of complex mode analysis. Matrix A and matrix B are both of $2N \times 2N$ size. For free vibration, F becomes zero matrix. Assume root of the equation is

$$X = \psi e^{\lambda t},$$
$$\dot{X} = \psi \lambda e^{\lambda t}, \tag{5}$$

where ψ is

$$\psi = [\phi_1, \ldots, \phi_N]. \tag{6}$$

So (4) can be solved, and $2N$ eigenvalues and eigenvectors can be acquired.

$$\lambda_1, \lambda_1^*, \lambda_2, \lambda_2^*, \ldots, \lambda_N, \lambda_N^*,$$

$$\begin{Bmatrix} \psi_1 \\ \psi_1 \lambda_1 \end{Bmatrix}, \begin{Bmatrix} \psi_1^* \\ \psi_1^* \lambda_1^* \end{Bmatrix}, \ldots, \begin{Bmatrix} \psi_N \\ \psi_N \lambda_N \end{Bmatrix}, \begin{Bmatrix} \psi_N^* \\ \psi_N^* \lambda_N^* \end{Bmatrix}. \tag{7}$$

Plug eigenvectors to original vibration equation (4),

$$(A\lambda + B)\overline{\psi} = 0, \quad \overline{\psi} = \begin{Bmatrix} \psi \\ \psi\lambda \end{Bmatrix} \tag{8}$$

for No. r mode; it can be shown as

$$(A\lambda_r + B)\overline{\psi}_r = 0. \tag{9}$$

Combine with eigenvector of No. s mode. It can be inferred that

$$\overline{\psi}_s^T (A\lambda_r + B)\overline{\psi}_r = 0,$$
$$\overline{\psi}_r^T (A\lambda_s + B)\overline{\psi}_s = 0. \tag{10}$$

Because A and B are both symmetric,

$$\overline{\psi}_r^T A\overline{\psi}_s = \overline{\psi}_s^T A\overline{\psi}_r,$$
$$\overline{\psi}_r^T B\overline{\psi}_s = \overline{\psi}_s^T B\overline{\psi}_r. \tag{11}$$

So, it is evident that

$$\overline{\psi}_r^T A\overline{\psi}_s (\lambda_r - \lambda_s) = 0. \tag{12}$$

Therefore, orthogonal property of complex mode analysis is

$$\overline{\psi}_s^T A\overline{\psi}_r = 0,$$
$$\overline{\psi}_s^T B\overline{\psi}_r = 0. \tag{13}$$

λ_r is the complex modal frequency of the vibration system. $\overline{\psi}_r$ is the complex modal eigenvector. These two parameters always come in conjugate pairs. According to the complex mode properties, a system with N degrees of freedom has $2N$ eigenvalues and eigenvectors.

3.2. FEM Analysis with Complex Mode Method. It is vital to calculate undamped vibration frequency and vibration modes before any complex frequency analysis in ABAQUS software. Besides, static analysis step must be added to exert actual working conditions on the entire model, such as rotation speed and pressure, so accuracy can be ensured. Therefore, analysis steps of the entire simulation are listed below.

Step one is the static mechanic analysis. By exerting pressure on outward stator surface, brake force can be simulated, and contact interactions among parts are established. Moreover, stress status under actual working condition can be achieved. Stress may alter the stiffness matrix and influence subsequent frequency simulation.

Step two is generating rotating velocity. In ABAQUS static mechanic analysis, there are no rotating freedoms for nodes of 3D solid objects, so it is unable to exert rotation on rotors as boundary condition as usual. So, special keywords must be added to the model to solve this problem. Categorize every node of rotors under one set, named as "rotor." Then use keywords "*Motion, Rotation" to establish relevant rotation axis and velocity.

Step three is the extraction of natural frequency of undamped model. This step is the foundation of complex mode analysis.

Step four is the complex mode analysis. Complex frequency and complex eigenvalues are calculated and recorded.

Complex eigenvalues are all complex numbers. The imaginary part of complex eigenvalue indicates the complex vibration frequency, while real part of eigenvalue denotes stability of vibration. Complex modal damping equals to opposite number of real part divided by imaginary part. Therefore, a positive real part means a negative modal damping, which ultimately leads to unstable vibration.

There is another factor that needs to be taken into consideration. The existing material damping and structural damping are not included in the model, so total damping effect in the vibration system is smaller than it should be; this may cause the result to be more unstable. As a result, considering existing practical analysis experience, vibration modes with a modal damping less than −0.01 are determined to be unstable complex modes, which lead to brake squeal phenomenon.

In this analysis, eigenvalue data are extracted by history output and mapped in Figure 7.

The slope in the figure represents a −0.01 damping. Dots above the slope mean unstable vibration mode. Corresponding frequency can be seen in the figure as well. There are 36 unstable vibration modes in the analysis totally.

There has been no certain conclusion that the stronger squeal is, the more serious negative damping exists. And theoretical improvement of the relationship between squeal extent and negative damping value has not been conducted yet. However, based on past experiment experience and complex mode simulation results, the tendency of serious brake squeal phenomenon does relate to the real part value of corresponding eigenvalue. Thus it is practical to determine the brake squeal extent by analyzing real part of the eigenvalue.

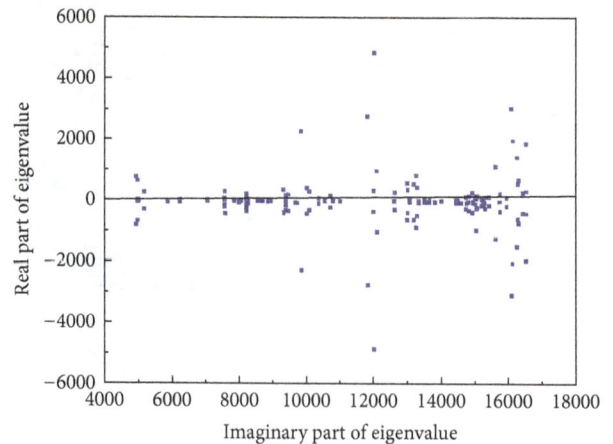

FIGURE 7: Complex eigenvalue scattergram.

In the analysis conducted in this chapter, unstable vibration mode firstly emerges at 4930 Hz. It is clear that unstable modes do not distribute uniformly. In fact, they cluster around several certain values. 5 kHz, 8 kHz, 10 kHz, 12 kHz, 13.5 kHz, and 16.2 kHz are intervals where most unstable modes exist. As a result, the frequencies of brake squeal also exist in those intervals. So far, the frequency of brake squeal phenomenon has been calculated from complex mode analysis.

3.3. Properties of Unstable Vibration Modes. To take a deeper look at the unstable vibration property, it is vital to analyze specific orders of unstable vibration modes.

There are two major vibration mode categories, in-plane mode and out-of-plane mode [19]. According to past research experience, each mode is mainly dominated by only one category of vibration, while the other category has little effect.

Vibration modes dominated by out-of-plane motion possess obvious out-of-plane vibration tendency. This tendency can be described by static diameter and static circle. As for modes dominated by in-plane motion, circumferential direction and radial direction are analyzed. There will at least be one static diameter along circumferential direction motion, while there is no static diameter or static circle along radial direction. Some typical unstable vibration modes are shown in Figure 8. Contour shows the displacement along its major direction of every vibration mode. F_r represents the real part of corresponding eigenvalue. It is noteworthy that the value of displacement has been normalized so it does not reflect any actual length.

Properties of unstable vibration modes can be concluded from the figures. Most of the unstable modes are dominated by out-of-plane vibration; only a few exceptions exist. Among these exceptions, radial direction vibration plays the main character. In terms of vibration modes whose frequency is under 10 kHz, brake discs have similar vibration tendency. On the other hand, if vibration frequency exceeds 10 kHz, different brake discs vibrate differently, with more static diameters. Moreover, brake discs near the baseplate vibrate fiercer under high frequency.

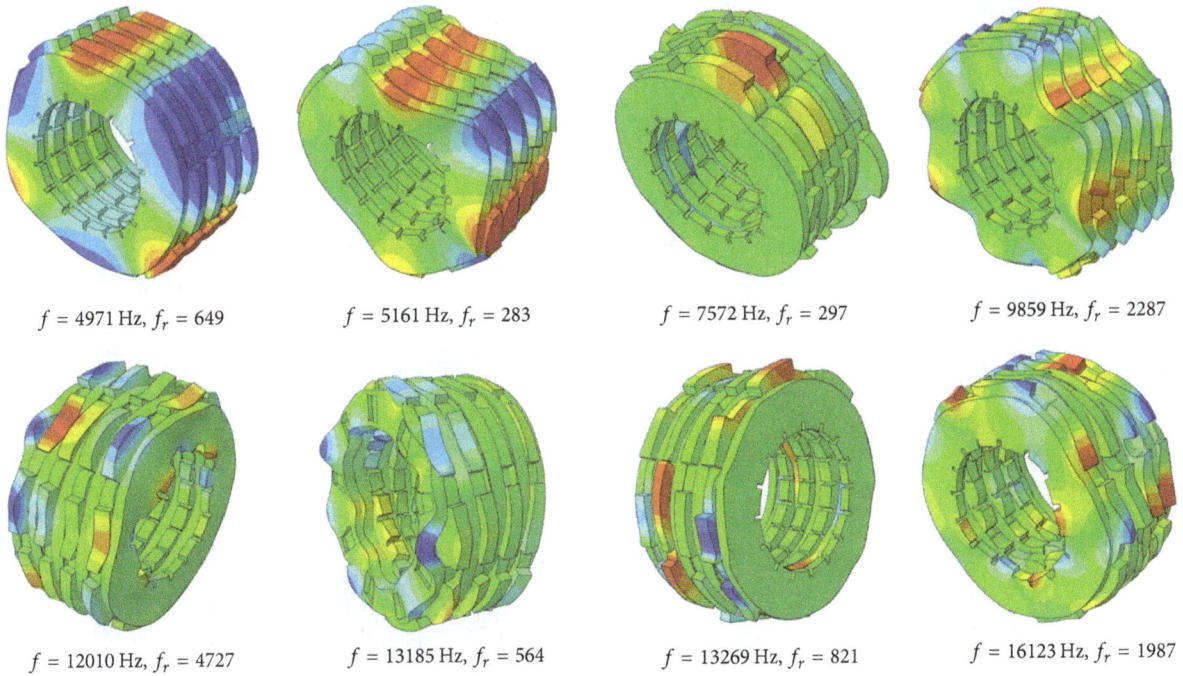

$f = 4971\,\text{Hz}, f_r = 649$ $f = 5161\,\text{Hz}, f_r = 283$ $f = 7572\,\text{Hz}, f_r = 297$ $f = 9859\,\text{Hz}, f_r = 2287$

$f = 12010\,\text{Hz}, f_r = 4727$ $f = 13185\,\text{Hz}, f_r = 564$ $f = 13269\,\text{Hz}, f_r = 821$ $f = 16123\,\text{Hz}, f_r = 1987$

FIGURE 8: Typical unstable vibration modes.

4. Transient Dynamic Analysis

In order to study brake squeal problem more precisely, non-linear properties must be taken into consideration. Explicit dynamic method is applied for transient dynamic analysis on brake squeal problem. Fourier transformation is also applied to analyze vibration property in the frequency domain.

4.1. Theory of Explicit Dynamic. Explicit dynamic analysis focuses on differences about time. No simultaneous equations need to be constructed or solved. Convergence is not a problem either. Dynamic analysis always offers a result, no matter how rough the model is. However, the increment step length is heavily influenced by the FEM element quality. The number of increment steps is also very tremendous, so it takes a lot of time and resource to conduct the analysis.

ABAQUS applies central difference method for solving transient dynamic problems. The FEM process can be shown as follows:

$$M\ddot{x}^{(t)} = f_{\text{ex}}^{(t)} - f_{\text{in}}^{(t)}. \tag{14}$$

In the equation, $\ddot{x}$ stands for acceleration vector and M is the mass matrix. f_{ex} represents external load, and f_{in} represents internal load. During explicit simulations, there are certain relations between velocity and displacement:

$$\dot{x}^{(t+0.5\Delta t)} = \dot{x}^{(t-0.5\Delta t)} + \frac{\Delta t^{(t+\Delta t)} + \Delta t^{(t)}}{2}\ddot{x}^{(t)},$$

$$x^{(t+\Delta t)} = x^{(t)} + \Delta t^{(t+\Delta t)}\dot{x}^{(t+0.5\Delta t)}. \tag{15}$$

In the equation, $t \pm 0.5\Delta t$ stands for the middle increment step during calculations and Δt is the step length. Difference,

instead of differential, is applied in explicit dynamic analysis, so step length must be short enough to ensure sufficient accuracy.

4.2. Explicit Dynamic Analysis. Explicit analysis step is unable to coexist with previous static analysis steps, so a whole new set of analysis steps are needed. During the linear perturbation steps in the last chapter, the entire system is stable in every increment step, and time does not participate in the actual simulation, while, in dynamic analysis, time is an important parameter which heavily influences the analysis procedure.

Therefore, the new analysis steps need further modification. Keywords are not able to exert rotation to the rotors, so a reference point is established at the axis of rotation. Then couple every node of rotor model with the reference point, so that boundary conditions can be exerted at the reference point as well as the rotors. Rotation and brake force are both gradually applied according to time, as shown in Figure 9.

After finishing applying rotation speed, simulation continues for 0.01 seconds to observe and analyze the vibration, which grants a 100 Hz lower limit for frequency domain analysis.

Change contact algorithm into finite sliding method, because the relative sliding is large compared to the size of single element. According to the purpose of the transient dynamic analysis, extra history output must be added. Pick a certain observation point on the rotor and record the displacement data every 5 ms, which ensures a high sampling density. After completing the simulation, map the curves of displacements along three directions, as shown in Figures 10 and 11. In this analysis, X is the axial direction and Y and Z are

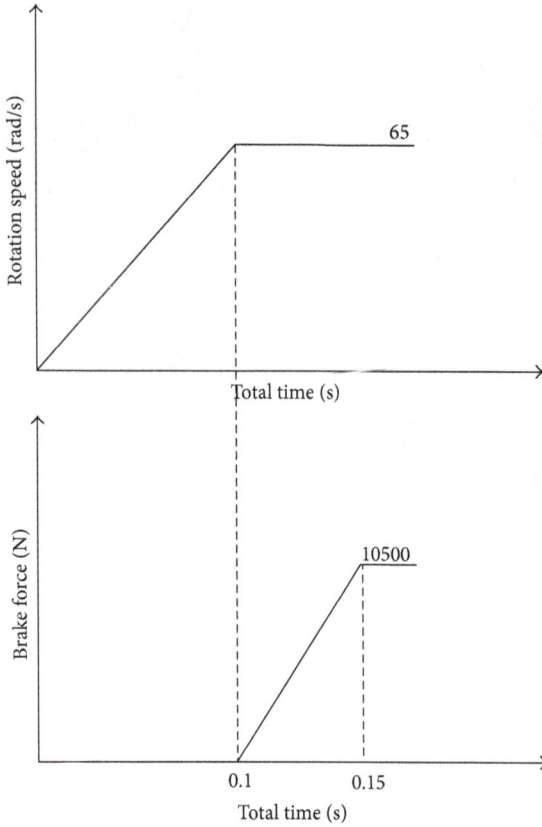

FIGURE 9: Applying process of rotation speed and brake force.

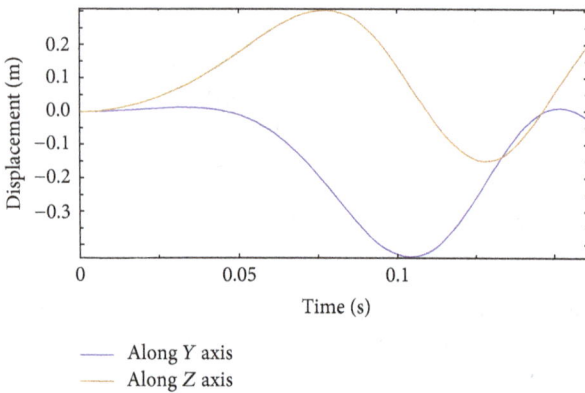

FIGURE 10: Displacement curves of Y and Z direction.

the directions in the disc plane. In Figure 10, the blue curve indicates displacement along Y direction and brown curve reflects Z direction.

From these two figures, it can be concluded that rotation dominates the motion along Y and Z axis; no obvious high frequency vibration takes place. This matches the result from complex mode analysis, where in-plane unstable vibration modes are minority. In terms of X axis direction, the amplitude of the displacement is merely among millimeters, which is very tiny compared with the size of brake disc. At 0.01 seconds when brake force is applied, high frequency vibration takes place. As brake force gradually increases, the vibration

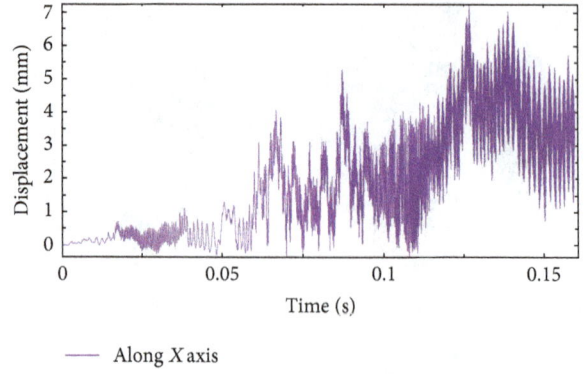

— Along X axis

FIGURE 11: Displacement curve of X direction.

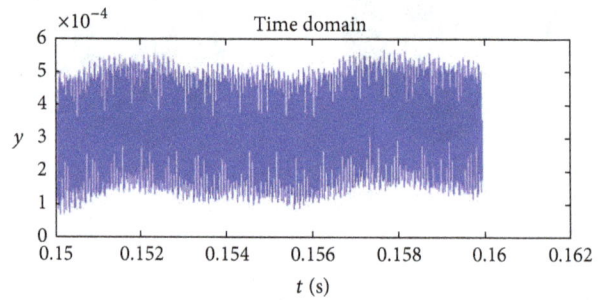

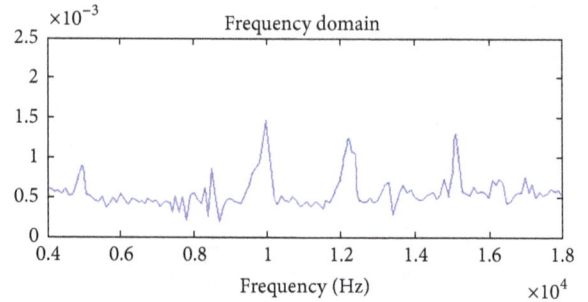

FIGURE 12: Displacement curves in time and frequency domain.

becomes more acute. To study the vibration property in frequency domain, extract the last 0.01-second curve and then use MATLAB to perform Fourier transformation; then Figure 12 can be achieved.

From Figure 12 it is evident that acute vibration takes place around 5 kHz, 8 kHz, 10 kHz, 12 kHz, and 15 kHz intervals. By comparing this result with result from last chapter, it can be concluded that squeal phenomenon around 5 kHz, 10 kHz, and 12 kHz has been accurately predicted by complex mode analysis, while 8 kHz, 13.5 kHz, and 15 kHz vibration frequency do not match the result from complex mode analysis. These phenomena are called "overprediction" and "missing order" by other researchers. The reason for such phenomena lies in the linear method used by complex mode analysis. As a linear perturbation analysis step, complex mode analysis uses linear algorithm to approach nonlinear problem in its essence. Although it saves a lot of time and resource, inaccuracy is inevitable.

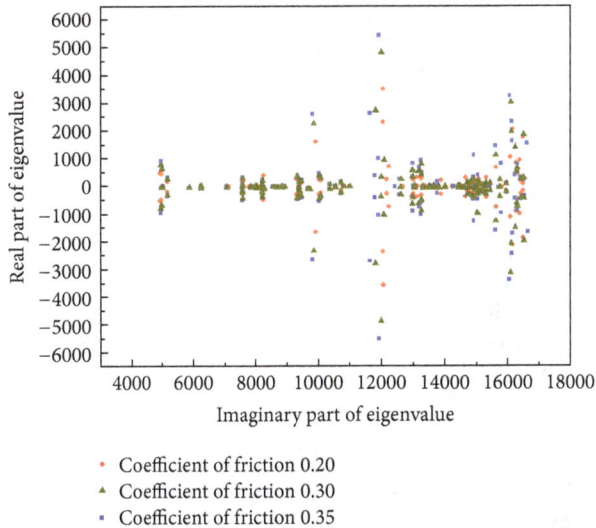

* Coefficient of friction 0.20
▲ Coefficient of friction 0.30
■ Coefficient of friction 0.35

FIGURE 13: Complex eigenvalue scattergram of different coefficient of friction.

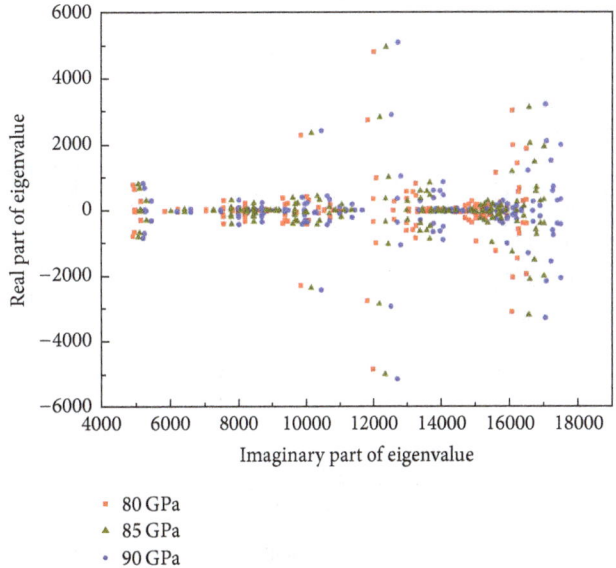

■ 80 GPa
▲ 85 GPa
• 90 GPa

FIGURE 14: Complex eigenvalue scattergram of different stiffness.

On the other hand, explicit dynamic analysis takes nonlinearity into consideration, so accuracy is guaranteed. Nevertheless, explicit dynamic analysis requires a considerable amount of time and resource. For instance, the simulation in this chapter is conducted using 16 cores of high-performance CPU and takes more than two hours to finish. Complex mode analysis merely takes less than 20 minutes to complete. As a result, explicit dynamic method is not practical to conduct squeal parameter study.

As a result, though complex mode analysis is less accurate than explicit dynamic method, it is still a handy tool to analyze brake squeal problem. Complex mode analysis consumes moderate resource and time and provides satisfactory accuracy, which is enough for subsequent simulations.

5. Parameter Study of Brake Squeal

As is analyzed in previous chapters, brake squeal is caused by friction-induced unstable modal coupling. Thus changing physical properties of the brake discs may be an effective way to ease the squeal phenomenon. In this chapter, three parameters are analyzed using complex mode analysis method.

5.1. Coefficient of Friction. After modifying the penalty function of the contact property, the effect of different coefficients of friction can be analyzed.

In Figure 13 it can be discovered that, as the coefficient of friction increases, the real part of complex eigenvalue rises as well, which means vibration becomes more unstable. Moreover, the increase of coefficient of friction also slightly alters the distribution of vibration frequency.

So, according to the analysis result, to ease the brake squeal phenomenon, coefficient of friction has to be as low as possible. However, a low coefficient of friction weakens the brake performance, which makes it difficult to stop the aircraft, because the brake time and brake length cannot be

extended at will. As a result, altering coefficient of friction is not a suitable option to solve the brake squeal problem.

5.2. Stiffness of Brake Disc. Stiffness matrix of the vibration equation gets changed while stiffness of brake disc changes. Let Young's modulus be 80 GPa, 85 GPa, and 90 GPa, respectively, map the complex eigenvalue scattergram in Figure 14.

It can be concluded that vibration frequency rises as the stiffness increases, which matches the common sense. The real part values have a positive correlation with stiffness, which means the harder the brake discs become, the more easily they squeal. So changing the stiffness of the brake disc for the lower is an auxiliary option for solving brake squeal problem and lowering the squeal frequency.

5.3. Brake Force Fluctuation. During taxiing procedure, the carbon brake discs have a complicated effect on the entire brake system [20–22]. For example, assembling clearance [23], eccentric wear, and wear particles may cause considerable brake force fluctuation. The fluctuation brings extra excitement for the vibration system, which heavily influences brake system performance. Steady-state dynamic analysis has been performed to study the effect of brake force fluctuation.

Steady-state dynamic is an important category of dynamic analysis. It focuses on researching dynamic response under steady simple harmonic load. By performing steady-state dynamic analysis, it is able to verify the system's ability to tolerate resonance and fatigue. Steady-state dynamic aims at acquiring the relationship between displacement and the frequency of external excitement, so subsequent structural optimization and modification can be carried out. Only steady-state response will be recorded and analyzed in steady-state analysis.

Delete original natural frequency and complex frequency analysis steps and set up a direct steady-state analysis step. Establish a load whose amplitude is 2 kN and use default

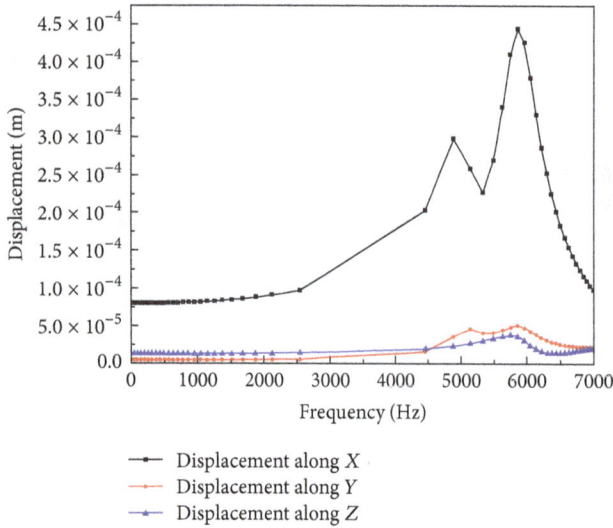

FIGURE 15: Steady-state displacement response and excitation frequency.

simple harmonic function to exert this load for steady-state analysis. The frequency range of the load is under 7 kHz.

After completing the simulation, extract displacement curves about frequency on the observation point, as shown in Figure 15.

It is obvious that displacement along X direction plays a main character compared to other directions. While the frequency of brake force excitation is under 2500 Hz, the steady-state displacement response is stable. At 4880 Hz, first displacement peak takes place, which is also the first unstable mode according to complex mode analysis. Between 4880 Hz and 5859 Hz, steady-state displacement response along Y and Z direction rises. As a result, external excitation fluctuation must be limited under 4880 Hz to prevent from acute vibration, which leads to severe squeal phenomenon.

6. Vibration Reduction Using Viscoelastic Damping Material

6.1. Vibration Reduction Principle Based on Viscoelastic Damping. According to the results above, modifying coefficient of friction, stiffness of brake disc, and brake force fluctuation frequency can improve brake squeal problem. But modifying coefficient of friction, stiffness is enslaved to the internal characteristics of carbon disc and the operability is limited. Brake force fluctuation frequency is related to many coefficients, such as manufacturing and installation error, using clearance, microscopic characteristics of disc surface, and so on. It is difficult to control the brake force fluctuation frequency. The viscoelastic damping material was introduced to the brake disc manufacturing and installation in the automotive industry, which has a good effect on the brake squeal problem. So the vibration reduction effect will be studied by using the viscoelastic damping material to the aircraft brake disc.

Dissipation factor is used to describe the damping. As the shearing deformation expends more energy than tensile deformation, the dissipation factor produced by complex shear modulus G is studied here.

$$G = G' + jG'' = G' \left(1 + j \tan \alpha \right),$$
$$\tau = \left(G' + jG'' \right) \gamma, \tag{16}$$

where G' is the real part of the complex shear modulus and G'' is the imaginary part of the complex shear modulus. α is the phase difference of the strain and stress after excitation. τ, γ are the shear stress and shear strain. Shear dissipation factor β can be described by

$$\beta = \frac{G''}{G'} = \tan \alpha. \tag{17}$$

The physical meaning of β is the ratio of the dissipation energy ΔW and the stored energy W by elastic deformation. Suppose the material excited by sine force. Because of hysteretic phenomena, ΔW can be derived:

$$\Delta W = \int_0^{2\pi} \tau \, d\gamma = \tau_0 \gamma_0 \pi \sin \alpha, \tag{18}$$

where τ_0, γ_0 are amplitude of shear stress and shear strain, respectively. Because

$$G' = \frac{\tau_0}{\gamma_0} \cos \alpha,$$
$$G'' = \frac{\tau_0}{\gamma_0} \sin \alpha, \tag{19}$$

the dissipation energy ΔW is

$$\Delta W = \pi G' \beta \gamma_0^2. \tag{20}$$

From formula (20) the key point to reduce vibration using the viscoelastic damping material is to maximize the product of the real part of the complex shear modulus G' and shear dissipation factor β. The two coefficients are affected by temperature and vibration frequency, so the ability of vibration reduction fluctuates under different work conditions and vibration frequency. The real part of the complex shear modulus G' and shear dissipation factor β can be determined by test [24, 25].

Considering the structure of the aircraft brake disc and work conditions, a ring-shaped layer of viscoelastic damping material is constrained to outer margin of the stator. Shear strain of the ring-shaped damping layer can absorb and dissipate the energy of the out-of-plane vibration to improve the brake squeal problem. The multiphase iron carbon alloy is used as the viscoelastic damping material in this paper. Because the dissipation factor β of the material cannot be directly defined in the software ABAQUS, the Rayleigh damping is used to define the damping. Under the circumstance of the material used in this paper, the dissipation factor satisfies

$$\eta = 2\zeta = \frac{\Delta}{\pi} = Q^{-1} = \frac{\text{SDC}}{200\pi}, \tag{21}$$

$$\frac{a}{2\omega} + \frac{b\omega}{2} = \xi, \tag{22}$$

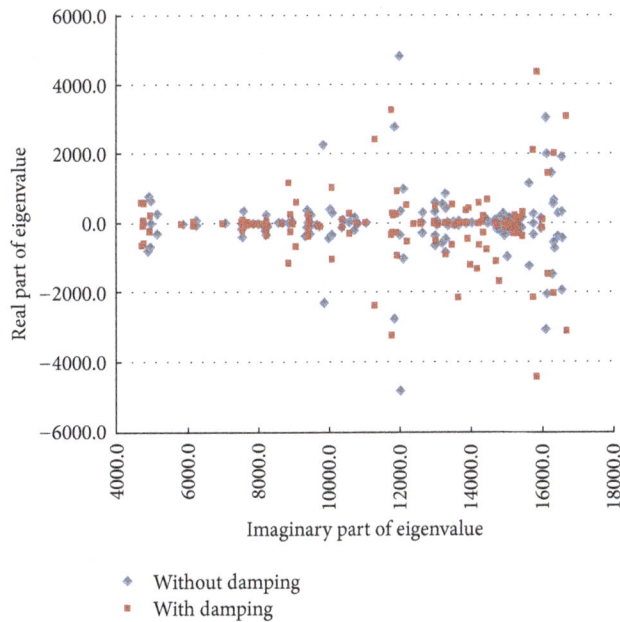

FIGURE 16: Complex eigenvalue scattergram comparison with or without damping.

where ζ is the damping ratio, Δ is the logarithm deletion rate, Q is quality factor, and SDC is the specific damping capacity. SDC of the multiphase iron carbon alloy is about 20%, so the damping ratio ζ can be derived by formula (21): 0.000159. a and b are the mass matrix coefficient and stiffness matrix coefficient, respectively. The CAE model of the ring-shaped damping layer is established separately and the normal frequency is computed. The first two frequencies are extracted, using formula (22); the Rayleigh damping factor can be obtained approximately. Then this factor is used to the complex eigenvalue analyzing.

6.2. Results of Vibration Reduction. After the ring-shaped damping layer constrained to outer margin of the stator and the damping factor set-up, the disc complex modal is computed and the complex eigenvalues distribution is obtained in Figure 16.

Under the effect of damping, the imaginary part of eigenvalue of every rank changes remarkably, which indicates the vibration frequency of the damping system changes. Meanwhile, the real part of eigenvalue decreases mostly. But in a few ranks, new unstable modal is produced by the damping material, so the damping method needs to be optimized further. In general, damping material changes the vibration nature and can restrain the unstable modal evidently. It is also an operable and economy method to improve the brake squeal problem.

7. Conclusions

(1) Aircraft brake noise problem contains a lot of nonlinear properties, which makes it a sophisticated engineering issue. A disc brake FEM model based on actual aircraft brake has

been established. By applying FEM analysis method, complex mode analysis has been performed on the brake model. The possibility of squeal has been analyzed according to the eigenvalues results, and vibration properties of typical unstable vibration modes which are highly likely to cause squeal have been studied thoroughly. Explicit dynamic method has been taken to conduct transient dynamic analysis. Brake disc's vibration properties have been researched in both time and frequency domain. Comparisons between these two methods have been performed. Complex mode method has been adopted for further parameter analysis.

(2) Parameter analysis has been conducted to study the influence of coefficient of friction, stiffness, and brake force fluctuation frequency on the brake squeal phenomenon. The results show that a small coefficient of friction, a low brake disc stiffness, or relatively low brake force fluctuation frequency all help improve the brake squeal problem.

(3) A ring-shaped layer of viscoelastic damping material is constrained to outer margin of the stator to restrain the unstable modal. The simulation results indicate that this method can change the vibration nature and improve the brake squeal problem.

Conflicts of Interest

The authors declare that there are no conflicts of interest regarding the publication of this paper.

Acknowledgments

This study was supported by the Fundamental Research Funds for the Central Universities (no. NS2016001), the National Natural Science Foundation of China (no. 51305198), and the Aero-Science Fund of China (no. 20142852025).

References

[1] J. J. Enright, "Reducing aircraft brake squeal with a damped brake-rod," in *Proceedings of the World Aviation Conference*, p. 5599, San Diego, Calif, USA, October 2000.

[2] M. Neubauer and R. Oleskiewicz, "Brake squeal control with shunted piezoceramics-efficient modelling and experiments," *Proceedings of the Institution of Mechanical Engineers, Part D: Journal of Automobile Engineering*, vol. 222, no. 7, pp. 1141–1151, 2008.

[3] D. Hochlenert, G. Spelsberg-Korspeter, and P. Hagedorn, "Friction induced vibrations in moving continua and their application to brake squeal," *Journal of Applied Mechanics*, vol. 74, no. 3, pp. 542–549, 2007.

[4] S. Y. Liu, J. T. Gordon, and M. A. Özbek, "Nonlinear model for aircraft brake squeal analysis: model description and solution methodology," *Journal of Aircraft*, vol. 35, no. 4, pp. 623–630, 1998.

[5] H. Hetzler, "Bifurcation analysis for brake squeal," in *Proceedings of ASME 10th Biennial Conference on Engineering Systems Design and Analysis, ESDA '10*, pp. 253–262, American Society of Mechanical Engineers, July 2010.

[6] H. D. Guan and X. D. Su, "An overview on brake vibration and noise," *Engineering Mechanics*, vol. 21, no. 4, pp. 150–155, 2004.

[7] P. Z. Yu, X. D. Yin, and J. L. Zhang, "A review on brake judder," *Automotive Engineering*, vol. 27, no. 3, pp. 372–376, 2005.

[8] F. Bergman, M. Eriksson, and S. Jacobson, "The effect of reduced contact area on the occurrence of disc brake squeals for an automotive brake pad," *Proceedings of the Institution of Mechanical Engineers, Part D: Journal of Automobile Engineering*, vol. 214, no. 5, pp. 561–568, 2000.

[9] H. M. Lu, L. J. Zhang, and Z. P. Yu, "A review of automotive disc brake squeal," *Journal of Vibration and Shock*, vol. 40, no. 4, pp. 1–7, 2011.

[10] F. Chen, "Disc brake squeal: an overview," SAE Technical Papers 2007-01-0587, 2007.

[11] G. Spelsberg-Korspeter, "Eigenvalue optimization against brake squeal: symmetry, mathematical background and experiments," *Journal of Sound and Vibration*, vol. 331, no. 19, pp. 4259–4268, 2012.

[12] J. Z. Zhuo, W. D. Xie, and B. X. Ning, "An overview on disc brake vibrations and noise," *Machinery Design & Manufacture*, vol. 11, pp. 215–217, 2007.

[13] O. Dessouki, G. Drake, B. Lowe, and W. K. Chang, "Disc brake squeal: diagnosis and prevention," SAE Technical Papers 2003-01-1618, 2003.

[14] F. Renaud, G. Chevallier, J.-L. Dion, and G. Taudire, "Motion capture of a pad measured with accelerometers during squeal noise in a real brake system," *Mechanical Systems and Signal Processing*, vol. 33, no. 2, pp. 155–166, 2012.

[15] S. Oberst and J. C. S. Lai, "Statistical analysis of brake squeal noise," *Journal of Sound and Vibration*, vol. 330, no. 12, pp. 2978–2994, 2011.

[16] D. M. Beloiu and R. A. Ibrahim, "Analytical and experimental investigations of disc brake noise using the frequency-time domain," *Structural Control and Health Monitoring*, vol. 13, no. 1, pp. 277–300, 2006.

[17] W. A. Wen, *Complex eigenvalue analysis of the squeal propensity of a railway vehicle disc brake system using the finite element method [Ph.D. Thesis]*, Southwest Jiaotong University, Chengdu, China, 2007.

[18] H. Chen and L. Y. Chen, "Study on automobile disc brake brakes shaking problem," *Internal Combustion Engine and Parts*, no. 2, pp. 11–13, 2013.

[19] K. Chen, *Experimental and computational research on disc brake squeal [Ph.D. Thesis]*, Harbin Institute of Technology, Harbin, China, 2014.

[20] T. J. Hutton, B. McEnaney, and J. C. Crelling, "Structural studies of wear debris from carbon-carbon composite aircraft brakes," *Carbon*, vol. 37, no. 6, pp. 907–916, 1999.

[21] L. Dagli and Y. Remond, "Identification of the non-linear behaviour a 4D carbon-carbon material designed for aeronautic application," *Applied Composite Materials*, vol. 9, no. 1, pp. 1–15, 2002.

[22] X. Xiong, B.-Y. Huang, J.-H. Li, and H.-J. Xu, "Friction behaviors of carbon/carbon composites with different pyrolytic carbon textures," *Carbon*, vol. 44, no. 3, pp. 463–467, 2006.

[23] G. Zhang, M. Xie, J. Li, G. Qi, and X. Pu, "Vehicle brake moan noise induced by brake pad taper wear," *Journal of Mechanical Engineering*, vol. 49, no. 9, pp. 81–86, 2013.

[24] X. T. Liu, M. Y. Shi, and X. H. Hua, "Current situation and prospect of active constrained layer damping vibration control technology," *Journal of Vibration and Shock*, vol. 20, no. 2, pp. 1–6, 2001.

[25] Q. Yang, X. Wang, W. W. Zhang et al., "Study on Noise Suppressing Performance of Constrained Layer Damping Structure," *Journal of Noise and Vibration Control*, no. 4, pp. 150–157, 2010.

14

The Application Research of Inverse Finite Element Method for Frame Deformation Estimation

Yong Zhao,[1] Hong Bao,[1] Xuechao Duan,[1] and Hongmei Fang[2]

[1]*Key Laboratory of Electronic Equipment Structure Design of Ministry of Education, Xidian University, Xi'an 710071, China*
[2]*Nanjing Research Institute of Electronics Technology, Nanjing 210039, China*

Correspondence should be addressed to Hong Bao; hbao@xidian.edu.cn

Academic Editor: Vaios Lappas

A frame deformation estimation algorithm is investigated for the purpose of real-time control and health monitoring of flexible lightweight aerospace structures. The inverse finite element method (iFEM) for beam deformation estimation was recently proposed by Gherlone and his collaborators. The methodology uses a least squares principle involving section strains of Timoshenko theory for stretching, torsion, bending, and transverse shearing. The proposed methodology is based on stain-displacement relations only, without invoking force equilibrium. Thus, the displacement fields can be reconstructed without the knowledge of structural mode shapes, material properties, and applied loading. In this paper, the number of the locations where the section strains are evaluated in the iFEM is discussed firstly, and the algorithm is subsequently investigated through a simple supplied beam and an experimental aluminum wing-like frame model in the loading case of end-node force. The estimation results from the iFEM are compared with reference displacements from optical measurement and computational analysis, and the accuracy of the algorithm estimation is quantified by the root-mean-square error and percentage difference error.

1. Introduction

Aircraft flexible wings with embedded conformal antennas and large frame structures that carry antennas require accurate real-time deformation estimation to provide feedback for their actuation and control systems [1–3]. Using the measured strain to reconstruct the shape of the structure is a key technology in the accurate real-time deformation estimation, which has been studied by many researchers.

The computation of the displacement field of the deformed structure is commonly performed on the basis of strain data measured in real time by a network of strain gauges [4–6]. For example, fiber Bragg grating (FBG) sensors have been extensively researched for deformation estimation due to their lightness, accuracy, and easy embedment. The strategies that reconstruct the deformed shape displacement field of the structure with in situ strain data can be divided into two kinds. One of the two kinds trains the mapping relation between the displacement field and the measured strain by model learning algorithms, for example, neural network model and fuzzy network algorithm [7, 8]. When the training

system has enough measured strains and displacement field data of the structure, a certain relation matrix can be determined and the stable relationship between the measured strain and the displacement field can be obtained. But the strategy requires a large number of training data, and the mapping relation is easy to fail when the actual loading is beyond the range of the training cases.

The other establishes the mapping relation between the displacement field and in situ strain data without the model learning. In the literature [9–11], the global or piecewise continuous basis function methods were employed to fit the surface-measured strain into the structure strain field, and then, the structure deformation displacement was obtained from the strain-displacement relationship. These methods are easy to implement, but the reconstruction accuracy of deformation estimation depends on the appropriate selection of basis function and weight coefficients. Mode shapes have been used as basis function in [12, 13]. The deformation displacements are reconstructed from measured strains by using the modal transformation method. However, there exist the following disadvantages in this method. (1) The detailed material elasticity and inertial parameters are needed to precisely construct mode shapes. (2) The accuracy of deformation reconstruction is severely limited to the modeling precision of the structure, and it is quite difficult to precisely model the complex structure. Maincon [14] developed a finite element-based methodology involving an inverse interpolation formulation that employs the surface-measured strain to determine the loads and structural response of aerospace vehicles, while this algorithm needs an appropriate quality function to adapt the different loading cases and this function is constructed based on a mass of computer simulation and experimental statistics.

Based on the Euler-Bernoulli beam equation, Jute et al. evaluated the deflection of beam by the integration of discretely measured strains directly [15]. The algorithm applied the classical beam equation and piecewise continuous polynomials to approximate beam curvature through integration. Derkevorkian et al. [16] compared the algorithm with the modal transformation method and demonstrated the additional benefits of the algorithm to achieve a robust method for monitoring ultralightweight flying wings or next-generation commercial airplanes. Though this one-dimensional scheme has displayed high accuracy in predicting deflection, it fails to estimate the element deformation under multidimensional complex loads.

Tessler and Spangler [17] proposed the inverse finite methodology (iFEM), which can be used to reconstruct the displacement field of shear-deformable structures, not only beam structure but also plate and shell structures. The main idea is reconstructing a three-dimensional displacement field of beam structure from the surface-measured strains according to a least squares approach. Due to the fact that only the displacement-strain relationship is used, the deformation reconstruction can be accomplished by the methodology without the prior knowledge of loads, materials, and inertial and damping properties. To model arbitrary plate and shell structures [18, 19], Tessler and his partners developed the iFEM algorithm using the first-order shear deformation theory and a three-node inverse shell element. FBG sensors were applied to measure the surface strains on the slender beams, and then, the deformed displacement was reconstructed by using an iFEM shell model.

The beam-deformed displacement and cross-sectional torsion were reconstructed by Gherlone et al. who employed the inverse finite element formulation to achieve high reconstruction accuracy of deformed displacement [20–21]. The authors first used the Timoshenko beam theory to model the beam kinematic accurately, then used the C^0 or C^1 inverse frame elements and least squares formulation to establish the relationship between the measured strain and the arbitrary node displacement field of the beam element with no prior knowledge of the finite element model and loads. The reconstruction equation is nonsingular when the boundary conditions are applied; that is, the status of one end node of the beam element must be known. As the displacement field of the beam element in iFEM is constructed based on an isotropic straight beam structure, whose cross section is invariable along the whole beam, the deformation of the tapered beam cannot be estimated. Meanwhile, in view

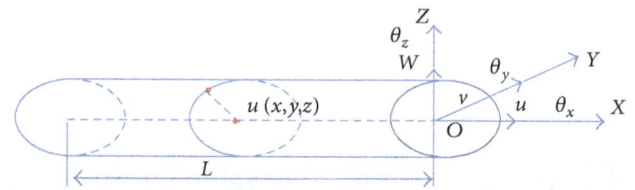

FIGURE 1: Beam geometry and kinematic variables.

of different load cases, the authors discussed different types of shape functions that are used to interpolate the kinematic variables in the beam element; then, the number of surface strain measurements and the deformation reconstruction equation are confirmed. Unfortunately, the minimum number of locations where the section strains are evaluated, which is critical for obtaining a correct solution to the algorithm, is not discussed clearly in different load cases. If the minimum number of the locations where the section strains are evaluated is inappropriately set, the solution will be nonunique; that is, the reconstructed displacement may be incorrect.

The contribution of this paper is twofold. (i) The relationship between the loading case and the number of the section strain locations is discussed in detail, and the following conclusions are verified; on the one hand, the minimum number of locations should be 2 in the loading case of end-node forces; on the other hand, the minimum number should be 3 under the uniform distributed loading. (ii) In order to examine the minimum number which is 2, estimation studies are carried on a beam structure and a wing-like three-dimensional frame structure, in the loading case of end-node force. Six fiber optic strain sensors are placed at two nodes of each beam to capture in situ strain dates. To assess the estimation error of the iFEM algorithm, the RMS and difference percent parameters are employed.

2. iFEM Algorithms for Beam Estimation

A beam deflection estimation algorithm was developed by Gherlone and his collaborators [20]. In the algorithm, the Timoshenko beam theory is first used to analyze the expression of the displacement field in a straight beam element (Figure 1).

$$u_x(x, y, z) = u(x) + z\theta_y(x) - y\theta_z(x),$$
$$v_y(x, y, z) = v(x) - z\theta_x(x), \qquad (1)$$
$$u_z(x, y, z) = w(x) + y\theta_x(x),$$

where u_x, v_y, and w_z are the point displacements along the x, y, and z axes, respectively. $u(x), v(x)$, and $w(x)$ denote the displacements at $y = z = 0$; $\theta_x(x)$, $\theta_y(x)$, and $\theta_z(x)$ are the rotations about the three coordinate axes. The six kinematic variables in the middle axes can be grouped in vector form as follows:

$$\mathbf{u} = \{u, v, w, \theta_x, \theta_y, \theta_z\}^T. \qquad (2)$$

The arbitrary section strains $\mathbf{e}(\mathbf{u}) = [e_1, e_2, e_3, e_4, e_5, e_6]^T$ can be obtained by (1).

$$e_1(x) = u_x(x),$$
$$e_2(x) = \theta_{y,x}(x),$$
$$e_3(x) = -\theta_{z,x}(x),$$
$$e_4(x) = w_x(x) + \theta_y(x), \tag{3}$$
$$e_5(x) = v_x(x) - \theta_z(x),$$
$$e_6(x) = \theta_{x,x}(x).$$

The six kinematic variables $\mathbf{u}$ can be interpolated by the right shape functions

$$\mathbf{u} = \mathbf{N}(x)\mathbf{u}^\mathbf{e}, \tag{4}$$

where $\mathbf{N}(x)$ and $\mathbf{u}^\mathbf{e}$ denote the shape function and nodal degrees of freedom, respectively. Substituting (4) into (3) gives arbitrary section strains in terms of the nodal degrees of freedom as follows:

$$\mathbf{e}(\mathbf{u}) = \mathbf{B}(x)\mathbf{u}^\mathbf{e}, \tag{5}$$

where the matrix $\mathbf{B}(x) = [B_1(\mathbf{x}), B_2(\mathbf{x}), ..., B_6(\mathbf{x})]$ contains the derivatives of the shape functions $\mathbf{N}(x)$. Once the section strains $\mathbf{e}(\mathbf{u})$ are obtained, the nodal displacement $\mathbf{u}^\mathbf{e}$ is determined; then, the kinematic variables $\mathbf{u}$ can be acquired by (4). However, the section strains $\mathbf{e}(\mathbf{u})$ are derived from the kinematic variables $\boldsymbol{u}$ theoretically, rather than the strain measurements. So, iFEM uses in situ section strains $\mathbf{e}^\varepsilon$ computed from the measured strains to replace $\mathbf{e}(\mathbf{u})$ when the least squares error function $\varphi(\mathbf{u})$ reaches the minimum.

$$\varphi(\mathbf{u}) = \|\mathbf{e}(\mathbf{u}) - \mathbf{e}^\varepsilon\|^2, \quad \mathbf{e}^\varepsilon = [e_1^\varepsilon, e_2^\varepsilon, e_3^\varepsilon, e_4^\varepsilon, e_5^\varepsilon, e_6^\varepsilon]^T. \tag{6}$$

In view of the effect of the axial stretching, bending, twisting, and transverse shearing, the improved least squares error functional $\varphi^\mathbf{e}(\mathbf{u})$ is obtained by the dot product of the weighting coefficient vector $\mathbf{W}$ and the original vector $\varphi(\mathbf{u}) = \{\varphi_k^e\}$.

$$\mathbf{W} = \{w_k\} = \left\{ w_1^0, w_2^0\left(\frac{I_y^e}{A^e}\right), w_3^0\left(\frac{I_z^e}{A^e}\right), w_4^0, w_5^0, w_6^0\left(\frac{I_p^e}{A^e}\right) \right\},$$

$$\varphi^\mathbf{e}(\mathbf{u}) = \mathbf{W} \cdot \varphi(\mathbf{u}),$$

$$\varphi_k^e = \frac{L}{n}\sum_{i=1}^{n} [e_k(x_i) - e_k^\varepsilon(x_i)]^2, \quad (k = 1, 2, ..., 6),$$

$$\tag{7}$$

where $w_k^0(k = 1, 2, ..., 6)$ denote dimensionless weighting coefficients whose initial values are identically set as 1; A^e, I_y^e and I_z^e, and I_p^e are, respectively, the cross-sectional area, second moments of the area according to the y- and z-axes, and polar moment of the area of the beam element. L is the length of the beam element; $x_i(0 \le x_i \le L)$ and n are, respectively, the axial coordinate of the locations where the section strains are evaluated and the number of locations, that is, the axial coordinate of sections where section strains are distributed in, and the number of sections.

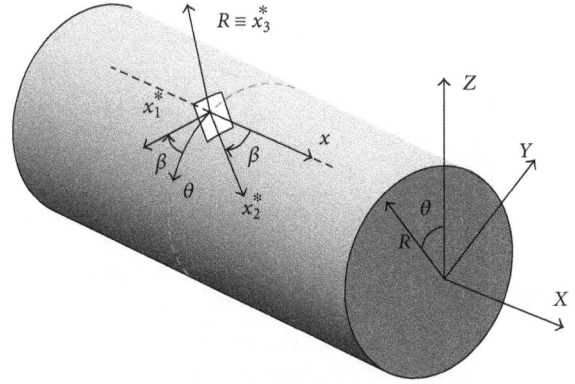

FIGURE 2: Location and coordinates of a strain gauge placed on the beam external surface.

For a straight beam member of constant circle cross section, the in situ section strains at $x = x_i$, $\mathbf{e}^\varepsilon(x_i) = [e_1^\varepsilon(x_i), e_2^\varepsilon(x_i), ..., e_6^\varepsilon(x_i)]^T$, could be derived from the measured strains of the beam surface with appropriate strain-tensor transformations from the (θ, x, r) to (x_X, x_Y, x_Z) coordinates [22].

$$\varepsilon_2^*(x_i, \theta, \beta) = e_1^\varepsilon(x_i)\left(c_\beta^2 - vs_\beta^2\right) + e_2^\varepsilon(x_i)\left(c_\beta^2 - vs_\beta^2\right)s_\theta R$$
$$+ e_3^\varepsilon(x_i)\left(c_\beta^2 - vs_\beta^2\right)c_\theta R + e_4^\varepsilon(x_i)c_\beta s_\beta c_\theta \tag{8}$$
$$- e_5^\varepsilon(x_i)c_\beta s_\beta s_\theta + e_6^\varepsilon(x_i)c_\beta s_\beta R,$$

with $c_\beta \equiv \cos \beta$, $s_\beta \equiv \sin \beta$, $c_\theta \equiv \cos \theta$, and $s_\theta \equiv \sin \theta$.

In (8), v is the Poisson ratio, $\varepsilon_2^*(x_i, \theta, \beta)$ is the in situ strains that are obtained from strain sensors. R denotes the radius of the beam cross section (Figure 2).

Substituting (5) into (7) results in the following quadratic form:

$$\varphi^\mathbf{e}(\mathbf{u}) = \frac{1}{2}(\mathbf{u}^\mathbf{e})^T\mathbf{k}^\mathbf{e}\mathbf{u}^\mathbf{e} - (\mathbf{u}^\mathbf{e})^T\mathbf{k}^\mathbf{e} + \mathbf{c}^\mathbf{e}. \tag{9}$$

Herein, $\mathbf{c}^\mathbf{e}$ is a constant, and $\mathbf{k}^\mathbf{e}$ and $\mathbf{f}^\mathbf{e}$ are indicated as follows:

$$\mathbf{k}^\mathbf{e} = \sum_{k=1}^{6} w_k k_k^e, \quad k_k^e = \frac{L}{n}\sum_{i=1}^{n}[B_k^T(x_i)B_k(x_i)],$$
$$\mathbf{f}^\mathbf{e} = \sum_{k=1}^{6} w_k f_k^e, \quad f_k^e = \frac{L}{n}\sum_{i=1}^{n}[B_k^T(x_i)e_k^\varepsilon(x_i)]. \tag{10}$$

Finally, the relationship between the deformation and in situ section strains, shown in (11), can be confirmed when the minimization of functional $\varphi^\mathbf{e}(\mathbf{u})$ is performed.

$$\mathbf{k}^\mathbf{e}\mathbf{u}^\mathbf{e} = \mathbf{f}^\mathbf{e}. \tag{11}$$

Once the appropriate shape functions and the problem-dependent displacement boundary conditions (e.g., setting the displacement of one end point to zero, which means that one of end nodes of the beam is fixed) are given, $\mathbf{u}^\mathbf{e}$ can be derived from a nonsingular system, and the vector $\mathbf{f}^\mathbf{e}$ depends on the measured strain values that change during

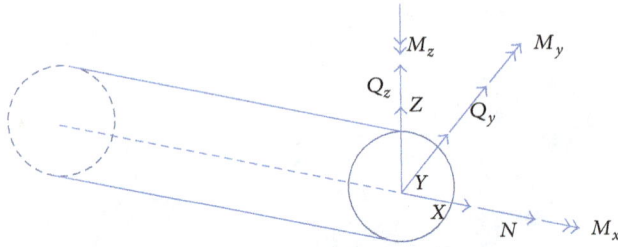

FIGURE 3: Beam section forces and moments.

deformation. Once the nodal degrees of freedom $\mathbf{u}^e$ are confirmed, the displacements and rotations of every node along the centroid axis of the beam element are obtained by (4) and the deformed shape of the whole beam can be reconstructed by (1).

3. Minimum Number of the Section Strain Location Selection

Although the displacement shape function and the number of surface strain measurements in different loading cases have been discussed in [20, 21], n, the number of the sections where the section strains are distributed in, is not shown clearly. In (11), once the problem-dependent displacement boundary conditions are determined, the nonsingularity of the equation will depend on the selection of n. In our investigation, it is found that n can be determined by a specific loading case. More specifically, the orders of the section forces and the moments can be obtained by the loading case, which determines the order of the section strains. With the value of the section strain order, n is determined immediately.

With the equilibrium equations (12), the section forces (N, Q_y, and Q_z) and moments (M_x, M_y, and M_z) can be obtained by the load concentrations $q_x(x)$, $q_y(x)$, and $q_z(x)$ along the x, y, and z directions (Figure 3).

$$
\begin{aligned}
N_x + q_x(x) &= 0, \\
Q_{y,x} + q_y(x) &= 0, \\
Q_{z,x} + q_z(x) &= 0, \\
M_{x,x} &= 0, \\
M_{y,x} - Q_z &= 0, \\
M_{z,x} - Q_y &= 0.
\end{aligned} \tag{12}
$$

The relationship between section forces and moments and the section strains e_i ($i = 1, 2, \ldots, 6$) can be interpreted as the following constitutive equations:

$$
\begin{aligned}
N &= A_x e_1, \\
M_x &= J_x e_6, \\
Q_y &= G_y e_5, \\
M_y &= D_y e_2, \\
Q_z &= G_z e_4, \\
M_z &= D_z e_3.
\end{aligned} \tag{13}
$$

where $A_x \equiv EA$ is the axial rigidity; $G_y \equiv k_y^2 GA$ and $G_z \equiv k_z^2 GA$ are the shear rigidities, with k_y^2 and k_z^2 denoting the shear correction factors; and $J_x \equiv GI_p$ and $D_y \equiv EI_y$ and $D_z \equiv EI_z$ are, respectively, the torsional rigidity and the bending rigidities. For the uniform section beam element, parameters mentioned above are constant. Then, the order of the section strains e_i will be identical to that of the section forces and moments.

Substituting (13) into (12) gives the relationship between the section strains and the loads as follows:

$$
\begin{aligned}
e_1 &= \frac{N}{A_x} = \frac{\int(-q_x(x))dx}{A_x}, \\
e_2 &= \frac{M_y}{D_y} = \frac{\int Q_z dx}{D_y} = \frac{\int\int(-q_z(x))dxdx}{D_y}, \\
e_3 &= \frac{M_z}{D_z} = \frac{\int Q_y dx}{D_z} = \frac{\int\int(-q_y(x))dxdx}{D_z}, \\
e_4 &= \frac{Q_z}{G_z} = \frac{\int(-q_z(x))dx}{G_z}, \\
e_5 &= \frac{Q_y}{G_y} = \frac{\int(-q_y(x))dx}{G_y}, \\
e_6 &= \frac{M_x}{J_x} = \frac{\int 0 dx}{J_x} = c_1.
\end{aligned} \tag{14}
$$

In most cases, the load in the x direction is zero, that is, $q_x(x) = 0$; then,

$$
e_1 = \frac{\int(-q_x(x))dx}{A_x} = \frac{\int 0 dx}{A_x} = c_2. \tag{15}
$$

The section strains e_1 and e_6 are constant in (14) and (15), and the order of the residual section strains, e_2, e_3, e_4, and e_5, will be discussed as the following two loading cases.

Consider a beam element loaded by the end-node forces, $q_y(x)$ and $q_z(x)$ are zero along the x-axis. Then, e_2, e_3, e_4, and e_5 are determined by (14).

$$
\begin{aligned}
e_2 &= \frac{\int\int(-q_z(x))dxdx}{D_y} = \frac{\int\int 0 dxdx}{D_y} = a_1 x + b_1, \\
e_3 &= \frac{\int\int(-q_y(x))dxdx}{D_z} = \frac{\int\int 0 dxdx}{D_z} = a_2 x + b_2, \\
e_4 &= \frac{\int(-q_z(x))dx}{G_z} = \frac{\int 0 dx}{G_z} = a_3, \\
e_5 &= \frac{\int(-q_y(x))dx}{G_y} = \frac{\int 0 dx}{G_y} = a_4.
\end{aligned} \tag{16}
$$

where a_1, a_2, a_3, a_4, b_1, and b_2 are unknown constant parameters.

As in (15), the highest order of the section strains $\mathbf{e(u)}$ is linear (e_2, e_3), which means that the distribution of the bending moments is a skew line, that the corresponding errors in (7) can be obtained from the shaded area between two skew lines, and that each line can be confirmed by two different nodes, that is, two section strains in different sections

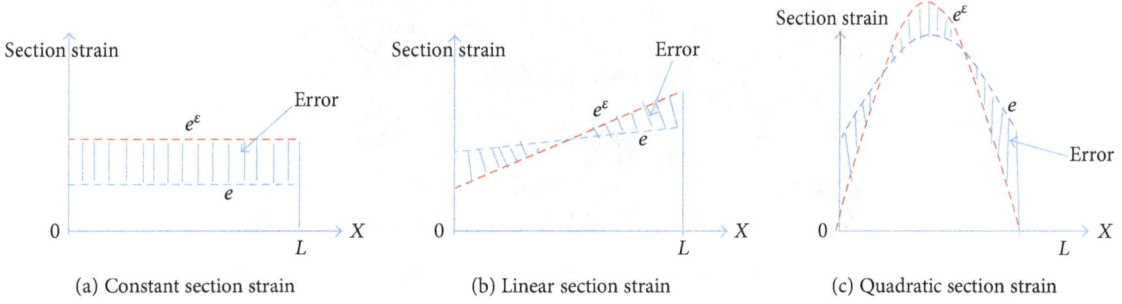

FIGURE 4: Distribution and error of section strains in different orders. e^ε is a section strain in $\mathbf{e}^\varepsilon$ computed from surface strain measurements by (8). e is a section strain in $\mathbf{e}(\mathbf{u})$ deduced by (3).

(Figure 4(a)). Thus, $n = 2$ for e_2 and e_3 in (7). Meanwhile, the other section strains, $e_1, e_4, e_5,$ and e_6, are constant, which means that their distributions are lines parallel to the x-axis, that the associative errors in (7) can be obtained from the shaded area between two parallel lines, and that every line can be confirmed by one node, that is, one section strain (Figure 4(b)). Thus, the number of the sections where section strains are distributed is $n = 1$. Finally, the number of the sections, n, is 2 in (7) and (10) in the loading case of end-node force and moments.

Another case is the beam element loaded by uniformly distributed forces, $q_y(x)$ and $q_z(x)$ are constant along the x-axis, and the order of the section strains $e_2, e_3, e_4,$ and e_5 can be deduced as follows.

Assumption: $q_y(x) = d_1$, $q_z(x) = d_2$.

Then,

$$
\begin{aligned}
e_2 &= \frac{\iint(-q_z(x))dxdx}{D_y} = \frac{\iint(-d_2)dxdx}{D_y} = \frac{-d_2}{2D_y}x^2 + d_3x + d_4, \\
e_3 &= \frac{\iint(-q_z(x))dxdx}{D_z} = \frac{\iint(-d_1)dxdx}{D_z} = \frac{-d_1}{2D_z}x^2 + d_5x + d_6, \\
e_4 &= \frac{\int(-q_z(x))dx}{G_z} = \frac{\int(-d_2)dx}{G_z} = \frac{-d_2x}{G_z} + d_7, \\
e_5 &= \frac{\int(-q_y(x))dx}{G_y} = \frac{\int(-d_1)dx}{G_y} = \frac{-d_1x}{G_y} + d_8.
\end{aligned}
\tag{17}
$$

Herein, d_i $(i = 1, ..., 8)$ are unknown constant parameters. The highest order of the section strains $\mathbf{e}(\mathbf{u})$ is quadratic (e_2, e_3), and then, the distribution of the bending moments is a parabola. The corresponding errors in (7) can be obtained from the shaded area between two parabolas which are confirmed by three different nodes (Figure 4(c)), that is, three section strains in three different sections. Thus, $n = 3$ for e_2 and e_3 in (7). Meanwhile, e_4 and e_5 are linear, where the distribution is a skew line, and e_1 and e_6 are constant. Similar to the loading case of end-node force, the corresponding numbers of sections are 1 and 2, respectively. Finally, the number of the sections, n, is 3 in (7) and (10) in the loading case of uniformly distributed forces.

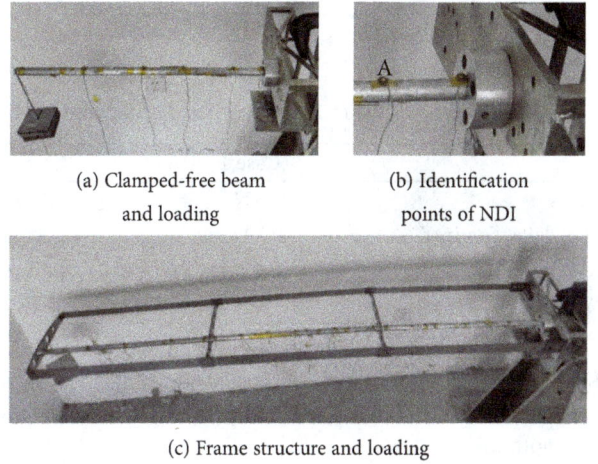

(a) Clamped-free beam and loading

(b) Identification points of NDI

(c) Frame structure and loading

FIGURE 5: Loading case and the distributions of NDI identification points.

4. Verification

A simple cantilevered solid beam and a three-dimensional frame structure were subjected to the end-node static loads to assess the iFEM potential for the flexible wing deformation estimation. The beam and frame structures were made of 6061-T6 aluminum alloy. The Young's modulus is $E = 73000$ Mpa, the Poisson ratio is $v = 0.3$, and its density is $\rho = 2712.63 \text{kg/m}^3$. The frame was composed of three solid beams and middle plates; the span of each beam $L = 640$ mm and the radius $R = 10$ mm (see Figure 5). For the solid circular cross section, the shear correction factors are $k_y^2 = k_z^2 = 0.887$ [21]. In our verification, every whole solid beam is regarded as one beam element. Accordingly, the principle axis of frame structure is divided into three elements.

The experimentally measured surface strains are obtained by fiber optic strain sensors. Displacement measurements are captured by 3D optical measurement instruction (NDI, Optotrak Certus), which determines the position of the identification point by using three CCD cameras to capture the infrared lights emitted by position sensors (Figure 6). The instruction is also used to assess the iFEM-recovered deflections, where the accuracy of the NDI is 0.1 mm. A position sensor was placed as close as possible to

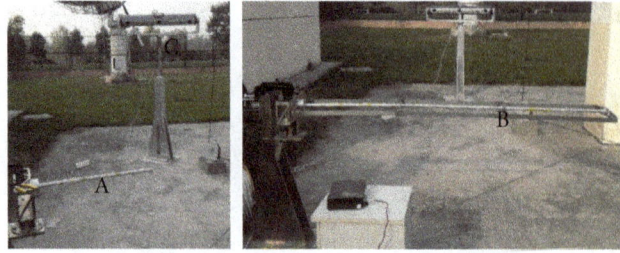

FIGURE 6: Static test setup. The single beam (A), frame structure (B), and NDI Optotrak Certus (C).

TABLE 1: Optical fiber sensor location in one element.

$e^{\varepsilon_1}(x,\theta,\beta)$	$e^{\varepsilon_2}(x,\theta,\beta)$	$e^{\varepsilon_3}(x,\theta,\beta)$	$e^{\varepsilon_4}(x,\theta,\beta)$	$e^{\varepsilon_5}(x,\theta,\beta)$	$e^{\varepsilon_6}(x,\theta,\beta)$
$(L/2,-2\pi/3,0)$	$(L/2,0,0)$	$(L/2,2\pi/3,0)$	$(4L/5,-2\pi/3,0)$	$(4L/5,0,\pi/4)$	$(4L/5,2\pi/3,0)$

TABLE 2: Loading case for the clamped-free beam. Max deflections are captured from NDI.

	Case 1	Case 2	Case 3	Case 4	Case 5
Loads	5 N	15 N	37 N	41 N	68 N
Max deflection in Y	−0.79 mm	−2.5 mm	−6.11 mm	−6.82 mm	−10.96 mm
RMS in Y	0.05 mm	0.16 mm	0.32 mm	0.33 mm	0.42 mm
%Diff(v) in Y	6%	6.4%	5.2%	4.8%	3.8%

the beam root to verify the effectiveness of the clamping arrangement. The force is achieved by placing several weights on a pothook.

For the frame structure, each beam is regarded as an element and the optical fiber sensor location scheme used for each element is identical. For the end-node static loads, the displacement field of the whole beam element is interpolated by C^0 continuous shape function and the number of required strain sensors is 6 (see [21]). As the radius of the beam element is small ($R = 10$ mm) and the grid length of every fiber grating sensor is 10 mm, it is difficult to stick the six strain gauges on one section; six fiber optic strain gauges are placed on two different sections along the beam (Table 1).

The accuracy of the reconstitution is assessed by root-mean-square (RMS) and percentage difference (%Diff).

$$\text{RMS} = \sqrt[2]{\sum_{i=1}^{m}\left(\tau^{\text{iFEM}}(x_i) - \tau^{\text{NDI}}(x_i)\right)^2},$$
$$\%\text{Diff}(\tau) = 100 * \left[\frac{\text{RMS}}{\max|\tau^{\text{NDI}}(x_i)|}\right], \quad (18)$$

where $\tau = (v, w)$ is the deformation displacement along the y- or z-axis and m is the number of deflection shape displacement measurement identification points of NDI (Figure 5(b)A) in the structure. For the beam, $m = 6$ (Figure 5(a)), and for the frame, $m = 14$ (Figure 5(c)). The superscript "iFEM" refers to the reconstitution by iFEM while "NDI" refers to the experimental measure from 3D optical measurement instruction; x_i is the ith location along the axis where the displacement u is measured.

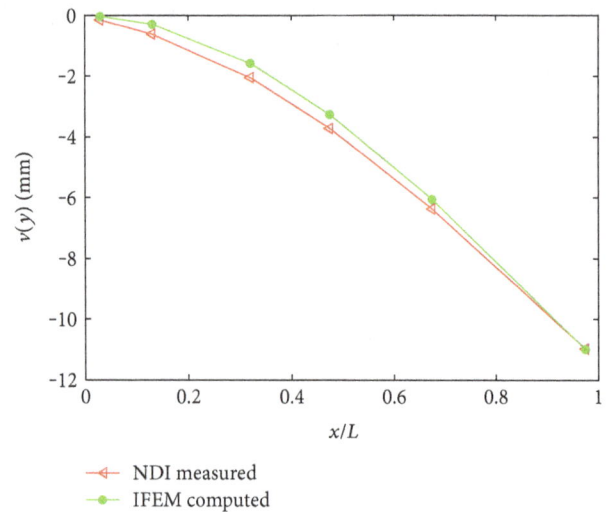

FIGURE 7: Comparison of iFEM reconstruction for a beam to NDI measurement in Y.

At first, five different vertical force cases (Table 2) were applied at the free end of the beam (Figure 5(a)). Figure 7 shows the comparison between iFEM reconstruction and NDI measurement along the beam axis in force direction, for the loading case 68 N. Symbol x/L indicates the location of identification point along the beam surface. Zero means identification point at the clamped end node, and 1 means identification point at the free end node. $v(y)$ indicates the corresponding displacement in force direction.

TABLE 3: Loading case for the frame structure. Max deflections are captured from NDI.

	Case 1	Case 2	Case 3	Case 4	Case 5
Loads	18 N	23 N	35 N	45.5 N	68 N
Max deflection in Y	16.31 mm	20.5 mm	31.2 mm	41.3 mm	62.8 mm
RMS in Y	0.22 mm	0.49 mm	2.0 mm	3.3 mm	6.8 mm
%Diff(v) in Y	1.3%	2.4%	6.4%	8.1%	10.9%
Max deflection in Z	2.53 mm	3.23 mm	5.1 mm	6.5 mm	10.2 mm
RMS in Z	0.32 mm	0.41 mm	0.73 mm	0.75 mm	1.1 mm
%Diff(w) in Z	12.6%	12.8%	14.6%	11.5%	10.8%

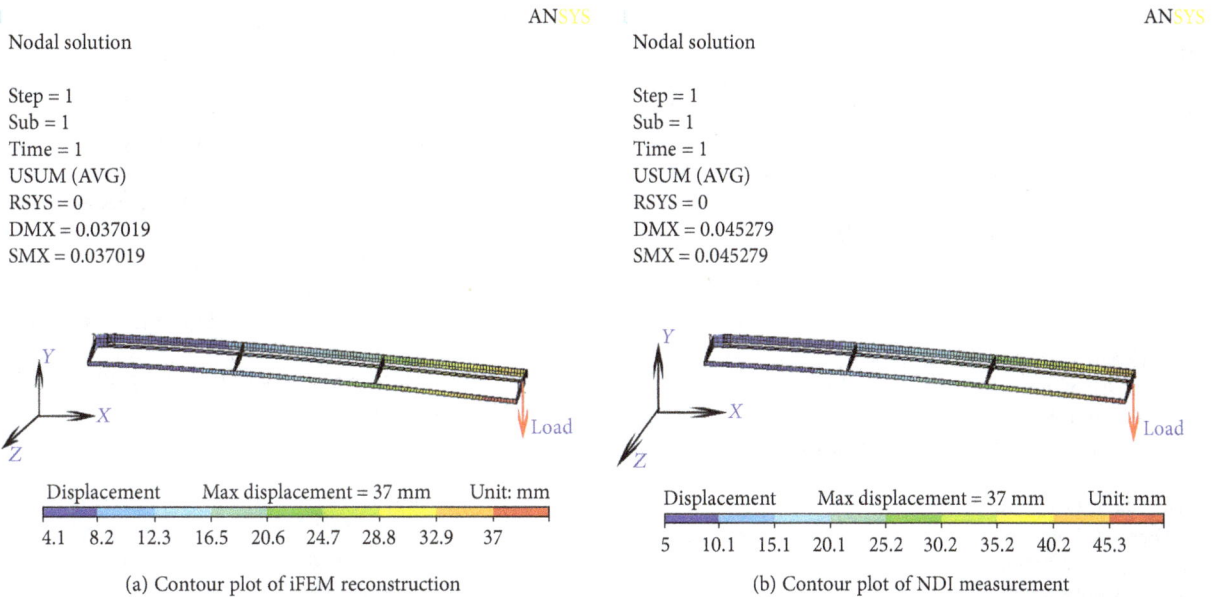

ANSYS

Nodal solution

Step = 1
Sub = 1
Time = 1
USUM (AVG)
RSYS = 0
DMX = 0.037019
SMX = 0.037019

Displacement   Max displacement = 37 mm   Unit: mm

4.1 8.2 12.3 16.5 20.6 24.7 28.8 32.9 37

(a) Contour plot of iFEM reconstruction

ANSYS

Nodal solution

Step = 1
Sub = 1
Time = 1
USUM (AVG)
RSYS = 0
DMX = 0.045279
SMX = 0.045279

Displacement   Max displacement = 37 mm   Unit: mm

5   10.1 15.1 20.1 25.2 30.2 35.2 40.2 45.3

(b) Contour plot of NDI measurement

FIGURE 8: Comparison of iFEM reconstruction for a frame to NDI measurement.

For the frame structure, five different vertical force cases (Table 3) were applied at the free end of the frame (Figure 5(c)). The two deformation contour plots of the frame are plotted by high-fidelity direct FE analysis (ANSYS 12.0), which shows the comparison between iFEM reconstruction and NDI measurement along the beam axis in force direction, for the loading case 45.5 N (Figure 8). The deformation unit in the two figures is meter.

It is seen from the preceding results that the iFEM methodology shows a good potential as a reliable estimation technique. The estimation results gained from the iFEM algorithm have a better approximation to the NDI measurement results. For the beam test, the accuracy of the reconstitution assessed by percentage difference does not exceed 6.4% and the error brings down with the loading increase. For the frame test in the loading case of 18 N, the percentage difference is 1.3%, but the difference grows up with the loading increase; especially from the loading case 23 N to 35 N, the difference increases 4%. The cause for this phenomenon is that the gripper of the frame structure is not very stable, which leads to the fact that the gap spacing between the frame and its carrier grows up with the loading increase that has a great influence on the boundary conditions.

5. Conclusion

This study investigates the application potential of the iFEM algorithm for the flexible wing and other frame structures. This method employs the in situ strain measurements and proper displacement shape function to estimate the deformation shape of the beam and frame structure, without the need to know the applied loads and material properties and to use modal shapes. When the boundary conditions of the reconstruction equation and the shape function are determined, the deformation shape will be accurately estimated in a nonsingular system. Although the boundary conditions are known in hypothesis, the number of location where the section strains are evaluated is not shown clearly; thereby, the singularity of the equation will be influenced.

This paper discusses the minimum number of location where the section strains are evaluated with the loading case, using equilibrium equations in detail, and verifies the result that for the loading case of end-node force, the number is 2. The results of the static loads in the beam and the frame model show that the iFEM algorithm has a good potential for the flexible frame structure estimation; especially for a simply supported beam, the accuracy of the estimation by iFEM is around 6%, and for the frame structure in small load (18 N), the accuracy reaches up to 1.3%, while as the gripper of the frame is not firm enough, the estimation accuracy brings down with the loading increase. So, the next work is to explore the technology that reduces the influence of the gap spacing of the frame gripper on accuracy of the frame deformation estimation.

Conflicts of Interest

The authors declare that there is no conflict of interest regarding the publication of this paper.

Acknowledgments

This work was financially supported by the National Natural Science Foundation of China (Grants 51675398 and 51775401), the National Key Basic Research Program of China (Grant 2015CB857100), and the CAS "Light of West China" Program (no. 2016-QNXZ-A-7).

References

[1] W. Akl, S. Poh, and A. Baz, "Wireless and distributed sensing of the shape of morphing structures," *Sensors and Actuators A*, vol. 140, no. 1, pp. 94–102, 2007.

[2] Z. Mao and M. Todd, "Comparison of shape reconstruction strategies in a complex flexible structure," *Proceedings of SPIE*, vol. 6932, article 69320H, 2008.

[3] P. David, J. Kenneth, and G. Vladimir, "Distortion measurement and compensation in a synthetic aperture radar phased-array antenna," in *2010 14th International Symposium on Antenna Technology and Applied Electromagnetics and the American Electromagnetics Conference (ANTEM-AMERRM)*, pp. 1–5, Ottawa, ON, Canada, 2010.

[4] W. Yin, T. Fu, J. Liu, and J. Leng, "Structural shape sensing for variable camber wing using FBG sensors," in *Proceedings of SPIE 7292, sensors and smart structures technologies for civil, mechanical, and aerospace systems*, vol. 7292, p. 72921H, San Diego, CA, USA, March 2009.

[5] S. Rapp, L. H. Kang, J. H. Han, U. C. Mueller, and H. Baier, "Displacement field estimating for a two-dimensional structure using fiber Bragg grating sensors," *Smart Materials and Structures*, vol. 18, no. 2, article 025006, 2009.

[6] L. Kang, D. Kim, and J. Han, "Estimation of dynamic structural displacements using fiber Bragg grating strain sensors," *Journal of Sound and Vibration*, vol. 305, no. 3, pp. 534–542, 2007.

[7] D. Ganotra, J. Joseph, and K. Singh, "Object reconstruction in multilayer neural network based profilometry using grating structure comprising two regions with different spatial periods," *Optics and Lasers in Engineering*, vol. 42, no. 2, pp. 179–192, 2004.

[8] Z. Yong, B. Hong, D. Xuechao, and F. Hongmei, "Research of situ strain of measuring flexible truss deformation based on fuzzy network method," in *2016 International Conference on Computer Engineering, Information Science & Application Technology (ICCIA 2016)*, pp. 401–410, Guilin, China, 2016.

[9] G. Foss and E. Haugse, "Using modal test results to develop strain to displacement transformation," in *Proc. 13th The International Society for Optical Engineering*, pp. 112–118, Nashville, TN, February 1995.

[10] S. L. Peter, J. A. Mauro, and W. H. Nesbitt, "Investigation of filtering techniques applied to the dynamic estimation problem," *Smart Materials and Structures*, vol. 10, no. 2, pp. 264–272, 2001.

[11] M. D. Todd and S. T. Vohra, "Shear deformation correction to transverse shape reconstruction from distributed strain measurements," *Journal of Sound and Vibration*, vol. 225, no. 3, pp. 581–594, 1999.

[12] B. B. Philip, H. Eric, and E. G. Ralph, "Structural shape identification from experimental strains using a modal transformation technique," in *4th AIAA/ASME/ASCE/AHS Structures, Structural Dynamics, and Materials Conference*, pp. 7–10, Norfolk, Virginia, April 7, 2003.

[13] H. I. Kim, L. H. Kang, and J. H. Han, "Shape estimation with distributed fiber Bragg grating sensors for rotating structures," *Smart Materials and Structures*, vol. 20, no. 3, pp. 35011–35021, 2011.

[14] P. Maincon, "2004 inverse FEM I: load and response estimates from measurements," in *Proceedings of 2nd International Conference on Structural Engineering, Mechanics and Computation*, pp. 967–971, Cape Town, 2004.

[15] C. Jute, W. Ko, and C. Stephens, "Deformed shape calculation of a full-scale wing using fiber optic strain data from a ground loads test," NASA Langley Research Center, Rept. TP-215975, Hampton, VA, USA, 2011.

[16] A. Derkevorkian, S. F. Masri, J. Alvarenga, H. Boussalis, J. Bakalyar, and W. L. Richards, "Strain-based deformation shape-estimation algorithm for control and monitoring applications," *AIAA Journal*, vol. 51, no. 9, pp. 2231–2240, 2013.

[17] A. Tessler and J. Spangler, "A variational principle for reconstruction of elastic deformations in shear deformable plates and shells," NASA Langley Research Center TM-212445, Hampton, VA, USA, 2003.

[18] A. Tessler and J. Spangler, "A least-squares variational method for full-field reconstruction of elastic deformations in shear-deformable plates and shells," *Computer Methods in Applied Mechanics and Engineering*, vol. 194, no. 2–5, pp. 327–339, 2005.

[19] C. C. Quach, S. L. Vazquez, A. Tessler, and J. P. Moore, "Structural anomaly detection using fiber optic sensors and inverse finite element method," in *AIAA Guidance, Navigation, and Control Conference and Exhibit*, San Francisco, CA, USA, 2005.

[20] M. Gherlone, P. Cerracchio, M. Mattone, M. Di Sciuva, and A. Tessler, "Dynamic shape reconstruction of three-dimensional frame structures using the inverse finite element method," NASA Langley Research Center TM-217315, Hampton, VA, USA, 2011.

[21] M. Gherlone, P. Cerracchio, M. Mattone, M. Di Sciuva, and A. Tessler, "Shape sensing of 3D frame structures using an inverse finite element method," *International Journal of Solids and Structures*, vol. 49, no. 22, pp. 3100–3112, 2012.

[22] A. I. Lurie, *Theory of Elasticity*, Springer-Verlag Berlin Heidelberg, New York, NY, USA, 2005.

15

A Two-Phased Guidance Law for Impact Angle Control with Seeker's Field-of-View Limit

Haoqiang Zhang◉,[1] Shengjing Tang◉,[1] Jie Guo◉,[1] and Wan Zhang◉[2]

[1]*Key Laboratory of Dynamic and Control of Flight Vehicle of the Ministry of Education, School of Aerospace Engineering, Beijing Institute of Technology, Beijing 100081, China*
[2]*Beijing Electro-Mechanical Engineering Institute, Beijing 100071, China*

Correspondence should be addressed to Jie Guo; guojie1981@bit.edu.cn

Academic Editor: Hikmat Asadov

A two-phased guidance problem with terminal impact angle constraints and seeker's field-of-view limit is addressed in this paper for a missile against a nonmaneuvering incoming target. From the conventional PN guidance without any constraints, it is found that satisfying the impact angle constraint causes a more curved missile trajectory requiring a large look angle. To avoid the look angle exceeding the seeker's physical limit, a two-phased look angle control guidance scheme with the terminal constraint is introduced. The PN-typed guidance law is designed for each guidance phase with a specific switching condition of line-of-sight. The proposed guidance law is comprised of two types of acceleration commands: the one in the initial phase which aims at controlling the missile's look angle to reach the limit and the other for final phase which is produced by switching the navigation gain. The monotonicity of the line-of-sight angle and look angle is analyzed and proved to support the proposed method. To evaluate the specific navigation gains for both initial and final phases, the scaling coefficient between them is discussed by solving a quadratic equation with respect to the initial navigation gain. To avoid a great abrupt acceleration change at the switching instant, a minimum coefficient is chosen. Extensive simulations are performed to validate the efficiency of the proposed approach.

1. Introduction

The impact angle control guidance laws (IACGL) have been widely studied for several years [1], for both stationary and moving targets. Practically, the engagement cannot be implemented successfully for a homing missile without the target information from a seeker. The seeker's field-of-view (FOV) is a critical limit for missile engagement, especially when the missile utilizes a seeker with narrow FOV and intercepts a high-speed target with terminal angle constraints [2, 3]. In such a strict condition, the line-of-sight (LOS) violates the FOV limit more easily. Meanwhile, the missile's detection process might be within a short period of time for moving targets.

A considerable number of previous works focusing on the impact angle guidance used the optimal control theory or the proportional navigation (PN) guidance method. The optimal control theory is mainly used to design those guidance laws in the assumption of linear engagement model. This method is typically applied with a class of cost functions involving the quadratic term of the control input which needs to be minimized [4]. By using the Schwartz inequality, the guidance laws which considered the terminal impact angle constraints (TIAC) were developed with the estimated time-to-go [5, 6], especially for a type of linear quadratic optimal control problem. The optimal guidance laws (OGL) for impact angle control which were proposed in [7, 8] also paid attention to the other performance of the shaping trajectory, such as the target observability and the acceleration constraints. Because of its efficiency and ease of implementation, PN guidance law is more popular and widely used [9]. There are also many modified PN-typed laws related to the impact angle control problem. A biased PN (BPN) guidance law was adopted in [10] to control the missile to impact a target with terminal angular constraints. This guidance law is a variation of the conventional PN guidance law, which is combined with a supplementary time-varying bias to control the impact angle. Lu et al. [11] introduced a three-dimensional guidance law, which was also based on PN with adaptive guidance parameters, to control a hypersonic vehicle in its terminal phase to impact a stationary target. Ratnoo and Ghose [12, 13] proposed a two-phased variable navigation gain PN guidance law for intercepting stationary and moving (nonmaneuvering) targets with all possible impact angles, but not considering the FOV limitation. On the other hand, sliding mode control (SMC) theory becomes more and more popular to be applied to design IACGL. A finite-time convergent sliding mode guidance law with terminal impact angle constraint was presented in [14], which was mainly addressed by the finite-time convergence stability theory. Zhao et al. [15] designed a SMC-based guidance law for an unpowered lifting reentry vehicle against a stationary target, to satisfy the TIAC with high guidance accuracy. Recently, the interception against maneuvering target was proposed based on several new SMC methods, for example, the nonsingular fast sliding mode (NFSM), with consideration of TIAC [16, 17].

Furthermore, several works are introduced to design the guidance laws considering the FOV limit as well as the TIAC. Park et al. [18] proposed an optimal impact angle control guidance law with the seeker's FOV limit for missiles with strapdown seekers. By using the optimal control theory, the look angle which was regarded as a new state for inequality constraint was introduced during the homing phase. In [19], a two-phased scheme was developed in which BPN was used to shape the missile trajectory by making the integral value of the bias to reach a certain value at first and then switched to PN in the second phase. The integral value was calculated from initial engagement conditions and desired impact angle. Based on the same strategy of [19], a few of two-phased BPN (TPBPN) guidance laws were proposed in [20–25] for attacking a stationary target. A TPBPN was proposed in [20] based on the bias-shaping method, which can satisfy both the terminal-angle constraint and the FOV limitation to maintain the seeker's lock-on condition. The TPBPN in [21] utilized the unbiased and biased pure proportional navigation laws and applied seeker's FOV maximum value to calculate the bias in the initial phase. A time-varying BPN guidance law was designed by Yang et al. [22] to achieve the angular constraints without violating the look angle. Two time-varying biases were adopted in [22] to keep the seeker's look angle and to ensure the terminal impact angle, respectively. Along with the thoughts of TPBPN and composite guidance law, Tekin and Erer [23] proposed the PN-gain-switched strategy, which can admit the allowed look angle and acceleration values to meet the demand of impact angle and at the same time, have a relative simple structure for implementation because of the PN form. Based on a two-stage PN guidance law, Ratnoo [24] derived a closed-form solution for the choice of navigation gains to satisfy the look angle and impact angle limit. Wen et al. [25] proposed a new guidance parameter design strategy based on the classical time-to-go weighted impact angle

optimal guidance law. A robust guidance law which was based on the switching logic was designed in [26] by an additional term and combined the sliding mode control. This kind of guidance problem was also solved by SMC [27] and handled without any additional switching logic. However, since most of these works dealt with stationary targets, the desired impact angle may not be achieved against moving targets when applying these works to homing missiles. A composite guidance scheme was presented considering the case of a nonmaneuvering moving target, which was composed of modified deviated pure pursuit (DPP) with error feedback loop of look angle for initial guidance phases and PN with $N \geq 3$ for final guidance phases [28]. Park et al. [29] addressed the similar guidance problem and strategy in [28] and extended to consider the command limit.

Based on the studies of previous works, this article draws on the experience of PN-gain-switched strategy from [23] to extend the investigated case to a nonmaneuvering moving target. The algorithm procedures and switching logic are introduced to achieve the guidance law proposed in this work. The main contributions of this work are summarized as follows:

(1) We have studied the scaling factor between the different navigation gains of the initial and final guidance phases. The minimum value of this scaling factor is chosen to reduce the abrupt acceleration change at switching instant. Therefore, the specific navigation gain values for both two guidance phases can be calculated according to the scaling factor and the desired impact angle, which is different from [23, 28].

(2) Two propositions are presented to prove the monotonicity of the line-of-sight angle and look angle. The switching of guidance phases occurs when a specific LOS angle similar to [28] is satisfied. However, the look angle reaches the FOV limit only at the switching instant, which reduces the required load in initial guidance phase compared with [28].

(3) Different from [20–22], the integral biased term is not needed in the proposed method, which indicates that we do not need to estimate the value of time-to-go. It results in a convenient implementation in real missile model.

2. Problem Statement

Consider the planar homing guidance geometry of a missile with narrow FOV against a nonmaneuvering incoming target as shown in Figure 1, where $X_I - O - Y_I$ is a Cartesian inertial reference frame. The relative distance between the missile and the target is r. The subscripts M and T denote the missile and the target, respectively. The acceleration vector **a** is applied perpendicularly to the velocity vector **V**. γ, λ, and σ represent the flight-path angle, LOS angle, and look angle. Angles are positive in the counterclockwise direction.

As characteristics of the surface-to-surface engagement, the flight-path angle of target equals to π which is under the condition of an incoming target. The initial condition

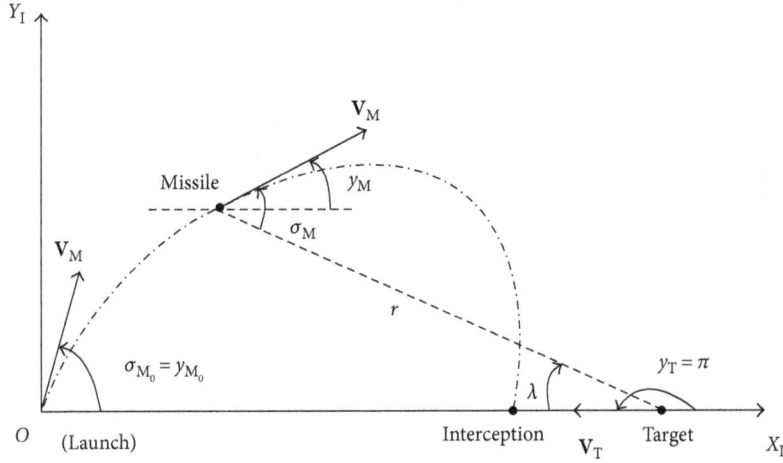

FIGURE 1: Engagement geometry.

on the launch point is also characterized with $\gamma_{M_0} = \sigma_{M_0}$. These conditions imply an approximate head-on case with initial heading errors. The nonlinear equations of motion in this homing problem are given by

$$\dot{r} = -\mathbf{V}_T \cos \lambda - \mathbf{V}_M \cos \sigma_M, \tag{1}$$

$$r\dot{\lambda} = \mathbf{V}_T \sin \lambda - \mathbf{V}_M \sin \sigma_M, \tag{2}$$

$$\dot{\gamma}_M = \frac{\mathbf{a}_M}{\mathbf{V}_M}, \tag{3}$$

$$\sigma_M = \gamma_M - \lambda, \tag{4}$$

where σ_M can be defined as the seeker's look angle in the assumption that the angle of attack (AOA) of the missile is small.

If the missile follows PN guidance law with a navigation gain N, the flight-path angle rate can be written as

$$\dot{\gamma} = N\dot{\lambda}. \tag{5}$$

Integrating (5) with boundary conditions of γ_{M_f} and λ_f yields

$$\gamma_{M_f} - \gamma_{M_0} = N(\lambda_f - \lambda_0), \tag{6}$$

where the subscripts 0 and f denote the initial and final conditions, respectively. Combining (4) and (5) to give

$$\dot{\sigma}_M = \dot{\gamma}_M - \dot{\lambda} = (N-1)\dot{\lambda}. \tag{7}$$

Also integrating (7) with boundary conditions of σ_{M_f} and λ_f, we can obtain

$$\sigma_{M_f} - \sigma_{M_0} = (N-1)(\lambda_f - \lambda_0). \tag{8}$$

Considering the collision triangle condition, we enable the LOS angle rate to be zero to intercept a moving target. Equation (2) should satisfy the following relationship:

$$\mathbf{V}_M \sin\left(\gamma_{M_f} - \lambda_f\right) = \mathbf{V}_T \sin \lambda_f. \tag{9}$$

Expanding both sides of (9), the final LOS angle which will keep the collision triangles can be expressed as (10) with some algebraic and derivations.

$$\lambda_f = \tan^{-1}\left(\frac{\sin \gamma_{M_f}}{\cos \gamma_{M_f} + \eta}\right), \tag{10}$$

where $\eta = \mathbf{V}_T / \mathbf{V}_M$ and the function of $\tan^{-1}$ should be calculated from atan2, which is a four-quadrant inverse tangent function.

Expressing (6) with respect to λ_0 leads to

$$\lambda_0 = \lambda_f - \frac{\gamma_{M_f} - \gamma_{M_0}}{N}. \tag{11}$$

3. Analysis for the Switched-Gain Guidance

Consider a two-phased guidance scheme illustrated in Figure 2, replacing λ_0 with λ_s which is the LOS angle at the specific switching instant. Substituting (4) and (10) into (11), the relationship can be expressed as

$$\lambda_s = \left[\tan^{-1}\left(\frac{\sin \gamma_{M_f}}{\cos \gamma_{M_f} + \eta}\right) - \frac{\gamma_{M_f} - \sigma_{M_s}}{N_F}\right] \frac{N_F}{N_F - 1}, \tag{12}$$

where N_F and σ_{M_s} denote the navigation gain in the final phase and the look angle at the switching instant. And we define N_I as the navigation gain in the initial phase which will be used next. Meanwhile, (12) means that, to be able to reach a desired impact angle when attacking a moving target, the LOS angle should be a certain value at some specific instant with a desired look angle. A switching condition λ_s depends on the three main parameters, which are the desired look angle, the impact angle, and the navigation gain in the final phase.

Proposition 1. *Considering the variation of look angle under the PN guidance in the initial phase, the look angle should satisfy the condition:* $N_I < 2$, $\sin^{-1} \eta < \sigma_M < \cos^{-1}(2\eta/N_I - 2)$,

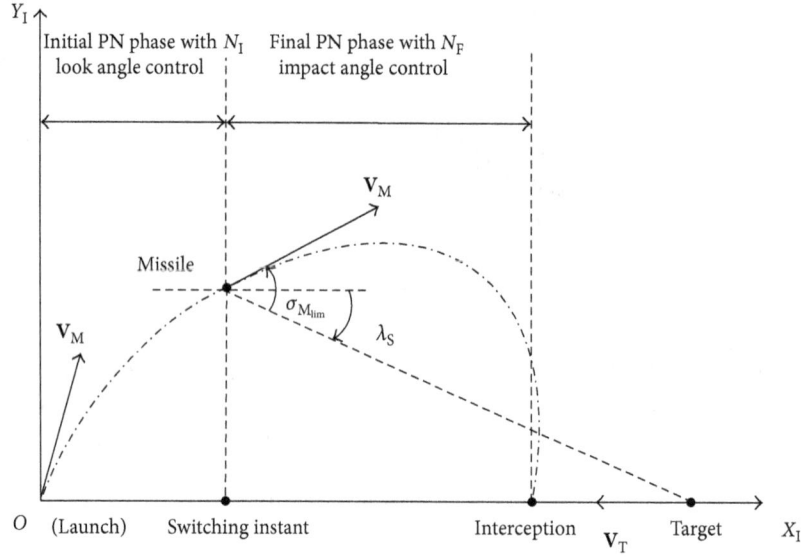

FIGURE 2: Two-phased guidance scheme.

or $N_I > 2$, $\sigma_M > \max(\sin^{-1}\eta, \cos^{-1}(2\eta/2 - N_I))$. *Hence, the LOS rate $\dot\lambda$ with negative values strictly decreases until the switching condition λ_s is reached, which is $\ddot\lambda < 0 \ \forall \ t \in [t_0, t_s]$ (where t_s is the switching time of the initial phase and final phase).*

Proof. Differentiating (2) with respect to t, we can obtain

$$\ddot\lambda = \frac{(\mathbf{V}_T \cos\lambda - \mathbf{V}_M \cos\sigma_M(N-1) - \dot r)\dot\lambda}{r}. \qquad (13)$$

Substituting (1) and (2) into (13) leads to

$$\ddot\lambda = \frac{\mathbf{V}_M^2}{r^2}[2\eta\cos\lambda + (2-N)\cos\sigma_M](\eta\sin\lambda - \sin\sigma_M). \quad (14)$$

To guarantee that the LOS angle decreases to the switching value monotonously, the LOS angle rate should keep in a negative value until switching, which is equivalent to

$$\begin{aligned} &\dot\lambda(t_0) < 0, \\ &\ddot\lambda(t) < 0, \qquad\qquad\qquad (15) \\ &\qquad \forall t \in [t_0, t_s]. \end{aligned}$$

Substituting the initial condition into (2), we can get

$$\dot\lambda(t_0) = -\frac{\mathbf{V}_M \sin\sigma_{M_0}}{r_0}, \qquad (16)$$

when $\sigma_{M_0} \in [0, \pi/2]$, $\dot\lambda(t_0) < 0$. Since the value of $\mathbf{V}_M^2/r^2$ must be positive to ensure $\ddot\lambda < 0$ and $\dot\lambda < 0$, the following conditions should be satisfied:

$$\begin{aligned} &2\eta\cos\lambda + (2-N)\cos\sigma_M > 0, \\ &\qquad\qquad\qquad\qquad\qquad\qquad\qquad (17) \\ &\eta\sin\lambda - \sin\sigma_M < 0. \end{aligned}$$

As we know that $|\sin\lambda|_{\max} = 1$, $|\cos\lambda|_{\max} = 1$ for $\lambda \in [-\pi, 0]$, the inequalities in (17) can be expressed as

$$\begin{aligned} &\sin^{-1}\eta < \sigma_M < \cos^{-1}\left(\frac{2\eta}{N-2}\right), \quad N < 2, \\ &\qquad\qquad\qquad\qquad\qquad\qquad\qquad\qquad\qquad (18) \\ &\sigma_M > \max\left(\sin^{-1}\eta, \cos\left(\frac{2\eta}{2-N}\right)\right), \quad N > 2. \end{aligned}$$

Replacing the navigation gain N with N_I in (18), it satisfies the condition proposed in Proposition 1.

Proposition 2. *For initial guidance phases, the condition of $N_I < 1$ and $\sigma_{M_0} > \sin^{-1}\eta$, $\sigma_{M_{lim}} < \cos^{-1}(2\eta/N_I - 2)$ can ensure that the look angle will increase from σ_{M_0} to the limit value $\sigma_{M_{lim}}$ during the time of $(t_s - t_0)$, monotonously.*

Proof. To guarantee that the look angle increases monotonously, the look angle rate $\dot\sigma_M$ should be positive. As proven in Proposition 1, $\dot\lambda < 0$ when the relationship of (18) is satisfied. Combined with (7), it should meet the condition that is $N_I < 1$ to keep $\dot\sigma_M > 0$ with $\dot\lambda < 0$. This condition means that

$$\begin{aligned} &\sigma_{M_{min}} = \sigma_{M_0}, \\ &\sigma_{M_{max}} = \sigma_{M_{lim}}, \qquad\qquad (19) \\ &\qquad \forall t \in [t_0, t_s]. \end{aligned}$$

Substituting (19) into (18) considering the monotonicity of the look angle, we can get the inequalities as

$$\sigma_{M_0} > \sin^{-1}\eta,$$

$$\qquad\qquad\qquad\qquad\qquad\qquad\qquad (20)$$

$$\sigma_{M_{lim}} < \cos^{-1}\left(\frac{2\eta}{N_I - 2}\right).$$

Integrating (7) with the boundary conditions of σ_{M_s} and λ_s yields

$$\sigma_{M_s} - \sigma_{M_0} = (N_I - 1)(\lambda_s - \lambda_0). \tag{21}$$

4. Implementation of the Two-Phased Guidance Law

To avoid violating the look angle limitation, the switching instant that look angle increases monotonically according to Proposition 2 is designed to reach the limit value exactly. According to (21), the navigation gain of initial can be expressed as

$$N_I = \frac{\sigma_{M_{lim}} - \sigma_{M_0}}{\lambda_s} + 1. \tag{22}$$

In (22), λ_s should be calculated by (12) and actually decided by N_F with certain values of γ_{M_f} and σ_{M_s}. Establish the proportion relation between N_F and N_I, which yields

$$k = \frac{N_F}{N_I}. \tag{23}$$

Substituting (12) and (23) into (22) leads to

$$N_I = \frac{\sigma_{M_{lim}} - \sigma_{M_0}}{\left[\tan^{-1}\left(\sin \gamma_{M_f}/\cos \gamma_{M_f} + \eta\right) - \gamma_{M_f} - \sigma_{M_{lim}}/kN_I\right](kN_I/kN_I - 1)} + 1. \tag{24}$$

Expanding (24), we can obtain a quadratic equation with respect to N_I as shown in (25), where the proportionality coefficient should be regarded as a known parameter. To solve the value of unknown parameter N_I of (25), a simple algebraic method is applied in (26).

$$k\lambda_f N_I^2 - \left[k\left(\lambda_f + \sigma_{M_{lim}} - \sigma_{M_0}\right) + \gamma_{M_f} - \sigma_{M_{lim}}\right]N_I$$
$$+ \gamma_{M_f} - \sigma_{M_0} = 0, \tag{25}$$

$$a = \lambda_f,$$
$$b = -(kb_1 + b_2), \tag{26}$$
$$c = \gamma_{M_f} - \sigma_{M_0},$$

where $b_1 = \lambda_f + \sigma_{M_{lim}} - \sigma_{M_0}$ and $b_2 = \gamma_{M_f} - \sigma_{M_{lim}}$. The solution of (25) should be

$$N_I = \frac{(kb_1 + b_2) \pm \sqrt{b_1^2 k^2 + (2b_1 b_2 - 4ac)k + b_2^2}}{2ak}, \tag{27}$$

and the discriminant should be nonnegative as

$$b_1^2 k^2 + (2b_1 b_2 - 4ac)k + b_2^2 \geq 0. \tag{28}$$

Therefore, to guarantee the real solution of (25), (28) should be satisfied and the inequality relation merely yields

$$k \geq \frac{(4ac - 2b_1 b_2) + 4\sqrt{ac(ac - b_1 b_2)}}{2b_1^2},$$
$$\text{or } k \leq \frac{(4ac - 2b_1 b_2) - 4\sqrt{ac(ac - b_1 b_2)}}{2b_1^2}, \tag{29}$$

where a and c should be in same sign and satisfy $ac \geq b_1 b_2$. These parameters in (29) are only determined by some boundary conditions and constraints, such as σ_{M_0}, $\sigma_{M_{lim}}$, and γ_{M_f}.

As expressed in (27), N_I has two values that will be discussed. The one(s) less than 1 should be considered according to Proposition 2. If there is no appropriate value to satisfy Proposition 2, the proportionality coefficient k needs to be chosen again. In addition, note that the guidance law for moving target requires two additional measurements or estimates of $\mathbf{V}_T$ and γ_T for implementation.

In summary, the proposed method can be described in procedures as:

Step 1. Give launch conditions, the initial position, velocity, and path angle of missile and target.

Step 2. Give the desired impact angle and calculate the parameters a, b_1, b_2, c according to (26), then judge if the conditions of (29) are satisfied or not.

Step 3. If satisfied, calculate the range of parameter k by (29). If not, change the given launch conditions.

Step 4. Calculate N_I using (27), in which there are parameters calculated in all the steps above, $N_F = kN_I$.

Step 5. Confirm the switching condition λ_s according to N_F and (12).

The proposed guidance loop is depicted in Figure 3. The switching logic is to compare the value of λ and λ_s, then to decide which navigation gain is utilized.

5. Numerical Simulations

To investigate the characteristics of the proposed guidance law, a number of nonlinear simulations are performed in a surface-to-surface scenario. For a given desired impact angle, a limit value of look angle and final navigation gain, the switched LOS angle conditions in (12), is applied.

In the given launch conditions of simulation, a missile has a constant speed of 200 m/s with the initial position $(x_{M_0}, y_{M_0}) = (0, 0)m$ and initial path angle $\gamma_{M_0} = 15°$. A moving but nonmaneuvering target has a constant speed of 50 m/s with the initial position $(x_{T_0}, y_{T_0}) = (5000, 0)m$ and constant path angle $\gamma_T = 180°$. The constraint of look angle is given as $\sigma_{M_{lim}} = 30°$. According to the given launch

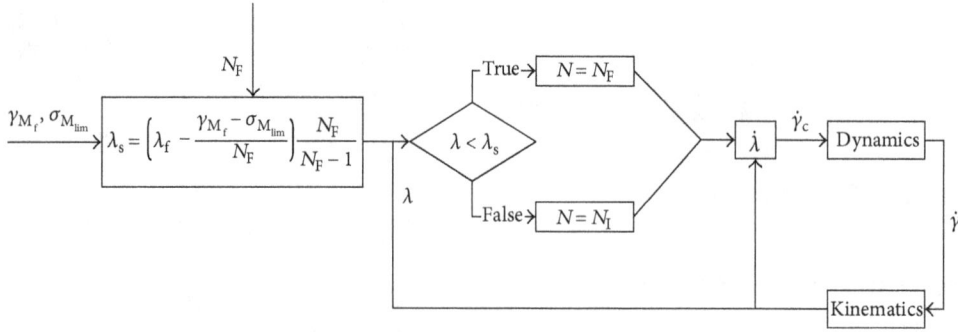

FIGURE 3: Switching logic of the guidance.

conditions, some important parameters and criterions for different impact angles are summarized in Table 1.

5.1. The Choice of N_I with Different k.

From Table 1, to intercept a moving target with desired TIAC, a suitable k, which has two values, should be chosen to design N_I and N_F. The basic principles for choosing such a value come from Proposition 2, and $N_I < 1$ should be fulfilled. The larger value of k is chosen to satisfy the relation $N_F > N_I$ [18]. In Table 2, the values are presented where N_I is chosen based on different k with specific terminal constraint $\gamma_{M_f} = -90°$.

As shown in Table 2, all the values of N_I satisfy Proposition 2. Figure 4 displays the scenarios under the given launch conditions. In Figure 4(a), how the trajectories reach the interception point can be observed, as they have the same impact angle but using different navigation gains (both N_I and N_F). The black dash line represents the trajectory of the moving target. Figures 4(b) and 4(c) show the variations of path angle and look angle in all scenarios. It is easy to see that each curve of path angle reaches the desired impact angle in an approaching case. Figure 4(c) shows that, under the guidance law proposed in this work, the look angles can vary in the same trend without violating the FOV limitation, while switching at different time. As calculated and shown in Table 2, 3.56 is the critical value of k, which leads to two solutions of N_I that are almost the same. On the contrary, if we choose k out of the critical value, it will correspond with two quite different solutions of N_I. N_F is calculated by (23), with the result that one value is larger than 2 and the other less and equal. As revealed in Figure 4(d), the scenarios of $k = 4, N_I = 0.49$ and $k = 5, N_I = 0.37$ which simultaneously lead to $N_F < 2$ trend to make the missile's accelerations diverge. Besides, the scenarios of $k = 4, N_I = 0.71$ and $k = 5$, $N_I = 0.75$ make the missile's accelerations sudden change from near zero to a much larger value. When $k = 3.65$ is applied, the acceleration profile depicted in Figure 4(d) looks to be implemented easier.

5.2. Simulations of Constant Velocity Missile Model.

All the simulation cases in this subsection are performed with the constant velocity missile model.

Case 1 (simulations for various impact angles). As discussed in the last subsection, the minimum value of k can make a minimum abrupt acceleration change of the missile. In this

TABLE 1: Summary of the parameters and criterions.

γ_{M_f} (deg)	a	b_1	b_2	c	k
-50	-0.71	-0.45	-1.40	-1.13	$k \geq 8.72$ or $k \leq 1.12$
-70	-1.01	-0.75	-1.75	-1.48	$k \geq 4.95$ or $k \leq 1.10$
-90	-1.33	-1.06	-2.09	-1.83	$k \geq 3.56$ or $k \leq 1.09$

TABLE 2: Values of N_I with different k.

k	N_I
3.56	$N_{I,1} = 0.62, N_{I,2} = 0.63$
4.00	$N_{I,1} = 0.49, N_{I,2} = 0.71$
5.00	$N_{I,1} = 0.37, N_{I,2} = 0.75$

subsection, Figure 5 reflects the performance of the proposed guidance law with various impact angles, using the critical value of k from Table 1 for every scenario. Figure 5(a) exhibits the trajectories resulting from the two-phased guidance strategy. The trajectories satisfy the different impact angle constraints and engage the curve of incoming target. Satisfying the impact angle constraint, which can be observed clearly in Figure 5(b), also influences the switching condition between the initial phase and final phase. The switching condition is obviously revealed from the trend variation of the look angle curves in Figure 5(c). It is evident that when the look angle increases to the limit, the navigation gain will switch to make the look angle decrease. Precisely because of the variation of navigation gain at switching instant, a sudden change appears in each acceleration history as shown in Figure 5(d).

Case 2 (comparative cases with other guidance laws). The performance of the proposed guidance scheme, which is illustrated in Figure 6 with the impact angle constraint of $\gamma_{M_f} = -90°$, is compared to the BPN guidance law in [10] and a two-phased guidance scheme described in [28]. The implemented acceleration in [28] is summarized as

$$\mathbf{a}_M = \begin{cases} \mathbf{V}_M \dot{\lambda} + K(\sigma_d - \sigma), & \text{for } |\lambda| < |\lambda_s|, \\ N \mathbf{V}_M \dot{\lambda}, & \text{for } |\lambda| \geq |\lambda_s|. \end{cases} \tag{30}$$

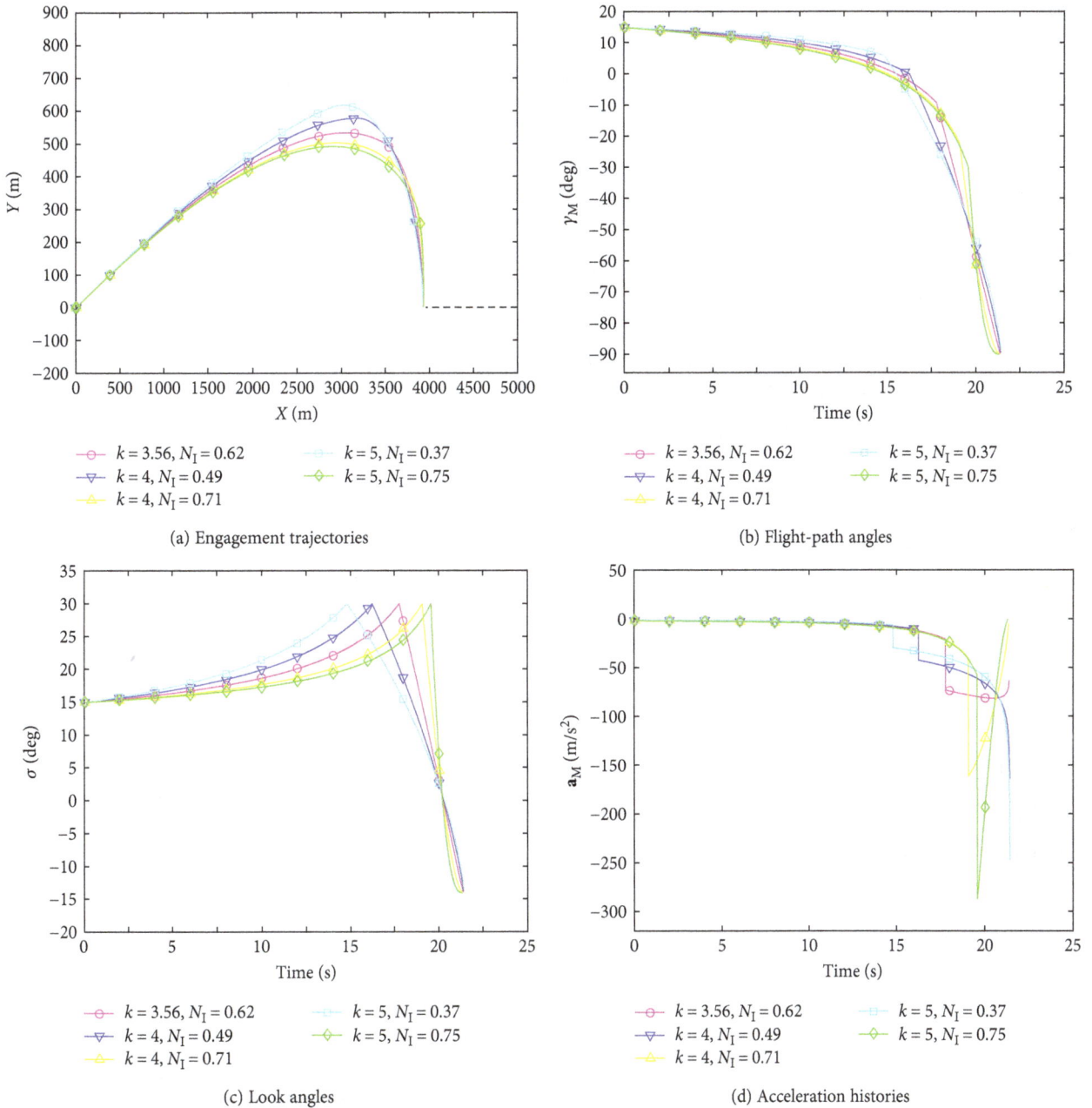

(a) Engagement trajectories

(b) Flight-path angles

(c) Look angles

(d) Acceleration histories

FIGURE 4: The results for the different choice of N_I with $\gamma_{M_f} = -90°$.

In Figure 6(a), it is obvious that the trajectories of BPN and [28] appear more curved than the proposed one. For the terminal impact constraint, all the methods can satisfy the demand as shown in Figure 6(b). However, the look angle variations are quite distinct from each other as seen in Figure 6(c). The BPN guidance law violates the seeker's FOV limitation at a certain time. The method in [28] leads the look angle increase to the limit rapidly and then maintain at this value until switching. Similarly, the method presented in this work controls the look angle to increase to the limit, but slowly and monotonously until switching. In Figure 6(d), compared with the scheme in [28], the acceleration history produced

by the proposed method appears more smoothly especially at initial time.

Case 3 (comparative cases considering first-order lag). Several simulation cases for the guidance law in [28] and the proposed one in this work are carried out to compare the performance with regard to autopilot delay. Consider a first-order lag system in this case, which can be described with a time constant T_m

$$\frac{\mathbf{a}_M}{\mathbf{a}_C} = \frac{1}{T_m s + 1}, \qquad (31)$$

(a) Engagement trajectories

(b) Flight-path angles

(c) Look angles

(d) Acceleration histories

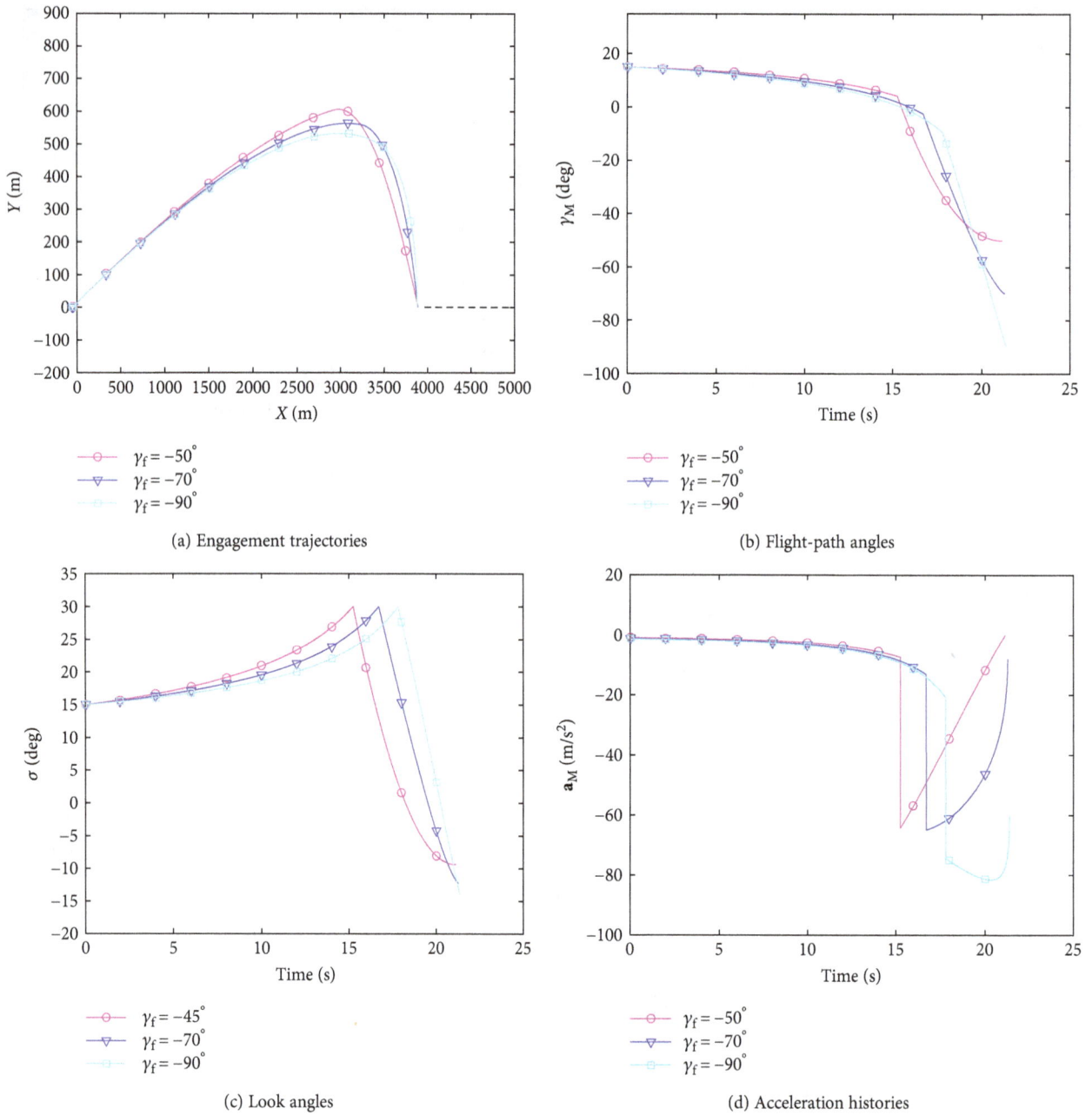

FIGURE 5: The results for the different terminal impact angles.

where $\mathbf{a}_M$ and $\mathbf{a}_C$ represent the lagged system response and the acceleration command, respectively.

To observe the autopilot delay influence on the guidance accuracy, T_m is set to vary from 0 to 0.3 with the TIAC of $-70°$. Comparative results of $T_m = 0.25$ are illustrated in Figure 7. When the autopilot lag is considered, the differences between Figures 6(c) and 7(b) show that the look angle of [17] exceeds the FOV limit at first in the initial phase. The impact angle error (IAE) is described in Figure 8 with the variation of T_m. The IAE of the proposed guidance law in this work is less than that of [28], for each investigated value of T_m. The different trends of the two curves shown in Figure 8 indicate that,

compared with [28], the influence on the proposed method varies not obviously with the increasing of T_m. The IAE of the proposed guidance law can keep in a small value.

5.3. Simulations of Realistic Missile Model with Uncertain Target Velocity. In this subsection, further simulations are carried out with a realistic missile model considering the uncertainty of target's velocity.

Case 4 (realistic model with uncertainty cases). The aerodynamics drag and gravity are taken into account in the realistic

(a) Engagement trajectories

(b) Flight-path angles

(c) Look angles

(d) Acceleration histories

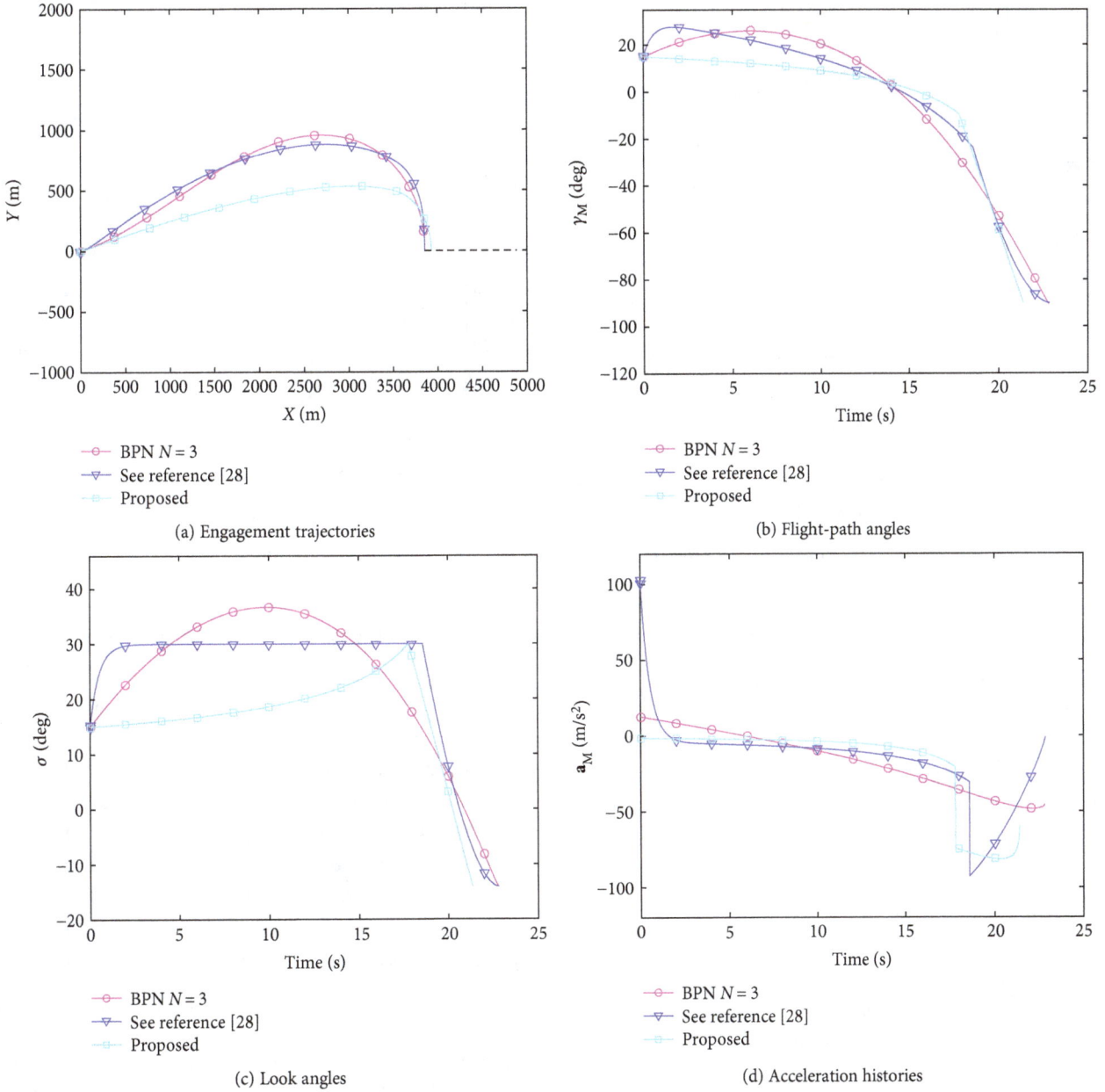

FIGURE 6: The results for the comparative cases with $\gamma_{M_f} = -90°$.

missile model, which is taken from [30]. The nonlinear dynamics equations are given by

$$\dot{x}_M = V_M \cos \gamma_M,$$

$$\dot{y}_M = V_M \sin \gamma_M,$$

$$\dot{V}_M = \frac{T_M - D_M}{m} - g \sin \gamma_M,$$

$$\dot{\gamma}_M = \frac{a_M - g \cos \gamma_M}{V_M}.$$

(32)

Without loss of generality, we assume that $T_M = 0$ in the terminal engagement case. Furthermore, the gravity compensation term should be added to the original acceleration

command, which is based on the five-step procedure proposed in this work, to offset the gravity influence on the flight-path angle rate. For realistic implementation, the original acceleration command needs to be changed as

$$a_M = a_M^* + g \cos \gamma_M,$$

(33)

where a_M^* represents the acceleration command for constant speed missile model. The target's velocity uncertainty is also considered in the simulations. The constant target speed 50 m/s is regarded as a nominal value to design the proposed guidance law; however, the simulations are carried out by the percentage perturbation values which are disturbed by 20% (positive and negative) deviation about the nominal value.

(a) Flight-path angles

(b) Look angles

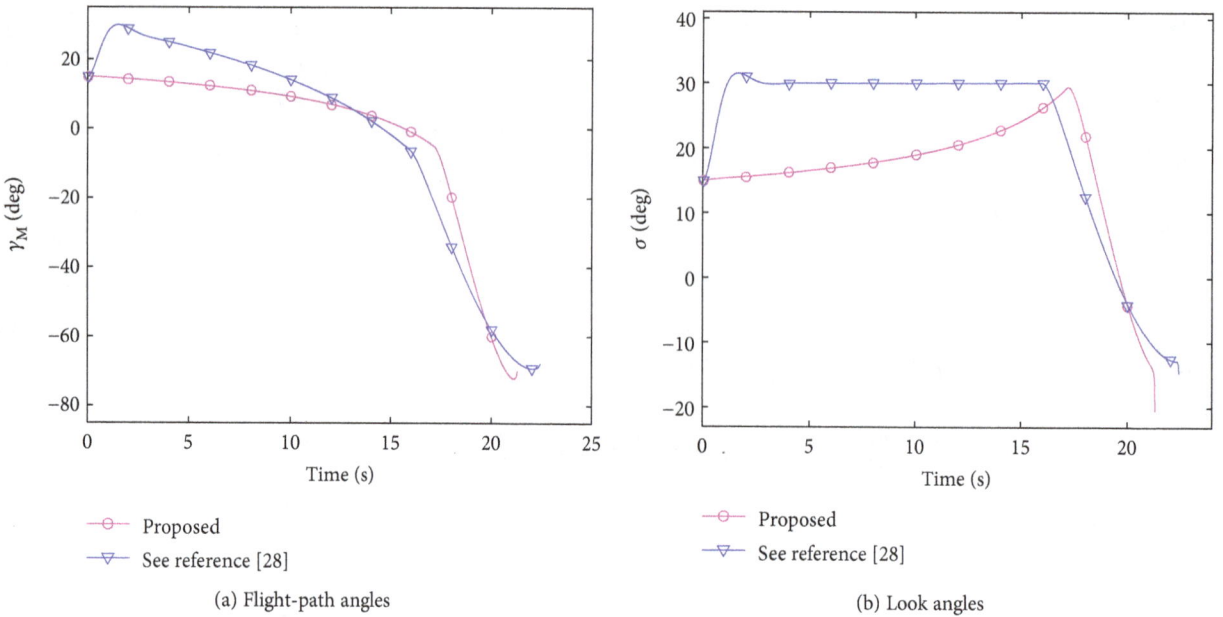

FIGURE 7: Comparative results with system lag $T_m = 0.25$ and $\gamma_{M_f} = -70°$.

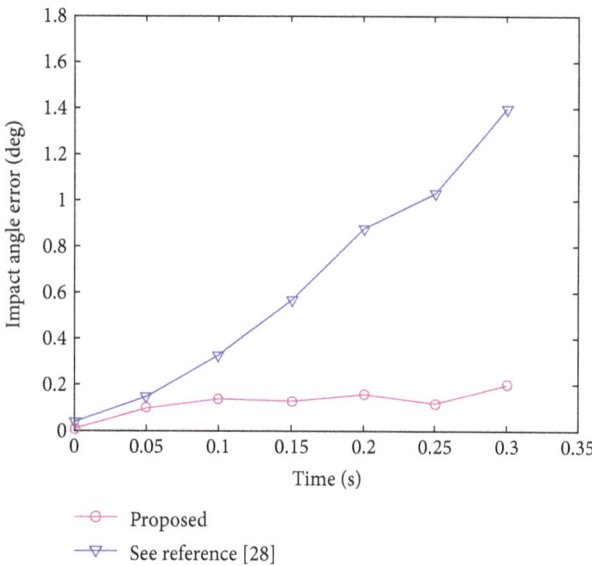

FIGURE 8: Impact angle error with the variation of system lag.

TABLE 3: Angle constraints of different conditions.

	TIAE (deg)	Look angle at switching instant (deg)
$\gamma_f = -50°$ nominal	4.82	27.35
$\gamma_f = -50°$ +20% deviation	0.82	27.08
$\gamma_f = -50°$ −20% deviation	8.68	27.41
$\gamma_f = -70°$ nominal	4.70	23.29
$\gamma_f = -70°$ +20% deviation	−1.24	23.10
$\gamma_f = -70°$ −20% deviation	9.43	20.87
$\gamma_f = -90°$ nominal	4.67	29.13
$\gamma_f = -90°$ +20% deviation	−2.33	29.25
$\gamma_f = -90°$ −20% deviation	8.12	22.74

The detailed simulation results are displayed in Table 3 and Figure 9.

Figure 9(a) shows that the missile can successfully intercept the target with realistic missile model under the condition of uncertain target velocities. In Figure 9(b), the missile velocities vary due to the gravity and air drag. Both the nominal and disturbed target velocity cases are simulated and compared. The main influences on LOS angles, flight-path angles, and look angles are exhibited in Figures 9(c)–9(e), respectively. The detailed numerical results are listed in Table 3 for comparison. Both the realistic decreasing velocity of missile and the uncertain target velocity bring the terminal impact angle errors (TIAE) in engagement simulations. The influence of target's −20% deviation of velocity

is added on that of decreasing missile velocity, which leads to a larger TIAE. On the contrary, the target's +20% deviation of velocity can partly offset the influence of decreasing missile velocity, which can be observed for all the impact angles in Table 3. The decreasing missile velocity and uncertain target velocity also effect the look angles. The look angle at switching instant is not equal to the limit FOV angle but does not exceed the limitation yet. The acceleration histories are plotted in Figure 9(f). It is clear that high impact angle constraint needs large maximum acceleration effect on the missile especially for the realistic missile model.

6. Conclusions

The look angle control strategy introduced in this study is motivated by the physical limit of the missile's seeker. It leads to a convenient design approach similar to the PN guidance law but taking the FOV limitation into consideration. The proposed two-phased scheme admits the allowed look angle

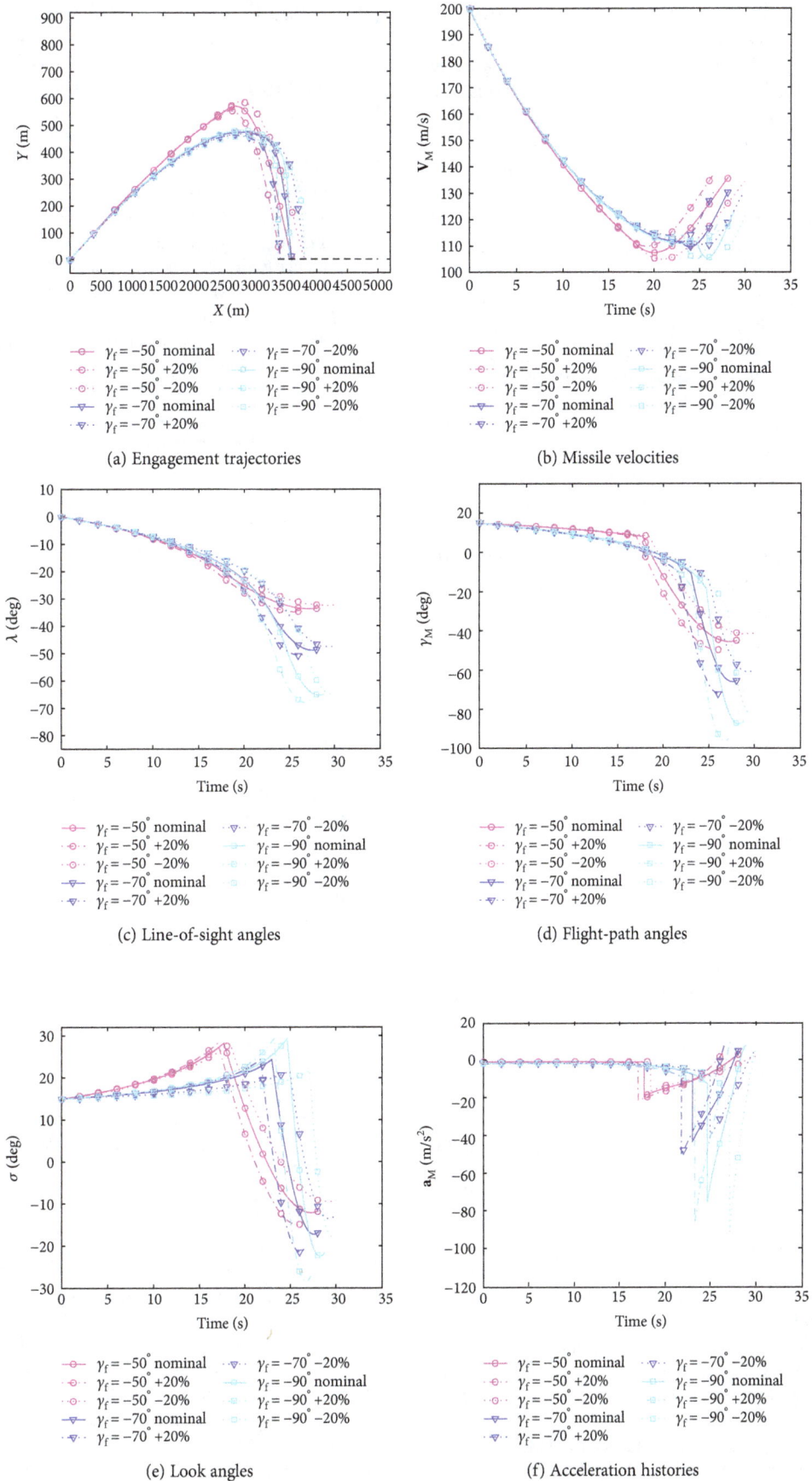

(a) Engagement trajectories

(b) Missile velocities

(c) Line-of-sight angles

(d) Flight-path angles

(e) Look angles

(f) Acceleration histories

FIGURE 9: The results for realistic missile model.

along with the desired impact angle constraint and switches the proportional navigation gain at an appropriate instant in the case for engagement. In this article, to evaluate the specific navigation gains for the initial and final phases, the scaling factor between them is discussed. A quadratic equation regarding to the navigation gain of the initial phase is presented and solved with consideration of the minimum jump amplitude at the switching instant. Actually, the abrupt acceleration change cannot be eliminated completely because of the switching of navigation gain at a certain instant. In spite of the acceleration histories with abrupt change, this proposed method also has a simple and feasible structure for implementation because only the LOS and LOS rate information are required to shape the trajectory and decide when to switch. Simulation results indicate the performance and feasibility of the proposed guidance strategy, in terms of satisfying the terminal impact angle constraint and FOV limit. In addition, under the lag system, the proposed method can also provide a satisfactory guidance performance for TIAC without exceeding the FOV limit. When applying the proposed guidance law, the decreasing missile velocity and uncertainties in the target speed may result in some unavoidable TIAEs compared to the ideal missile and target.

Conflicts of Interest

The authors declare that there is no conflict of interests regarding the publication of this paper.

Acknowledgments

This work is supported by the National Natural Science Foundation of China (Grant no. 11202024).

References

[1] M. Kim and K. V. Grider, "Terminal guidance for impact attitude angle constrained flight trajectories," *IEEE Transactions on Aerospace and Electronic Systems*, vol. AES-9, no. 6, pp. 852–859, 1973.

[2] P. L. Vergez and J. R. Mcclendon, "Optimal control and estimation for strapdown seeker guidance of tactical missiles," *Journal of Guidance, Control, and Dynamics*, vol. 5, no. 3, pp. 225-226, 1982.

[3] R. K. Mehra and R. D. Ehrich, "Air-to-air missile guidance for strapdown seekers," in *The 23rd IEEE Conference on Decision and Control*, pp. 1109–1115, Las Vegas, NV, USA, 2007, IEEE.

[4] A. E. Bryson, *Applied Optimal Control: Optimization, Estimation, and Control*, CRC Press, 1975.

[5] P. Zarchan, *Tactical and Strategic Missile Guidance*, American Institute of Aeronautics & Astronautics Inc, Washington, DC, USA, Sixth edition, 1990.

[6] C.-K. Ryoo, H. Cho, and M.-J. Tahk, "Time-to-go weighted optimal guidance with impact angle constraints," *IEEE Transactions on Control Systems Technology*, vol. 14, no. 3, pp. 483–492, 2006.

[7] C. H. Lee, M. J. Tahk, and J. I. Lee, "Generalized formulation of weighted optimal guidance laws with impact angle constraint," *IEEE Transactions on Aerospace and Electronic Systems*, vol. 49, no. 2, pp. 1317–1322, 2013.

[8] C. H. Lee, T.-H. Kim, M.-J. Tahk, and I.-H. Whang, "Polynomial guidance laws considering terminal impact angle and acceleration constraints," *IEEE Transactions on Aerospace and Electronic Systems*, vol. 49, no. 1, pp. 74–92, 2013.

[9] N. A. Shneydor, *Missile Guidance and Pursuit: Kinematics, Dynamics and Control*, Elsevier, 1998.

[10] B. S. Kim, J. G. Lee, and H. S. Han, "Biased PNG law for impact with angular constraint," *IEEE Transactions on Aerospace and Electronic Systems*, vol. 34, no. 1, pp. 277–288, 1998.

[11] P. Lu, D. B. Doman, and J. D. Schierman, "Adaptive terminal guidance for hypervelocity impact in specified direction," *Journal of Guidance, Control, and Dynamics*, vol. 29, no. 2, pp. 269–278, 2006.

[12] A. Ratnoo and D. Ghose, "Impact angle constrained interception of stationary targets," *Journal of Guidance, Control, and Dynamics*, vol. 31, no. 6, pp. 1817–1822, 2008.

[13] A. Ratnoo and D. Ghose, "Impact angle constrained guidance against nonstationary nonmaneuvering targets," *Journal of Guidance, Control, and Dynamics*, vol. 33, no. 1, pp. 269–275, 2010.

[14] Y. Zhang, M. Sun, and Z. Chen, "Finite-time convergent guidance law with impact angle constraint based on sliding-mode control," *Nonlinear Dynamics*, vol. 70, no. 1, pp. 619–625, 2012.

[15] Y. Zhao, Y. Sheng, and X. Liu, "Sliding mode control based guidance law with impact angle constraint," *Chinese Journal of Aeronautics*, vol. 27, no. 1, pp. 145–152, 2014.

[16] L. Sun, W. Wang, R. Yi, and S. Xiong, "A novel guidance law using fast terminal sliding mode control with impact angle constraints," *ISA Transactions*, vol. 64, pp. 12–23, 2016.

[17] S. Lyu, Z. H. Zhu, S. Tang, and X. Yan, "Prescribed performance slide mode guidance law with terminal line-of-sight angle constraint against maneuvering targets," *Nonlinear Dynamics*, vol. 88, no. 3, pp. 2101–2110, 2017.

[18] B. G. Park, T. H. Kim, and M. J. Tahk, "Optimal impact angle control guidance law considering the seeker's field-of-view limits," *Proceedings of the Institution of Mechanical Engineers, Part G: Journal of Aerospace Engineering*, vol. 227, no. 8, pp. 1347–1364, 2012.

[19] K. S. Erer and O. Merttopçuoglu, "Indirect impact-angle-control against stationary targets using biased pure proportional navigation," *Journal of Guidance, Control, and Dynamics*, vol. 35, no. 2, pp. 700–704, 2012.

[20] T. H. Kim, B. G. Park, and M. J. Tahk, "Bias-shaping method for biased proportional navigation with terminal-angle constraint," *Journal of Guidance, Control, and Dynamics*, vol. 36, no. 6, pp. 1810–1816, 2013.

[21] K. S. Erer, R. Tekin, and M. K. Ozgoren, "Look angle constrained impact angle control based on proportional navigation," in *AIAA Guidance, Navigation, and Control Conference*, pp. 91–97, Kissimmee, FL, USA, 2015.

[22] Z. Yang, H. Wang, and D. Lin, "Time-varying biased proportional guidance with seeker's field-of-view limit," *International Journal of Aerospace Engineering*, vol. 2016, Article ID 9272019, 11 pages, 2016.

[23] R. Tekin and K. S. Erer, "Switched-gain guidance for impact angle control under physical constraints," *Journal of Guidance, Control, and Dynamics*, vol. 38, no. 2, pp. 205–216, 2015.

[24] A. Ratnoo, "Analysis of two-stage proportional navigation with heading constraints," *Journal of Guidance, Control, and Dynamics*, vol. 39, no. 1, pp. 156–164, 2016.

[25] Q. Wen, Q. Xia, and S. Weixia, "A parameter design strategy for seeker's field-of-view constraint in impact angle guidance," *Proceedings of the Institution of Mechanical Engineers, Part G: Journal of Aerospace Engineering*, vol. 229, no. 13, pp. 2389–2396, 2015.

[26] S. He and D. Lin, "A robust impact angle constraint guidance law with seeker's field-of-view limit," *Transactions of the Institute of Measurement and Control*, vol. 37, no. 3, pp. 317–328, 2015.

[27] X. Wang, Y. Zhang, and H. Wu, "Sliding mode control based impact angle control guidance considering the seeker's field-of-view constraint," *ISA Transactions*, vol. 61, pp. 49–59, 2016.

[28] B. G. Park, H.-H. Kwon, Y.-H. Kim, and T.-H. Kim, "Composite guidance scheme for impact angle control against a nonmaneuvering moving target," *Journal of Guidance, Control, and Dynamics*, vol. 39, no. 5, pp. 1132–1139, 2016.

[29] B. G. Park, T. H. Kim, and M. J. Tahk, "Biased PNG with terminal-angle constraint for intercepting nonmaneuvering targets under physical constraints," *IEEE Transactions on Aerospace and Electronic Systems*, vol. 53, no. 3, pp. 1562–1572, 2017.

[30] P. Kee, D. Li, and S. Chai, "Near optimal midcourse guidance law for flight vehicle," in *36th AIAA Aerospace Sciences Meeting and Exhibit*, pp. 583–590, Reno, NV, USA, 2006.

16

A Bayesian Classifier for X-Ray Pulsars Recognition

Hao Liang, Yafeng Zhan, and Chaowei Duan

School of Aerospace Engineering and Space Centre, Tsinghua University, Beijing 100084, China

Correspondence should be addressed to Yafeng Zhan; zhanyf@tsinghua.edu.cn

Academic Editor: Paul Williams

Recognition for X-ray pulsars is important for the problem of spacecraft's attitude determination by X-ray Pulsar Navigation (XPNAV). By using the nonhomogeneous Poisson model of the received photons and the minimum recognition error criterion, a classifier based on the Bayesian theorem is proposed. For X-ray pulsars recognition with unknown Doppler frequency and initial phase, the features of every X-ray pulsar are extracted and the unknown parameters are estimated using the Maximum Likelihood (ML) method. Besides that, a method to recognize unknown X-ray pulsars or X-ray disturbances is proposed. Simulation results certificate the validity of the proposed Bayesian classifier.

1. Introduction

PULSARS have been recognized as nature's "most stable clock" [1, 2] and could emit ideal period pulse signals in multielectromagnetic bands [3]. Since Chester and Butman firstly proposed using X-ray pulsars for navigation [4], many scholars focus on this completely autonomous celestial navigation method [3, 5–8]. Recognition for X-ray pulsars is one of the key technologies in the X-ray Pulsar Navigation (XPNAV) system. When XPNAV system is failure or restarted, the spacecraft rolls over and the attitude of the spacecraft is unknown. At this time, the detector receives one X-ray pulsar signal in a very short time. If we recognize which X-ray pulsar emits the signal, we can extract the observed pulsar's direction vector from the XPNAV Database [3] to speed up the spacecraft's attitude redetermination [6].

At present, most literatures proposed to recovery the X-ray pulsar's profile [9] and use transforming method to extract the features of profile for pulsars recognition, such as the Selected Line Spectra Transform [10, 11], the Bispectra-Mellin Transform [12], the S Transform (ST) [13, 14], and the Wavelet Transform [15]. Besides that, the neural network has been used for pulsars recognition at the Pulsar Arecibo L-band Feed Array (PALFA) [16]. The major disadvantage of the above classifiers is that they have to recovery the X-ray pulsar's profile. As the X-ray pulsar is always far away from the solar system (thousands of light years), the received X-ray pulsar signal is very weak. Therefore, unless the observation

time for X-ray pulsars could last hundreds of seconds, it is too hard to build the X-ray pulsar's profile well. In consequence, the existing classifiers are inefficient and even useless for the scenario with short observation time.

XPNAV could only use the well-known X-ray pulsars for navigation. The XPNAV Database stores each alternative X-ray pulsar's prior knowledge, which includes the X-ray pulsar's direction vector, the X-ray background noise's rate, and the X-ray pulsar signal's flux, period, and standard profile [3]. Based on the XPNAV Database, the X-ray pulsar signal photons' Joint Probability Density Function (JPDF) could be built [17]. By using the JPDF, we propose a Bayesian classifier for X-ray pulsars recognition with the minimum recognition error criterion [18]. Unlike the traditional transforming method, the proposed Bayesian classifier could extract the features of pulsars without recovering the profile. In the condition of short observation time, the proposed Bayesian classifier shows high recognition efficiency.

This paper is organized as follows: Section 2 introduces the X-ray pulsar signal model; Section 3 explains the proposed Bayesian classifier in detail; Computer simulations for proving the classifier's validity are shown in Section 4, and some conclusions are given in the final part.

2. X-Ray Pulsar Signal Model

Unlike the radio signal, the X-ray pulsar signal shows the particle property. The X-ray detector onboard aims at the

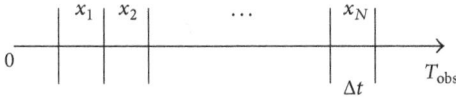

FIGURE 1: The segments of the observation time.

X-ray pulsar and records the arriving time of photons. The number of received photons is a stochastic process which obeys the nonhomogeneous Poisson process [17]; the Probability Density Function (PDF) is shown in

$$P\left(N_{T_{obs}} = k\right) = \frac{\left(\int_0^{T_{obs}} \lambda(t)\, dt\right)^k \exp\left(-\int_0^{T_{obs}} \lambda(t)\, dt\right)}{k!}, \quad (1)$$

where T_{obs} is the observation time, $N_{T_{obs}}$ is the number of received photons, and $\lambda(t)$ is the arriving rate of the photons. $\lambda(t)$ is composed of the X-ray pulsar signal and the X-ray background noise:

$$\lambda(t) = \lambda_b + \lambda_s h\left(\theta(t)\right), \quad (2)$$

where λ_b is the X-ray background noise's rate which is almost time-invariant in the cosmos environment, λ_s is the X-ray pulsar signal's flux, $h(\theta(t))$ is the X-ray pulsar signal's normalization standard profile, and $\theta(t)$ is the X-ray pulsar signal's phase. As the X-ray pulsar signal is periodic, $h(\theta(t) + T) = h(\theta(t))$, where T is the period.

3. Bayesian Classifier for X-Ray Pulsar Recognition

3.1. Classifier Description. Divide the observation time into N segments, which is shown in Figure 1. Each segment time is Δt. If Δt is short enough, the X-ray photons' arriving rate in the nth segment $\lambda_n(\theta)$ could be expressed as [17]

$$\lambda_n(\theta) \approx \frac{1}{\Delta t} \int_{t_0+(n-1)\Delta t}^{t_0+n\Delta t} \lambda(t; \theta)\, dt. \quad (3)$$

The number of photons in the nth segment x_n is a random variable which obeys the Poisson distribution. The PDF of x_n could be expressed as

$$p(x_n = k; \theta) = \frac{\left(\lambda_n(\theta)\Delta t\right)^k}{k!} \exp\left(-\lambda_n(\theta)\Delta t\right). \quad (4)$$

As the random variables in each segment is independent, the Joint Probability Density Function (JPDF) could be expressed as

$$p(\mathbf{X}; \theta) = \prod_{n=1}^{N} p(x_n; \theta)$$
$$= \prod_{n=1}^{N} \frac{\left(\lambda_n(\theta)\Delta t\right)^{x_n}}{x_n!} \exp\left(-\lambda_n(\theta)\Delta t\right), \quad (5)$$

where $\mathbf{X} = [x_1 \; x_2 \; \cdots \; x_N]^{\mathrm{T}}$.

In order to simplify the analysis, we suppose recognizing two X-ray pulsars. The JPDF of these X-ray pulsar signals could be, respectively, expressed as

$$H_1 : p_1(\mathbf{X}; \theta_n)$$
$$= \prod_{n=1}^{N} \frac{\left(\left(\lambda_b + \lambda_s^1 h_1(\theta_n)\right)\Delta t\right)^{x_n}}{x_n!} \exp\left(-\left(\lambda_b + \lambda_s^1 h_1(\theta_n)\right)\Delta t\right),$$

$$H_2 : p_2(\mathbf{X}; \theta_n)$$
$$= \prod_{n=1}^{N} \frac{\left(\left(\lambda_b + \lambda_s^2 h_2(\theta_n)\right)\Delta t\right)^{x_n}}{x_n!} \exp\left(-\left(\lambda_b + \lambda_s^2 h_2(\theta_n)\right)\Delta t\right). \quad (6)$$

Based on the Bayesian theory, the probability of the error recognition is [18]

$$p_e = p(H_1 \mid H_2)\, p(H_2) + p(H_2 \mid H_1)\, p(H_1)$$
$$= p(H_2) \int_{R_1} p(\mathbf{X} \mid H_2)\, d\mathbf{X}$$
$$\quad + p(H_1) \int_{R_2} p(\mathbf{X} \mid H_1)\, d\mathbf{X}$$
$$= p(H_2) \int_{R_1} p(\mathbf{X} \mid H_2)\, d\mathbf{X} \quad (7)$$
$$\quad + p(H_1)\left(1 - \int_{R_1} p(\mathbf{X} \mid H_1)\, d\mathbf{X}\right)$$
$$= p(H_1)$$
$$\quad + \int_{R_1} \left\{p(H_2)\, p(\mathbf{X} \mid H_2) - p(H_1)\, p(\mathbf{X} \mid H_1)\right\} d\mathbf{X},$$

where R_1 is the decision region of H_1 and R_2 is the decision region of H_2.

Our goal is to minimize p_e. Then, we decide that $\mathbf{X}$ belong to H_1, if and only if

$$p(H_2)\, p(\mathbf{X} \mid H_2) - p(H_1)\, p(\mathbf{X} \mid H_1) < 0. \quad (8)$$

Assume $p(H_1) = p(H_2)$ and take logarithm to the two sides of inequality (8). Then, we could get the decision criterion:

$$\ln \frac{p(\mathbf{X} \mid H_1)}{p(\mathbf{X} \mid H_2)} > 0, \quad \mathbf{X} \in H_1,$$
$$\ln \frac{p(\mathbf{X} \mid H_1)}{p(\mathbf{X} \mid H_2)} \leq 0, \quad \mathbf{X} \in H_2. \quad (9)$$

Taking (6) into the proposed decision criterion (9),

$$\ln \frac{p_1(\mathbf{X}; \theta_n)}{p_2(\mathbf{X}; \theta_n)} = \sum_{n=1}^{N} \left\{x_n \left[\ln\left(\lambda_b + \lambda_s^1 h_1(\theta_n)\right)\right.\right.$$
$$\left.\left. - \ln\left(\lambda_b + \lambda_s^2 h_2(\theta_n)\right)\right]\right\} - \sum_{n=1}^{N} \left\{\left[\lambda_s^1 h_1(\theta_n)\right.\right. \quad (10)$$
$$\left.\left. - \lambda_s^2 h_2(\theta_n)\right] \Delta t\right\}.$$

In (10), $\sum_{n=1}^{N} \{[\lambda_s^1 h_1(\theta_n) - \lambda_s^2 h_2(\theta_n)]\Delta t\}$ is constant in the condition of a fixed observation time. Assume $-\sum_{n=1}^{N} \{[\lambda_s^1 h_1(\theta_n) - \lambda_s^2 h_2(\theta_n)]\Delta t\} = W_0$, $\ln(\lambda_b + \lambda_s^1 h_1(\theta_n)) - \ln(\lambda_b + \lambda_s^2 h_2(\theta_n)) = w_n$, and $\mathbf{W} = [w_1 \; w_2 \; \cdots \; w_N]^{\mathrm{T}}$, and then the classifier using the Bayesian minimum recognition error criterion for X-ray pulsar recognition could be expressed as

$$g(\mathbf{X}) = \mathbf{W}^{\mathrm{T}}\mathbf{X} + W_0. \tag{11}$$

From (11), $g(\mathbf{X}) = 0$ determines a decision surface, and the surface is a hyperplane. The hyperplane divides the hyperspace into two parts, where $R_1 = \{g(\mathbf{X}) \geq 0\}$ and $R_2 = \{g(\mathbf{X}) < 0\}$.

3.2. Performance Estimation. The probability of the error recognition is shown in (7). In order to estimate p_e, we assume $p(H_1) = p(H_2) = 0.5$, without generality, and calculate $\int_{R_1} p(\mathbf{X} \mid H_2)d\mathbf{X}$ and $\int_{R_2} p(\mathbf{X} \mid H_1)d\mathbf{X}$.

As x_n obeys the Poisson distribution with the arriving rate $\lambda_n(\theta)\Delta t$, the PDF of x_n is shown in (4). Then, the average of x_n is $\varepsilon_n = E(x_n) = \lambda_n(\theta)\Delta t$ and the variance of x_n is $\sigma_n^2 = \mathrm{var}(x_n) = \lambda_n(\theta)\Delta t$. Each element of $\mathbf{X}$ is independent; based on the central limit theorem, $g(\mathbf{X})$ obeys the Gaussian distribution with the average $\varepsilon_{g(\mathbf{X})}$ and variance $\sigma_{g(\mathbf{X})}^2$. The PDF of $g(\mathbf{X})$ could be expressed as

$$P(g(\mathbf{X})) = \frac{1}{\sqrt{2\pi}\sigma_{g(\mathbf{X})}} \exp\left(-\frac{(g(\mathbf{X}) - \varepsilon_{g(\mathbf{X})})^2}{2\sigma_{g(\mathbf{X})}^2}\right),$$

$$\varepsilon_{g(\mathbf{X})} = E(g(\mathbf{X})) = E\left(\mathbf{W}^{\mathrm{T}}\mathbf{X} + W_0\right)$$

$$= \sum_{n=1}^{N} w_n \varepsilon_n + W_0, \tag{12}$$

$$\sigma_{g(\mathbf{X})}^2 = \mathrm{var}(g(\mathbf{X})) = \mathrm{var}\left(\mathbf{W}^{\mathrm{T}}\mathbf{X} + W_0\right)$$

$$= \sum_{n=1}^{N} w_n^2 \sigma_n^2;$$

then,

$$\int_{R_2} p(\mathbf{X} \mid H_1)d\mathbf{X} = P(g(\mathbf{X} \mid H_1) < 0)$$

$$= \int_{-\infty}^{0} \frac{1}{\sqrt{2\pi}\sigma_{g(\mathbf{X}|H_1)}} \exp\left(-\frac{(g(\mathbf{X} \mid H_1) - \varepsilon_{g(\mathbf{X}|H_1)})^2}{2\sigma_{g(\mathbf{X}|H_1)}^2}\right)dg \tag{13}$$

$$\cdot(\mathbf{X} \mid H_1) = Q\left(\frac{g(\mathbf{X} \mid H_1) - \varepsilon_{g(\mathbf{X}|H_1)}}{\sigma_{g(\mathbf{X}|H_1)}}\right),$$

where $Q(\cdot)$ is the q-function, $Q(x) = \int_x^{+\infty}(1/\sqrt{2\pi})\exp(-(1/2)t^2)dt$.

The recognition probability of H_1 could be expressed as

$$P_c^1 = 1 - \int_{R_2} p(\mathbf{X} \mid H_1)d\mathbf{X}$$

$$= 1 - Q\left(\frac{g(\mathbf{X} \mid H_1) - \varepsilon_{g(\mathbf{X}|H_1)}}{\sigma_{g(\mathbf{X}|H_1)}}\right). \tag{14}$$

Similarly, the recognition probability of H_2 is

$$P_c^2 = 1 - \int_{R_1} p(\mathbf{X} \mid H_2)d\mathbf{X}$$

$$= 1 - Q\left(\frac{g(\mathbf{X} \mid H_2) - \varepsilon_{g(\mathbf{X}|H_2)}}{\sigma_{g(\mathbf{X}|H_2)}}\right). \tag{15}$$

3.3. Recognition for Multiple X-Ray Pulsars. From (11), two X-ray pulsars determine a decision surface. We expand the number of X-ray pulsars to K; the number of decision surfaces is the combination C_K^2. To decrease the complexity of the Bayesian classifier, we deform (10) and calculate the features of each X-ray pulsar:

$$\ln \frac{p_1(\mathbf{X}; \theta_n)}{p_2(\mathbf{X}; \theta_n)}$$

$$= \sum_{n=1}^{N} \left\{x_n \ln\left(\lambda_b + \lambda_s^1 h_1(\theta_n)\right) - \lambda_s^1 h_1(\theta_n)\Delta t\right\} \tag{16}$$

$$- \sum_{n=1}^{N} \left\{x_n \ln\left(\lambda_b + \lambda_s^2 h_2(\theta_n)\right) - \lambda_s^2 h_2(\theta_n)\Delta t\right\}.$$

The first item of (16) is only with a matter of the first X-ray pulsar; the second item of (16) is only with a matter of the second X-ray pulsar. Then, we could get the features of the *i*th X-ray pulsar:

$$J_i(\mathbf{X}) = \sum_{n=1}^{N} \left\{x_n \ln\left(\lambda_b + \lambda_s^i h_i(\theta_n)\right) - \lambda_s^i h_i(\theta_n)\Delta t\right\}. \tag{17}$$

By calculating $J(\mathbf{X})$ of every X-ray pulsar, the determined X-ray pulsar is the one with the $\max(J_i(\mathbf{X}))$, where $\max(\cdot)$ is the maximum function.

3.4. Classifier with Unknown Parameters. As the spacecraft's position and the velocity are unknown in XPNAV, $\theta(t)$ in (2) would be influenced by the Doppler frequency f_d and the initial phase ϕ_0:

$$\theta(t) = (f_0 + f_d) \cdot t + \phi_0,$$

$$f_d = \frac{\mathbf{v} \cdot \mathbf{r}}{c} f_0, \tag{18}$$

where f_0 is the frequency of the X-ray pulsar, $\mathbf{v}$ is the spacecraft's velocity vector, $\mathbf{r}$ is the X-ray pulsar's direction vector, and c is the light velocity. Before recognition, we have to evaluate f_d and ϕ_0 first. In this paper, we use the Maximum Likelihood (ML) estimation to deal with this

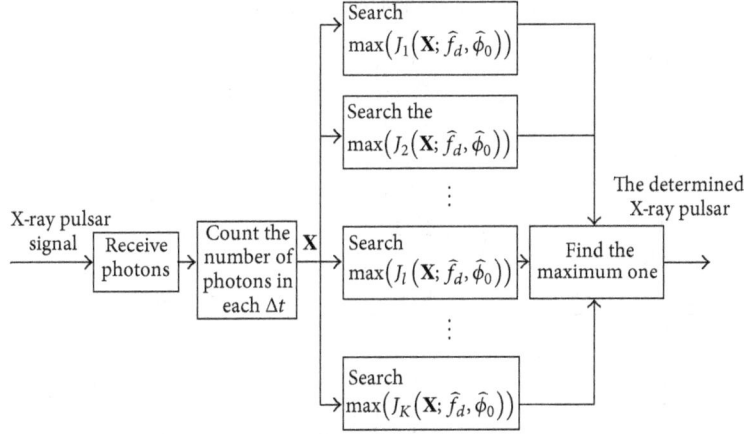

FIGURE 2: The whole recognition process for multiple X-ray pulsars with unknown parameters Bayesian classifier.

problem, because ML is asymptotic-efficient to the Cramer-Rao Low Bound (CRLB) [17, 19].

Define $\max(J_i(\mathbf{X}; \widehat{f}_d, \widehat{\phi}_0))$ as the maximum of $J_i(\mathbf{X}; \widehat{f}_d, \widehat{\phi}_0)$, and $\max(J_i(\mathbf{X}; \widehat{f}_d, \widehat{\phi}_0))$ is selected as the features of the ith X-ray pulsar for comparison, where $\widehat{f}_d$ is the evaluation of f_d and $\widehat{\phi}_0$ is the evaluation of ϕ_0. We do the two-dimensional grid searches for f_d and ϕ_0 to find the maximum value. The whole recognition process for multiple X-ray pulsars with unknown parameters Bayesian classifier is shown in Figure 2.

From (17), ignoring the computation burden for the part in $\{\cdot\}$, we need N times additions for calculating the features of one X-ray pulsar. For K X-ray pulsars, the computation is $K \times N$ times additions. If the searching interval of f_d and ϕ_0 is divided into G grids. the whole number of additions is $G \times K \times N$.

3.5. Recognition for Unknown X-Ray Pulsars. There are hundreds of X-ray pulsars in universe. The XPNAV Database may only store some regular used X-ray pulsars' prior information. Therefore, we have to find a way to identify whether the received X-ray signal comes from the X-ray pulsars stored in the XPNAV Database or not.

From (17), $\{x_n \ln(\lambda_b + \lambda_s^i h_i(\theta_n)) - \lambda_s^i h_i(\theta_n)\Delta t\}$ obeys the Poisson distribution. When N is large enough, $J_i(\mathbf{X})$ obeys the Gaussian distribution with the PDF:

$$P\left(J_i(\mathbf{X})\right) = \frac{1}{\sqrt{2\pi}\sigma_{J_i(\mathbf{X})}} \exp\left(-\frac{\left(J_i(\mathbf{X}) - \varepsilon_{J_i(\mathbf{X})}\right)^2}{2\sigma_{J_i(\mathbf{X})}^2}\right), \quad (19)$$

$$\varepsilon_{J_i(\mathbf{X})} = E\left(J_i(\mathbf{X})\right)$$

$$= \sum_{n=1}^{N} \left\{\left(\lambda_b + \lambda_s^i h_i(\theta_n)\right) \ln\left(\lambda_b + \lambda_s^i h_i(\theta_n)\right) \quad (20)\right.$$

$$\left. - \lambda_s^i h_i(\theta_n)\Delta t\right\},$$

$$\sigma_{J_i(\mathbf{X})}^2 = \text{var}\left(J_i(\mathbf{X})\right) = \text{var}\left(\mathbf{W}^{\mathrm{T}}\mathbf{X} + W_0\right)$$

$$= \sum_{n=1}^{N} \left\{\left(\lambda_b + \lambda_s^i h_i(\theta_n)\right)\left[\ln\left(\lambda_b + \lambda_s^i h_i(\theta_n)\right)\right]^2\right\}. \quad (21)$$

We choose $(\varepsilon_{J_i(\mathbf{X})} - 3\sigma_{J_i(\mathbf{X})}, \varepsilon_{J_i(\mathbf{X})} + 3\sigma_{J_i(\mathbf{X})})$ as the detection range of $J_i(\mathbf{X})$, as $P(\varepsilon_{J_i(\mathbf{X})} - 3\sigma_{J_i(\mathbf{X})} < J_i(\mathbf{X}) \leq \varepsilon_{J_i(\mathbf{X})} + 3\sigma_{J_i(\mathbf{X})}) \approx$ 99.7%. If $J_i(\mathbf{X})$ is out of $(\varepsilon_{J_i(\mathbf{X})} - 3\sigma_{J_i(\mathbf{X})}, \varepsilon_{J_i(\mathbf{X})} + 3\sigma_{J_i(\mathbf{X})})$, we judge that the received X-ray signal comes from other X-ray pulsars which are not included in the XPNAV Database or some X-ray background noise in space.

4. Computer Simulations

4.1. Simulation for Two X-Ray Pulsars. Six X-ray pulsars from [7] are provided for Computer simulations. The parameters are shown in Table 1 [3, 7], and the normalization standard profiles are shown in Figure 3 [7, 20].

In order to certify the theory in Sections 3.1 and 3.2, we simulate the Bayesian classifier for two X-ray pulsars' (B1821-24, B1937+21) recognition. The time segment Δt is set as 1 ms for simulation simplicity. In the condition of increasing observation time, we compare the recognition probability estimation from (14) with the 100 times Monte Carlo simulations. The photons generating algorithm refers to [17]. Simulation results are shown in Figures 4 and 5.

From Figures 4 and 5, the recognition probabilities increase with the increasing observation time. Besides that, the results of the Monte Carlo simulations are in accordance with the theory performance estimation from (14).

4.2. Simulation for Multiple X-Ray Pulsars. We simulate the Bayesian classifier for six X-ray pulsars' recognition with unknown parameters and without unknown parameters. The initial phase ϕ_0 is set as a random number obeying the uniform distribution with the range [0, 1). The Doppler frequency f_d obeys the uniform distribution with the range $(-0.0023, 0.0023)$ Hz. The other simulation conditions are the same as those in Section 4.1. In the condition of increasing

TABLE 1: The parameters of X-ray pulsars.

ID	Name	Period (sec)	Galactic longitude (degree)	Galactic latitude (degree)	λ_b (cnts/s)	λ_s (cnts/s)
1	B0531+21	0.0335	184.56	−5.78	1540	15400
2	B0540-69	0.0504	279.72	−31.52	5.15	51.5
3	B0833-45	0.0893	263.55	−2.79	1.59	15.9
4	B1509-58	0.1502	320.32	−1.16	16.2	162
5	B1821-24	0.0031	7.8	−5.58	1	1.93
6	B1937+21	0.0016	57.51	−0.29	1	0.499

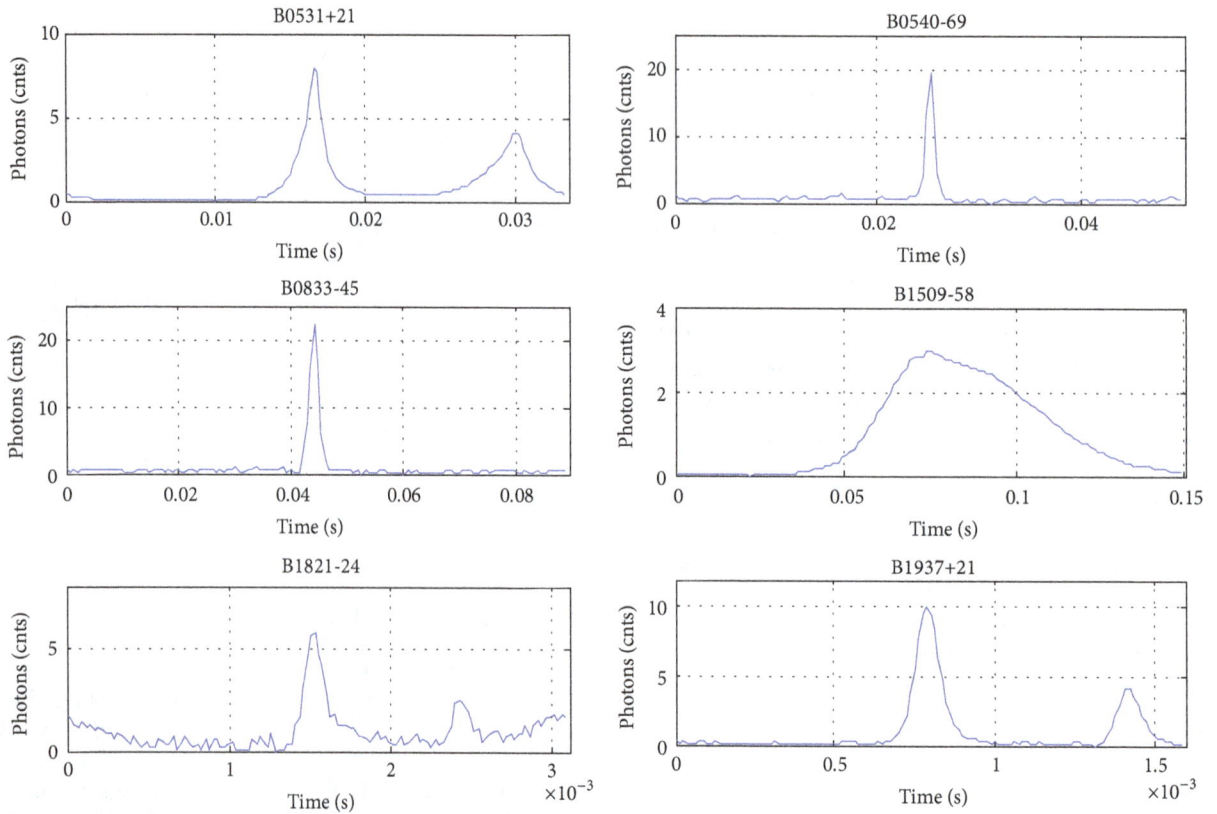

FIGURE 3: The normalization standard profile of six X-ray pulsars.

observation time, we simulate the recognition process presented in Figure 2. The searching range of ϕ_0 is [0, 1), and the searching range of f_d is (−0.0023, 0.0023) Hz. The searching step of ϕ_0 is 0.02, and the searching step of f_d is $4.6e − 4$ Hz. The number of the Monte Carlo simulations is 100 times.

Besides that, as the ST method has both higher recognition rate and fast processing speed compared to the Bispectra-Mellin Transform and the Wavelet Transform, we simulate the ST method for comparison [14]. The ST of the four X-ray pulsars' normalization standard profiles are shown in Figure 6. The detailed steps of the ST for X-ray pulsars recognition refer to [13, 14]. The simulations for ST method ignore the impact of the unknown parameters for simplicity.

The recognition probability of each X-ray pulsar is shown in Figures 7–12.

From Figures 7–12, the recognition probabilities of six X-ray pulsars utilizing the Bayesian classifier increase with the increasing observation time. Compared with the ST classifier, the Bayesian classifier obviously shows higher recognition probability, which certificates the effectiveness of the proposed Bayesian classifier. However, compared to the Bayesian classifier without unknown parameters, the one with unknown parameters shows lower recognition probabilities because of the limited estimation accuracy for ϕ_0 and f_d.

The first four X-ray pulsars' (B0531+21, B0540-69, B0833-45, B1509-58) recognition probabilities could reach 100% in no more than 1 s, and the last two X-ray pulsars (B1821-24, B1937+21) need tens of seconds to get the same effect because of their lower flux. Therefore, we may select X-ray pulsars in

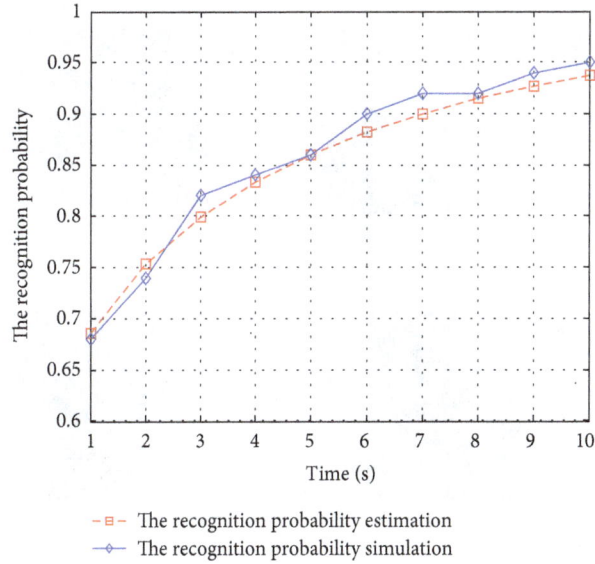

FIGURE 4: The recognition probability (two pulsars) for B1821-24.

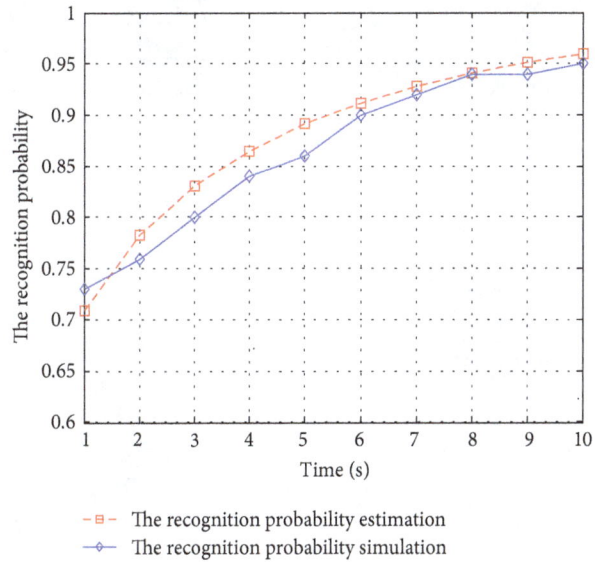

FIGURE 5: The recognition probability (two pulsars) for B1937+21.

the XPNAV Database with high flux for recognition and treat the X-ray pulsars with low flux as background noise.

4.3. Simulation for Unknown X-Ray Pulsars. We simulate the Bayesian classifier to recognize unknown X-ray pulsars. Two weak X-ray pulsars (B1821-24, B1937+21) are selected as the unknown X-ray pulsars. Besides that, the signal only with background noise is also included in the simulation. We calculate $J_i(\mathbf{X})$, $\varepsilon_{J_i(\mathbf{X})}$, and $\sigma^2_{J_i(\mathbf{X})}$ of the remaining four X-ray pulsars in Table 1. $J_i(\mathbf{X})$ is calculated from (17), $\varepsilon_{J_i(\mathbf{X})}$ is calculated from (20), and $\sigma^2_{J_i(\mathbf{X})}$ is calculated from (21). If $J_i(\mathbf{X})$ is out of $(\varepsilon_{J_i(\mathbf{X})} - 3\sigma_{J_i(\mathbf{X})}, \varepsilon_{J_i(\mathbf{X})} + 3\sigma_{J_i(\mathbf{X})})$, we judge that

the received X-ray signal comes from other X-ray pulsars or X-ray disturbances. The other simulation conditions are the same as those in Section 4.1. The number of the Monte Carlo simulations is 1000.

The recognition probabilities of unknown X-ray pulsars are shown in Figures 13–15.

From Figures 13–15, the recognition probabilities for unknown X-ray pulsars or disturbance increase with the increasing observation time, and the recognition probabilities for two unknown X-ray pulsars and background noise could reach 100% in about 1.2 s. Comparing Figures 7–10 with 13–15, 1.2 s is long enough for the first four X-ray pulsars' recognition process.

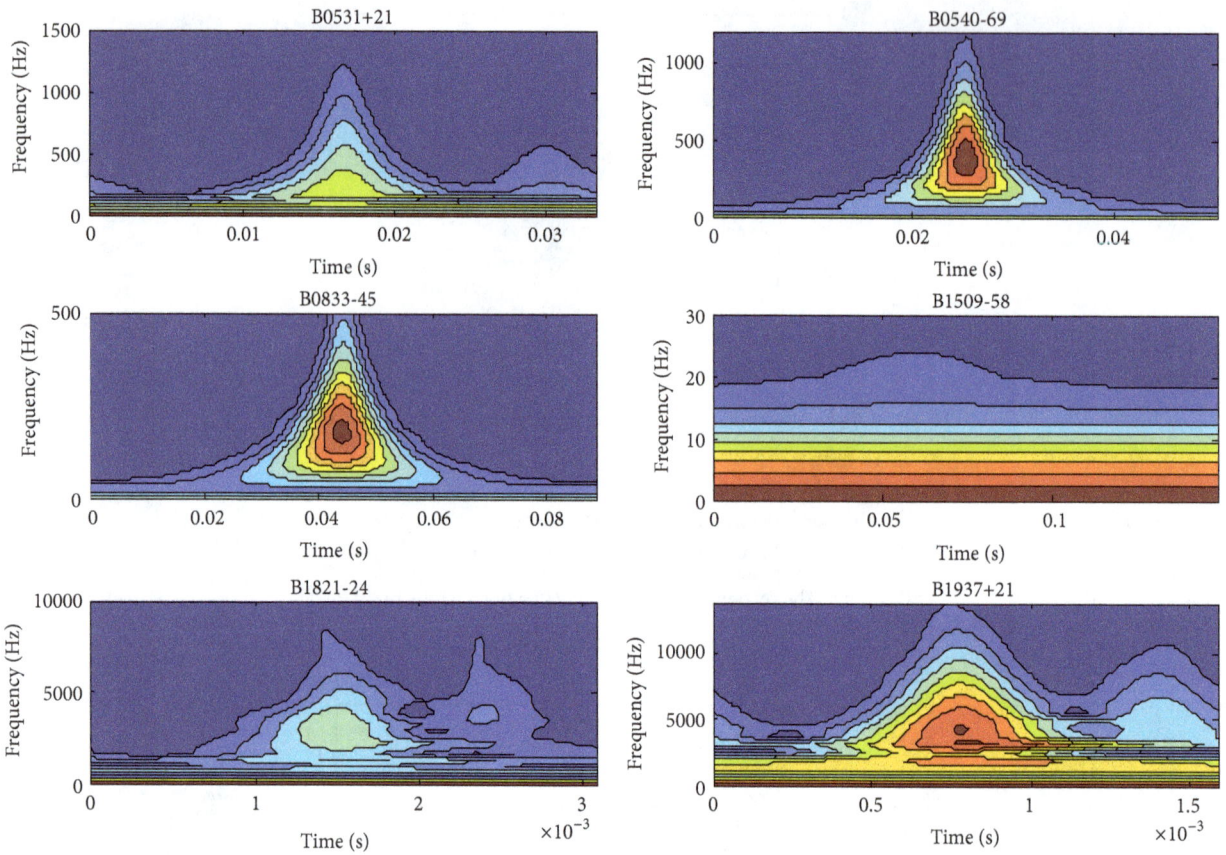

FIGURE 6: The S Transform for six X-ray pulsars.

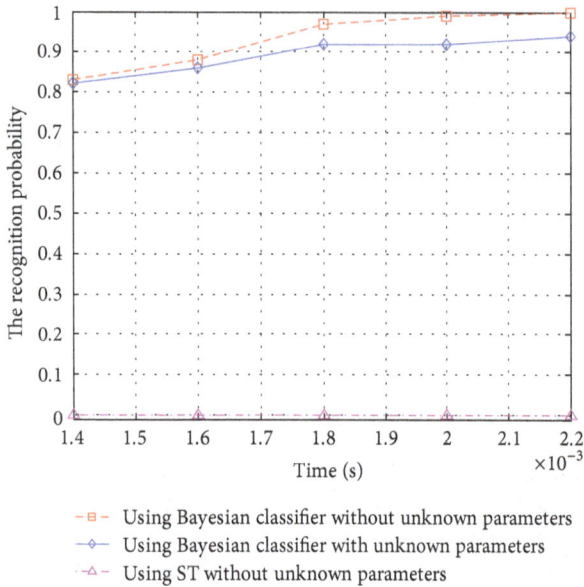

- ▫ - Using Bayesian classifier without unknown parameters
- ◇ - Using Bayesian classifier with unknown parameters
- △ - Using ST without unknown parameters

FIGURE 7: The recognition probability for B0531+21.

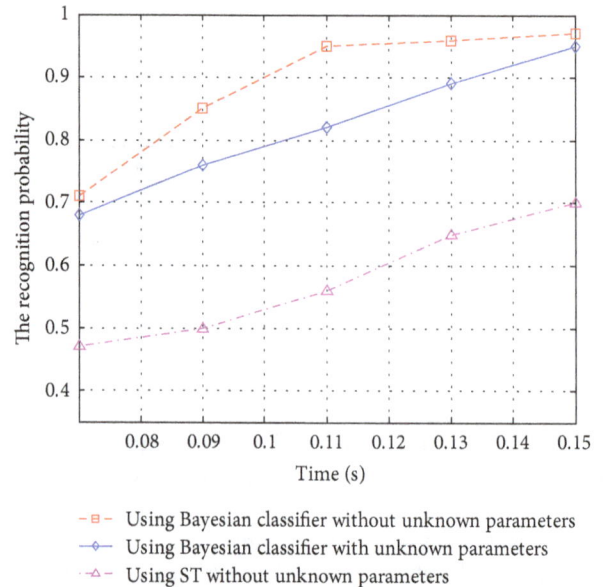

- ▫ - Using Bayesian classifier without unknown parameters
- ◇ - Using Bayesian classifier with unknown parameters
- △ - Using ST without unknown parameters

FIGURE 8: The recognition probability for B1509-58.

5. Conclusion

Recognition for X-ray pulsars is important in XPNAV system, especially for spacecraft's attitude determination. In this paper, we propose a decision surface of every two X-ray pulsars' recognition based on the Bayesian theory firstly and derive the recognition performance. Simulation results show that the theory performance estimations are in accordance

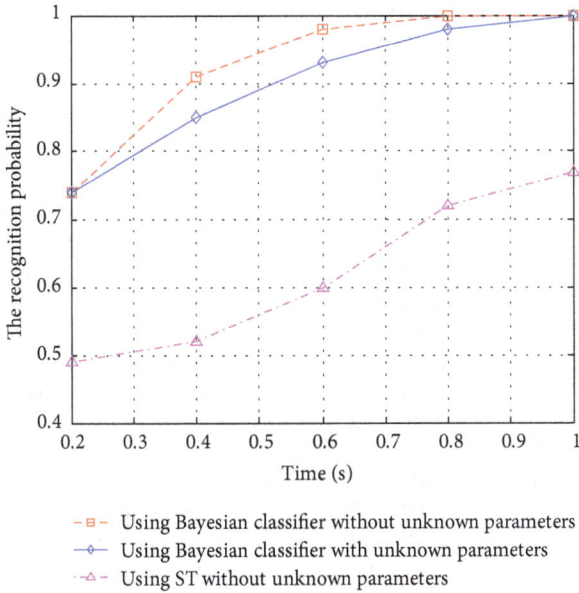

FIGURE 9: The recognition probability for B0833-45.

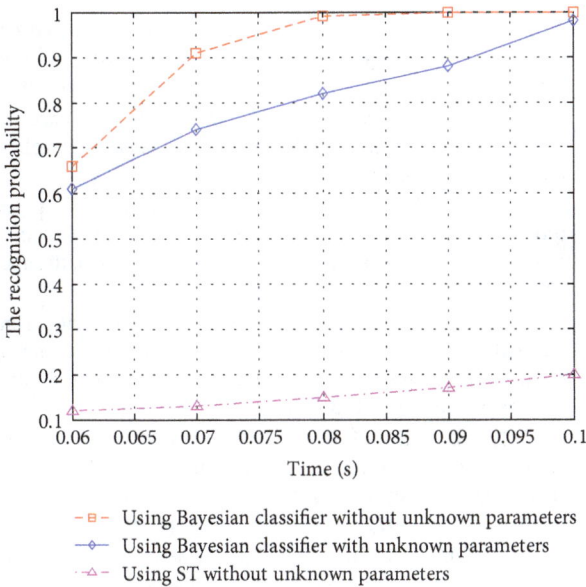

FIGURE 11: The recognition probability for B1821-34.

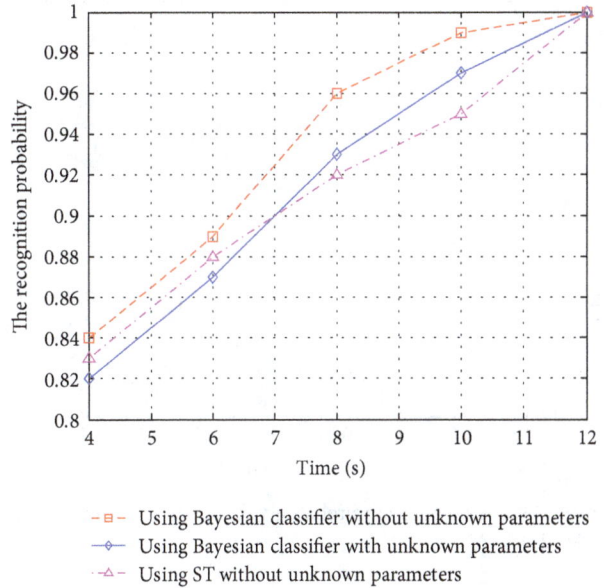

FIGURE 10: The recognition probability for B1509-58.

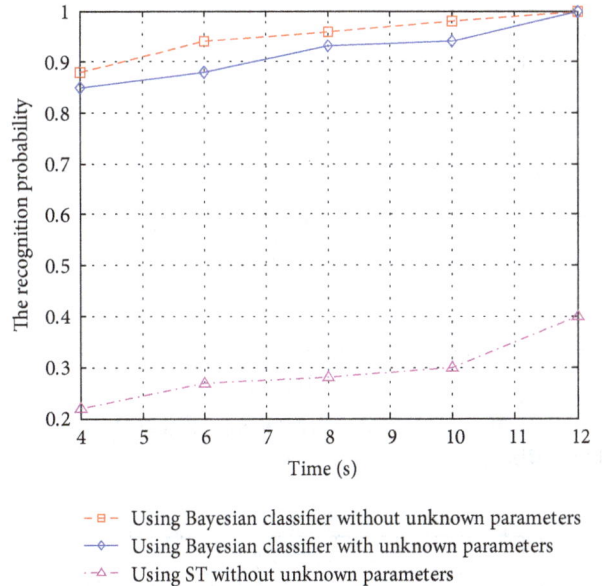

FIGURE 12: The recognition probability for B1937+21.

with the Monte Carlo simulations. We also extract the features of X-ray pulsars for multiple X-ray pulsars recognition and use the ML algorithm to estimate the unknown Doppler frequency and initial phase. Simulation results show that, in the condition of short observation time, the Bayesian classifier's recognition probability is obviously higher than that of the S Transform method. Besides that, we propose a method to recognize the unknown X-ray pulsars or X-ray disturbances. For some weak X-ray pulsars or background noise, the recognition time would be less than 1.2 s, which could fulfill the practical requirements.

We believe the proposed Bayesian classifier may also be extended to other pattern recognition problems with the probability density obeying the Poisson distribution.

Competing Interests

The authors declare that they have no competing interests.

Acknowledgments

This work was supported in part by National Natural Science Foundation of China under Grants 61271265 and

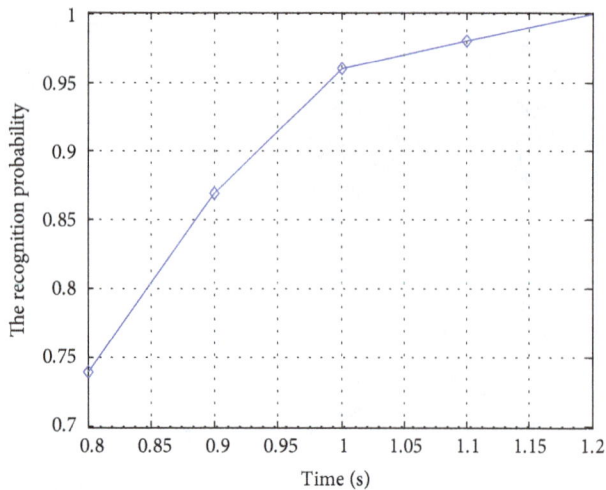

FIGURE 13: The recognition probability for unknown X-ray pulsars (B1821-24).

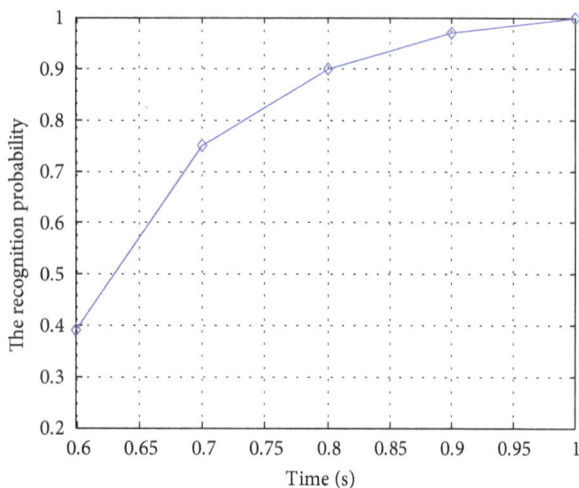

FIGURE 14: The recognition probability for unknown X-ray pulsars (B1937+21).

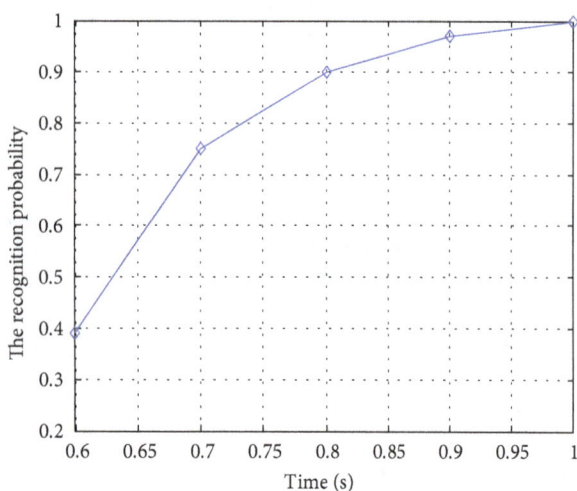

FIGURE 15: The recognition probability for unknown X-ray pulsars (background noise).

61671263 and in part by Tsinghua University Initiative Scientific Research Program under Grants 2013089244 and 20161080057.

References

[1] J. H. Taylor Jr., "Millisecond pulsars: Nature's most stable clocks," *Proceedings of the IEEE*, vol. 79, no. 7, pp. 1054–1062, 1991.

[2] D. N. Matsakis, J. H. Taylor, and T. M. Eubanks, "A statistic for describing pulsar and clock stabilities," *Astronomy and Astrophysics*, vol. 326, no. 3, pp. 924–928, 1997.

[3] S. I. Sheikh, *The use of variable celestial X-ray sources for spacecraft navigation [Ph.D. dissertation]*, Department of Aerospace Engineering, University of Maryland, College Park, Md, USA, 2005.

[4] T. J. Chester and S. A. Butman, "Navigation using X-ray pulsars," Tech. Rep. 81N27129, JPL, NASA, Pasadena, Calif, USA, 1981.

[5] J. E. Hanson, *Principles of X-ray navigation [Ph.D. thesis]*, Department of Aeronautics and Astronautics, Stanford University, Stanford, Calif, USA, 1996.

[6] S. I. Sheikh, D. J. Pines, P. S. Ray, K. S. Wood, M. N. Lovellette, and M. T. Wolff, "Spacecraft navigation using x-ray pulsars," *Journal of Guidance, Control, and Dynamics*, vol. 29, no. 1, pp. 49–63, 2006.

[7] A. A. Emadzadeh and J. L. Speyer, "Relative navigation between two spacecraft using X-ray pulsars," *IEEE Transactions on Control Systems Technology*, vol. 19, no. 5, pp. 1021–1035, 2011.

[8] S. I. Sheikh, J. E. Hanson, P. H. Graven et al., "Spacecraft navigation and timing using X-ray pulsars," *Navigation*, vol. 58, no. 2, pp. 165–186, 2011.

[9] D. Zhang, W. Zheng, Y. Wang, and L. Zhang, "X-ray pulsar profile recovery based on tracking-differentiator," *Mathematical Problems in Engineering*, vol. 2016, Article ID 4238165, 10 pages, 2016.

[10] X.-D. Zhang, Y. Shi, and Z. Bao, "A new feature vector using selected bispectra for signal classification with application in radar target recognition," *IEEE Transactions on Signal Processing*, vol. 49, no. 9, pp. 1875–1885, 2001.

[11] Z.-H. Xie, L.-P. Xu, G.-R. Ni, and Y. Wang, "A new feature vector using selected line spectra for pulsar signal bispectrum characteristic analysis and recognition," *Chinese Journal of Astronomy and Astrophysics*, vol. 7, no. 4, pp. 565–571, 2007.

[12] Z. Su, Y. Wang, L.-P. Xu, and N. Luo, "A new pulsar integrated pulse profile recognition algorithm," *Journal of Astronautics*, vol. 31, no. 6, pp. 1563–1568, 2010.

[13] R. G. Stockwell, L. Mansinha, and R. P. Lowe, "Localization of the complex spectrum: the S transform," *IEEE Transactions on Signal Processing*, vol. 44, no. 4, pp. 998–1001, 1996.

[14] L. Wang, L.-P. Xu, H. Zhang, and N. Luo, "A new algorithm of pulsar signal recognition," *Journal of Astronautics*, vol. 33, no. 10, pp. 1460–1465, 2012.

[15] H. Shan, X. Wang, X. Chen et al., "Wavelet based recognition for pulsar signals," *Astronomy and Computing*, vol. 11, pp. 55–63, 2015.

[16] W. W. Zhu, A. Berndsen, E. C. Madsen et al., "Searching for pulsars using image pattern recognition," *The Astrophysical Journal*, vol. 781, no. 2, pp. 1–12, 2014.

[17] A. A. Emadzadeh and J. L. Speyer, "On modeling and pulse phase estimation of X-ray pulsars," *IEEE Transactions on Signal Processing*, vol. 58, no. 9, pp. 4484–4495, 2010.

[18] A. R. Webb and K. D. Copsey, *Statistical Pattern Recognition*, John Wiley & Sons, West Sussex, UK, 2011.

[19] N. Ashby and A. R. Golshan, "Minimum uncertainties in position and velocity determination using x-ray photons from millisecond pulsars," in *Proceedings of the ION NTM Conference*, pp. 110–118, San Diego, Calif, USA, January 2008.

[20] The EPN (European Pulsar Network) Database Browser, http://www.jb.man.ac.uk/~pulsar/Resources/epn/browser.html.

17

Design of Three-Dimensional Path Following Guidance Logic

Sungsu Park⬤

Department of Aerospace Engineering, Sejong University, Seoul 05006, Republic of Korea

Correspondence should be addressed to Sungsu Park; sungsu@sejong.ac.kr

Academic Editor: Christian Circi

This paper presents a three-dimensional path following guidance logic. The proposed guidance logic is composed of the guidance law and the motion strategy of virtual target along the desired path. The guidance law makes a vehicle purse the virtual target, and the motion strategy explicitly specifies the motion of virtual target by introducing the concept of the projection point and the tangentially receding distance. The proposed logic is simple and efficient and yet provides precise path following. Numerical simulations are performed to demonstrate the effectiveness of the proposed guidance logic.

1. Introduction

Path following problem is to design a guidance law that will make a vehicle to follow a desired path. The desired path usually consists of consecutive segments which are comprised of circular arc and straight-line path. Path following problem has been a research topic over the last two decades, and many methods have been developed in response to the need for efficient and reliable path following guidance systems. Most of the methods have been developed for the paths in planar motion.

The path following guidance law specifies command acceleration that is applied to the vehicle, and the control law of the vehicle follows the command generated by the guidance law. There are many control methods [1] for implementing these commands in the vehicle. Although guidance loop and control loop can be designed separately, there are alternative methods for combining both guidance and control loops [2, 3]. However, these methods are usually more complex to use in real applications [4]. This paper is concerned only with guidance loop.

Conventional approach for path following is based on the proportional-integral-derivative (PID) control. This approach relies on the cross-track error, and if the error is small, then linear feedback on the cross-track error provides good performance. Other approach is to construct vector fields surrounding the path so that the vehicle converges to the desired path along the vector field [5, 6]. Global

convergence is proved, but this method is not applicable to three-dimensional space paths. Recent common approach [7–13] is rewriting the path following problem as the classical line-of-sight guidance problem by specifying a moving virtual target along the desired path. This approach generates a simple guidance law, yet can compensate a large deviation from the desired path. However, the position of virtual target is not easy to determine, and the initial distance between the vehicle and the desired path should be less than the specified distance. Some methods in this approach are still for the paths in planar motion [9–13].

Motivated by these observations, a new three-dimensional path following guidance logic is proposed in this paper. The proposed guidance logic is equally applicable to two primitive path types in three-dimensional space: straight-line path and circular path, and might be extendable to any smooth path. Similar to the methods in [7–13], the proposed guidance logic also uses a moving virtual target on the desired path; however, the proposed logic explicitly specifies the motion of virtual target by introducing the concept of the projection point and the tangentially receding distance. The proposed logic is simple and efficient, yet provides precise path following.

This paper is organized as follows. In the next section, a three-dimensional engagement geometry of a vehicle and virtual target is presented to formulate the problem, and the guidance law is proposed to pursue this virtual target. In Section 3, the motion of virtual target is specified explicitly

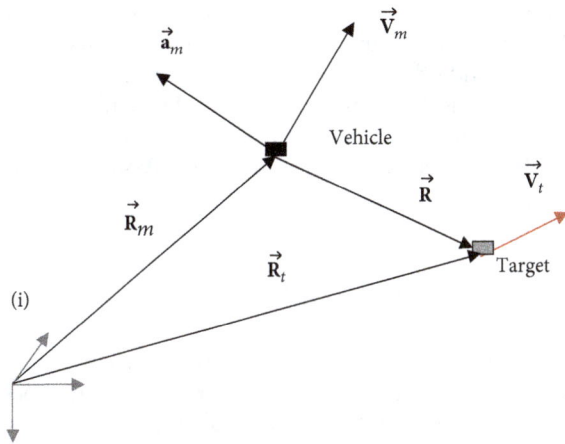

FIGURE 1: Three-dimensional vehicle/target engagement.

according to the position and velocity vectors of the vehicle and the geometry of the desired path. Some linearized stability analyses are also performed. In Section4, numerical simulation results are provided to demonstrate the performance of the proposed guidance logic. Conclusions are given in Section 5.

2. Design of Guidance Law

The proposed path following guidance logic is composed of the guidance law and the motion strategy of virtual target. In this section, the guidance law is firstly described. The guidance law generates command acceleration that is applied to the vehicle.

2.1. Engagement Kinematics. Figure 1 illustrates the overall kinematics of three-dimensional engagement between the vehicle and the virtual target on the desired path. The inertial frame is represented by $\{i\}$. $\vec{\mathbf{R}}_m$ and $\vec{\mathbf{V}}_m$ are the position and velocity vectors of the vehicle, respectively; $\vec{\mathbf{a}}_m$ represents the command acceleration vector which is assumed to be perpendicular to the velocity vector of vehicle; and $\vec{\mathbf{R}}$ is the relative position vector or line-of-sight (LOS) vector from the vehicle to the target. The position vector $\vec{\mathbf{R}}_t$ and velocity vector $\vec{\mathbf{V}}_t$ of the virtual target will be determined in further sections.

In this framework, the path following guidance law leads the vehicle to pursue a moving virtual target, while the interception of a target is not a goal as in the missile guidance law.

The guidance law in this paper issues an acceleration command which is perpendicular to the velocity vector. Therefore, the speed of the vehicle is not to be constrained by the guidance law but allowed to be controlled independently. This is because path following is required for a vehicle to follow a desired path without any temporal specifications. Note that, on the other hand, trajectory tracking is required to track a time-parameterized trajectory [14].

2.2. Guidance Law. In a missile community, pure proportional navigation guidance (PPNG) law has been widely used for targeting an enemy target [15]. PPNG tends to

keep a constant LOS angle and eventually delivers a missile into a target if the missile speed is sufficiently larger than that of the target. However, PPNG is not considered in path following problem since a constant LOS angle does not guarantee path following, and moreover the relative distance between the target and vehicle is typically kept constant in path following problem.

Pursuit guidance (PG) law has also been used in the missile community. Although it is known to be suboptimal guidance law for intercept purpose [15], PG can direct the vehicle along the LOS irrespective of the vehicle and target velocities. Therefore, many PG and PG variants are used in recent path following studies with a moving virtual target concept.

In this paper, however, the combined PPNG and PG law is proposed as a path following guidance law as in (1).

$$\vec{\mathbf{a}}_m = N\frac{\left(\vec{\mathbf{R}} \times \vec{\mathbf{V}}\right)}{R^2} \times \vec{V}_m - hN\frac{\left(\vec{\mathbf{R}} \times \vec{V}_m\right)}{R^2} \times \vec{V}_m, \quad (1)$$

where $N > 0$ and $h > 0$ are proportional and pursuit gains, respectively; R represents the distance between the vehicle and target; and $\vec{\mathbf{V}} = \vec{V}_t - \vec{V}_m$ is the relative velocity vector. The first term of (1) is PPNG law and the second is PG law. Therein, the gain terms N and h determine the performance of path following and will be further investigated in the next section by using a linear analysis.

The reason of the addition of PPNG to PG is two folds. One is for enhancing damping for a straight-line path and the other is for providing the centripetal acceleration for a circular path. In this way, both paths are successfully followed.

The proof of convergence of PPNG and PG law in three-dimensional space is very difficult and complex mainly due to the much higher complexity of the three-dimensional pursuit dynamics than that of two-dimensional dynamics. Some sufficient conditions can be found in [16, 17] under which a vehicle guided by the three-dimensional guidance law can always intercept a target maneuvering arbitrarily with time-varying normal acceleration.

The next section determines the explicit motion of virtual target.

3. Design of Virtual Target Motion

In this section, the motion of virtual target is explicitly determined for two cases, straight-line path and circular path. The proposed positioning method for the virtual target is conceptually simple and extendible to any smooth path.

3.1. Straight-Line Path. The straight-line path is specified by two waypoint vectors $\vec{\mathbf{W}}_i$ and $\vec{\mathbf{W}}_{i+1}$ and a unit direction vector $\hat{\mathbf{e}}_{t_i}$ along the straight-line segment connecting waypoints i and $i + 1$ as shown in Figure 2. The waypoint vector is a position vector to a waypoint from the origin of inertial frame. The point D is the projection of the vehicle on the path, and R_0 is the receding distance of the virtual target computed along the path starting from D. In this way, the

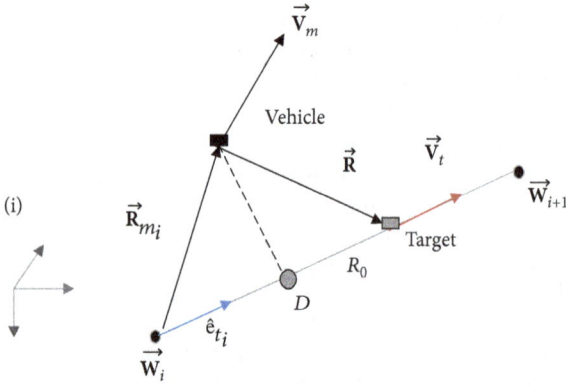

FIGURE 2: Straight path following.

virtual target is always ensured to be located on the path. In Figure 2, $\vec{R}_{m_i} = \vec{R}_m - \vec{W}_i$ is the relative position vector of a vehicle from the waypoint i.

The speed of the virtual target is determined by the projection of the current velocity vector of a vehicle on the path. Therefore, the position and velocity vectors of the virtual target are computed as

$$\vec{R}_t = \vec{W}_i + \left(\vec{R}_{m_i} \cdot \hat{e}_{t_i} \right) \hat{e}_{t_i} + R_0 \hat{e}_{t_i}, \qquad (2)$$
$$\vec{V}_t = \dot{\vec{R}}_t = \left(\vec{V}_m \cdot \hat{e}_{t_i} \right) \hat{e}_{t_i}.$$

The receding distance R_0 between the virtual target and the projection point is the design choice of the proposed guidance logic. The influence of this value can be investigated by the linear analysis using similar method to [10]. Figure 3 shows a linearization situation and the relationship between the vehicle velocity vector $\vec{V}_m$ and the relative velocity vector $\vec{V}$, where d denotes the cross-track error.

Then, the guidance law of (1) becomes

$$a_m \approx N \frac{V_m^2}{R} \cos \alpha_2 \sin \alpha_1 + hN \frac{V_m^2}{R} \sin \alpha. \qquad (3)$$

Assuming that the magnitudes of d/R_0 and $\alpha = \alpha_1 + \alpha_2$ are small,

$$R \approx R_0,$$
$$\cos \alpha_2 \approx 1,$$
$$\sin \alpha_1 \approx \frac{\dot{d}}{V_m}, \qquad (4)$$
$$\sin (\alpha_1 + \alpha_2) \approx \frac{\dot{d}}{V_m} + \frac{d}{R_0},$$

then (3) can be written as

$$a_m \approx N \frac{V_m}{R_0} \dot{d} + hN \frac{V_m}{R_0} \dot{d} + hN \frac{V_m^2}{R_0^2} d. \qquad (5)$$

Equation (5) resembles the proportional and derivative (PD) control and indicates that the two gains N and h and

the ratio of vehicle speed V_m and the distance R_0 for the virtual target behave as the PD control gains. The first term of the right-hand side of the above equation is due to pure proportional navigation guidance (PPNG) law and the second and third terms pursuit guidance (PG) law. Thus, the PPNG law enhances the damping performance. Assuming further $a_m \approx -\ddot{d}$, (5) reduces to

$$\ddot{d} + N(1+h) \frac{V_m}{R_0} \dot{d} + hN \frac{V_m^2}{R_0^2} d = 0. \qquad (6)$$

Equation (6) shows that the cross-track error dynamics is a second-order system, and the track error eventually goes to zero. The damping ratio and natural frequency are determined by PD control gains, which are $(N(1+h))/(2\sqrt{hN})$ and $\sqrt{hN}(V_m/R_0)$, respectively. For example, if $N = 1$ and $h = 2$ are chosen, then the damping ratio is 1.06 and the natural frequency is $\sqrt{2}(V_m/R_0)$.

The receding distance R_0 also plays a role as a look-ahead distance to detect the end of the path segment. As the vehicle moves, the virtual target also moves along the desired path which consists of consecutive segments. When the receding distance R_0 reaches the end of the current segment, a new virtual target position must be selected on the next segment to proceed to the next waypoint $\vec{W}_{i+2}$.

In this segment switching situation as shown in Figure 4, a new projection point D' on the next segment is firstly determined, and a new virtual target position is computed at a point on the next path segment that is R_0 distant from the new projection point D'. This segment switching logic is very simple and practical compared to the previous methods in the literatures since no additional methods are required to ensure that the virtual target always exists on the new path segment.

3.2. Circular Path. The circular path is specified by a unit vector $\hat{e}_{rot}$ for indicating the direction of rotation, a position vector $\vec{P}_c$ pointing to the centre of the circle, and the circle radius R_c, as shown in Figure 5. The two waypoint vectors $\vec{W}_i$ and $\vec{W}_{i+1}$ indicate the start and end points of the circular path.

For the circular path following case, a virtual target is proposed to be positioned on the tangent line at the point D with a tangential distance R_0 ahead of that point, where the point D is the projection point of the vehicle position onto the circular path. Note that the virtual target is not positioned on the circular path. In this way, the positioning method of the virtual target maintains consistency with the straight-line path case. Furthermore, this positioning method is very simple and practical compared to the previous methods in which the target is positioned at a point on the circular path, which is usually selected as an intersection point of the circular path and the reference distance from the vehicle [9–11].

In Figure 5, a unit direction vector $\hat{e}_t$ represents a direction of tangent line at the projection point D, and $\vec{R}_{cm}$ is a position vector from the centre of circle to the vehicle.

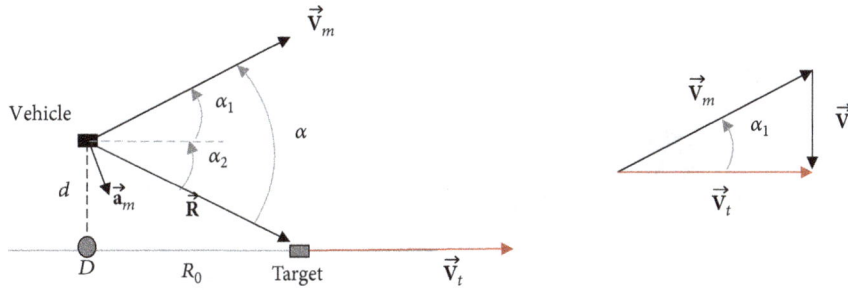

FIGURE 3: Linear model for straight-line path following.

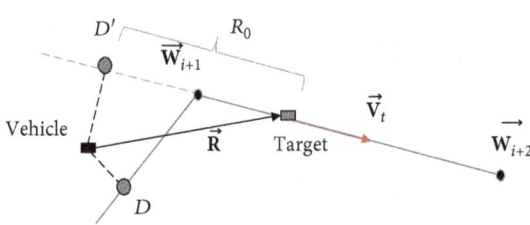

FIGURE 4: Path segment switching.

Then, the position and velocity vectors of the virtual target are expressed as

$$\overrightarrow{\mathbf{R}}_t = R_c \hat{e}_d + R_0 \hat{e}_t + \overrightarrow{P}_c, \tag{7}$$

$$\overrightarrow{\mathbf{V}}_t = \dot{\overrightarrow{R}}_t = R_c \dot{\hat{e}}_d + R_0 \dot{\hat{e}}_t. \tag{8}$$

In (8), a unit direction vector $\hat{\mathbf{e}}_d$ represents a direction from the centre of the circle to the projection point D and is given by

$$\overrightarrow{R}_{cm}^{\perp} = \overrightarrow{R}_{cm} - \left(\overrightarrow{R}_{cm} \cdot \hat{e}_{rot}\right) \hat{e}_{rot}, \tag{9}$$

$$\hat{\mathbf{e}}_d = \frac{\overrightarrow{R}_{cm}^{\perp}}{\left| \overrightarrow{R}_{cm}^{\perp} \right|}, \tag{10}$$

where $\overrightarrow{R}_{cm}^{\perp}$ represents a projection of the vector $\overrightarrow{\mathbf{R}}_{cm}$ onto the circle contained plane. The time derivative terms for the two unit direction vectors in (8) can be calculated by differentiating (10) as

$$\dot{\hat{e}}_d = \frac{\dot{\overrightarrow{R}}_{cm}^{\perp}}{\left| \overrightarrow{R}_{cm}^{\perp} \right|} - \frac{\left(\overrightarrow{R}_{cm}^{\perp} \cdot \dot{\overrightarrow{R}}_{cm}^{\perp}\right) \overrightarrow{R}_{cm}^{\perp}}{\left| \overrightarrow{R}_{cm}^{\perp} \right|^3}, \tag{11}$$

$$\dot{\overrightarrow{R}}_{cm}^{\perp} = \overrightarrow{V}_m - \left(\overrightarrow{V}_m \cdot \hat{e}_{rot}\right) \hat{e}_{rot},$$

which can be further simplified as follows:

$$\dot{\hat{e}}_d = \frac{1}{\left| \overrightarrow{R}_{cm}^{\perp} \right|} \left(\overrightarrow{V}_m \cdot \hat{e}_t\right) \hat{e}_i. \tag{12}$$

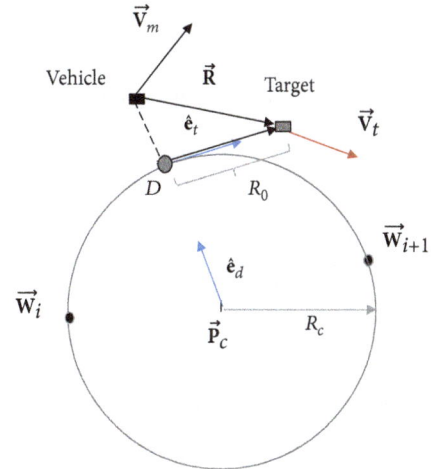

(a)

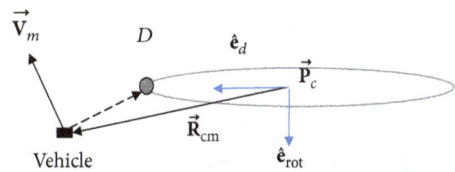

(b)

FIGURE 5: Circular path following. (a) Top view, (b) side view.

Since $\hat{\mathbf{e}}_t = \hat{e}_{rot} \times \hat{e}_d$, differentiating it gives

$$\dot{\hat{e}}_t = \hat{e}_{rot} \times \dot{\hat{e}}_d = -\frac{1}{\left| \overrightarrow{R}_{cm}^{\perp} \right|} \left(\overrightarrow{V}_m \cdot \hat{e}_t\right) \hat{e}_d. \tag{13}$$

Substituting (12) and (13) to (8), the velocity vector of the virtual target can be computed as

$$\overrightarrow{V}_t = \frac{R_c}{\left| \overrightarrow{R}_{cm}^{\perp} \right|} \left(\overrightarrow{V}_m \cdot \hat{e}_t\right) \hat{e}_t - \frac{R_0}{\left| \overrightarrow{R}_{cm}^{\perp} \right|} \left(\overrightarrow{V}_m \cdot \hat{e}_t\right) \hat{e}_d. \tag{14}$$

Note that unlike the straight-line path case, the velocity vector of the virtual target is not along the $\hat{\mathbf{e}}_t$ direction.

As a special case, if the motion of the vehicle initially begins on the circular path with the direction along the tangential unit vector $\hat{\mathbf{e}}_t$, that is,

$$\overrightarrow{\mathbf{R}}_m = R_c \hat{e}_d + \overrightarrow{P}_c, \ \overrightarrow{V}_m = V_m \hat{e}_t, \tag{15}$$

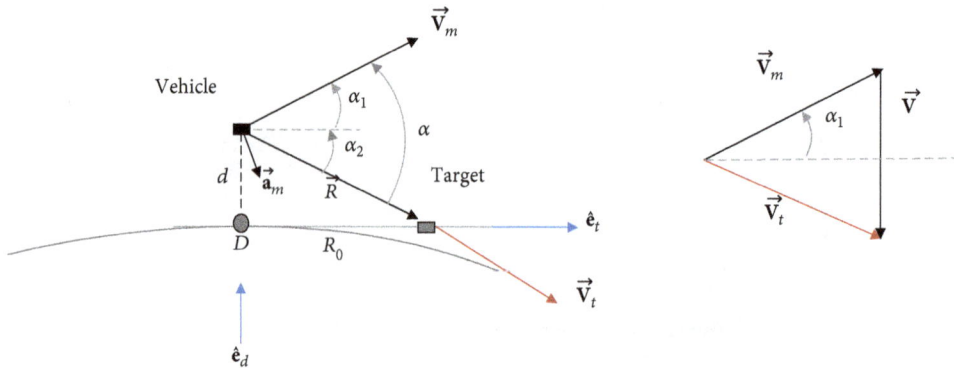

FIGURE 6: Linear model for circular path following.

then the target velocity vector and the relative position and velocity vectors are reduced to

$$\vec{V}_t = V_m \hat{e}_t - \frac{R_0 V_m}{R_c}\hat{e}_d,$$

$$\vec{R} = R_0\hat{e}_t, \qquad (16)$$

$$\vec{V} = -\frac{R_0 V_m}{R_c}\hat{e}_d,$$

and the guidance law in (1) becomes

$$\vec{a}_m = -N\frac{V_m^2}{R_c}\hat{e}_d. \qquad (17)$$

Therefore, with a proportional gain of $N = 1$, the guidance law produces the command acceleration that is the same with the centripetal acceleration which is appropriate to follow the circle with radius R_c. Note that pursuit guidance (PG) law generates zero command acceleration, which means that PG law makes no contribution in this special case.

The tangentially receding distance R_0 between the virtual target and the projection point is also the design choice of the proposed guidance logic. The influence of this value can be investigated by the linear analysis. Figure 6 shows a linearization situation with assumption that the vehicle motion is perturbed from the initial circular path following motion. In Figure 6, d denotes the cross-track error, and the relationship between the vehicle velocity vector $\vec{V}_m$ and the relative velocity vector $\vec{V}$ is also shown.

Assuming that all the vectors are almost in the same plane which contains a circular path and the magnitudes of d/R_c and α_1 are small, then $\vec{R}_{cm}^{\perp} \approx \vec{R}_{cm}$ and $\dot{\vec{R}}_{cm}^{\perp} \approx \vec{V}_m$ are satisfied, and the relative velocity vector is approximately expressed as

$$\vec{V} \approx -\frac{R_0 V_m}{R_c}\hat{e}_d. \qquad (18)$$

Therefore, the guidance law of (1) becomes

$$a_m \approx N\frac{V_m^2 R_0}{RR_c}\cos\alpha_2 + hN\frac{V_m^2}{R}\sin\alpha. \qquad (19)$$

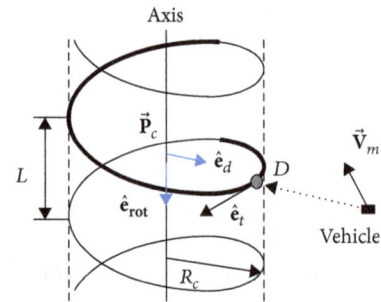

FIGURE 7: Helical path following.

Assuming also that the magnitudes of d/R_0 and $\alpha = \alpha_1 + \alpha_2$ are small,

$$R \approx R_0,$$

$$\cos\alpha_2 \approx 1, \qquad (20)$$

$$\sin(\alpha_1 + \alpha_2) \approx \frac{\dot{d}}{V_m} + \frac{d}{R_0},$$

then (19) can be written as

$$a_m \approx N\frac{V_m^2}{R_c} + hN\frac{V_m}{R_0}\dot{d} + hN\frac{V_m^2}{R_0^2}d. \qquad (21)$$

Equation (21) resembles proportional and derivative (PD) control with the forcing term of $N(V_m^2/R_c)$. The first term of the right-hand side of the above equation is due to pure proportional navigation guidance (PPNG) law, and the second and third terms are due to pursuit guidance (PG) law. Thus, the PPNG law provides the centripetal acceleration, and the PG law plays a role as a PD control. Since the vehicle is approximately in circular motion, $a_m \approx (V_m^2/R_c) - \ddot{d}$ holds and (21) is reduced to

$$\frac{V_m^2}{R_c} = N\frac{V_m^2}{R_c} + \ddot{d} + hN\frac{V_m}{R_0}\dot{d} + hN\frac{V_m^2}{R_0^2}d. \qquad (22)$$

Equation (22) implies that with a proportional gain of $N = 1$, the cross-track error dynamics is a second-order system and the track error eventually goes to zero. The damping

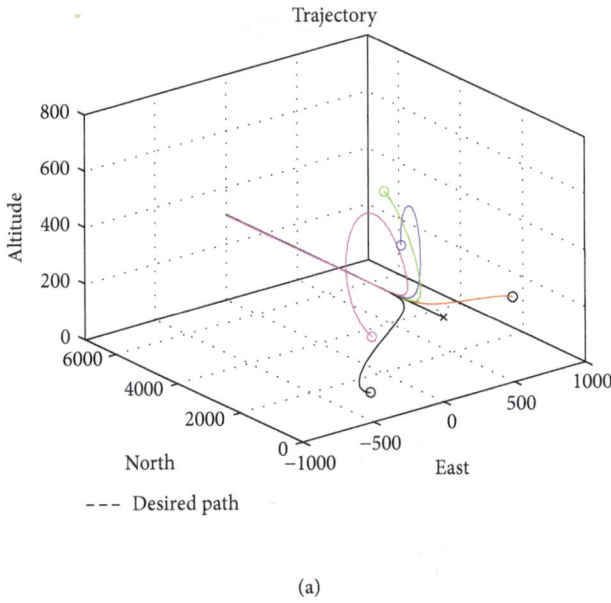

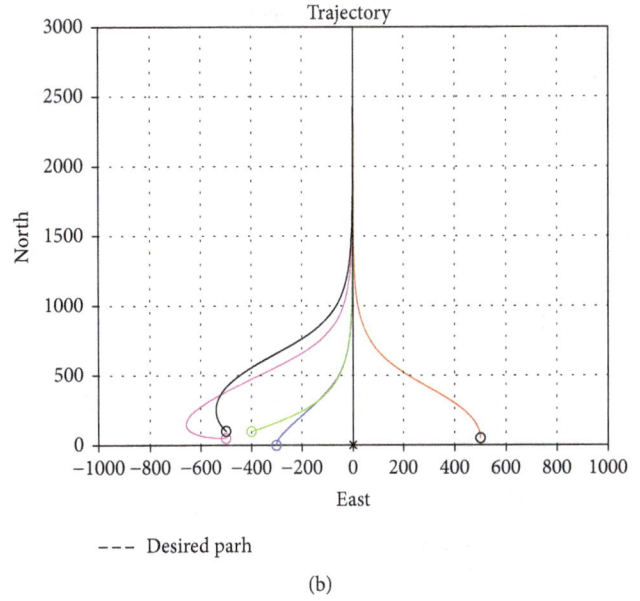

FIGURE 8: Straight-line path following for various initial conditions. (a) Three-dimensional view, (b) planar view.

ratio and natural frequency are determined by two guidance gains and the ratio of the vehicle speed and the tangentially receding distance R_0, which are $\sqrt{hN}/2$ and $\sqrt{hN}(V_m/R_0)$, respectively. For example, if $N = 1$ and $h = 2$ are chosen, then the damping ratio is $1/\sqrt{2}$ and the natural frequency is $\sqrt{2}(V_m/R_0)$.

The proposed strategy for specifying the motion of virtual target involves computations of position and velocity of the projection point and the receding point on the tangent line at the projection point, which depend on both the velocity of vehicle and the geometry of the desired path. Although more complex computations are expected than the straight-line or circular path, this is conceptually possible to any smooth path.

For example, in case of helical path defined on a vertical cylinder of radius R_c as shown in Figure 7, the position and velocity vectors of virtual target in (8) are modified as

$$\overrightarrow{\mathbf{R}}_t = R_c \widehat{e}_d + R_0 \widehat{e}_t + \overrightarrow{\mathbf{P}}_c, \tag{23}$$
$$\overrightarrow{\mathbf{V}}_t = R_c \dot{\widehat{e}}_d + R_0 \dot{\widehat{e}}_t + \dot{\overrightarrow{\mathbf{P}}}_c,$$

where $\overrightarrow{\mathbf{P}}_c$ is now a position vector pointing to the instantaneous centre of curvature at the point D. The time derivative of $\overrightarrow{\mathbf{P}}_c$ is given by a vertical component of the velocity vector of a vehicle as follows:

$$\dot{\overrightarrow{P}}_c = \left(\overrightarrow{V}_m \cdot \widehat{e}_{\text{rot}} \right) \widehat{e}_{\text{rot}}. \tag{24}$$

The time derivative of a unit direction vector $\widehat{e}_d$ in (12) is modified as

$$\dot{\widehat{e}}_d = \frac{1}{\left| \overrightarrow{R}_{\text{cm}}^{\perp} \right|} \left(\overrightarrow{V}_m \cdot \widehat{e}_n \right) \widehat{e}_n, \tag{25}$$

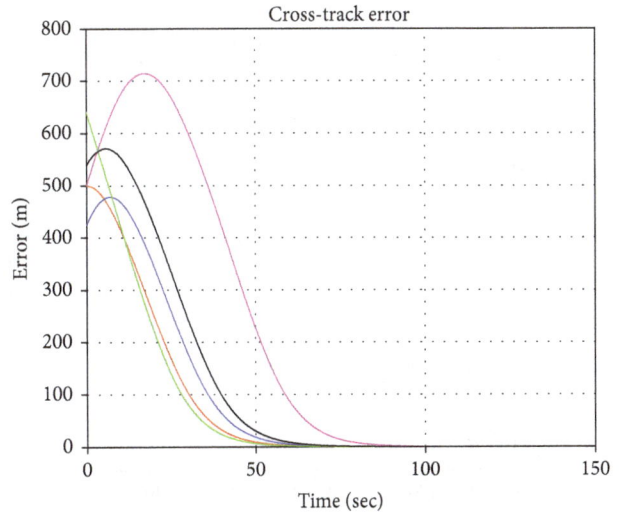

FIGURE 9: Cross-track errors.

where $\widehat{e}_n = \widehat{e}_{\text{rot}} \times \widehat{e}_d$ is a unit normal vector which is different from the unit tangential vector $\widehat{e}_t$ of the path at the point D. The time derivative of $\widehat{e}_t$ in (13) is also modified as

$$\dot{\widehat{e}}_t = \frac{1}{\left| \widehat{e}_n + (L/2\pi R_c)\widehat{e}_{\text{rot}} \right|} \left(-\frac{1}{\left| \overrightarrow{R}_{\text{cm}}^{\perp} \right|} \left(\overrightarrow{V}_m \cdot \widehat{e}_n \right) \widehat{e}_d \right), \tag{26}$$

where L is a pitch of helix.

In this way, the motion of virtual target can be computed according to the geometry of the desired path. Therefore, the proposed guidance logic might be extendable to any smooth path.

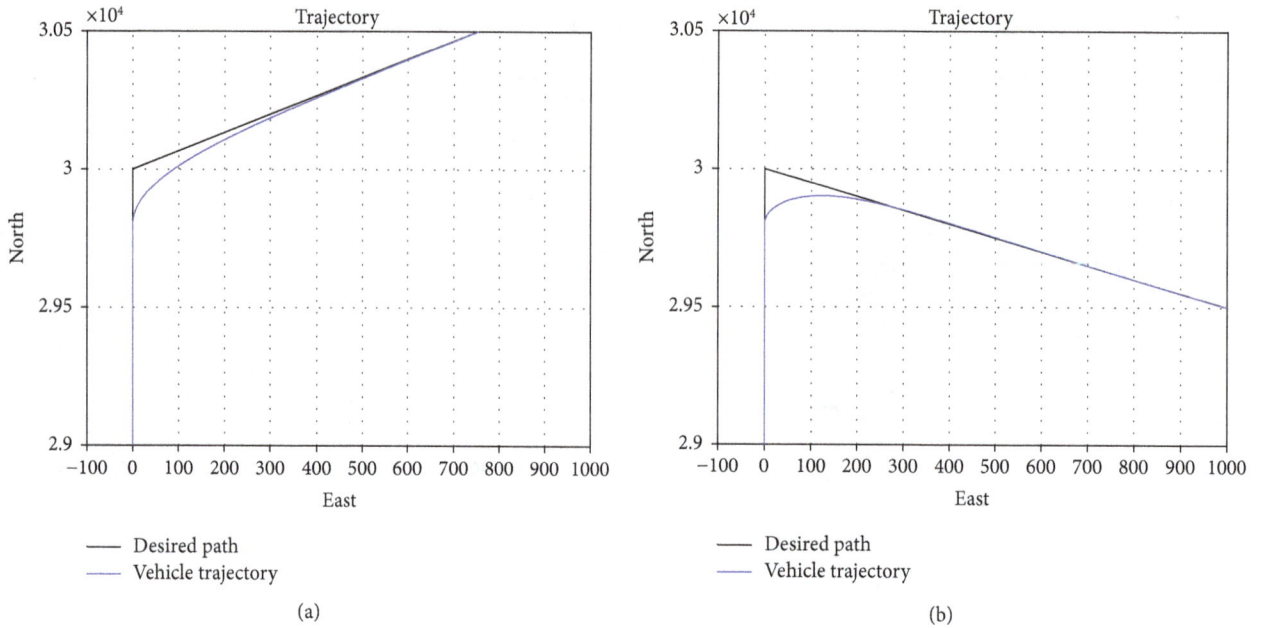

(a)

(b)

FIGURE 10: Path transition between straight-line segments. (a) Obtuse angle, (b) acute angle.

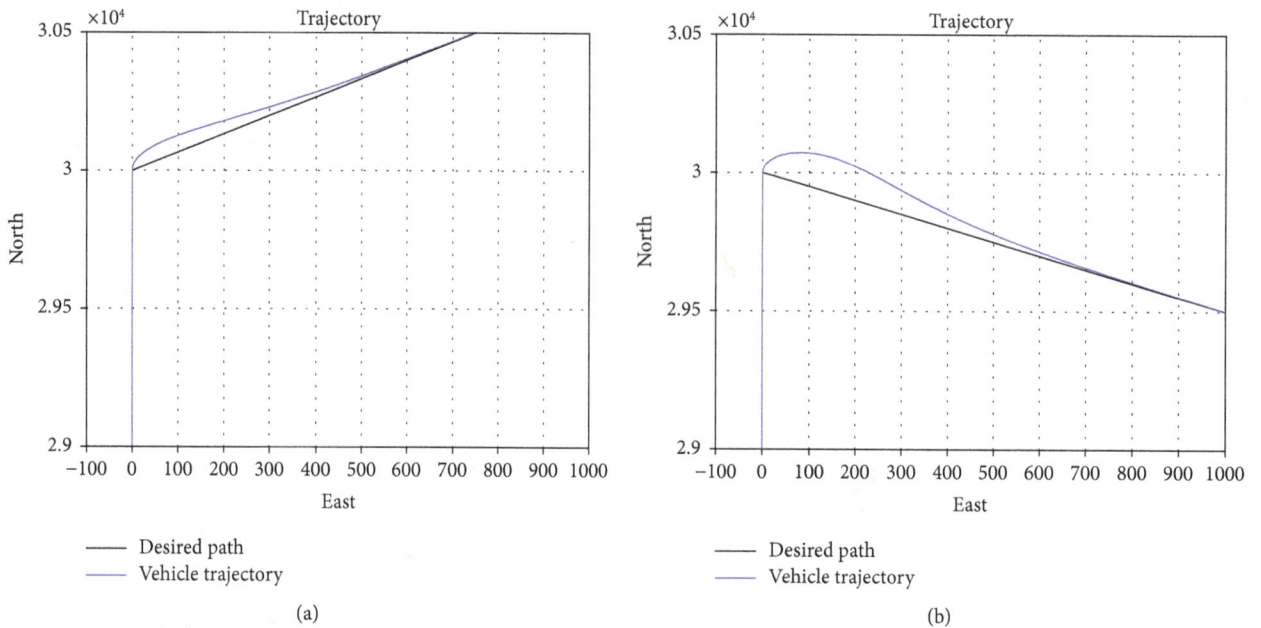

(a)

(b)

FIGURE 11: Path transition over the waypoint. (a) Obtuse angle, (b) acute angle.

TABLE 1: Initial conditions used in simulations.

Line colour	Initial positions (x: north, y: east, h: altitude in meters)	Initial headings ψ and flight path angles γ (deg)
Orange	$x = 50, y = 500, h = 300$	$\psi = 0, \gamma = 0$
Magenta	$x = 50, y = -500, h = 300$	$\psi = -90, \gamma = 30$
Blue	$x = 0, y = -500, h = 600$	$\psi = 0, \gamma = 60$
Green	$x = 100, y = -400, h = 800$	$\psi = 60, \gamma = -30$
Black	$x = 100, y = -500, h = 100$	$\psi = -30, \gamma = 0$

4. Simulations

In this section, the numerical simulations are performed to demonstrate the effectiveness of the proposed path following guidance logic. The path of a vehicle is defined in three dimensions and is assumed to be given to the vehicle a priori. It is also assumed that the vehicle is a point mass and is flying at a constant speed of 25 m/sec. The tangentially receding distance R_0 is chosen as 200 m, and $N = 1$ and $h = 2$ are used in simulations.

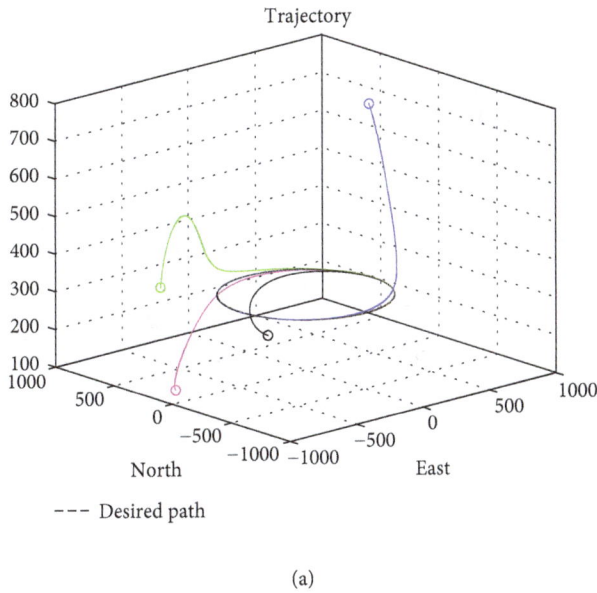

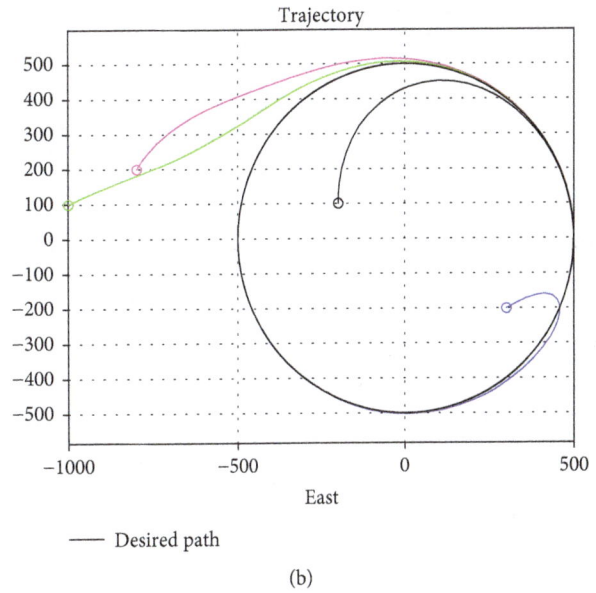

FIGURE 12: Circular path following. (a) Three-dimensional view, (b) planar view.

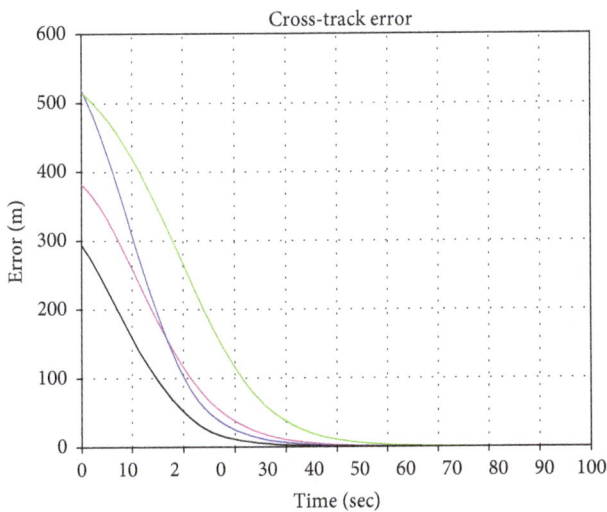

FIGURE 13: Cross-track errors.

To illustrate straight-line path following with the proposed guidance logic, three scenarios are considered, and the results are shown in Figures 8–11. In the first scenario, the vehicle begins at various initial positions and poses and is supposed to fly north along the desired straight-line path at an altitude of 300 m. The initial conditions used in the first scenario are summarized in Table 1. Figure 8 shows that the proposed logic can accommodate large deviations from the desired path, and the vehicle trajectory converges to the path without any following errors. The trajectories of cross-track error are shown in Figure 9.

As for the next two scenarios, Figure 10 illustrates the ability of the proposed guidance logic to follow straight-line path segments with both obtuse and acute angles. The receding distance R_0 detects the end of path segment, and the vehicle performs segment switching maneuver.

If the receding distance is longer, the vehicle will start its segment switching earlier. On the contrary, instead of using the receding distance, if the projection point D is used to detect the end of path segment, then over-waypoint transition is possible as shown in Figure 11. Although the transitions show some deviations around the waypoint, the vehicle returns to the next path segment without any following errors.

To illustrate circular path following with the proposed guidance logic, three scenarios are considered and the results are shown in Figures 12–15. In the first scenario, the vehicle begins at various initial positions and poses and is supposed to fly along circular path with the circle radius of 500 m at an altitude of 300 m. The initial conditions used in the first scenario are summarized in Table 2. Figure 12 shows that the vehicle trajectory converges to the desired circular path with any following errors. The trajectories of cross-track error are shown in Figure 13.

To further test the circular path following capabilities of the proposed guidance logic, two more scenarios are simulated. One is for oblique circular paths defined in three-dimensional space and the other is for path transitions between concentric circular paths. In Figure 14, it can be seen that the vehicle trajectories converge to an oblique circular path without any following errors, and Figure 15 shows the vehicle successfully transits and follows two concentric circular paths with varying radii and inclination angles.

Figure 16 illustrates how the proposed guidance logic can track a helical path with any following error. The helical path is defined on a vertical cylinder of radius 500 m, having pitch of $L = 10(2\pi)$ m.

5. Conclusions

This paper presents a new simple and efficient three-dimensional path following guidance logic. The proposed

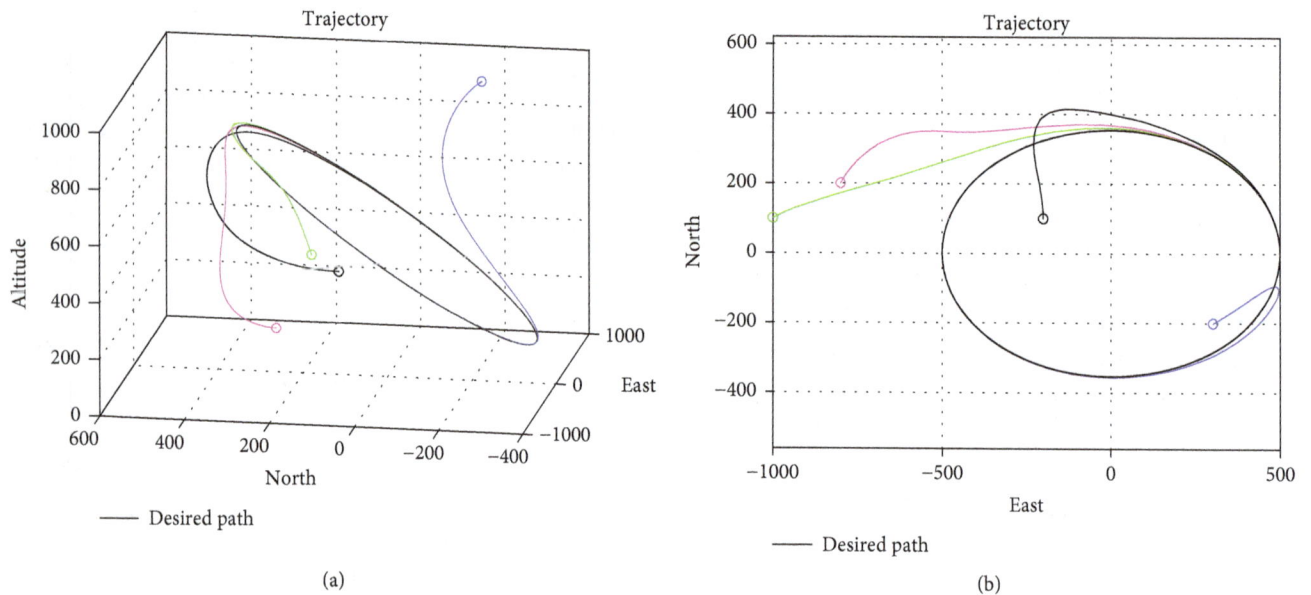

(a)

(b)

FIGURE 14: Path following with an oblique circle (with inclination angle of 45 degree). (a) Three-dimensional view, (b) planar view.

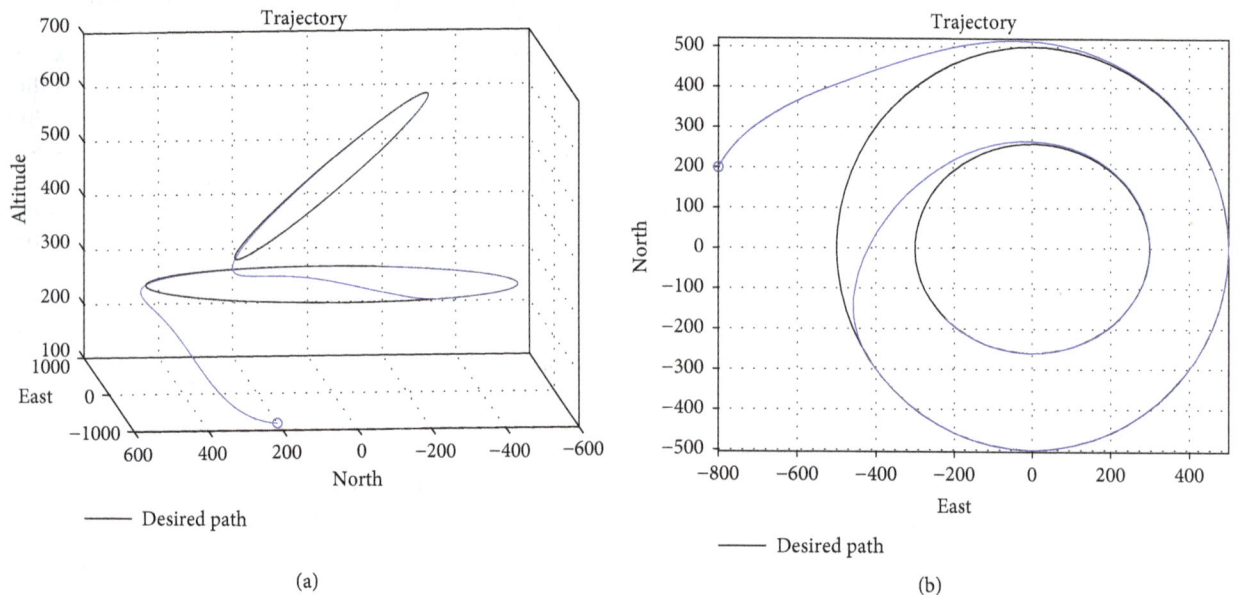

(a)

(b)

FIGURE 15: Circular path transition with circle radii of 500 m and 300 m and inclination angles of 0 and 30 degrees, respectively. (a) Three-dimensional view, (b) planar view.

TABLE 2: Initial conditions used in simulations.

Line colour	Initial positions (x: north, y: east, h: altitude in meters)	Initial headings ψ and flight path angles γ (deg)
Magenta	$x = 200, y = -800, h = 100$	$\psi = 30, \gamma = 0$
Blue	$x = -200, y = 300, h = 800$	$\psi = 60, \gamma = -30$
Green	$x = 100, y = -1000, h = 400$	$\psi = 60, \gamma = 60$
Black	$x = 100, y = -200, h = 200$	$\psi = 0, \gamma = 0$

guidance logic is composed of the guidance law and the motion strategy of the virtual target along the desired path.

As for the guidance law, the combined pure proportional navigation guidance (PPNG) and pursuit guidance (PG) are proposed. The combined law enhances damping for a straight-line path and provides the centripetal acceleration for a circular path. This guidance law makes a vehicle pursue the virtual target efficiently. As for the motion strategy of the virtual target, the motion is explicitly specified by introducing the concept of the projection point and the tangentially receding distance. Although more complex computations are expected than the straight-line or circular path, the

(a)

(b)

(c)

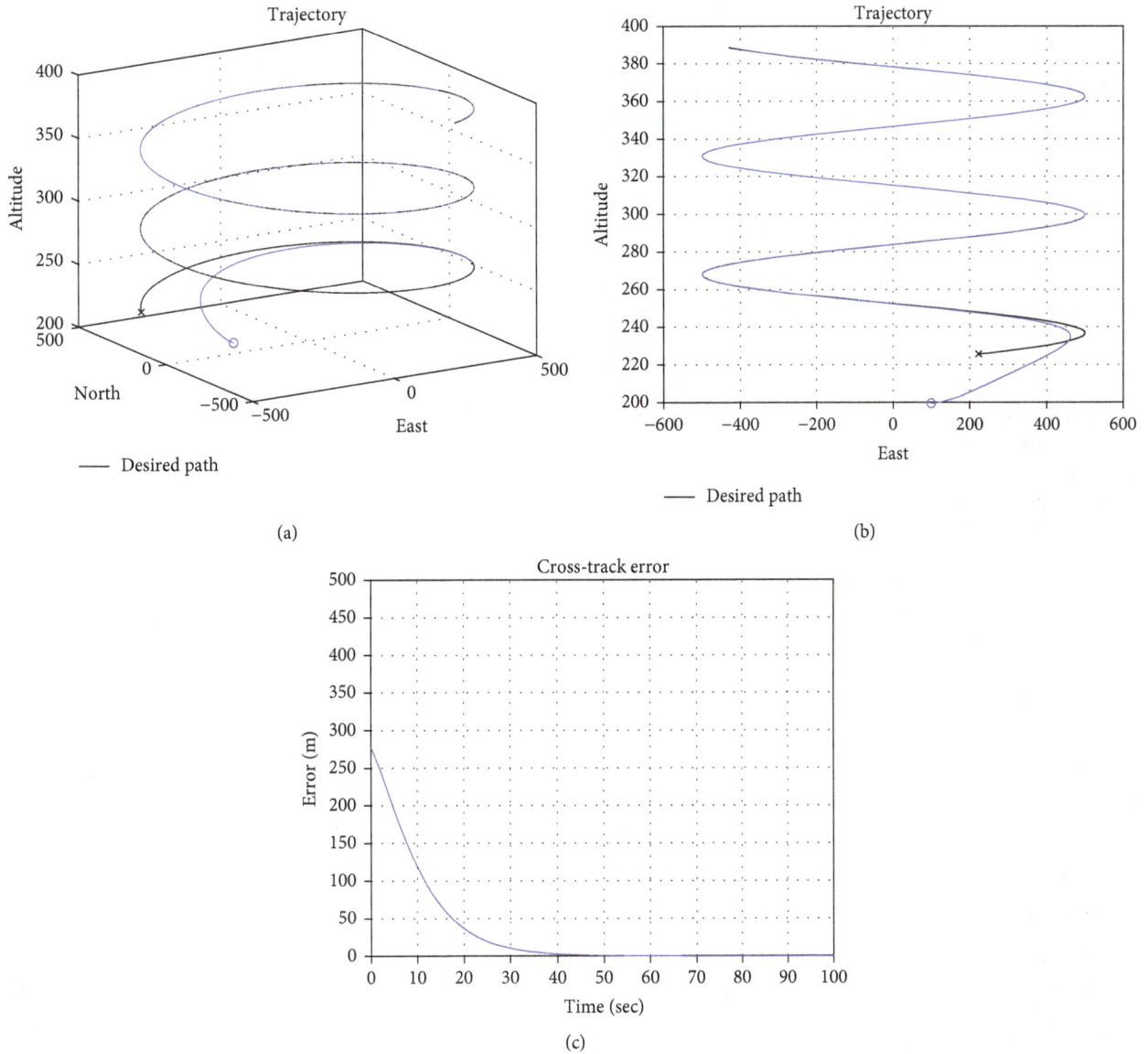

FIGURE 16: Helical path following. (a) Three-dimensional view, (b) side view, (c) cross-track error.

proposed guidance logic is conceptually extendable to any smooth path.

The effectiveness of the proposed logic has been demonstrated by the numerical simulations with various scenarios. For straight-line, circular, and helical paths defined in three-dimensional space, the proposed logic provides precise path following.

Conflicts of Interest

The authors declare that there is no conflict of interest regarding the publication of this paper.

Acknowledgments

This work was supported by Sejong University.

References

[1] G. E. Franklin, J. D. Powell, and A. Emami-Naeini, *Feedback Control of Dynamic Systems*, Prentice–Hall, 2002.

[2] M. Rathinam and R. M. Murray, *Configuration Flatness of Lagrangian Systems Underactuated by Prentice–Hall*, Upper Saddle River, NJ, 2002.

[3] E. Johnson, A. Calise, and E. Corban, "A six degree-of-freedom adaptive flight control architecture for trajectory following," in *AIAA Guidance, Navigation, and Control Conference and Exhibit*, Monterey, California, August 2002.

[4] D. J. Gates, "Nonlinear path following method," *Journal of Guidance, Control, and Dynamics*, vol. 33, no. 2, pp. 321–332, 2010.

[5] D. R. Nelson, D. B. Barber, T. W. McLain, and R. W. Beard, "Vector field path following for miniature air vehicles," *IEEE Transactions on Robotics*, vol. 23, no. 3, pp. 519–529, 2007.

[6] D. A. Lawrence, E. W. Frew, and W. J. Pisano, "Lyapunov vector fields for autonomous unmanned aircraft flight con-

trol, *Journal of Guidance, Control, and Dynamics*, vol. 31, no. 5, pp. 1220–1229, 2008.

[7] V. Cichella, I. Kaminer, V. Dobrokhodov, E. Xargay, N. Hovakimyan, and A. Pascoal, "Geometric 3D path-following control for a fixed-wing UAV on SO(3)," in *AIAA Guidance, Navigation, and Control Conference*, Portland, Oregon, August 2011.

[8] T. Yamasaki, H. Takano, and Y. Baba, *Robust Path-Following for UAV Using Pure Pursuit Guidance*, Aerial Vehicles, InTech, 2009.

[9] S. Park, *Avionics and Control System Development for Mid-Air Rendezvous of Two Unmanned Aerial Vehicles, [Ph.D. thesis]*, MIT, 2004.

[10] S. Park, J. Deyst, and J. P. How, "A new nonlinear guidance logic for trajectory tracking," in *AIAA Guidance, Navigation, and Control Conference and Exhibit*, Providence, Rhode Island, August 2004.

[11] G. Ducard, K. C. Kulling, and H. P. Geering, "A simple and adaptive on-line path planning system for a UAV," in *2007 Mediterranean Conference on Control & Automation*, pp. 1–6, Athens, Greece, June 2007.

[12] E. D. B. Medagoda and P. W. Gibbens, "Synthetic-waypoint guidance algorithm for following a desired flight trajectory," *Journal of Guidance, Control, and Dynamics*, vol. 33, no. 2, pp. 601–606, 2010.

[13] J.-m. Zhang, Q. Li, N. Cheng, and B. Liang, "Nonlinear path-following method for fixed-wing unmanned aerial vehicles," *Journal of Zhejiang University Science C*, vol. 14, no. 2, pp. 125–132, 2013.

[14] P. Encarnacao and A. Pascoal, "Combined trajectory tracking and path following: an application to the coordinated control of autonomous marine craft," in *Proceedings of the 40th IEEE Conference on Decision and Control (Cat. No.01CH37228)*, vol. 1, pp. 964–969, Orlando, FL, USA, December 2001.

[15] P. Zarchan, *Tactical and Strategic Missile Guidance*, vol. 176, AIAA, 1997.

[16] S.-H. Song and I.-J. Ha, "A Lyapunov-like approach to performance analysis of 3-dimensional pure PNG laws," *IEEE Transactions on Aerospace and Electronic Systems*, vol. 30, no. 1, pp. 238–248, 1994.

[17] J.-H. Oh and I.-J. Ha, "Capturability of the 3-dimensional pure PNG law," *IEEE Transactions on Aerospace and Electronic Systems*, vol. 35, no. 2, pp. 491–503, 1999.

Permissions

List of Contributors

Tania Savitri, Youngjoo Kim and Hyochoong Bang
Department of Aerospace Engineering, Korea Advanced Institute of Science and Technology (KAIST), Daejeon 305-701, Republic of Korea

Sujang Jo
Korea Aerospace Research Institute (KARI), Daejeon 305-701, Republic of Korea

Massimo Gennaretti, Federico Porcacchia, Simone Migliore and Jacopo Serafini
Department of Engineering, Roma Tre University, Rome, Italy

Xiangzhen Xue and Sanmin Wang
School of Mechanical Engineering, Northwestern Polytechnical University, Xi'an 710072, China

Jie Yu and Liyun Qin
35th Department, Xi'an Space Engine Factory, Xi'an 710061, China

Zhi-yong Liu
The Key Laboratory of Electronic Equipment Structure Design of Ministry of Education, Xidian University, Xian 710071, China
Xian Yang Vocational & Technical College, Xianyang 712000, China

Jing-li Du and Hong Bao
The Key Laboratory of Electronic Equipment Structure Design of Ministry of Education, Xidian University, Xian 710071, China
Collaborative Innovation Center of Information Sensing and Understanding, Xidian University, Xian 710071, China

Qian Xu
Xinjiang Astronomical Observatories, CAS, Urumqi 830011, China

Xiangyang Zhou
School of Instrumentation Science & Opto-Electronics Engineering, Beihang University, Beijing 100191, China
State Key Laboratory for Manufacturing Systems Engineering, Xi'an Jiaotong University, Xi'an 710049, China

Yating Li
School of Instrumentation Science & Opto-Electronics Engineering, Beihang University, Beijing 100191, China

Yuan Jia
China Aerospace Academy of Electronic Technology Beijing Institute of Aerospace Micro-Electromechanical Technology, Beijing 100094, China

Libo Zhao
State Key Laboratory for Manufacturing Systems Engineering, Xi'an Jiaotong University, Xi'an 710049, China

Changxin Luo, Humin Lei, Dongyang Zhang and Xiaojun Zou
Air and Missile Defence College, Air Force Engineering University, Xi'an 710051, China

Adel Ghenaiet
Faculty of Mechanical and Process Engineering, University of Sciences and Technology (USTHB), BP 32, El-Alia, Bab-Ezzouar, 16111 Algiers, Algeria

Liang Lv, Jianguo Tan and Yue Hu
Science and Technology on Scramjet Laboratory, College of Aerospace Science and Engineering, National University of Defense Technology, Changsha, Hunan 410073, China

F. Caputo and A. De Luca
Department of Engineering, University of Campania Luigi Vanvitelli, Aversa, Italy

F. Marulo and M. Guida
Department of Industrial Engineering, University of Naples "Federico II", Napoli, Italy

B. Vitolo
Geven, Zona Asi Boscofangone, Nola, Napoli, Italy

M. Carminati and S. Ricci
Department of Aerospace Science and Technology, Politecnico di Milano, Via La Masa 34, 20156 Milano, Italy

Weiwei Zhang, Shengyuan Jiang, Dewei Tang, Huazhi Chen and Jieneng Liang
State Key Laboratory of Robotics and System, Harbin Institute of Technology, Harbin 150001, China

Chang-sheng Gao, Jian-qing Li and Wu-xing Jing
Department of Aerospace Engineering, Harbin Institute of Technology, Harbin, China

Ming Zhang and Ran Xu
Key Laboratory of Fundamental Science for National Defense-Advanced Design Technology of Flight Vehicle, Nanjing University of Aeronautics and Astronautics, Nanjing 210016, China

Hong Nie
Key Laboratory of Fundamental Science for National Defense-Advanced Design Technology of Flight Vehicle, Nanjing University of Aeronautics and Astronautics, Nanjing 210016, China
State Key Laboratory of Mechanics and Control of Mechanical Structures, Nanjing University of Aeronautics and Astronautics, Nanjing 210016, China

Yong Zhao, Hong Bao and Xuechao Duan
Key Laboratory of Electronic Equipment Structure Design of Ministry of Education, Xidian University, Xi'an 710071, China

Hongmei Fang
Nanjing Research Institute of Electronics Technology, Nanjing 210039, China

Haoqiang Zhang, Shengjing Tang and Jie Guo
Key Laboratory of Dynamic and Control of Flight Vehicle of the Ministry of Education, School of Aerospace Engineering, Beijing Institute of Technology, Beijing 100081, China

Wan Zhang
Beijing Electro-Mechanical Engineering Institute, Beijing 100071, China

Hao Liang, Yafeng Zhan and Chaowei Duan
School of Aerospace Engineering and Space Centre, Tsinghua University, Beijing 100084, China

Sungsu Park
Department of Aerospace Engineering, Sejong University, Seoul 05006, Republic of Korea

Index

www.ingramcontent.com/pod-product-compliance
Lightning Source LLC
Chambersburg PA
CBHW082031190326
41458CB00010B/3330